Infrared and Raman
Spectra of Inorganic and
Coordination Compounds

Infrared and Raman Spectra of Inorganic and Coordination Compounds

Part B: Applications in Coordination, Organometallic, and Bioinorganic Chemistry

Fifth Edition

Kazuo Nakamoto

A WILEY-INTERSCIENCE PUBLICATION

John Wiley & Sons, Inc.

NEW YORK / CHICHESTER / WEINHEIM / BRISBANE / SINGAPORE / TORONTO

Copyright © 1997 by John Wiley & Sons, Inc.

Library of Congress Cataloging-in-Publication Data:
Nakamoto, Kazuo, 1922–
    Infrared and Raman spectra of inorganic and coordination compounds
  / Kazuo Nakamoto.—5th ed.
      p.   cm.
    "A Wiley-Interscience publication."
    Includes bibliographical references.
    Contents: pt. A. Theory and applications of inorganic chemistry—
  pt. B. Applications in coordination, organometallic, and
  bioinorganic chemistry.
    ISBN 0-471-16394-5 (pt. A: cloth: alk. paper).—ISBN
  0-471-16392-9 (pt. B: cloth: alk. paper)
    1. Infrared spectroscopy. 2. Raman spectroscopy. I. Title.
  QD96.I5N33   1997
  543'.08583—dc20   96-33456
    ISBN 0-471-16394-5 (pt. A)
    ISBN 0-471-16392-9 (pt. B)
    ISBN 0-471-19406-9 (set)

Contents

Section IV. Applications in Organometallic Chemistry   **257**

Section V. Applications in Bioinorganic Chemistry   **319**

Index   **379**

Contents of Part A

Section II. Applications in Inorganic Chemistry 153

Appendices 321

Preface

The first edition of this book, entitled *Infrared Spectra of Inorganic and Coordination Compounds*, was published in 1963. Since then, it has been revised in 1970, 1978, and 1986 to keep up with ever-increasing new literature. The preparation of the fifth edition was begun in the fall of 1990 and completed in the spring of 1996.

As I emphasized in the Preface of the previous editions, this book is intended to describe fundamental theories of vibrational spectroscopy in a condensed form (Section I) and to illustrate their applications in inorganic (Section II), coordination (Section III), organometallic (Section IV) and bioinorganic chemistry (Section V), using typical examples. In the fifth edition, all of these sections have been updated by adding many new references while omitting those shown to be erroneous by later studies. Furthermore, several subsections have been added in all sections to cover new topics. In particular, I have included many infrared and Raman spectral charts of typical compounds because they provide real "feel," which cannot be expressed by tables. The book has been divided into two volumes: Part A covers Sections I and II, and Part B covers Sections III–V. I felt that this division was justified, because Part A covers basic theory and its applications to relatively simple inorganic compounds, whereas Part B is focused on applications of basic theory to large and complex molecules.

As in the past editions, I have tried to give a broad and balanced coverage of the field. It was clearly impossible, however, to include all significant work in the limited space available. I only hope that any unbalanced presentation will be compensated for by the review articles and reference books which are abundantly quoted throughout this book.

I wish to express my sincere thanks to all who helped me in preparing this edition. I am particularly indebted to Professors D. P. Strommen (Idaho State University), R. A. Condrate (Alfred University), R. S. Czernuszewicz (University of Houston), J. R. Kincaid (Marquette University), and T. Kitagawa (Institute for Molecular Science, Okazaki, Japan). Special thanks are given to Prof. R. S. Czernuszewicz, who kindly drew some figures in Sections I, IV, and V. As I mentioned earlier, this edition contains many spectral charts, which could not have been included without the cooperation of many authors and their publishers. For the sake of uniformity, I simply used reference numbers in the figure captions to cite the source of each chart. I would like to acknowledge the cooperation I received by listing here published and publications quoted in this edition: Academic Press (monographs), American Chemical Society (*Journal of the American Chemical Society, Journal of Physical Chemistry, Inorganic Chemistry*, and *Analytical Chemistry*), American Institute of Physics (*Journal of Chemical Physics*), Cornell University Press (monograph), Elsevier Science (*Chemical Physics, Chemical Physics Letters, Coordination Chemistry Review, Journal of Molecular Structure, Journal of Organometallic Chemistry, Solid State Communications*, and *Spectrochimica Acta*), John Wiley (monographs and *Journal of Raman Spectroscopy*), Material Research Society (*MBS Bulletin*), Macmillan Magazines (*Nature*), National Academy of Science (monograph), Plenum Press (monograph), The Royal Society of Chemistry (*Journal of the Chemical Society, Dalton Transactions*), The Society for Applied Spectroscopy (*Applied Spectroscopy*), and Verlag der Zeitschrift der Naturforschung (*Zeitschrift für Naturforschung*).

Finally, I would like to thank the staffs of the Science Library and the Chemistry Department of Marquette University for their help in preparing this book.

KAZUO NAKAMOTO

Milwaukee, Wisconsin

Abbreviations

IR, infrared; R, Raman; RR, resonance Raman; p, polarized; dp, depolarized; ap, anomalous polarization; i.a., inactive.

ν, stretching; δ, in-plane bending or deformation; ρ_w, wagging; ρ_r, rocking; ρ_t, twisting; π, out-of-plane bending. Subscripts a, s, and d denote antisymmetric, symmetric, and degenerate modes, respectively. Approximate normal modes of vibration corresponding to these vibrations are given in Figs. I-25 and I-26.

NCA, normal coordinate analysis.

GVF, generalized valence force field; UBF, Urey-Bradley force field.

M, metal; L, ligand; X, halogen; R, alkyl group or cyclopentadienyl (Cp) or other ring compound.

g, gas; l, liquid; s, solid; m or mat, matrix; sol'n or sl, solution.

Me, methyl; Et, ethyl; Pr, propyl; Bu, butyl; Ph, phenyl; Cp, cyclopentadienyl; OAc, acetate ion; py, pyridine; pic, pycoline; en, ethylenediamine. Abbreviations of other ligands are given when they appear in the text.

In the tables of observed frequencies, values in parentheses are calculated or estimated values unless otherwise stated.

III

APPLICATIONS IN COORDINATION CHEMISTRY

III-1. AMMINE, AMIDO, AND RELATED COMPLEXES

(1) Ammine (NH₃) Complexes

Vibrational spectra of metal ammine complexes have been studied extensively, and these are reviewed by Schmidt and Müller.[1] Figure III-1 shows the infrared spectra of typical hexammine complexes in the high-frequency region. To assign these NH_3 group vibrations, it is convenient to use the six normal modes of vibration of a simple 1 : 1 (metal/ligand) complex model such as that shown in Fig. III-2. Table III-1 lists the infrared frequencies and band assignments of hexammine complexes. It is seen that the antisymmetric and symmetric NH_3 stretching, NH_3 degenerate deformation, NH_3 symmetric deformation, and NH_3 rocking vibrations appear in the regions of 3400–3000, 1650–1550, 1370–1000, and 950–590 cm⁻¹, respectively. These assignments have been confirmed by NH_3/ND_3 and $NH_3/^{15}NH_3$ isotope shifts.

The NH_3 stretching frequencies of the complexes are lower than those of the free NH_3 molecule for two reasons.[18] One is the effect of coordination. Upon coordination, the N–H bond is weakened and the NH_3 stretching frequencies are lowered. The stronger the M–N bond, the weaker is the N–H bond and the lower are the NH_3 stretching frequencies if other conditions are equal. Thus the NH_3 stretching frequencies may be used as a rough measure of the M–N bond strength. The other reason is the effect of the counterion. The NH_3 stretching frequencies of the chloride are much lower than those of the perchlorate,

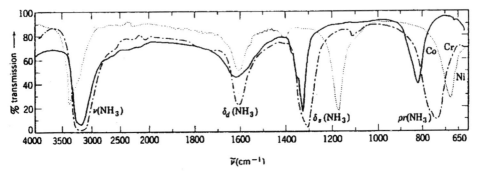

Fig. III-1. Infrared spectra of hexammine complexes: $[Co(NH_3)_6]Cl_3$ (solid line), $[Cr(NH_3)_6]Cl_3$ (dotted–dashed line), and $[Ni(NH_3)_6]Cl_2$ (dotted line).

for example. This is attributed to the weakening of the N–H bond due to the formation of the N–H ... Cl-type hydrogen bond in the former.

The effects of coordination and hydrogen bonding mentioned above shift the NH_3 deformation and rocking modes to higher frequencies. Among them, the NH_3 rocking mode is most sensitive, and the degenerate deformation is least sensitive, to these effects. Thus the NH_3 rocking frequency is often used to compare the strength of the M–N bond in a series of complexes of the same type and anion.[18] As will be shown in the next subsection, a simple 1 : 1 complex such as that shown in Fig. III-2 has been prepared in inert gas matrices.[19]

To assign the skeletal modes such as the MN stretching and NMN bending modes, it is necessary to consider the normal modes of the octahedral MN_6 skeleton (O_h symmetry). The MN stretching mode in the low-frequency region is of particular interest since it provides direct information about the structure of the MN skeleton and the strength of the M–N bond. The octahedral MN_6 skeleton exhibits two ν(M–N) (A_{1g} and E_g) in Raman and one ν(M–N) (F_{1u}) in infrared spectra (Sec. II-8 of Part A). Most of these vibrations have been assigned based on observed isotope shifts (including metal isotopes, NH_3/ND_3 and $NH_3/^{15}NH_3$) and normal coordinate calculations. Although the assignment of the ν(Co–N) in the infrared spectrum of $[Co(NH_3)_6]Cl_3$ had been controversial, Schmidt and Müller[5] confirmed the original assignments made by Nakamoto et al.; the three weak bands at 498, 477, and 449 cm^{-1} are the split components of the triply degenerate F_{1u} mode (Fig. III-3).[20] According to Nakagawa and Shimanouchi,[21] the intensity of the MN stretching mode in the infrared increases as the M–N bond becomes more ionic and as the MN stretching frequency becomes lower. Relative to the Co(III)–N bond of the $[Co(NH_3)_6]^{3+}$ ion, the Co(II)–N bond of the $[Co(NH_3)_6]^{2+}$ ion is more ionic, and its stretching frequency is much lower (325 cm^{-1}). This may be responsible for the strong appearance of the Co(II)–N stretching band in the infrared.

As listed in Table III-1, two Raman-active MN stretching modes (A_{1g} and E_g) are observed for the octahedral hexammine salts. In general, $\nu(A_{1g})$ is higher than $\nu(E_g)$. However, the relative position of $\nu(F_{1u})$ with respect to these two

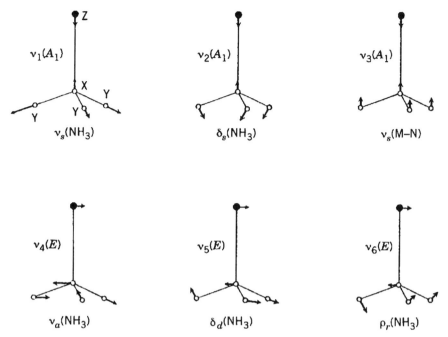

Fig. III-2. Normal modes of vibration of tetrahedral ZXY_3 molecules. (The band assignment is given for an $M-NH_3$ group.)

vibrations changes from one compound to another. Another obvious trend in $\nu(MN)$ is $\nu(M^{4+}-N) > \nu(M^{3+}-N) > \nu(M^{2+}-N)$. This holds for all symmetry species. Table III-1 shows that the NH_3 rocking frequency also follows the same trend as above.

Normal coordinate analyses on metal ammine complexes have been carried out by many investigators. Among them, Nakagawa, Shimanouchi, and co-workers.[9,17,21] have made the most comprehensive study, using the UBF field. The MN stretching force constants of the hexammine complexes follow this order:

$$Pt(IV) \gg Co(III) > Cr(III) > Ni(II) \approx Co(II)$$
$$2.13 \qquad 1.05 \qquad 0.94 \qquad 0.34 \qquad 0.33 \quad mdyn/\text{Å}$$

Terrasse et al.[14] report a value of 1.6 mdyn/Å for the Rh-N stretching force constant of the $[Rh(NH_3)_6]^{3+}$ ion in the UBF field. Recently, Acevedo and co-workers carried out normal coordinate calculations on the $[Cr(NH_3)_6]^{3+}$ and $[Ni(NH_3)_6]^{2+}$ ions.[22,23] On the other hand, Schmidt and Müller[5,6] and other workers[11] calculated the GVF constants of a number of ammine complexes by using the point mass model (i.e., the NH_3 ligand is regarded as a single atom having the mass of NH_3), and refined their values with isotope shift data

TABLE III-1. Infrared Frequencies of Octahedral Hexammine Complexes (cm^{-1})^a

Complex	ν_a(NH₃)	ν_s(NH₃)	δ_a(HNH)	δ_s(HNH)	ρ_r(NH₃)	ν(MN) IR	ν(MN) Raman	δ(NMN)	Refs.
[Mg(NH₃)₆]Cl₂	3353	3210	1603	1170	660	363	335 (A_{1g}), 243 (E_g)	198	2
[Cr(NH₃)₆]Cl₃	3257	3185, 3130	1630	1307	748	495, 473, 456	465 (A_{1g}), 412 (E_g)	—	3, 4
[⁵⁰Cr(NH₃)₆](NO₃)₃	3310	3250, 3190	1627	1290	770	471	—	270	5
[Mn(NH₃)₆]Cl₂	3340	3160	1608	1146	592	302	330 (A_{1g})	165	1, 6
[Fe(NH₃)₆]Cl₂	3335	3175	1596	1156	633	315	—	170	1, 6
[Ru(NH₃)₆]Cl₂	3315	3210	1612	1220	763	409	—	—	7
[Ru(NH₃)₆]Cl₃	3077		1618	1368, 1342	788	463	500 (A_{1g}), 475 (E_g)	283, 263	8
[Os(NH₃)₆]OsBr₆	3330	3125	1595	1339	818	452	—	256	8
[Co(NH₃)₆]Cl₂	3330	3250	1602	1163	654	325	357 (A_{1g}), 255 (E_g)	92	6, 9, 10
[Co(NH₃)₆]Cl₃	3240	3160	1619	1329	831	498, 477, 449	500 (A_{1g}), 445 (E_g)	331	11–13
[Co(ND₃)₆]Cl₃	2440	2300	1165	1020	667	462, 442, 415	—	294	5
[Rh(NH₃)₆]Cl₃		3200	1618	1352	845	472	515 (A_{1g}), 480 (E_g)	302	8, 14
[Ir(NH₃)₆]Cl₃		3155	1587	1350, 1323	857	475	527 (A_{1g}), 500 (E_g)	279, 264	8, 14
[⁵⁸Ni(NH₃)₆]Cl₂	3345	3190	1607	1176	685	335	370 (A_{1g}), 265 (E_g)	217	11, 15
[Zn(NH₃)₆]Cl₂	3350	3220	1596	1145	645	300	—	—	1, 10
[Cd(NH₃)₆]Cl₂	—	—	1585	1091	613	298	342 (A_{1g})	—	6, 10
[Pt(NH₃)₆]Cl₄	3150	3050	1565	1370	950	530, 516	569 (A_{1g}), 545 (E_g)	318	16, 17

^{a}All infrared frequencies are those of the F_{1u} species.

(H/D, ^{14}N/^{15}N, and metal isotopes). For the hexammine series, they obtained the following order:

$$Pt^{4+} > Ir^{3+} > Os^{3+} > Rh^{3+} > Ru^{3+} > Co^{3+} >$$
$$\quad\;\; 2.75 \quad 2.28 \quad 2.13 \quad 2.10 \quad 2.01 \quad 1.86$$
$$Cr^{3+} > Ni^{2+} > Co^{2+} > Fe^{2+} \sim Cd^{2+} > Zn^{2+} > Mn^{2+}$$
$$1.66 \quad 0.85 \quad 0.80 \quad\; 0.73 \qquad\;\; 0.69 \quad 0.67 \;\; \text{mdyn/Å}$$

For a series of divalent metals, the above order is parallel to the Irving–Williams series ($Mn^{2+} < Fe^{2+} < Co^{2+} < Ni^{2+} < Cu^{2+} > Zn^{2+}$). Schmidt and Müller[1] discussed the relationship between the MN stretching force constant and the stability constant or the bond energy.

Table III-2 lists the observed infrared frequencies and band assignments of tetrahedral, square-planar, and linear metal ammine complexes. The Raman-active MN stretching frequencies are also included in Table III-2. Normal coordinate analyses have been made by Nakagawa et al.[9,17,21] by using the UBF field; the following values were obtained for the MN stretching force constants:

$$Hg^{2+} > Pt^{2+} > Pd^{2+} > Cu^{2+}$$
$$2.05 \quad 1.92 \quad 1.71 \quad 0.84 \;\; \text{mdyn/Å}$$

Normal coordinate calculations have also been made by Tellez[35] on the tetrahedral $[Zn(NH_3)_4]^{2+}$ and $[Cd(NH_3)_4]^{2+}$ ions. Using the GVF field and the point mass approximation, Schmidt and Müller[6] obtained the following values;

$$Pt^{2+} > Pd^{2+} \gg Co^{2+} \sim Zn^{2+} \sim Cu^{2+} > Cd^{2+}$$
$$2.54 \quad 2.15 \quad 1.44 \quad 1.43 \quad 1.42 \quad 1.24 \;\; \text{mdyn/Å}$$

(2)   Ammine Complexes in Inert Gas Matrices

Ault[19] has measured the infrared spectra of cocondensation products of alkali halide (MX) vapors with NH_3 diluted in argon. In the case of KCl, for example, the bands at 3365, 3177, and 1103 cm^{-1} have been assigned to the $\nu_a(NH_3)$, $\nu_s(NH_3)$, and $\delta_s(HNH)$, respectively, of the 1:1 ion pair of the type I shown below.

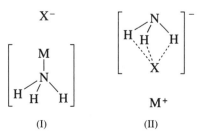

(I)   (II)

TABLE III-2. Infrared Frequencies of Other Ammine Complexes (cm^{-1})

Complex	ν_a(NH$_3$)	ν_s(NH$_3$)	δ_a(HNH)	δ_s(HNH)	ρ_r(NH$_3$)	ν(MN) IR	ν(MN) Raman	δ(NMN)	Refs.
Tetrahedral									
[Co(NH$_3$)$_4$](ReO$_4$)$_2$	3340	3260	1610	1240	693	430	405 (A_1)	195	24
[^{64}Zn(NH$_3$)$_4$]I$_2$	3275	3150	1596	1253	685	—	432 (A_1)	156	25
	3233			1239			412 (F_2)		
[Cd(NH$_3$)$_4$](ReO$_4$)$_2$	3354	3267	1617	1176	670	370	—	166	1, 26
								160	
Square-planar									
[^{104}Pd(NH$_3$)$_4$]Cl$_2$·H$_2$O	3270	3170	1630	1279	849	495	502 (A_{1g})	325	5, 27, 28
					802		482 (B_{1g})	300	
[Pt(NH$_3$)$_4$]Cl$_2$	3236	3156	1563	1325	842	510	543 (A_{1g})	301	17, 29, 28
							522 (B_{1g})		
[Cu(NH$_3$)$_4$]SO$_4$·H$_2$O	3327	3169	1669	1300	735	426	420 (A_{1g})	256	5, 30
	3253		1639	1283			375 (B_{1g})	227	
[Au(NH$_3$)$_4$](NO$_3$)$_3$	3490	3105	1571	1331	936	555	566	327	31
	3220				914		544	307	
								272	
Linear									
[Ag(NH$_3$)$_2$]$_2$SO$_4$	3320	3150	1642	1236	740	476	372 (A_1)	221	32, 33
	3230		1626	1222	703	400		211	
[Hg(NH$_3$)$_2$]Cl$_2$	3265	3197	1605	1268	719	513	412	—	33, 34

The type II structure was ruled out because of the following reasons. First, the $\delta_s(HNH)$ frequency should be sensitive to the metal ion in (I) and to the anion in (II). The fact that it shows relatively large shifts by changing the metal ion, but almost no shifts by changing the anion, supports (I), Second, the $\nu_a(NH_3)$ and $\nu_s(NH_3)$ in (II) are expected to be highly sensitive to the anion, owing to formation of the $N-H\cdots X$ hydrogen bonds; this is not the case in (I). The fact that they show only small shifts in going from CsCl to CsI supports (I). Further supports for structure (I) are given by the appearance of the $\rho_r(NH_3)$ and $\nu(M-N)$ at 458 and 232 cm^{-1} (KCl), respectively. These frequencies are much lower than those of transition metal complexes discussed earlier, because their M–N bonds are much weaker (more ionic).

Süzer and Andrews[36] studied the IR spectra of cocondensation products of alkali metal (M) vapors with NH_3/Ar. They assigned the following bands:

	Li	Na	K	Cs	
$\nu_s(NH_3)$	3277	3294	3292	3287	(all in cm^{-1})
$\delta_s(NH_3)$	1133	1079	1064	1049	

to the $1:1$ adduct of C_{3v} symmetry which is similar to that of the $M(NH_3)^+$ cation discussed earlier. The $M-NH_3$ bonding has been attributed to a small charge transfer from NH_3 to M in the case of Li and Na, and to a reverse charge transfer in the case of K and Cs. At high concentrations of M and NH_3, large aggregates of undefined stoichiometries were formed. Similar work including Fe and Cu was carried out by Szczepanski et al.[37] Loutellier et al.[38] have made the most extensive IR study on the $Li(K)/NH_3$/Ar system. By varying the concentrations and relative ratios of M/NH_3 in a wide range, they were able to observe bands characteristic of the $1:1$, $1:2$, ... $1:n$, and $2:1$, $3:1$ ... $m:1$ adducts. As an example, Fig. III-3 shows the IR spectra of the Li/NH_3/Ar system in the $\delta_s(NH_3)$ region. The molar ratios (Li/NH_3) and the peaks characteristic of each species are indicated in the figure. In general, the $1:1$ adduct is formed when the concentrations of Li and NH_3 are close. If the concentration of NH_3 is high relative to Li, the $1:n$ ($n = 2, 3, 4...$) adducts are formed. On the other hand, the $m:1$ ($m = 2, 3, 4...$) adducts result when the concentration of Li is high relative to NH_3. For the $1:1$ adduct of Li, the bands at 381 and 320 cm^{-1} have been assigned to the $\rho_r(NH_3)$ and $\nu(Li-N)$, respectively. The Li–N stretching force constant was found to be 0.3 mdyn/Å.

(3)   Halogenoammine Complexes

If the NH_3 groups of a hexammine complex are partly replaced by other groups, the degenerate vibrations are split because of lowering of symmetry, and new bands belonging to other groups appear. Here we discuss only halogenoammine complexes. The infrared spectra of $[Co(NH_3)_5X]^{2+}$- and *trans*-$[Co(NH_3)_4X_2]^+$-type complexes have been studied by Nakagawa and

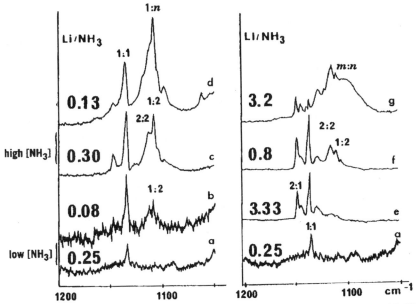

Fig. III-3. IR spectra (1200–1050 cm^{-1}) of cocondensation products of Li atoms with NH$_3$ molecules in Ar matrices. Left column: (a) Li/NH$_3$/Ar = 0.1/0.4/1000, (b) 0.1/1.2/1000, (c) 3/10/1000, and (d) 2.5/20/1000. Right column: (a) Li/NH$_3$/Ar = 0.1/0.4/1000, (e) 2/0.6/1000, (f) 2/2.5/1000, and (g) 8/2.5/1000 (reproduced with permission from Ref. 38).

TABLE III-3   Skeletal Vibrations of Pentammine and *trans*-Tetrammine Co(III) Complex (cm^{-1})[9,21]

Complex	ν(CoN)	ν(CoX)	Skeletal Bending
Pentammine (**C**$_{4v}$)			
[Co(NH$_3$)$_5$F]$^{2+}$			
A_1	480, 438	343	308
E	498	—	345, 290, 219
[Co(NH$_3$)$_5$Cl]$^{2+}$			
A_1	476, 416	272	310
E	498	—	292, 287, 188
[Co(NH$_3$)$_5$Br]$^{2+}$			
A_1	475, 410	215	287
E	497	—	290, 263, 146
[Co(NH$_3$)$_5$I]$^{2+}$			
A_1	473, 406	168	271
E	498	—	290, 259, 132
trans-Tetrammine (**D**$_{4h}$)			
[Co(NH$_3$)$_4$Cl$_2$]$^+$			
A_{2u}	—	353	186
E_u	501	—	290, 167
[Co(NH$_3$)$_4$Br$_2$]$^+$			
A_{2u}	—	317	227
E_u	497	—	280, 120

Shimanouchi.[9,21] Table III-3 lists the observed frequencies and band assignments obtained by these workers. The infrared spectra of some of these complexes in the CoN stretching region are shown in Fig. III-4. Normal coordinate analyses on these complexes[9] have yielded the following UBF stretching force constants (mdyn/Å): K(Co—N), 1.05; K(Co—F), 0.99; K(Co—Cl), 0.91; K(Co—Br), 1.03; and K(Co—I), 0.62. Raman spectra of some chloroammine Co(III) complexes have been assigned.[39] In the series of the $[Cr(NH_3)_5X]^{2+}$ ions, the ν(Cr–N) are in the 475–400 cm^{-1} region, and the ν(Cr–X) are at 540, 302, 264, and 184 cm^{-1}, respectively, for X = F, Cl, Br, and I.[40] For more information on halogenoammine complexes of Cr(III), see Refs. 41 and 3. Detailed vibrational assignments are available for halogenoammine complexes of Os(III)[42] and of Ru(III), Rh(III), Os(III), and Ir(III).[43]

In regard to $M(NH_3)_4X_2$- and $M(NH_3)_3X_3$-type complexes, the main interest has been the distinction of stereoisomers by vibrational spectroscopy. As shown in Appendix V of Part A, *trans*-MN_4X_2 (D_{4h}) exhibits one MN stretch-

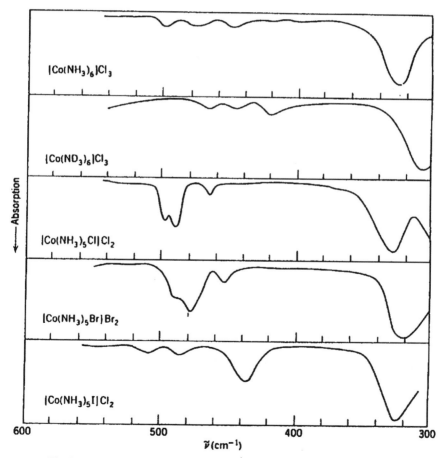

Fig. III-4. Infrared spectra (600–300 cm^{-1}) of Co(III) ammine complexes.[20]

ing (E_u) and one MX stretching (A_{2u}), while *cis*-MN$_4$X$_2$ ($\mathbf{C}_{2v}$) shows four MN stretching (two A_1, B_1, and B_2) and two MX stretching (A_1 and B_1) vibrations in the infrared. For *mer*-MN$_3$X$_3$ ($\mathbf{C}_{2v}$), three MN stretching and three MX stretching vibrations are infrared active, whereas only two MN stretching and two MX stretching vibrations are infrared active for *fac*-MN$_3$X$_3$ ($\mathbf{C}_{3v}$). Nolan and James[16] have measured and assigned the Raman spectra of a series of [Pt(NH$_3$)$_n$Cl$_{6-n}$]$^{(n-2)+}$-type complexes. Li et al.[44] carried out normal coordinate analysis on *cis*-Pt(NH$_3$)$_2$Cl$_4$.

Vibrational spectra of the planar M(NH$_3$)$_2$X$_2$-type complexes [M = Pt(II) and Pd(II)] have been studied by many investigators. Table III-4 summarizes the observed frequencies and band assignments of their skeletal vibrations, including those of "*cis*-platin"—the well-known anticancer drug. Figure III-5 shows the infrared spectra of *cis*- and *trans*-[Pd(NH$_3$)$_2$Cl$_2$] obtained by Layton et al.[49] As expected, both the PdN and PdCl stretching bands split into two in the *cis*-isomer. Durig et al.[50] found that the PdN stretching frequencies range from 528 to 436 cm^{-1}, depending on the nature of other ligands in the complex. In general, the PtN stretching band shifts to a lower frequency as a ligand of stronger *trans*-influence is introduced in the position *trans* to the Pt–N bond.[51] Using infrared spectroscopy, Durig and Mitchell[52] studied the isomerization of *cis*-[Pd(NH$_3$)$_2$X$_2$] to its *trans*-isomer.

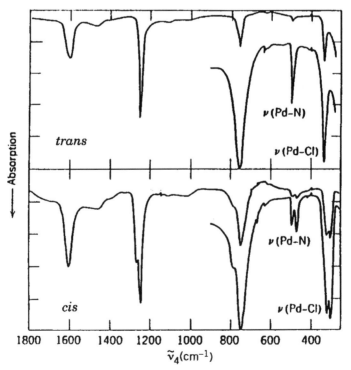

Fig. III-5. Infrared spectra of *trans*- and *cis*-[Pd(NH$_3$)$_2$Cl$_2$].[49]

TABLE III-4. Skeletal Frequencies of Square-Planar M(NH$_3$)$_2$X$_2$-Type Complexes (cm^{-1})^a

Complex	ν(MN)	ν(MX)	Bending	Refs.
trans-[Pd(NH$_3$)$_2$Cl$_2$]				
IR	496	333	245, 222, 162, 137	45, 28
R	492	295	224	
cis-[Pd(NH$_3$)$_2$Cl$_2$]				
IR	495, 476	327, 306	245, 218, 160, 135	45
trans-[Pd(NH$_3$)$_2$Br$_2$]				
IR	490	—	220, 220, 122, 101	45
R	483	182	172	28
cis-[Pd(NH$_3$)$_2$Br$_2$]				
IR	480, 460	258	225, 225, 120, 100	45
trans-[Pd(NH$_3$)$_2$I$_2$]				
IR	480	191	263, 218, 109	45
trans-[Pt(NH$_3$)$_2$Cl$_2$]				
IR	572	365	220, 195	46, 47
R	538	334	—	28, 46
cis-[Pt(NH$_3$)$_2$Cl$_2$]				
IR	510	330, 323	250, 198, 155, 123	47
R	507	253	160	28
trans-[Pt(NH$_3$)$_2$Br$_2$]				
IR	504	260	230	46, 47
R	535	206	—	46
trans-[Pt(NH$_3$)$_2$I$_2$]				
R	532	153	—	46

^{a}For band assignments, see also Refs. 17 and 48.

(4) Linear-Chain Ammine Complexes

Mixed-valence compounds such as Pd^{II}Pt^{IV}(NH$_3$)$_4$Cl$_6$ and Pd^{II}Pd^{IV}(NH$_3$)$_4$Cl$_6$ take the form of a chain structure as shown on the next page. Both compounds exhibit an intense, extremely broad spectrum in the visible region. The IR spectra of these mixed-valence compounds are approximately superpositions of those of each of the components. However, the RR spectra (Secs. I-22 and 23 of Part A) obtained by using exciting lines in this region are markedly different from the IR spectra. In the case of the Pd–Pt complex, RR spectra involving the progressions of three totally symmetric metal–chlorine stretching vibrations were observed. Thus, the visible spectrum was attributed to a metal–metal mixed-valence transition. On the other hand, the Pd–Pd complex exhibits a RR spectrum involving several stretching and bending fundamentals and their combinations and overtones which originate in the Pd(NH$_3$)$_2$Cl$_4$ component only. Thus, Clark and Trumble[53] attributed the visible spectrum to the metal–ligand charge-transfer transitions within this component. Later, this work was extended to the Ni–Pt complex of ethylenediamine [Sec. III-2(3)].

In the Magnus green salt, [Pt(NH$_3$)$_4$] [PtCl$_4$], the Pt(II) atoms form a linear chain structure with relatively short Pt–Pt distances (~3.3 Å). Originally,

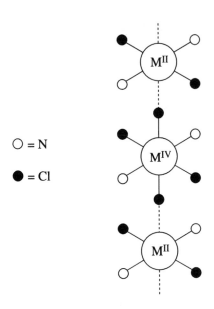

$\bigcirc$ = N

$\bullet$ = Cl

Hiraishi et al.[17] assigned the infrared band at 200 cm^{-1} to a lattice mode which corresponds to the stretching mode of the Pt–Pt–Pt chain. This high frequency was justified on the basis of the strong Pt–Pt interaction in this salt. Adams and Hall,[54] on the other hand, assigned this mode at 81 cm^{-1}, and the 201 cm^{-1} band to a NH$_3$ torsion. In fact, the latter is shifted to 158 cm^{-1} by the deuteration of NH$_3$ ligands.[55] Different from the mixed-valence complexes, the Raman spectrum of the Magnus green salt obtained by excitation in the visible absorption band does not display long overtone series.[55] This is expected since it has no axial bonds which would change the bond lengths upon electronic excitation. Resonance Raman spectra of these and other linear-chain complexes are reviewed by Clark.[56]

(5)   Lattice Vibrations of Ammine Complexes

Vibrational spectra of metal ammine complexes in the crystalline state exhibit lattice vibrations below 200 cm^{-1}. Assignments of lattice modes have been made for the hexammine complexes of Mg(II),[2] Co(II),[10,57] Ni(II),[10,21,57,58] [Co(NH$_3$)$_6$] [Co(CN)$_6$],[59] and [Pt(NH$_3$)$_4$]Cl$_2$.[60] Lattice modes and low-frequency internal modes of hexammine complexes have also been studied by Janik et al.[61,62] using the inelastic neutron-scattering technique.

(6)   Amido (NH$_2$) Complexes

The vibrational spectra of amido complexes may be interpreted in terms of the normal vibrations of a pyramidal ZXY$_2$-type molecule. Mizushima et al.[63] and Niwa et al.[64] carried out normal coordinate analysis on the [Hg(NH$_2$)$_2$]$_\infty^+$ ion

TABLE III.5. Infrared Frequencies and Band Assignments of Amido Complexes (cm^{-1})[64]

Compound	$\nu(NH_2)$	$\delta(NH_2)$	$\rho_w(NH_2)$	$\rho_r(NH_2)$	$\nu(HgN)$
$[Hg(NH_2)]_\infty^+(Cl)_\infty^-$	3200 ⎱ 3175 ⎰	1540	1025	673	573
$[Hg(NH_2)]_\infty^+(Br)_\infty^-$	3220 ⎱ 3180 ⎰	1525	1008	652	560

(infinite-chain polymer); the results of the latter workers are given in Table III-5. Brodersen and Becher[65] studied the infrared spectra of a number of compounds containing Hg–N bonds and assigned the HgN stretching bands at 700–400 cm^{-1}. The infrared spectrum of the NH_2^- ion in alkali-metal salts has been measured.[66] Alkylamido complexes of the type $M(NR_2)_{4,5}$ (M = Ti, Zr, Hf, V, Nb, and Ta) exhibit their MN stretching bands in the 700–530 cm^{-1} region.[67]

(7)   Amine(RNH$_2$) Complexes

Infrared spectra of methylamine complexes, $[Pt(CH_3NH_2)_2X_2]$ (X: a halogen), have been studied by Watt et al.[68] and Kharitonov et al.[69] Far-infrared spectra of $[M(R_2NH)_2X_2]$- [M = Zn(II) or Cd(II); R = ethyl or n-propyl; X = Cl or Br] type complexes have also been reported.[70] Chatt and co-workers[71] studied the effect of hydrogen bonding on the NH stretching frequencies of $trans$-$[Pt(RNH_2)Cl_2L]$-type complexes (R = Me, Et, etc.; L = C_2H_4, PEt$_3$, etc.) in organic solvents such as chloroform and dioxane. Their study revealed that the complexes of primary amines have a strong tendency to associate through intermolecular hydrogen bonds of the NH$\cdots$Cl type, whereas those of secondary amines have little tendency to associate. Later, this difference was explained on the basis of steric repulsion and intramolecular interaction between the NH hydrogen and the nonbonding d-electrons of the metal.[72] The $\nu(Pt–I)$ vibrations of Pt(RNH$_2$)$_2$I$_2$-type complexes are in the 200–150 cm^{-1} region.[73]

IR and Raman spectra of metal complexes of aniline have been reviewed by Thornton.[73a]

(8)   Complexes of Hydrazine and Hydroxylamine

Hydrazine ($H_2N–NH_2$) coordinates to a metal as a unidentate or a bridging bidentate ligand. No chelating (bidentate) hydrazines are known. For example, the hydrazine ligands in $[M(N_2H_4)_2]Cl_2$ [M(II) = Mn, Fe, Co, Ni, Cu, Zn, and Cd] are bridging bidentate (polymeric):

On the other hand, all hydrazine ligands in $[Co(N_2H_4)_6]Cl_2$ are coordinated to the Co atom as a unidentate ligand. According to Nicholls and Swindells,[74] the complexes of the former type exhibit the $\nu(N-N)$ near 970 cm^{-1}, whereas those of the latter type show it near 930 cm^{-1}. The IR spectra of hydrazine complexes of Hg(II),[75] M(II)(M = Ni, Co, Zn, and Cd),[76] Os(II),[77] and Ln(III) (Ln = Pr, Nd, and Sm)[78] have been reported. In these compounds, hydrazine acts as a unidentate or bridging bidentate ligand. The $\nu(M-N)$ are assigned empirically in the 440–330 cm^{-1} region.[79]

The vibrational spectra of hydroxyalmine(NH$_2$OH) have been reported by Kharitonov et al.[80]

III-2. COMPLEXES OF ETHYLENEDIAMINE AND RELATED LIGANDS

(1)   Chelating Ethylenediamine

When ethylenediamine(en) coordinates to a metal as a chelating ligand, it may take a *gauche* (δ and λ) or a *cis* conformation, as shown in Fig. III-6. Then, eight different conformations are probable for the $[M(en)_3]^{n+}$ ion if we consider all possible combinations of conformations of the three chelate rings (δ or λ) around the chiral metal center. They are designated as $\Lambda(\delta\delta\delta)$, $\Lambda(\delta\delta\lambda)$, $\Lambda(\delta\lambda\lambda)$, $\Lambda(\lambda\lambda\lambda)$, $\Delta(\lambda\lambda\lambda)$, $\Delta(\lambda\lambda\delta)$, $\Delta(\lambda\delta\delta)$, and $\Delta(\delta\delta\delta)$. According to X-ray

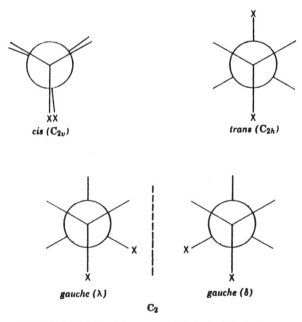

Fig. III-6. Rotational isomers of 1,2-disubstituted ethane.

analysis, all the en ligands in the $[Co(en)_3]^{3+}$ ion take the *gauche* conformation (δ), and the configuration of the whole ion is $\Lambda(\delta\delta\delta)$.[81,82] Although it is rather difficult to obtain such information from vibrational spectra, Cramer and Huneke[83] have shown that some of these conformers can be distinguished by the number of IR-active C–C stretching vibrations. For example, $[Cr(en)_3]Cl_3 \cdot 3.5H_2O$ [$\Lambda(\delta\delta\delta)$, D_3 symmetry] exhibits only one band at 1003 cm^{-1} whereas $[Cr(en)_3][Ni(CN)_5] \cdot 1.5H_2O$ [$\Lambda(\delta\delta\lambda, \delta\lambda\lambda)$, C_2 symmetry] exhibits three bands at 1008, 1002 (shoulder), and 995 cm^{-1}. Gouteron has shown[84] that racemic (*dl*) and optically active (*d*) forms of $[Co(en)_3]Cl_3$ can be distinguished in the crystalline state by comparing vibrational spectra below 200 cm^{-1}.

Normal coordinate analyses on metal complexes of ethylenediamine have been made by several groups of workers. Fleming and Shepherd[85] carried out normal coordinate calculations on the 1:1(Cu/en) model of the $[Cu(en)_2]^{2+}$ ion. These workers considered a 9-atom system of C_{2v} symmetry, assuming that the two hydrogen atoms bonded to the C and N atoms are single atoms having the double mass of hydrogen. The IR bands at 410 and 360 cm^{-1} have been assigned to the $\nu(Cu–N)$ which are coupled with other skeletal modes. The corresponding Cu–N stretching force constant (GVF) was 1.25 mdyn/Å. Borch and co-workers[86–88] have carried out more complete calculations by considering all the 37 atoms of the $[Rh(en)_3]^{3+}$ ion ($[\Lambda(\delta\delta\delta)]$ configuration of D_3 symmetry), and the force constants (GVF) have been refined by using the vibrational frequencies obtained for the N–d_{12}, C–d_{12}, N,C–d_{24}, and their ^{15}N analogues. In total, 38 force constants were employed, including the Rh–N stretch of 1.607 mdyn/Å. Three $\nu(Rh–N)$ vibrations are at 545 (A_1), 445 (A_2) and 506 (E), although they are strongly coupled with other skeletal bending modes. Figure III-7 shows the IR and Raman spectra of (N–d_{12}) $[Rh(en)_3]Cl_3 \cdot D_2O$ obtained by Borch et al.[88] Later, their calculations (E modes) were improved by Williamson et al.,[89] who assigned the polarized Raman spectra of tris(ethylenediamine) complexes of Co(III) and Rh(III) based on similar calculations.

Empirical assignments of $\nu(M–N)$ have been reported for $[M(en)_3]^{3+}$ (M = Cr and Co),[90] $[M(en)_3]^{2+}$ (M = Zn, Cd, Fe, etc.),[91] and $[M(en)_2]^{2+}$ (M = Cu, Pd, and Pt).[92] Bennett et al.[93,94] found that, in a series of the $M(en)_3SO_4$ complexes, the $\nu(M–N)$ frequencies follow the order:

M =	Mn(II)		Fe(II)		Co(II)		Ni(II)		Cu(II)		Zn(II)	
ν_4	391	<	397	<	402	<	410	<	485	>	405	(cm^{-1})
ν_5	303	<	321	≈	319	<	334	<	404	>	291	

As mentioned in Sec. III-1(1), this is the order of stability constants known as the Irving–Williams series. These assignments have been confirmed by extensive isotope substitutions including metal isotopes.

Stein et al.[95] observed that the Raman intensities of the totally symmetric stretching and chelate deformation modes of the $[Co(en)_3]^{3+}$ ion at 526 and 280 cm^{-1}, respectively, display minima near 21.5 KK where the *d–d* transition shows its absorption maximum. Figure III-8 shows the excitation profiles

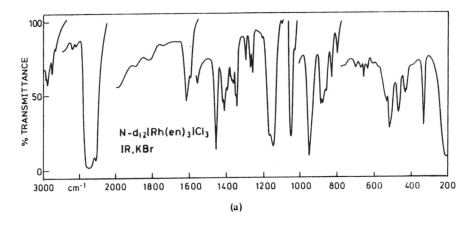

(a)

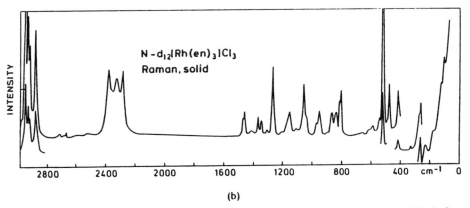

(b)

Fig. III-7. Infrared (top) and Raman spectra (bottom) of N-deuterated (N-d$_{12}$) [Rh(en)$_3$]Cl$_3$·D$_2$O (reproduced with permission from Ref. 88).

of these totally symmetric vibrations, as well as that of non-totally symmetric ν(Co–N)(E_g) at 444 cm^{-1}. Since this result is opposite to what one expects from resonance Raman spectroscopy (Sec. I-22 of Part A), it is called "antiresonance." These workers attributed its origin to "interference" between the weak scattering from the ligand-field state and strong preresonance scattering from higher energy allowed electronic states. For more theoretical study on this phenomenon, see Ref. 96.

Lever and Mantovani[97] assigned the MN stretching bands of M(N—N)$_2$X$_2$-[M = Cu(II), Co(II), and Ni(II); N—N = en, dimethyl-en, etc.; X = Cl, Br, etc.] type complexes by using the metal isotope technique. For these compounds, the CoN and NiN stretching bands have been assigned to 400–230 cm^{-1}[97] and the CuN stretching vibrations have been located in the 420–360 cm^{-1} range.[98] A straight-line relationship between the square of the CuN stretching frequency and the energy of the main electronic *d–d* band was found,[99] with some exceptions.[98]

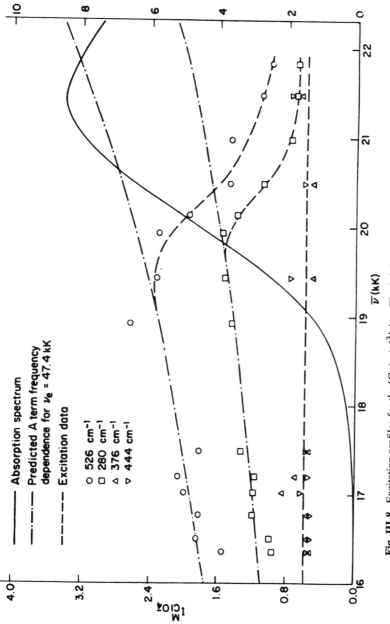

Fig. III-8. Excitation profiles for the $[Co(en)_3]^{3+}$ ion. The left-hand scale pertains to the excitation data, and the right-hand scale to the absorption spectrum. $I_{ClO_4^-}^M$ is the molar intensity relative to that of the ν_1 band of ClO_4^-: $(I_{Co}/C_{Co})(C_{ClO_4^-}/I_{ClO_4^-})$. The theoretical curves (----) are calculated with the A term frequency dependence given by A. C. Albrecht and M. C. Hutley (*J. Chem. Phys.*, **55**, 4438 (1971) (reproduced with permission from Ref. 95).

17

The infrared spectra of *cis*- and *trans*-[M(en)$_2$X$_2$]$^+$ [M = Co(III), Cr(III), Ir(III), and Rh(III); X = Cl, Br, etc.] have been studied extensively.[100–103] These isomers can be distinguished by comparing the spectra in the regions of 1700–1500 (NH$_2$ bending), 950–850 (CH$_2$ rocking), and 610–500 cm^{-1} (MN stretching).

(2)   Bridging Ethylenediamine

Ethylenediamine takes the *trans* form when it functions as a bridging group between two metal atoms. Powell and Sheppard[104] were the first to suggest that ethylenediamine in (C$_2$H$_4$)Cl$_2$Pt(en)PtCl$_2$(C$_2$H$_4$) is likely to be *trans*, since the infrared spectrum of this compound is simpler than that of other complexes in which ethylenediamine is *gauche*. However, a recent NMR study on this complex ruled out this possibility.[105] The *trans* configuration of ethylenediamine has been found in (AgCl)$_2$en,[106] (AgSCN)$_2$en,[107] (AgCN)$_2$en,[107] Hg(en)Cl$_2$,[108] and M(en)Cl$_2$ (M = Zn or Cd).[106] The structure of these complexes may be depicted as follows:

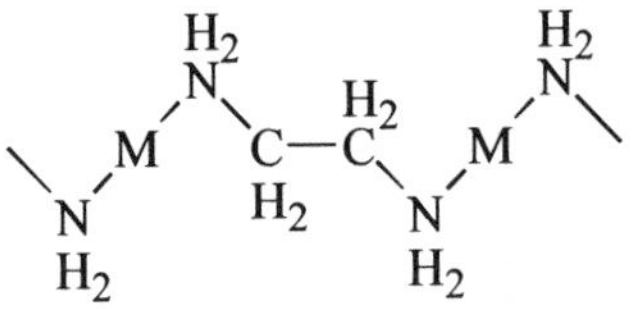

A more complete study, including the infrared and Raman spectra, of M(en)X$_2$-type complexes [M = Zn(II), Cd(II), and Hg(II); X = Cl, Br, and SCN] has been made by Iwamoto and Shriver.[109] Mutual exclusion of infrared and Raman spectra, along with other evidence, supports the C$_{2h}$ bridging structure of the en ligand in the Cd and Hg complexes (see Fig. III-9).

(3)   Mixed-Valence Complexes

In Sec. III-1(3), we discussed the RR spectra of mixed-valence complexes such as PdPt(NH$_3$)$_4$Cl$_6$ and Pd$_2$(NH$_3$)$_4$Cl$_6$. Analogous complexes can be prepared by changing the metal and the N-donor ligand. For example, Clark and Croud[110] measured the RR spectra of single crystals of [Ni(en)$_2$] [Pt(en)$_2$X$_2$] (ClO$_4$)$_4$ (X = Cl, Br, and I) in which a linear chain such as

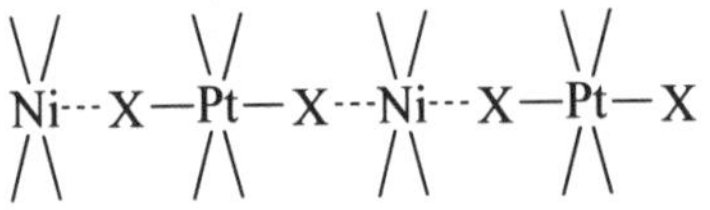

is formed via halogen bridges. These complexes exhibit strong, broad bands due to the Ni(II)–Pt(IV) charge-transfer transition in the visible region. Figure III-10 shows the RR spectra of the chloro complex obtained by 488-nm excitation at 20 K. It is seen that the complex exhibits a series of over-

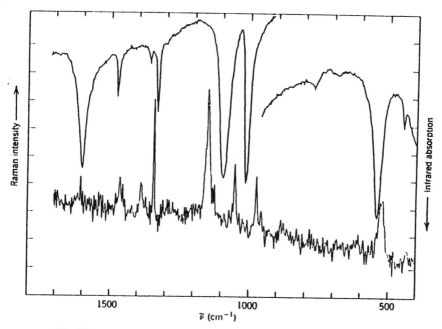

Fig. III-9. Infrared (top) and Raman (bottom) spectra of [Cd(en)Br$_2$].[109]

tones of the symmetric ν_s(Cl–Pt–Cl) vibration up to 6ν which split into three peaks due to mixing of ^{35}Cl/^{37}Cl isotopes. Similar overtone series were observed for X = Br and I. Using these data, the frequencies corrected for anharmonicity and anharmonicity constants have been calculated (Sec. I-23 of Part A). Polarized RR studies show that these vibrations are completely polarized along the Ni$\cdots$X–Pt–X$\cdots$Ni axis. Similar work is reported for [Pt(en)$_2$][Pt(en)$_2$Cl$_2$](ClO$_4$)$_4$.[111] In the [Pt(en)$_2$][Pt(en)$_2$X$_2$](ClO$_4$)$_4$ series, the IR-active chain phonon frequencies (antisymmetric stretching) are 359.1, 238.7, and 184.2 cm^{-1}, respectively, for X = Cl, Br, and I.[112]

Omura et al.[113] reported the far-IR spectra of Magnus-type salts [M(en)$_2$]-[M'Cl$_4$] [M, M' = Pt(II) or Pd(II)]. Berg and Rasmussen[114] measured the IR and far-IR spectra of the analogous complexes [M(en)$_2$][M'Br$_4$] [M, M' = Pt(II) or Pd(II)] and [M(en)$_2$][HgI$_4$]. No bands assignable to the metal–metal stretching were observed in these complexes.

(4) Complexes of Polyamines

Polyamines such as these shown below coordinate to a metal as tridentate or tetradentate ligands:

Diethylenetriamine(dien)

$$H_2N-(CH_2)_2-\overset{\overset{\displaystyle H}{|}}{N}-(CH_2)_2-NH_2$$

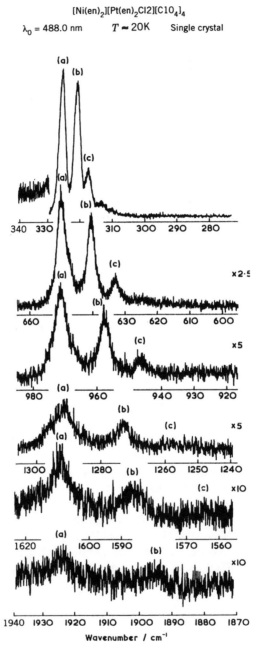

Fig. III-10. Isotopomer band intensities for ν_1–$6\nu_1$ of a single crystal of $[Ni(en)_2][Pt(en)_2Cl_2]$-$(ClO_4)_4$. For ease of presentation, the $^{35}ClPt^{35}Cl$ component (a) of each harmonic is lined up with the same abscissa value; (b) and (c) refer to the $^{35}ClPt^{37}Cl$ and $^{37}ClPt^{37}Cl$ components, respectively (reproduced with permission from Ref. 110).

Triaminotriethylamine(tren):

$$N \begin{cases} (CH_2)_2{-}NH_2 \\ (CH_2)_2{-}NH_2 \\ (CH_2)_2{-}NH_2 \end{cases}$$

Triethylenetetramine(trien):

$$H_2N{-}(CH_2)_2{-}\underset{H}{N}{-}(CH_2)_2{-}\underset{H}{N}{-}(CH_2)_2{-}NH_2$$

The infrared spectra of diethylenetriamine (dien) complexes have been reported for [Pd(dien)X]X (X = Cl, Br, and I)[115] and [Co(dien)(en)Cl]$^{2+}$.[116] The latter exists in the four isomeric forms shown in Fig. III-11. Their infrared spectra revealed that the ω- and κ-isomers contain dien in the *mer*-configuration; the π- and ε-isomers contain dien in the *fac*-configuration. The *mer*- and *fac*-isomers of [M(dien)X$_3$] [M = Cr(III), Co(III), and Rh(III); X: a halogen] can also be distinguished by infrared spectra.[117]

The infrared spectra of β, β', β''-triaminotriethylamine(tren) complexes with Co(III)[118] and lanthanides[119] have been reported. Buckingham and Jones[120]

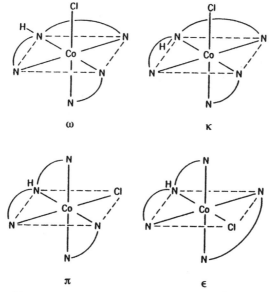

Fig. III-11. Structures of the [Co(dien)(en)Cl]$^{2+}$ ion.

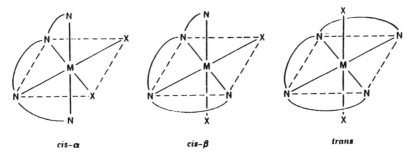

Fig. III-12. Structures of the [M(trien)X$_2$]$^+$ ions.

measured the infrared spectra of [M(trien)X$_2$]$^+$, where trien is triethylene-tetramine, M is Co(III), Cr(III), or Rh(III), and X is a halogen or an acido anion. These compounds give three isomers (Fig. III-12) which can be distinguished, for example, by the CH$_2$ rocking vibrations in the 920–869 cm^{-1} region. For [Co(trien)Cl$_2$]ClO$_4$, *cis-α*-isomer exhibits two strong bands at 905 and 871 cm^{-1} and *cis-β*-isomer shows four bands at 918, 898, 868, and 862 cm^{-1}; *trans*-isomer gives only one band at 874 cm^{-1} with a weak band at 912 cm^{-1}. Far-infrared spectra of some of these trien complexes have been reported.[121]

(5)   Complexes of 1,2-Disubstituted Ethanes

As is shown in Fig. III-6, 1,2-disubstituted ethane may exist in the *cis*, *trans*, or *gauche* form, depending on the angle of internal rotation. The *cis* form may not be stable in the free ligand because of steric repulsion between two X groups. The *trans* form belongs to point group C$_{2h}$, in which only the *u* vibrations (antisymmetric with respect to the center of symmetry) are infrared-active. On the other hand, both *gauche* forms belong to point group C$_2$, in which all the vibrations are infrared-active. Thus the *gauche* form exhibits more bands than the *trans* form. Mizushima and co-workers[122] have shown that 1,2-dithiocyana-toethane (NCS—CH$_2$—CH$_2$—SCN) in the crystalline state definitely exists in the *trans* form, because no infrared frequencies coincide with Raman frequencies (mutual exclusion rule). By comparing the spectrum of the crystal with that of a CHCl$_3$ solution, they concluded that several extra bands observed in solution can be attributed to the *gauche* form. Table III-6 summarizes the infrared frequencies and band assignments obtained by Mizushima et al. It is seen that the CH$_2$ rocking vibration provides the most clear-cut diagnosis of conformation: one band (A$_u$) at 749 cm^{-1} for the *trans* form, and two bands (A and B) at 918 and 845 cm^{-1} for the *gauche* form.

The compound 1,2-dithiocyanatoethane may take the *cis* or *gauche* form when it coordinates to a metal through the S atoms. The chelate ring formed will be completely planar in the *cis*, and puckered in the *gauche*, form. The *cis* and *gauche* forms can be distinguished by comparing the spectrum of a

TABLE III-6. Infrared Spectra of 1,2-Dithiocyanatoethane and Its Pt(II) Complex (cm^{-1})[122]

Ligand			
Crystal *trans*	CHCl$_3$ Solution (*gauche* + *trans*)	Pt Complex (*gauche*)	Assignment
—	2170(g)	2165(g) ⎫	
2155(t)	2170(t)	— ⎬	$\nu(C\equiv N)$
1423(t)	1423(t)	— ⎫	
—	1419(g)	1410(g) ⎬	$\delta(CH_2)$
1291^a(t)	—	— ⎭	
—	1285(g)	1280(g) ⎫	
1220(t)	1215(t)	— ⎬	$\rho_w(CH_2)$
1145(t)	1140(t)	— ⎫	
—	1100(g)	1110(g) ⎭	$\rho_t(CH_2)$
—	—(g)^b	1052(g) ⎫	
1037^a	—	— ⎬	$\nu(CC)$
—	918(g)	929(g) ⎫	
—	845(g)	847(g) ⎬	$\rho_r(CH_2)$
749(t)	—^{b}	— ⎭	
680(t)	677(t)	— ⎫	
660(t)	660(t)	— ⎬	$\nu(CS)$

^{a}Raman frequencies in the crystalline state.
^{b}Hidden by CHCl$_3$ absorption.

metal chelate with that of the ligand in CHCl$_3$ solution (*gauche* + *trans*). Table III-6 compares the infrared spectrum of 1,2-dithiocyanatoethanedichloroplatinum(II) with that of the free ligand in a CHCl$_3$ solution. Only the bands characteristic of the *gauche* form are observed in the Pt(II) complex. This result definitely indicates that the chelate ring in the Pt(II) complex is *gauche*. The method described above has also been applied to the metal complexes of 1,2-dimethylmercaptoethane (CH$_3$S—CH$_2$—CH$_2$—SCH$_3$).[123] In this case, the free ligand exhibits one CH$_2$ rocking at 735 cm^{-1} in the crystalline state (*trans*), whereas the metal complex always exhibits two CH$_2$ rockings at 920–890 and 855–825 cm^{-1} (*gauche*). In the case of ethylenediamine complexes discussed earlier, the CH$_2$ rocking mode does not provide a clear-cut diagnosis since it couples strongly with the NH$_2$ rocking and C–N stretching modes.

III-3. COMPLEXES OF PYRIDINE AND RELATED LIGANDS

(1) Complexes of Pyridine

Upon complex formation, the pyridine (py) vibrations in the high-frequency region are not shifted appreciably, whereas those at 604 (in-plane ring deformation) and 405 cm^{-1} (out-of-plane ring deformation) are shifts to higher frequen-

TABLE III-7. Vibrational Frequencies of Pyridine Complexes (cm^{-1})[124]

Complex	Structure	py[a]	py[a]	ν(M—py)
Co(py)$_2$Cl$_2$	Monomeric, tetrahedral	642	422	253[b]
Ni(py)$_2$I$_2$	Monomeric, tetrahedral	643	428	240
Cr(py)$_2$Cl$_2$	Polymeric, octahedral	640	440	219
Cu(py)$_2$Cl$_2$	Polymeric, octahedral	644	441	268
Co(py)$_2$Cl$_2$	Polymeric, octahedral	631	429	243, 235[b]
mer-[Rh(py)$_3$Cl$_3$]	Monomeric, octahedral	650	468	265, 245, 230
fac-[Rh(py)$_3$Cl$_3$]	Monomeric, octahedral	643	464	266, 245
trans-[Ni(py)$_4$Cl$_2$]	Monomeric, octahedral	626	426	236
trans-Ir(py)$_4$Cl$_2$]Cl	Monomeric, octahedral	650	469	260, (255)
cis-[Ir(py)$_4$Cl$_2$]Cl	Monomeric, octahedral	656	468	287, 273
trans-[Pt(py)$_2$Br$_2$]	Monomeric, square-planar	656	476	297
cis-[Pt(py)$_2$Br$_2$]	Monomeric, square-planar	659	448	260, 234
		644		

[a]For band assignments of pyridine, see Refs. 125 and 126.
[b]Assignments made by Postmus et al.[127]

cies. Clark and Williams[124] carried out an extensive far-infrared study on metal pyridine complexes. Table III-7 lists the observed frequencies of these metal-sensitive py vibrations and metal–py stretching vibrations. Clark and Williams showed that ν(M—py) and ν(MX) (X: a halogen) are very useful in elucidating the stereochemistry of these py complexes. For example, *fac*-[Rh(py)$_3$Cl$_3$] exhibits two ν(Rh—py) (C_{3v} symmetry), whereas *mer*-Rh(py)$_3$Cl$_3$ shows three ν(Rh—py) (C_{2v} symmetry) near 250 cm^{-1}. The infrared spectra of these two compounds are given in Fig. III-13.

The metal isotope technique has been used to assign the ν(M—py) and ν(MX) vibrations of Zn(py)$_2$X$_2$[128] and Ni(py)$_4$X$_2$.[129] The former vibrations have been located in the 225–160 and 250–225 cm^{-1} regions, respectively, for the Zn(II) and Ni(II) complexes. Figure III-14 shows the infrared and Raman spectra of [^{64}Zn(py)$_2$Cl$_2$] and its ^{68}Zn analog. As expected from its C_{2v} symmetry, two ν(Zn—py) and two ν(ZnCl) are metal-isotope sensitive.

Thornton and co-workers[130,131] carried out an extensive study on a variety of pyridine complexes with emphasis on band assignments based on isotope shift data (py-d_5, ^{15}N-py, and metal isotopes). As an example, Fig. III-15 illustrates the IR spectra and band assignments of a series of the M(py)$_2$Cl$_2$-type complexes.[131] The shaded bands (ν_3 and ν_4) are assigned to the ν(M—N) while the solid bands (ν_1 and ν_2) are due to the ν(M—Cl). The ν_5 is assigned to a bending mode. In a series of polymeric octahedral complexes shown in Fig. III-15, both ν(M—N) and ν(M—Cl) follow the Irving–Williams order shown in Sec. III-2(1).

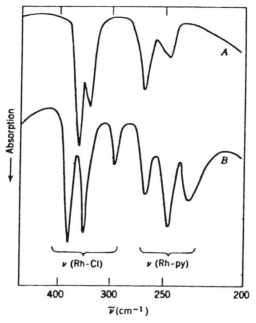

Fig. III-13. Far-IR spectra of (A) *fac-* and (B) *mer-*[Rh(py)₃Cl₃].[124]

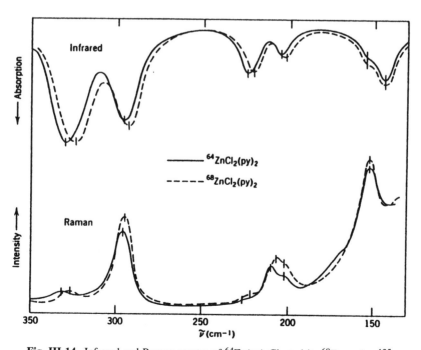

Fig. III-14. Infrared and Raman spectra of ^{64}Zn(py)₂Cl₂ and its ^{68}Zn analog.[125]

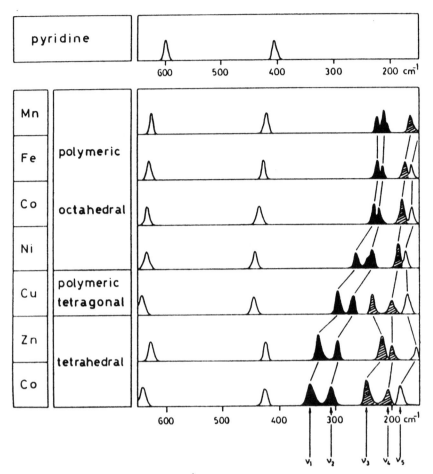

Fig. III-15. Infared spectra (650–150 cm^{-1}) of the M(py)$_2$Cl$_2$ complexes: solid bands, ν(M–Cl) and shaded bands, ν(M–N) (reproduced with permission from Ref. 131).

Two ν(M—Cl) and one ν(M—N) are expected in IR spectra of polymeric octahedral complexes (**C**$_1$ symmetry) whereas two ν(M—Cl) and two ν(M—N) are expected in IR spectra of tetrahedral complexes (**C**$_{2v}$ symmetry). The latter also holds for polymeric tetragonal Cu(II) complex. It should be noted that the violet Co(II) complex is polymeric octahedral while the blue Co(II) complex is monomeric tetrahedral. An extensive review on IR and Raman spectra of metal pyridine complexes has been made by Thornton.[131] Far-infrared spectra of metal pyridine nitrate complexes, M(py)$_x$(NO$_3$)$_y$, have been reported.[132,133]

(2) Surface-Enhanced Raman Spectra of Pyridine

In 1974, Fleischmann et al.[134] made the first observation of surface-enhanced Raman spectra (SERS) of pyridine adsorbed on a silver electrode. Figure III-16

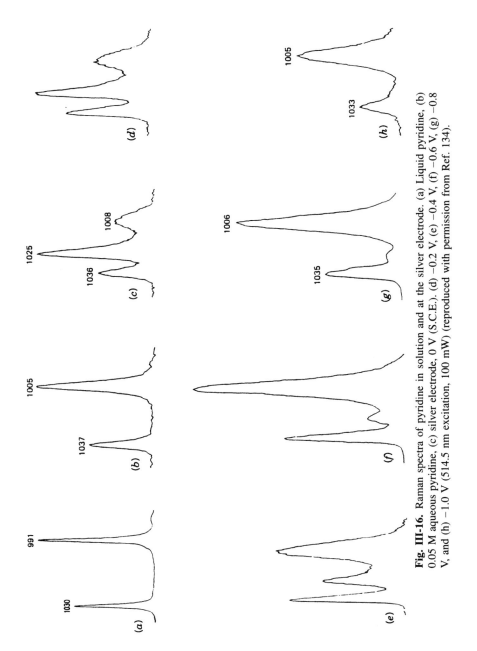

Fig. III-16. Raman spectra of pyridine in solution and at the silver electrode. (a) Liquid pyridine, (b) 0.05 M aqueous pyridine, (c) silver electrode, 0 V (S.C.E.). (d) −0.2 V, (e) −0.4 V, (f) −0.6 V, (g) −0.8 V, and (h) −1.0 V (514.5 nm excitation, 100 mW) (reproduced with permission from Ref. 134).

shows the Raman spectra of pyridine in solution (a and b) and SERS of pyridine on a Ag electrode which has the potential 0 to -1.0 V (c–h) relative to the SCE. It is seen that the intensities of the bands at 1037, 1025, and 1008 cm^{-1} (ring stretch) are changed markedly by changing the potential. The 1025 cm^{-1} band was assigned to the uncharged species, that is, pyridine bonded directly on the electrode surface (Lewis acid site) since its intensity is maximized near zero potential. The remaining two bands were attributed to the pyridine which is hydrogen bonded to water molecules on the electrode surface. These two environments of pyridine are illustrated in Fig. III-17. As expected, the relative intensity of the 1025 cm^{-1} band decreases and those of the 1037 and 1008 cm^{-1} bands increase as the negative potential increases.

While verifying these results in 1977, Jeanmaire and Van Duyne[135,136] noted that the Raman signals (3067, 1036, and 1008 cm^{-1}) from pyridine on Ag electrodes are enhanced by a huge factor (10^4–10^6) relative to normal Raman spectra in solution. In addition, they noted that the Raman intensity not only depends on the electrode potential but also on several other factors such as electrode surface preparation, concentration of pyridine in solution, and the nature and concentration of the supporting electrolyte anion. Almost simultaneously, Albrecht and Creighton[137] noted anomalous enhancements of Raman bands of

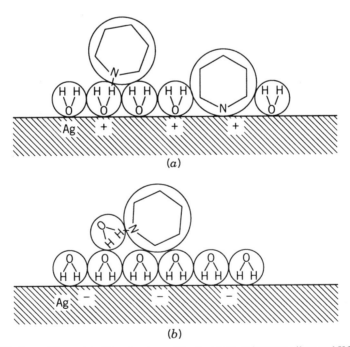

Fig. III-17. A possible model of the structure of the double layer between silver and KCl solution containing pyridine. (a) At positive potentials, showing pyridine adsorbed to silver via nitrogen and in an "aqueous acidic environment." (b) At negative potentials, showing adsorbed pyridine in "aqueous environment" (reproduced with permission from Ref. 134).

pyridine adsorbed on a Ag electrode. These bands include those at 3076 (CH stretch), 1600 and 1050–1000 (ring stretch), 669 (ring deformation), and 239 cm^{-1} [Ag–N(py) stretch].[138]

Yamada and Yamamoto[139,140] measured the SERS of pyridine adsorbed on metal oxides with UV excitation (363.8 nm). These workers were able to distinguish three types of adsorbed pyridine, as shown in Fig. III-18.

Type H   Pyridine hydrogen-bonded to the surface OH group (~3075 and 999 cm^{-1})

Type L   Pyridine adsorbed on a Lewis acid site (~3075 and 1025 cm^{-1})

Type B   Pyridine adsorbed on a Bronsted acid site (3090 and 1005 cm^{-1})

The frequencies in the brackets are those observed for γ-alumina. SERS of pyridine adsorbed on rhodium oxide have also been reported.[141]

The SERS of the [Os(NH$_3$)$_5$py]$^{n+}$ ($n = 2, 3$) and [Ru(NH$_3$)$_5$py]$^{2+}$ ions adsorbed at the silver–aqueous interface exhibit internal modes of pyridine as well as metal–ligand modes, although the lone-pair electrons of pyridine nitrogen atoms in these ions are not available for bonding to the surface silver atoms. This result demonstrates the utility of SERS in obtaining vibrational data of coordination compounds which are sometimes difficult to obtain in bulk media [e.g., Os(II) complex].[142]

(3)   Complexes of Pyridine Derivatives and Related Ligands

The infrared spectra of metal complexes with alkyl pyridines have been studied extensively.[143–147] Using the metal isotope technique, Lever and Ramaswamy[148] assigned the M—pic stretching bands of M(pic)$_2$X$_2$ [M = Ni(II) and Cu(II); pic = picoline; X = Cl, Br, and I] in the 300–230 cm^{-1} region. The infrared spectra of metal complexes with halogenopyridines have been reported.[144,149,150] Infrared spectra have been used to determine whether coordination occurs through the nitrile or the pyridine nitrogen in cyanopyridine complexes. It was found that 3- and 4-cyanopyridines coordinate to the metal via the pyridine

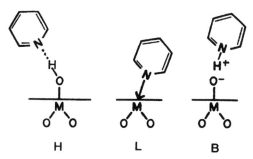

Fig. III-18. Three types of adsorbed pyridines on metal oxide surfaces.[140]

nitrogen,[151–153] whereas 2-cyanopyridine coordinates to the metal via the nitrile nitrogen.[151,154]

Infrared spectra of aromatic amine N-oxides and their metal complexes have been reviewed by Garvey et al.[155] The $N{=}O$ stretching band of pyridine N-oxide (1265 cm^{-1}) is shifted by 70–30 cm^{-1} to a lower frequency upon complexation. The following references are given for three complexes: Fe(II),[156] Hg(II),[157] and Fe(III).[158]

Imidazole (ImH) and its derivatives form complexes with a number of transition-metal ions. Infrared spectra are reported for metal complexes of

imidazole,[159–161] 2-methylimidozole,[162–163] 1 (or N)-methylimidazole,[164] 4- and 5-bromoimidazole,[165] and benzimidazole.[166,167] Among them, imidazole is biologically most important since imidazole nitrogens of histidyl residues coordinate to metal ions in many metalloproteins. Thus, the identification of M–N(Im) vibrations in biological systems provides valuable information about the structure of the active site of a metalloprotein (Sec. V-1). Using metal isotope techniques, Cornilsen and Nakamoto[161] assigned and M–N stretching vibrations of 16 imidazole complexes of Ni(II), Cu(II), Zn(II), and Co(II) in the 325–210 cm^{-1} region. Hodgson et al.[168] also assigned these vibrations in the same region. Salama and Spiro[169] were first to assign the Co–N stretching vibrations in resonance Raman spectra of Co(ImH)$_2$Cl$_2$ (274 and 232 cm^{-1}), [Co(ImH)$_4$]$^{2+}$ (301 cm^{-1}), and [Co(Im$^-$)$_4$]$^{2-}$ (306 cm^{-1}).

Caswell and Spiro[170] studied excitation profiles of imidazole, histidine, and related ligands including the [Cu(ImH)$_4$]$^{2+}$ ion in the UV region. These compounds exhibit maxima near 218 and 204 nm where the π–π^* transitions of the heterocyclic rings occur.

III-4. COMPLEXES OF BIPYRIDINE AND RELATED LIGANDS

(1) Complexes of 2,2′-Bipyridine

Infrared spectra of metal complexes of 2,2′-bipyridine (bipy) have been studied extensively. In general, the bands in the high-frequency region are not metal sensitive since they originate in the heterocyclic or aromatic ring of the ligand. Thus, the main interest has been focused on the low-frequency region, where ν(MN) and other metal-sensitive vibrations appear. It has been difficult, however, to assign ν(MN) empirically since several ligand vibrations also appear in the same frequency region. This difficulty was overcome by using the metal isotope technique. Hutchinson et al.[171] first applied this method to the tris-bipy complexes of Fe(II), Ni(II), and Zn(II). Later, this work was extended to other metals in various oxidation states.[172] Table III-8 lists $\nu(MN)$, magnetic moments, and the electronic configuration of these tris-bipy complexes. The results revealed several interesting relationships between $\nu(MN)$ and the electronic structure:

1. In terms of simple MO theory, Cr(III), Cr(II), Cr(I), Cr(0), V(II), V(0), Ti(0), Ti(−I), Fe(III), Fe(II), and Co(III) have filled or partly filled t_{2g} (bonding) and empty e_g (antibonding) orbitals. The $\nu(MN)$ of these metals (Group A) are in the 300–390 cm^{-1} region.

2. On the other hand, Co(II), Co(I), Co(0), Mn(II), Mn(0), Mn(−I), Ni(II), Cu(II), and Zn(II) have filled or partly filled e_g orbitals. The $\nu(MN)$ of these metals (Group B) are in the 180–290 cm^{-1} region.

3. Thus no marked changes in frequencies are seen in the Cr(III)–Cr(0) and Co(II)–Co(0) series, although a dramatic decrease in frequencies is observed in going from Co(III) to Co(II).

4. The fact that the $\nu(MN)$ do not change appreciably in the former two series indicates that the M–N bond strength remains approximately the same.

These results also suggest that, as the oxidation state is lowered, increasing numbers of electrons of the metal reside in essentially ligand orbitals which do not affect the M–N bond strength.

Other work on bipy complexes includes a far-infrared study of tris-bipy complexes with low-oxidation-state metals [Cr(0), V(−I), Ti(0), etc.],[173] the assignments of infrared spectra of M(bipy)Cl$_2$ (M = Cu, Ni, etc.),[174] normal coordinate analysis on Pd(bipy)Cl$_2$ and its bipy-d_8 analog.[175]

The [Fe(bipy)$_3$]$^{2+}$ ion and its analogs exhibit strong absorption near 520 nm which is due to Fe($3d$)–ligand(π) CT transition. When the laser wavelength is tuned in this region, a number of bipy vibrations (all totally symmetric) are strongly resonance-enhanced, as shown in Fig. III-19.[176] Excitation profile studies show that the intensities of all these bands are maximized at the main absorption maximum at 19,100 cm^{-1} (524 nm) and that no maxima are present at the side band near 20,500 cm^{-1} (488 nm). Thus, Clark et al.[176] concluded that the latter band is due to a vibronic transition. The resonance Raman spectrum of the [Fe(bipy)$_3$]$^{2+}$ ion was also observed near the iron electrode surface in borate buffer solution containing bipy.[177] The electronic spectrum of the singly reduced [Fe(bipy)$_3$]$^+$ ion has also been assigned based on excitation profile studies.[178]

(2) Time-Resolved Resonance Raman (TR^3) Spectra

Recently, the [Ru(bipy)$_3$]$^{2+}$ ion and related complexes have attracted much attention as potential compounds of solar-energy-conversion devices because

TABLE III-8. MN Stretching Frequencies and Electronic Structures in [M(bipy)$_3$]$^{n+}$-Type Compounds (cm^{-1})[a]

	-1	0	I	II	III
d^3					Cr (3.78) 385 349 $(t_{2g})^3$
d^4		Ti (0) 374 339 $(t_{2g})^4$-ls		V (3.67) 374 335 $(t_{2g})^3$	
d^5	Ti (1.74) 365 322 $(t_{2g})^5$-ls	V (1.68) 371 343 $(t_{2g})^5$-ls	Cr (2.0) 371 343 $(t_{2g})^5$-ls	Mn (5.95) 224 191 $(t_{2g})^3(e_g)^2$-hs	Fe (?) 384 367
d^6		Cr (0) 382 308 $(t_{2g})^6$		Fe (0) 386 376 $(t_{2g})^6$	Co (0) 378[b] 370 $(t_{2g})^6$
d^7		Mn (4.10) 258 227 $(t_{2g})^5(e_g)^2$		Co (4.85) 266 228 $(t_{2g})^5(e_g)^2$	
d^8	Mn (3.71) 235 184 $(t_{2g})^6(e_g)^2$		Co (3.3) 244 194 $(t_{2g})^6(e_g)^2$	Ni (3.10) 282 258 $(t_{2g})^6(e_g)^2$	
d^9		Co (2.23) 280 257 $(t_{2g})^6(e_g)^3$		Cu (?) 291 268 $(t_{2g})^6(e_g)^3$	
d^{10}				Zn (0) 230 184 $(t_{2g})^6(e_g)^4$	

[a]The numbers at the upper right of each group indicate the MN stretching frequencies (cm^{-1}). The number in parentheses gives the observed magnetic moment in Bohr magnetons. ls = low spin; hs = high spin.

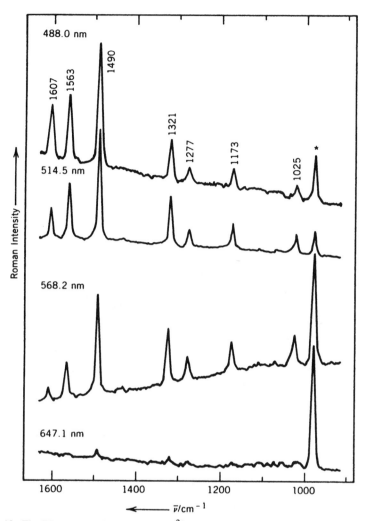

Fig. III-19. The RR spectra of the $[Fe(bipy)_3]^{2+}$ ion. The asterisk indicates the 981 cm^{-1} band of the SO_4^{2-} ion (internal standard).[177]

of their excited-state redox properties. When solutions of this ion are irradiated with 7-ns, high-intensity pulses from the third harmonic (354.5 nm) of a Nd–YAG laser, the irradiated volume can be saturated with the long-lived (~600 ns) triplet M–L CT state (A_3) via efficient ($\phi \cong 1$) and rapid ($\tau < 10$ ps) intersystem crossing ($A_2 \rightarrow A_3$) as shown in Fig. III-20. Since the A_3–A_4 (π–π^*) transition is close to the 354.5-nm exciting line, conditions are favorable for efficient resonance Raman scattering from the A_3 state; namely, it is possible to obtain the time-resolved resonance Raman (TR³) spectrum of the ion in the electronic excited state. The first observation of such spectra was made by Dallinger and Woodruff,[179] who were followed by many investigators.[180–183] These workers

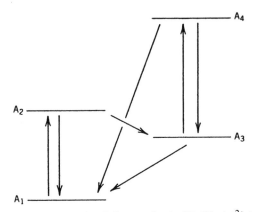

Fig. III-20. Energy-level diagram for the $[Ru(bipy)_3]^{2+}$ ion.

found that the TR^3 spectrum consists of two series of bipy vibrations; one series of bands are the same as those observed in the A_1 state and the other correspond to those of $Li^+(bipy^{\cdot -})$. Figure III-21 shows the spectra obtained by Mallick et al.[184] It is seen that the TR^3 spectrum (middle trace) is the addition of the RR spectra of the $[Ru(bipy)_3]^{2+}$ ion (top trace) and $Li^+(bipy^{\cdot -})$ (bottom trace). Thus, the triplet M–L CT state (A_3) is formulated as $[Ru(III)(bipy)_2(bipy^{\cdot -})]^{2+}$, that is, the electron is localized on one bipy rather than delocalized over all three ligands (at the least vibrational time scale). The RR spectra of electron-reduction products of several $[Ru(bipy)_3]^{2+}$ derivatives show similar electron localization.[185]

Kincaid and co-workers[185a,185b] carried out normal coordinate analyses on the 1 : 1 (metal/ligand) model of the $[Ru(bipy)_3]^{2+}$ ion in the ground and the M—L CT states. As expected, extensive vibrational couplings exist among those represented by local internal coordinates. The Ru–N stretching force constant was 2.192 mdyn /Å in both states.

In heteroleptic complexes, selective population of individual ligand-localized excited states is possible by using TR^3 spectroscopy. For example, Danzer and Kincaid[185c] have demonstrated that the triplet M—L CT state of the $[Ru(bipy)_2(bpz)]^{2+}$ ion (bpz: bipyrazine) should be formulated as $[Ru(III)(bipy)_2(bpz^{\cdot -})]^{2+}$, whereas that of the $[Ru(bipy)(bpz)_2]^{2+}$ ion should be formulated as $[Ru(III)(bipy)(bpz)(bpz^{\cdot -})]^{2+}$.

(3)   Complexes of Phenanthroline and Related Ligands

The metal isotope technique has been used to study the effect of magnetic crossover on the low-frequency spectrum of $Fe(phen)_2(NCS)_2$ (phen: 1,10-phenanthroline). This compound exists as a high-spin complext at 298 K and as a low-spin complex at 100 K. Figure III-22 shows the infrared spectra of $^{54}Fe(phen)_2(NCS)_2$ obtained by Takemoto and Hutchinson.[186] On the basis of observed isotopic shifts, along with other evidence, they made the following

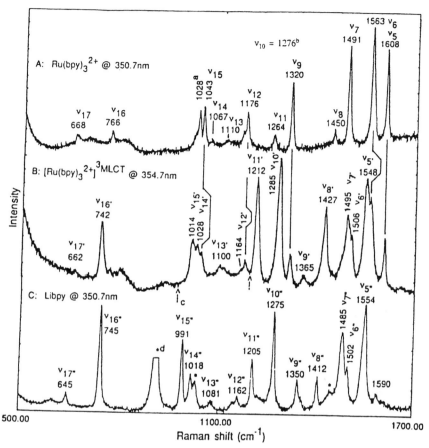

Fig. III-21. The RR and TR^3 spectra of the [Ru(bipy)$_3$]$^{2+}$ ion. (A) RR spectrum with 350.7 nm excitation. (B) TR^3 spectrum with 354.7 nm excitation. (C) RR spectrum of Li(bipy) with 350.7 nm excitation (reproduced with permission from Ref. 184).

assignments (cm^{-1}):

	ν(Fe—NCS)	ν(Fe—N(phen))
High spin	252(4.0)	222(4.5)
Low spin	532.6(1.6)	379(5.0)
	528.5(1.7)	371(6.0)

The numbers in parentheses indicate the isotope shift, ν(^{54}Fe)–ν(^{57}Fe). Both vibrations show large shifts to higher frequencies in going from the high- to the low-spin complex. This result suggests the marked strengthening of these coordinate bonds in going from the high- to the low-spin complex, as confirmed by X-ray analysis.[187] The work of Takemoto and Hutchinson has been extended to Fe(bipy)$_2$(NCS)$_2$ and Fe(phen)$_2$(NCSe)$_2$.[188] It has also been shown

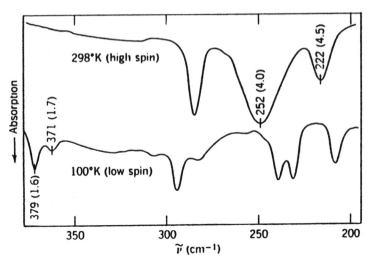

Fig. III-22. Infrared spectra of $^{54}Fe(phen)_2(NCS)_2$. The number in parentheses indicates the isotope shift due to $^{54}Fe/^{57}Fe$ substitution.

by infrared spectroscopy that a partial high- → low-spin conversion occurs under high pressure.[189] Barnard et al.[190] studied the vibrational spectra of bis[tri-(2-pyridyl)amine]Co(II) perchlorate in the high-spin (293 K) and low-spin (100 K) states. In the infrared, the CoN stretching band is at 263 cm^{-1} for the high-spin complex, whereas it splits and shifts to 312 and 301 cm^{-1} in the low-spin complex.

As stated above, $Fe(phen)_2(NCS)_2$ exhibits the room-temperature high-spin state (HS-1) and the low-temperature low-spin state (LS-1). Herber and Casson[190a] found that, when the latter is irradiated by white light below ~50 K, another high-spin state (HS-2) is obtained (light-induced excited-state trapping), and annealing of this HS-2 state above ~30 K produces another low-spin state (LS-2). These workers have shown that the ν(CN) of the NCS ligands near the 2100-cm^{-1} region are different among these four states. This work has been extended to $Fe(bt)_2(NCX)_2$ (bt, 2,2′-bi-2-thiazoline and X is S or Se),[190b] and to $Fe(5,6-dmp)_2(NCS)_2$ (dmp, dimethylphenanthroline).[190c]

The TR^3 spectrum of the [Cu(I)(DPP)_2]$^+$ ion (DPP: 2,9-diphenyl-phenanthroline) shows that its M—L CT state is formulated as [Cu(II)(DPP)(DPP$^{·-}$)]$^{+}$.[191]

(4) Complexes of Other Ligands

Simple α-diimines such as shown below form metal chelate compounds similar to bipy and phen, discussed earlier.

R = CH$_3$, R′ = H
Glyoxal-bis-methylimine(GMI)

R = R′ = CH$_3$
Biacetyl-bis-methylimine(BMI)

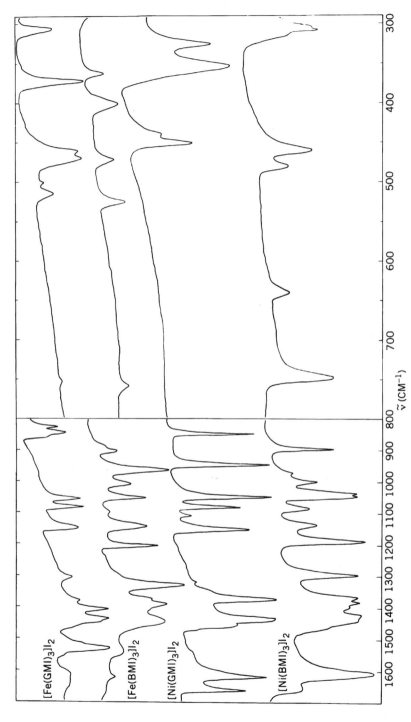

Fig. III-23. Infrared spectra of α-diimine complexes of Fe(II) and Ni(II).[192]

[Fe(GMI)₃]I₂

[Fe(BMI)₃]I₂

[Ni(GMI)₃]I₂

[Ni(BMI)₃]I₂

$\tilde{v}$ (CM⁻¹)

37

Figure III-23 shows the IR spectra of Fe(II) and Ni(II) complexes of these ligands.[192] Normal coordinate calculations indicate extensive vibrational couplings among those represented by individual internal coordinates.

Depending upon the nature of the alkyl group (R), an alkyl-substituted α-diimine (R—N=CH—CH=N—R) coordinates to the metal [Pt(II) or Pd(II)] as a unidentate or bidentate (chelating) ligand. van der Poel et al.[193] observed the C=N stretching bands at 1615–1624 and 1590–1604 cm^{-1} for the unidentate and bidentate coordination, respectively.

The metal isotope technique has been used to assign the MN vibrations of metal complexes with many other ligands. For example, Takemoto[194] assigned the NiN_2 and NiN_1 stretching vibrations of $[Ni(DAPD)_2]^{2-}$ at 416–341 and 276 cm^{-1}, respectively.

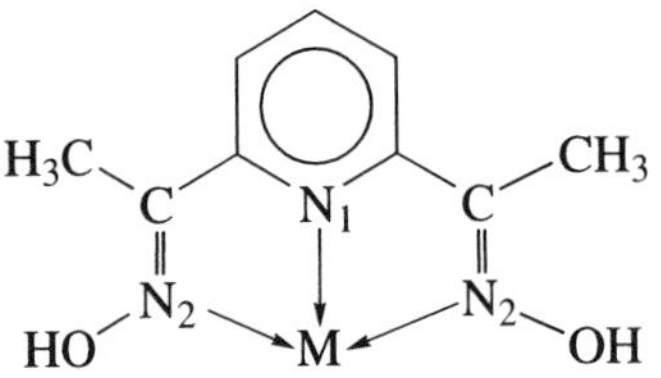

DAPD: 2,6-diacetylpyridine
dioxime

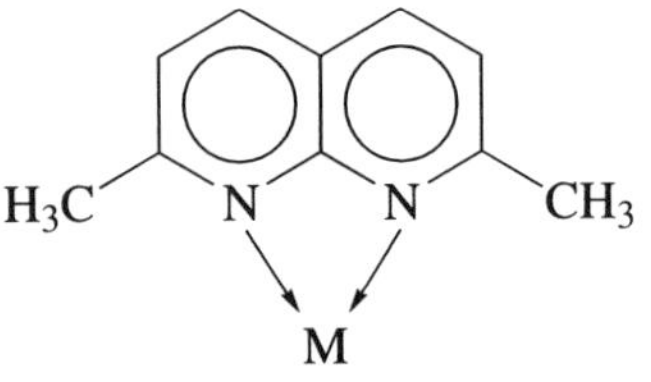

DMNAPY: 2,7-dimethyl-
1,8-naphthyridine

In $Ni(DAPD)_2$, where the Ni atom is in the +IV state, the NiN_2 and NiN_1 stretching bands are located at 509.8–472.0 and 394.8 cm^{-1}, respectively. These high-frequency shifts in going from Ni(II) (d^8) to Ni(IV), (d^6, diamagnetic) have also been observed for diarsine complexes (Sec. III-25(2)). Hutchinson and Sunderland[195] have noted that the MN stretching frequencies of the Ni(II) and Zn(II) complexes of 2,7-dimethyl-1,8-naphthyridine (DMNAPY), shown above, are lower than those of the corresponding tris-bipy complexes by 16 to 24%. This was attributed to the weakening of the M–N bond due to the strain in the four-membered chelate rings of the DMNAPY complexes. Normal coordinate analysis has been carried out on the $M(DMG)_2$ series (DMG: dimethylglyoximate ion)[196] and the MN stretching force constants (mdyn/Å) have been found to be as follows:

Pt(II)		Pd(II)		Cu(II)		Ni(II)	
3.77	>	2.84	>	1.92	>	1.88	(GVF)

This work was extended to bis(glyoximato) complexes of Pt(II), Pd(II), and Ni(II).[197] The Co–N(DMG) and Co–N(py) stretching bands of $Co(DMG)_2(py)X$ (X = a halogen) were assigned at 512 and 453 cm^{-1}, respectively, based on ^{15}N and py-d_5 isotope shifts.[198] The metal isotope technique has been used to assign the MN stretching vibrations of metal complexes with 8-hydroxyquinoline[199] and 1,8-naphthyridine.[200]

III-5. METALLOPORPHYRINS

Vibrational spectra of metalloporphyrins have been studied exhaustively because of their biological importance as prosthetic groups of a variety of heme proteins (Section V). Thus, many review articles have been published on this subject, and most of them discuss vibrational spectra of metalloporphyrins together with those of heme proteins. A review by Kitagawa and Ozaki,[201] however, is focused on metalloporphyrins.

(1) Normal Coordinate Analysis

Because of relatively high symmetry and biological significance, normal coordinate analyses on metalloporphyrins have been carried out by many investigators.[202–209] Figure III-24 shows the planar $\mathbf{D}_{4h}$ structure of a metalloporphyrin. The simplest porphyrin, porphin (Por), has 105 ($3 \times 37 - 6$) normal vibrations which are classified as shown in Table III-9.[202] As stated in Sec. I-23 of Part A, metalloporphyrins are ideal for resonance Raman (RR) studies because they exhibit strong absorption bands in the visible and near UV regions. It is well established that RR spectra obtained by excitation near the Q_0 (or α) band are dominated by those of the B_{1g} and B_{2g} species (dp) while those obtained by excitation between the Q_0 and Q_1 (or β) band are dominated by those of the A_{2g} species (ap). On the other hand, the RR spectra obtained by excitation near the **B** (or Soret) band are dominated by those of the A_{1g} species (p). This information, combined with isotope shift data (H/D, $^{14}N/^{15}N$, and metal isotopes), has been used extensively to refine the results of normal coordinate calculations.

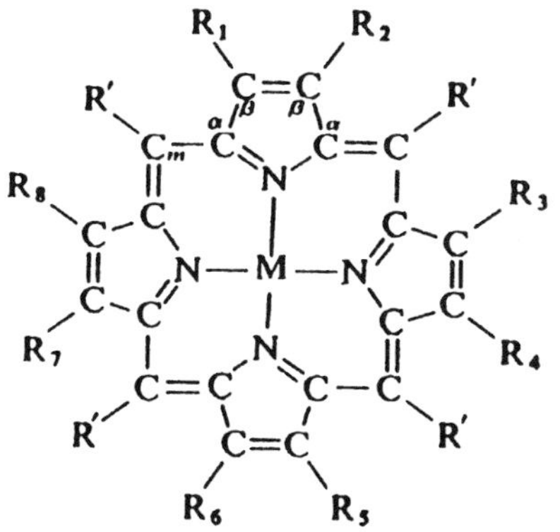

Fig. III-24. Structures of metalloporphyrins. Porphin (Por), R_1~R_8 = 8 and R′ = H. Octaethylporphyrin (OEP), R_1~R_8 = ethyl and R′ = H. Tetraphenylporphyrin (TPP), R_1~R_8 = H and R′ = phenyl. Tetramesitylporphyrin (TMP), R_1~R_8 = H and R′ = mesityl.

TABLE III-9. Classification of Normal Vibrations of a Metal Porphin Complex of D_{4h} Symmetry$^{a\,202}$

In-Plane Vibrations		Out-of-Plane Vibrations	
$A_{1g}(R)$	9	A_{1u}	3
$A_{2g}{}^b$	8	$A_{2u}(IR)$	6
$B_{1g}(R)$	9	B_{1u}	5
$B_{2g}(R)$	9	B_{2u}	4
$E_u(IR)$	18	$E_g(R)^c$	8

^{a}R = Raman-active; IR = IR-active.

^{b}The A_{2g} vibrations become Raman-active under resonance conditions (see Sec. I-23 of Part A).

^{c}The E_g vibrations are weak even under resonance conditions.

Thus far, Spiro and co-workers[207–209] have carried out the most extensive normal coordinate calculations on metalloporphyrins. As expected from their conjugated ring structures, strong vibrational couplings occur among vibrational modes represented by local internal coordinates. Thus, it is rather difficult to describe the normal modes by single internal coordinates. Table III-10 lists the observed frequencies and major local coordinates responsible for the IR/Raman-active in-plane skeletal modes of Ni(Por), Ni(OEP), and Ni(TPP) (the RR spectra of Ni(OEP) are shown in Fig. I-35). Figure III-25 shows the normal modes of the eight A_{1g} vibrations obtained for Ni(OEP) and Ni(Por). Table III-11 lists the normal modes to which the ν(Ni–N) coordinates make significant contributions. Previously, empirical assignments have been made for IR-active M–N stretching vibrations of a series of M(II)(TPP) complexes using metal isotopes such as ^{58}Ni/^{62}Ni, ^{63}Cu/^{65}Cu, and ^{64}Zn/^{68}Zn.[210] For normal coordinate analysis on the out-of-plane modes, see Ref. 209.

TABLE III-10. In-Plane Skeletal Mode Frequencies (cm^{-1}) and Local Mode Assignments for Ni(II) Complexes of OEP, Porphin, and TPP$^{207,\,208}$

Symmetry	ν_i	Description^a	NiOEP	NiPor	NiTPP
A_{1g}	ν_1	$\nu(Cm\text{-}X)$	$[3041]^b$	$[3042]$	1235
	ν_2	$\nu(C_\beta\text{-}C_\beta)$	1602	1579	1572
	ν_3	$\nu(C_\alpha\text{-}C_m)_{sym}$	1520	1463	1470
	ν_4	$\nu(\text{Pyr. half-ring})_{sym}$	1383	1380	1374
	ν_5	$\nu(C_\beta\text{-}Y)_{sym}$	1138	$[3097]$	$[3097]$
	ν_6	$\nu(\text{Pyr. breathing})$	804	999	1004
	ν_7	$\delta(\text{Pyr. def.})_{sym}$	674	735	889
	ν_8	$\nu(\text{Ni-N})$	360/343	372	402
	ν_9	$\delta(C_\beta\text{-}Y)_{sym}$	263/274	1070	1078

TABLE III-10. (*Continued*)

Symmetry ν_i		Description[a]	NiOEP	NiPor	NiTPP
B_{1g}	ν_{10}	$\nu(C_\alpha\text{-}C_m)_{asym}$	1655	1654	1594
	ν_{11}	$\nu(C_\beta\text{-}C_\beta)$	1577	1509	1504
	ν_{12}	$\nu(\text{Pyr. half-ring})_{sym}$	1331	1319	1302
	ν_{13}	$\delta(C_m\text{-}X)$	1220	1189	238
	ν_{14}	$\nu(C_\beta\text{-}Y)_{sym}$	1131	[3097]	[3097]
	ν_{15}	$\nu(\text{Pyr. breathing})$	751	1007	1004
	ν_{16}	$\delta(\text{Pyr. def.})_{sym}$	746	734	[900]
	ν_{17}	$\delta(C_b\text{-}Y)_{sym}$	305	1064	1084
	ν_{18}	$\nu(\text{Ni-N})$	168	241	277
A_{2g}	ν_{19}	$\nu(C_\alpha\text{-}C_m)_{asym}$	1603	1615	1550
	ν_{20}	$\nu(\text{Pyr. quarter-ring})$	1394	1358	1341
	ν_{21}	$\delta(C_m\text{-}X)$	1307	1143	[257]
	ν_{22}	$\nu(\text{Pyr. half-ring})_{asym}$	1121	1009	1016
	ν_{23}	$\nu(C_\beta\text{-}Y)_{asym}$	1058	[3087]	[3087]
	ν_{24}	$\delta(\text{Pyr. def.})_{asym}$	597	810	828
	ν_{25}	$\delta(\text{Pyr. rot.})$	551	433	560
	ν_{26}	$\delta(C_\beta\text{-}Y)_{asym}$	[243]	1321	1230
B_{2g}	ν_{27}	$\nu(C_m\text{-}X)$	[3040]	[3041]	1269
	ν_{28}	$\nu(C_\alpha\text{-}C_m)_{asym}$	1483	[1492]	[1481]
	ν_{29}	$\nu(\text{Pyr. quarter-ring})$	1407	1372	1377
	ν_{30}	$\nu(\text{Pyr. half-ring})_{asym}$	1160	1007	1004
	ν_{31}	$\nu(C_\beta\text{-}Y)_{asym}$	1015	[3088]	[3087]
	ν_{32}	$\delta(\text{Pyr. def.})_{asym}$	938	823	869
	ν_{33}	$\delta(\text{Pyr. rot.})$	493	439	450
	ν_{34}	$\delta(C_\beta\text{-}Y)_{asym}$	197	1197	1191
	ν_{35}	$\delta(\text{Pyr. transl.})$	144	201	109
E_u	ν_{36}	$\nu(C_m\text{-}X)$	[3040]	[3042]	
	ν_{37}	$\nu(C_\alpha\text{-}C_m)_{asym}$	[1637]	1624	
	ν_{38}	$\nu(C_\beta\text{-}C_\beta)$	1604	1547	
	ν_{39}	$\nu(C_\alpha\text{-}C_m)_{sym}$	1501	1462	
	ν_{40}	$\nu(\text{Pyr. quarter-ring})$	1396	1385	
	ν_{41}	$\nu(\text{Pyr. half-ring})_{sym}$	[1346]	1319	
	ν_{42}	$\delta(C_m\text{-}X)$	1231	1150	
	ν_{43}	$\nu(C_\beta\text{-}Y)_{sym}$	1153	[3097]	
	ν_{44}	$\nu(\text{Pyr. half-ring})_{asym}$	1133	1033	
	ν_{45}	$\nu(C_\beta\text{-}Y)_{asym}$	996	[3087]	
	ν_{46}	$\delta(\text{Pyr.})_{asym}$	927	806	
	ν_{47}	$\nu(\text{Pyr. breathing})$	766	995	
	ν_{48}	$\delta(\text{Pyr.})_{sym}$	[615]	745	
	ν_{49}	$\delta(\text{Pyr. rot.})$	[534]	366	
	ν_{50}	$\nu(\text{Ni-N})$	[358]	420	
	ν_{51}	$\delta(C_b\text{-}Y)_{asym}$	328	1064	
	ν_{52}	$\delta(C_\beta\text{-}Y)_{sym}$	263	1250	
	ν_{53}	$\delta(\text{Pyr. transl.})$	[167]	282	

[a]See Ref. 207 for definitions of local coordinates. X,Y = H,H for NiPor, H,C_2H_5 for NiOEP, and C_6H_5, H for NiTPP.

[b][], Calculated values.

A_{1g} Modes

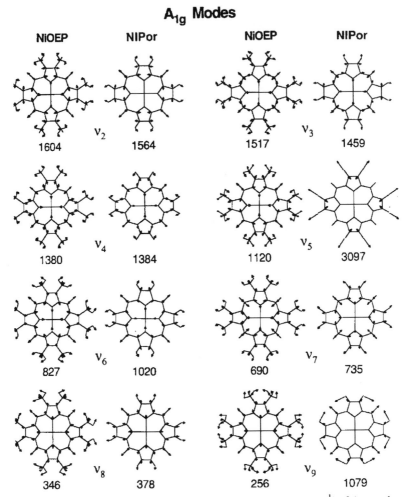

Fig. III-25. Atomic displacements and calculated skeletal frequencies (cm^{-1}) of A_{1g} modes compared for Ni(OEP) and Ni(Por) (reproduced with permission from Ref. 208).

TABLE III-11. Normal Modes Containing Ni—N Stretching Vibrations[a]

Complex	$\nu_8(A_{1g})$[b]	$\nu_{18}(B_{1g})$	$\nu_{50}(E_u)$	$\nu_{53}(E_u)$
Ni (Por)	369 (27%)	237 (64%)	420 (50%)	282 (44%)
Ni (OEP)	360/343 (7%)	168 (37%)	358* (30%)	328 (28%)[c]
Ni (TPP)	402 (24%)	277 (53%)	436 (21%)	306 (32%)

[a]The Ni—N stretching force constant of 1.68 mdyn/Å was used for all three porphyrins.
[b]Numbers in front of brackets are observed frequencies (cm^{-1}) except for that with an asterisk, which is calculated. Numbers in brackets indicate % PED.
[c]ν_{51}.

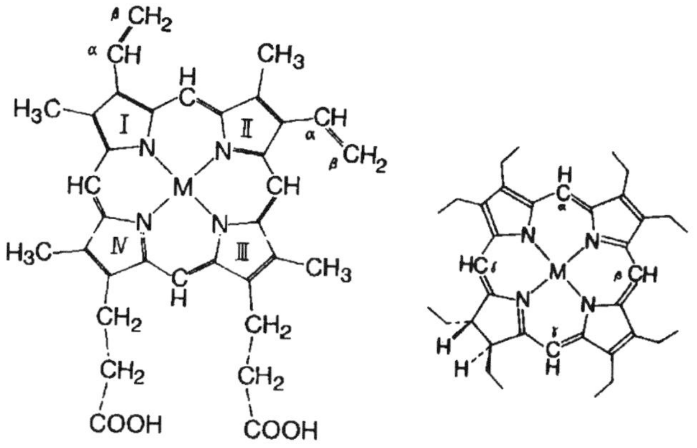

Fig. III-26. Structures of (a) M(PP) and (b) M(OEC).

It should be noted that the internal vibrations of the peripheral substituents such as the ethyl and phenyl groups also couple with the porphyrin core vibrations.[207,208] The vinyl group vibrations of Fe(PP) (PP:protoporphyrin IX, see Fig. III-26a), which are commonly found in natural heme proteins, can be resonance-enhanced by excitation near 200 nm (near the $\pi-\pi^*$ transition of the vinyl group).[211,212]

Normal coordinate analysis has also been made on metallochlorins in which one of the four pyrrole rings of a metalloporphyrin is saturated (e.g., octaethylchlorin (OEC), shown in Fig. III-26b).[213,214]

(2) Structure-Sensitive Vibrations

The RR spectra of iron porphyrins have been studied extensively because of their importance as models of heme proteins. If the frequencies of their skeletal modes above 1450 cm^{-1} (ν_2, ν_3, ν_{10}, ν_{11}, and ν_{19}) are plotted against the Ct–N distance (the center to pyrrole N distance in the porphyrin cavity), we obtain linear relationships such as shown in Fig. III-27.[215] Furthermore, the slope of each line increases in the order

$$\nu_{19} > \nu_{10} > \nu_3 > \nu_2 > \nu_{11}$$

which is parallel to the percent contribution to the $\nu(C_\alpha-C_m)$ coordinate in each normal mode. This result indicates that when the core size increases, the $C_\alpha-C_m$ bonds are weakened and the corresponding frequencies are lowered. In going from the low- to high-spin state, the porphyrin core tends to expand or be domed, and this results in weakening of the $C_\alpha-C_m$ bond. Thus, these vibrations

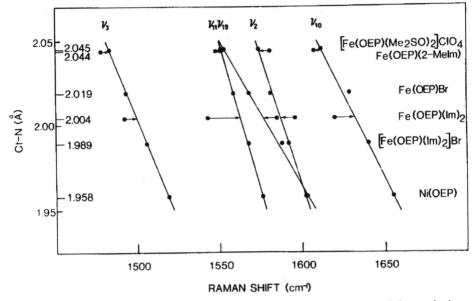

Fig. III-27. Correlations between the ν_{10}, ν_2, ν_{19}, ν_{11}, and ν_3 frequencies and the porphyrin core size (Ct–N, Å) for the Fe(OEP) complexes (reproduced with permission from Ref. 215).

serve as the spin-state marker bands.[216] They are also metal-sensitive because the core size varies with the nature of the metal ion.[217] The ν_{10} is sensitive to the number and the nature of axial ligands.[218] The ν_{11} is sensitive to the nature of the peripheral substituents since it is mainly due to $\nu(C_\beta C_\beta)$. Finally, ν_7 is a 16-membered porphyrin ring breathing motion. Thus, it is strong for planar complexes and weak for nonplanar (or domed) complexes.[219]

The totally symmetric breathing mode, ν_4, is known to be the best marker for the oxidation state. It is near 1360 cm^{-1} for Fe(II) complexes, and near 1375 cm^{-1} for Fe(III) complexes with relatively small dependence on the spin state. The frequency increase in going from Fe(II) to Fe(III) is attributed to the decrease of π-back bonding from the metal $d\pi$ orbital to the porphyrin π^* orbital. In Fe(IV) and Fe(V) porphyrins, the ν_4 are at 1379[220] and 1384 cm^{-1},[221] respectively.

Structure-sensitive bands of other metalloporphyrins have also been studied; Fe(OEC)[222] and water-soluble porphyrins such as Fe(TMpy-P2) [TMpy-P2: tetrakis(2-N-methylpyridyl)porphyrin, R_1~R_8 = H, R' = 2-N-methylpyridyl cation in Fig. III-24].[223,224]

According to X-ray analysis, triclinic and tetragonal crystals of Ni(OEP) contain "flat" and "ruffled" porphyrin rings, respectively. All the frequencies above 1400 cm^{-1} (in-plane skeletal modes) are lower in the "ruffled" form than in the "flat" form. The solution frequencies are intermediate between those of the triclinic and tetragonal forms, indicating that Ni(OEP) is definitely "ruffled" in solution.[225,226]

In Ni(OETPP) (R_1~R_8 = C_2H_5 and R′ = C_6H_5 in Fig. III-24), "saddle" distortion of the porphyrin ring occurs because of steric crowding of the peripheral substituents. Shelnutt and co-workers carried out RR as well as X-ray diffraction studies on this and related complexes.[227–229] Such distortion causes large downshifts (~70 cm^{-1}) relative to the planar porphyrin in a number of porphyrin skeletal modes, and activates three out-of-plane (γ_{15}, γ_{16}, and δ_4) vibrations. Among them, the γ_{16} (tilting of the pyrrole rings) becomes one of the strongest bands in the Soret-excited RR spectrum.[230]

(3) Axial Ligand Vibrations

Table III-12 lists the observed frequencies of Fe–L stretching vibrations of the iron porphyrins where L is an axial ligand. Polyatomic ligands such as N_3 and pyridine (py) exhibit their own internal modes as well. These axial ligand vibrations can be assigned by using isotopic ligands (H/D, $^{14}N/^{15}N$, $^{32}S/^{34}S$, etc.) and metal isotopes ($^{54}Fe/^{56}Fe$, etc.).

Kincaid and Nakamoto[235] observed the ν(Fe–F) of ^{54}Fe(OEP)F at 595 cm^{-1} with the 514.5 nm excitation. Kitagawa et al.[236] also observed the ν(Fe–X) of Fe(OEP)X at 364 and 279 cm^{-1} for X = Cl and Br, respectively, and the ν_s(L–Fe–L) of [Fe(OEP)L$_2$]$^+$ (L = ImH) at 290 cm^{-1} using the 488 nm excitation. These results show that the axial vibrations can be enhanced via resonance with in-plane π–π* transitions (α and β bands). According to the latter workers, vibrational coupling between these axial vibrations and totally symmetric in-plane porphyrin-core vibrations is responsible for their resonance enhancement. On the other hand, Spiro[237] prefers electronic coupling, namely, the π–π* transition induces the changes in the Fe–X (or L) distance, thus activating the

TABLE III-12. Observed Frequencies of Iron-Axial Ligand Stretching Vibrations

Complex[a]	Mode	Obs. Freq. (cm^{-1})	Ref.
Fe(OEP) (ImH)$_2^+$	ν_a(L—Fe—L)	385, 319[b]	231
Fe(PP) (ImH)$_2^+$	ν_s(L—Fe—L)	200	231
Fe(OEP) (γ-pic)$_2^+$	ν_a(L—Fe—L)	373	232
Fe(PP) (ImH)$_2$	ν_s(L—Fe—L)	200	231
Fe(MP) (py)$_2$	ν_s(L—Fe—L)	179	233
Fe(OEP)F	ν(Fe—L)	605.5	232
Fe(OEP)Cl	ν(Fe—L)	357	232
Fe(OEP)Br	ν(Fe—L)	270	232
Fe(OEP)I	ν(Fe—L)	246	232
Fe(OEP)NCS	ν(Fe—L)	315	232
Fe(OEP)N$_3$	ν(Fe—L)	420	233a
Fe(OEP) (SC$_6$H$_5$)	ν(Fe—L)	341	234
Fe(TPP) (SC$_6$H$_5$)$_2^-$	ν_a(L—Fe—L)	345	234
Fe(TPP) (THT)$_2^+$	ν_a(L—Fe—L)	328	234

[a]MP: mesoporphyrin IX dimethyl ester; THT: tetrahydrothiophene; pic: picoline.
[b]These two bands are due to ~50:50 mixing of two modes (antisymmetric Fe–L stretching and pyrrole tilting) (Ref. 231).

axial vibration. Direct excitation is possible if the metal–axial ligand CT transition is in the visible region. Thus, Asher and Sauer[238] observed the ν(Mn–X) of Mn(EP)X (X = F, Cl, Br, and I) with exciting lines in the 460–490 nm region where the Mn–X CT bands appear. Here EP denotes etioporphyrin ($R_1 = R_3 = R_5 = R_7 = CH_3$, $R_2 = R_4 = R_6 = R_8 = C_2H_5$ in Fig. III-24). Similarly, Wright et al.[233] were able to observe totally symmetric pyridine (py) vibrations as well as ν_s(Fe–N(py)) of Fe(MP)(py)$_2$ with exciting lines near 497 nm which are in resonance with the Fe($d\pi$)–py(π^*) CT transition. Here, MP denotes mesoporphyrin IX dimethylester ($R_1 = R_3 = R_5 = R_8 = CH_3$, $R_2 = R_4 = C_2H_5$, and $R_6 = R_7 = $ —CH_2—CH_2COOCH_3 in Fig. III-24).

Ogoshi et al.[239] reported the IR spectra of M(OEC) (M = Zn, Cu, and Ni), Mg(OEC)(py)$_2$, and Fe(OEC)X (X = F, Cl, Br, and I), and assigned ν(M–N(OEC)) and ν(Mg–N(py)) using metal isotope techniques. Ozaki et al.[240] report RR spectra of these and other OEC complexes.

Vibrations of axial ligands such as O_2, NO, CO, and so on are discussed in later sections. Axial ligand vibrations provide valuable information about the structure and bonding of heme proteins containing these ligands.

(4)   Metal–Metal Bonded Porphyrins

As shown in Sec. II-11(2) of Part A, RR spectra of metal–metal bonded complexes such as $Re_2F_8^{2-}$ exhibit the strong ν(Re–Re) at 320 cm^{-1} and a series of its overtones and combination bands. The metal–metal bonded porphyrin dimer, [Ru(OEP)]$_2^{n+}$, exhibits the ν(Ru–Ru) at 285, 301, and 310 cm^{-1} for $n = $ 0, 1, and 2, respectively.[241] The observed frequency increase indicates that the electron density of the Ru–Ru bond is removed from its antibonding π^* orbital as the complex ion is oxidized. However, this removal has almost no effect on the porphyrin core vibrations since the metal–metal bond is perpendicular to the porphyrin plane. Similar results are reported for the [Os(OEP)]$_2^{n+}$ series (233, 254, and 266 cm^{-1} for $n = $ 0, 1, and 2, respetively).[242] The RR spectra of asymmetric sandwich compounds such as Ce(OEP)(TPP),[243] triple-decker sandwich compounds such as Eu$_2$(OEP)$_3$, and their singly oxidized compounds are available.[244]

(5) π–π Complex Formation and Dimerization

The Cu(II) uroporphyrin I (R_1~$R_8 = $ —CH_2COO^- in Fig. III-24) forms molecular adducts with a variety of aromatic heterocyclic compounds in aqueous alkaline solution. Using Raman difference spectroscopy, Shelnutt[245] observed small shifts (+2.9 ~ −2.7 cm^{-1}) of the porphyrin skeletal vibrations resulting from the π–π charge-transfer (porphyrin to heterocycle), and showed that the planes of these two components are parallel to each other. This work has been extended to the study of dimerization of the Cu(II) and Ni(II) complexes of uroporphyrin

I. In this case, the porphyrin skeletal modes were upshifted by 1~3 cm^{-1} as a result of dimerization.[246]

(6) Metallophthalocyanines

Metallophthalocyanines, M(Pc), shown in Fig. III-28, are known for their exceptional thermal stability and extremely low solubility in any solvents. Under $\mathbf{D}_{4h}$ symmetry, their 165 (3 × 57 − 6) normal vibrations are classified into $14A_{1g} + 13A_{2g} + 14B_{1g} + 14B_{2g} + 13E_g + 6A_{1u} + 8A_{2u} + 7B_{1u} + 7B_{2u} + 28E_u$, of which the $A_{1g}, A_{2g}, B_{1g}, B_{2g}$, and E_u vibrations are in-plane modes, and the $A_{1u}, A_{2u}, B_{1u}, B_{2u}$ and E_g vibrations are out-of-plane modes. Only the A_{2u} and E_u vibrations are IR-active, whereas the A_{1g}, B_{1g}, B_{2g} and E_g vibrations are Raman-active. Similar to metalloporphyrins, the A_{2g} vibrations become Raman-active under resonance conditions. Melendres and Maroni[247] carried out normal coordinate analysis on Fe(Pc). As expected, extensive vibrational couplings occur among the local coordinates so that simple descriptions of normal modes by local coordinates cannot be justified. They estimated the Fe–N stretching force constant to be 1.00 mdyn/Å. Using metal isotope techniques, Hutchinson et al.[248] assigned primary ν(M–N) modes of M(Pc) at 240.7 (^{64}Zn), 284.0 (^{63}Cu), 376.0 and 317.8 (^{58}Ni), and 308.4 cm^{-1} (^{54}Fe). Infrared[249] and Raman spectra[250–253] of M(Pc) have been reported by many investigators. Infrared spectra of M(Pc) containing axial ligands (Cl, OH) are also available.[254,255]

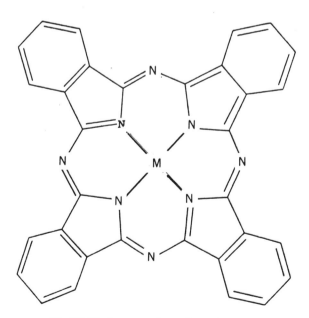

Fig. III-28. Structure of metallophthalocyanine.

III-6. NITRO AND NITRITO COMPLEXES

The NO_2^- ion coordinates to a metal in a variety of ways:

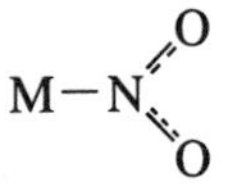

Nitro complex	Nitrito complex	Chelating nitrito complex
I	II	III

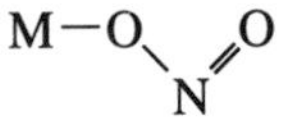

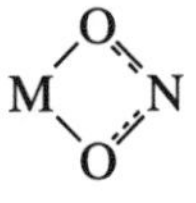

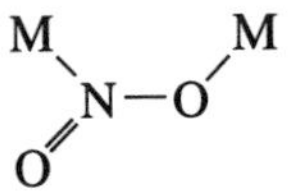

IV

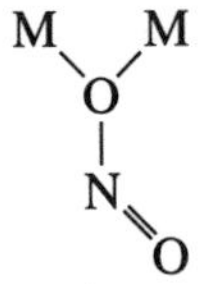

V

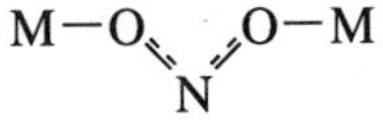

VI

Bridging complexes

Vibrational spectroscopy is very useful in distinguishing these structures.

(1) Nitro Complexes

The normal vibrations of the unidentate *N*-bonded nitro complex may be approximated by those of a planar ZXY_2 molecule, as shown in Fig. III-29. In addition to these modes, the NO_2 twisting and skeletal modes of the whole complex may appear in the low-frequency region. Table III-13 summarizes the observed frequencies and band assignments for typical nitro complexes. It is seen that these complexes exhibit $\nu_a(NO_2)$ and $\nu_s(NO_2)$ in the 1470–1370 and 1340–1320 cm^{-1} regions, respectively. On the other hand, the free NO_2^- ion exhibits these modes at 1250 and 1335 cm^{-1}, respectively. Thus $\nu_a(NO_2)$ shifts markedly to a higher frequency, whereas $\nu_s(NO_2)$ changes very little upon coordination.

Nakagawa and Shimanouchi[57] and Nakagawa et al.[256] carried out normal coordinate analyses to assign the infrared spectra of crystalline hexanitro cobaltic salts; both internal and lattice modes were assigned completely by factor group analysis. The results indicate that the complex ion takes the T_h symmetry in K, Rb, and Cs salts but the S_6 symmetry in the Na salt (see Fig. I-10 of Part A). The IR spectra of K and Na salts are compared in Fig. III-30. Kanamori et al.[262] obtained the RR spectra (632.8 nm excitation, ~80K) of these complexes, and showed that resonance enhancement occurs via the B-term since nontotally as well as totally symmetric vibrations are observed.

There are many nitro complexes containing other ligands such as NH_3 and Cl. In these cases, the main interest has been the distinction of stereoisomers

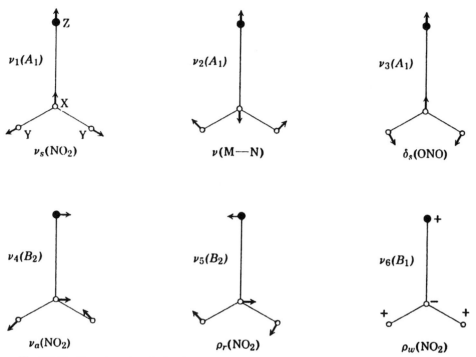

Fig. III-29. Normal modes of vibration of planar ZXY_2 molecules (the band assignment is given for an $M–NO_2$ group).

TABLE III-13. Observed Infrared Frequencies and Band Assignments of Nitro Complexes (cm^{-1})

Complex	$\nu_a(NO_2)$	$\nu_s(NO_2)$	$\delta(ONO)$	$\rho_w(NO_2)$	$\nu(MN)$	$\rho_r(NO_2)^a$	Refs.
$K_3[Co(NO_2)_6]$	1386	1332	827	637	416	293	256
$Na_3[Co(NO_2)_6]$	1425	1333	854 $\Big\}$ 831	623	449 $\Big\}$ 372	276 $\Big\}$ 249	256
$K_2Ba[Ni(NO_2)_6]$	1343	1306	838	433	291	255	256
$K_3[Ir(NO_2)_6]$	1395 $\Big\}$ 1375	1330	830	657	390	300	257
$K_3[Rh(NO_2)_6]$	1395	1340	833	627	386	283	257
$K_3[Ir(NO_2)Cl_5]$	1374	1315	835	644	325	288	258
$[Pt(NO_2)_6]^{4-}$	1488 $\Big\}$ 1458	1328	834	621	368	294	259
$K_2[Pt(^{15}NO_2)_4]$	1466 $\Big\}$ 1397	1343	847 $\Big\}$ 839 $\Big\}$ 833	640 $\Big\}$ 623	421		260, 261
$[Pd(NO_2)_4]^{2-\,b}$	1408	1364 $\Big\}$ 1320	834 $\Big\}$ 824	440	290		261
$K_2[Pt(NO_2)Cl_3]$	1401	1325	844	614	350	304	258

^{a}This mode may couple with other low-frequency modes.
^{b}Raman data in aqueous solution.

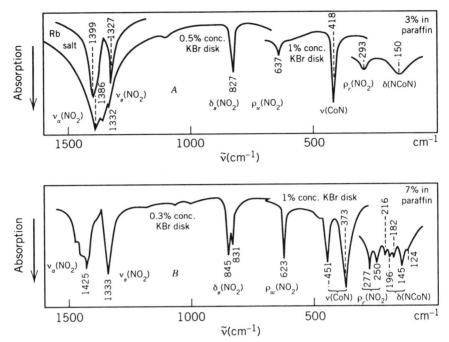

Fig. III-30. Infrared spectra of (A) $K_3[Co(NO_2)_6]$ and (B) $Na_3[Co(NO_2)_6]$.[262]

by the symmetry selection rules and the differences in frequency between iso-mers. It is possible to distinguish *cis*- and *trans*-[Co(NH_3)_4(NO_2)_2]^+ [263] by the rule that the *cis*-isomer exhibits more bands than the *trans*-isomer, and to distinguish *fac*- and *mer*-[Co(NH_3)_3(NO_2)_3][264] by the observation that $\delta(NO_2)$ and $\rho_w(NO_2)$ are higher for the *fac*-isomer (C_3, 832 and 625 cm^{-1}, respectively) than for the *mer*-isomer (C_{2v}, 825 and 610 cm^{-1}, respectively). Nakagawa and Shimanouchi[265] measured the infrared spectra of the $[Co(NO_2)_n(NH_3)_{6-n}]^{(3-n)+}$ series and carried out normal coordinate analysis on the mononitro and dini-tro complexes. Nolan and James[266] studied the infrared and Raman spectra of $[Pt(NO_2)_nCl_{6-n}]^{2-}$-type salts in the crystalline state.

(2)   Nitrito Complexes

If the NO_2 group is bonded to a metal through one of its O atoms, it is called a nitrito complex. Table III-14 lists the NO stretching frequencies of typical nitrito complexes. The two $\nu(NO_2)$ of nitrito complexes are well separated, $\nu(N{=}O)$ and $\nu(NO)$ being at 1485–1400 and 1110–1050 cm^{-1}, respectively. Distinction between the nitro and nitrito coordination can be made on this basis. It is to be noted that nitrito complexes lack the wagging modes near 620 cm^{-1} which appear in all nitro complexes. The $\nu(MO)$ of nitrito complexes were assigned in the 360–340 cm^{-1} region for metals such as Cr(III), Rh(III), and Ir(III).[258]

TABLE III-14. Vibrational Frequencies of Nitrito Complexes (cm^{-1})

Complex	$\nu(N{=}O)$	$\nu(NO)$	$\delta(ONO)$	Ref.
$[Co(NH_3)_5(ONO)]Cl_2$	1468	1065	825	267
$[Cr(NH_3)_5(ONO)]Cl_2$	1460	1048	839	267
$[Rh(NH_3)_5(ONO)]Cl_2$	1461 ⎱ 1445 ⎰	1063	830	258
$[Ni(py)_4(ONO)_2]$	1393	1114	825	268
trans-$[Cr(en)_2(ONO)_2]ClO_4$	1485 ⎱ 1430 ⎰	—	835 ⎱ 825 ⎰	269
$[Co(py)_4(ONO)_2](py)_2$	1405	1109	824	270

In many nitro complexes several types of nitro coordination are mixed. Goodgame, Hitchman, and their co-workers carried out an extensive study on vibrational spectra of nitro complexes containing various types of coordination. For example, all six nitro groups in $K_4[Ni(NO_2)_6]\cdot H_2O$ are coordinated through the N atom. However, its anhydrous salt exhibits the bands characteristic of nitro as well as nitrito coordination. From UV spectral evidence, Goodgame and Hitchman[271] suggested the structure $K_4[Ni(NO_2)_4(ONO)_2]$ for the anhydrous salt. Table III-15 lists the observed frequencies of two Ni(II) complexes containing both nitro and nitrito groups.

The red nitritopentammine complex, $[Co(NH_3)_5(ONO)]Cl_2$, is unstable and is gradually converted to the stable yellow nitro complex. The kinetics of this conversion can be studied by observing the disappearance of the nitrito bands,[273,274] and the rate of the photochemical isomerization in the solid state has been determined.[275] Burmeister[276] reviewed the vibrational spectra of these and other linkage isomers.

(3) Chelating Nitrito Complexes

If the nitrito group is chelating, the $\nu(N{=}O)$ and $\nu(N{-}O)$ of the nitrito group will be shifted to a lower and a higher frequency, respectively, relative to those of unidentate nitrito complexes. As a result, the separation between these two modes (Δ) becomes much smaller than those of unidentate complexes. Table III-16 lists the vibrational frequencies of chelating nitrito groups. It should be noted

TABLE III-15. Vibrational Frequencies of Ni(II) Complexes Containing Nitro and Nitrito Groups (cm^{-1})

Complex	Nitro Group			Nitrito Group		Ref.
	$\nu_a(NO_2)$	$\nu_s(NO_2)$	$\rho_w(ONO)$	$\nu(N{=}O)$	$\nu(NO)$	
$K_4[Ni(NO_2)_6]H_2O$	1346	1319	427	—	—	271
$K_4[Ni(NO_2)_4(ONO)_2]$	1347	1325	423 ⎱ 414 ⎰	1387	1206	271
Ni[2-(aminomethyl)-py]$_2$-(NO$_2$)(ONO)	1338	1318	—	1368	1251	272

TABLE III-16 Vibrational Frequencies of Chelating Nitrito Groups (cm^{-1})

Complex	$\nu_a(NO_2)$	$\nu_s(NO_2)$	$\delta(ONO)$	Δ^a	Ref.
Co(Ph$_3$PO)$_2$(NO$_2$)$_2$	1266	1199 ⎱ 1176 ⎰	856	78	277
Ni(α-pic)$_2$(NO$_2$)$_2$	1272	1199	866 ⎱ 862 ⎰	73	277
Re(CO)$_2$(PPh$_3$)$_2$ (NO$_2$)	1241	1180	887	61	277a
[Ni(N, N'-dimethyl-en)-(NO$_2$)] ClO$_4$	1300	1230	—	70	277b
Cs$_2$[Mn(NO$_2$)$_4$]	1302	1225	841	77	278
Co(Me$_4$-en)(NO$_2$)$_2$	1290	1207	850	83	270
Zn(py)$_2$(NO$_2$)$_2$	1351	1171	850	180	279
Zn(isoquinoline)$_2$(NO$_2$)$_2$	1370	1160	—	210	279
(o-cat) [Co(NO$_2$)$_4$]^b	1390	1191	—	199	278

$^a\Delta = \nu_a - \nu_s$.
^{b}o-cat = [o-xylylenebis(triphenylphosphonium)]$^{2+}$ ion.

that the Δ value depends on the degree of asymmetry of the coordinated nitrito group; it is expected that the Δ value is the smallest when the two N—O bonds are equivalent and increases as the degree of asymmetry increases. Relatively large Δ values observed for the last three compounds in Table III-16 may be accounted for on this basis.

(4)   Bridging Nitro Complexes

The nitro group is known to form a bridge between two metal atoms. Nakamoto et al.[267] suggested that among the three possible structures, IV, V, and VI, shown before, IV is most probable for

$$[(NH_3)_3Co\overset{\displaystyle OH}{\underset{\displaystyle NO_2}{\diagdown\!-OH\!-\!Co(NH_3)_3}}]^{3+}$$

since its NO$_2$ stretching frequencies (1516 and 1200 cm^{-1}) are markedly different from those of other types discussed thus far. Later, this structure was found by X-ray analysis of[280]

$$[(NH_3)_4Co\overset{\displaystyle NH_2}{\underset{\displaystyle NO_2}{\diagdown\!\diagup}}Co(NH_3)_4]Cl_4 \cdot 4H_2O$$

This compound exhibits the NO$_2$ stretching bands at 1492 and 1180 cm^{-1}. Upon

$^{16}O \rightarrow\ ^{18}O$ substitution of the bridging oxygen, the latter is shifted by -10 cm^{-1} while the former is almost unchanged. Thus, these bands are assigned to the $\nu(N{=}O)$ (outside the bridge) and $\nu(N{-}O)$ (bridge), respectively.[281] The $[Co_2\{NO_2(OH)_2\}(NO_2)_6]^{3-}$ ion exhibits the NO_2 bands at 1516, 1190, and 860 cm^{-1}, indicating the presence of a bridging nitro group:[282]

$$
\left[
\begin{array}{c}
\overset{\displaystyle O}{\underset{\displaystyle \diagdown}{}}\ \ \\
N{-}O \\
O_2N \diagup\ \ \ \ \ \diagdown\ NO_2 \\
O_2N{-}\underset{\diagup}{Co}{-}O{-}\underset{\diagdown}{Co}{-}NO_2 \\
O_2N\ \ \ \underset{O}{\overset{H}{|}}\ \ \ NO_2 \\
H
\end{array}
\right]^{3-}
$$

$[Ni(\beta\text{-pic})_2(NO_2)_2]_3\cdot C_6H_6$ exhibits a number of bands due to coordinated nitro groups. Goodgame et al.[283] suggested the presence of two different types of bridging nitro groups, IV, V, and III, on the basis of the crystal structure and infrared data for this compound: Type IV absorbs at 1412 and 1236, Type V at 1460 and 1019, and Type III at 1299 and 1236 cm^{-1}. Goodgame et al.[284] also studied the infrared spectra of other bridging nitro complexes of Ni(II). For example, they found that $Ni(en)(NO_2)_2$ contains a Type-IV bridge (1429 and 1241 cm^{-1}), while $Ni(py)_2(NO_2)_2(\frac{1}{3}C_6H_6)$ is similar to that of the analogous β-picoline complex.

III-7. LATTICE WATER AND AQUO AND HYDROXO COMPLEXES

Water in inorganic salts may be classified as lattice or coordinated water. There is, however, no definite borderline between the two. The former term denotes water molecules trapped in the crystalline lattice, either by weak hydrogen bonds to the anion or by weak ionic bonds to the metal, or by both:

$$
\begin{array}{ccc}
X^- & & X^- \\
\diagdown & & \diagup \\
H & & H \\
\diagdown & & \diagup \\
& O & \\
& \vdots & \\
& M &
\end{array}
$$

whereas the latter denotes water molecules bonded to the metal through partially covalent bonds. Although bond distances and angles obtained from X-ray and neutron-diffraction data provide direct information about the geometry of the water molecule in the crystal lattice, studies of vibrational spectra are also useful for this purpose. It should be noted, however, that the spectra of water molecules are highly sensitive to their surroundings.

(1) Lattice Water

In general, lattice water absorbs at 3550–3200 cm^{-1} (antisymmetric and symmetric OH stretchings) and at 1630–1600 cm^{-1} (HOH bending). If the spectrum is examined under high resolution, the fine structure of these bands is observed. For example, $CaSO_4 \cdot 2H_2O$ exhibits eight peaks in the 3500–3400 cm^{-1} region,[285] and its complete vibrational analysis can be made by factor group analysis (Sec. I-26 of Part A). In the low-frequency region (600~200 cm^{-1}) lattice water exhibits "librational modes" that are due to rotational oscillations of the water molecule, restricted by interactions with neighboring atoms. As are shown in Fig. III-31, they are classified into three types depending upon the direction of the principal axis of rotation. It should be noted, however, that these librational modes couple not only among themselves but also with internal modes of water (HOH bending) and other ions (SO_4^{2-}, NO_3^-, etc.) in the crystal. Tayal et al.[286] reviewed librational modes of water in hydrated solids.

The presence of the hydronium (H_3O^+) ion in crystalline acid hydrates is well established, and their spectra were discussed in Sec. II-3 of Part A. The existence of the $H_5O_2^+$ ion was first detected by X-ray analysis.[287] Pavia and Giguère[288] further confirmed its presence in $HClO_4 \cdot 2H_2O$ (namely, $[H_5O_2]ClO_4$) by the absence of some characteristic bands of the H_3O^+ and H_2O species. Its structure is suggested to be centrosymmetric H_2O—H—OH_2 of approximately C_{2h} symmetry. Both X-ray[289] and neutron-diffraction[290] studies suggest the presence of the $H_5O_2^+$ ion in $trans$-$[Co(en)_2Cl_2]Cl \cdot HCl \cdot 2H_2O$. Thus it should be formulated as $trans$-$[Co(en)_2Cl_2]Cl \cdot [H_5O_2]Cl$. The existence of the $H_7O_3^+$ ion in crystalline $HNO_3 \cdot 3H_2O$ and $HClO_4 \cdot 3H_2O$ was confirmed by infrared studies.[291] The spectra are consistent with a structure in which two of the hydrogens of the H_3O^+ ion are bonded to two H_2O molecules through short, asymmetrical hydrogen bonds.

(2) Aquo (H_2O) Complexes

In addition to the three fundamental modes of the free water molecule, coordinated water exhibits other modes, such as those shown in Fig. III-29. Nakagawa and Shimanouchi[292] carried out normal coordinate analyses on $[M(H_2O)_6]$- (T_h symmetry) and $[M(H_2O)_4]$- (D_{4h} symmetry) type ions to assign these low-frequency modes. Table III-17 lists the frequencies and band assignments, and Fig. III-32 illustrates the far-infrared spectra of aquo complexes obtained by these authors. According to Stefov et al.,[293] $[Cr(H_2O)_6]Cl_3$ exhibits the rocking

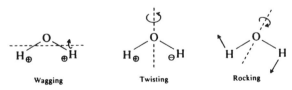

Wagging   Twisting   Rocking

Fig. III-31. The three rotational modes of H_2O in the solid state.

TABLE III-17. Observed Frequencies, Band Assignments, and MO Stretching Force Constants of Aquo Complexes[292]

Compound	$\rho_r(H_2O)$	$\rho_w(H_2O)$	$\nu(MO)$	$K(M—O)^a$
$[Cr(H_2O)_6]Cl_3$	800	541	490	1.31
$[Ni(H_2O)_6]SiF_6$	$(755)^b$	645	405	0.84
$[Ni(D_2O)_6]SiF_6$	—	450	389	0.84
$[Mn(H_2O)_6]SiF_6$	$(655)^c$	560	395	0.80
$[Fe(H_2O)_6]SiF_6$	—	575	389	0.76
$[Cu(H_2O)_4]SO_4·H_2O$	887, 855	535	440	0.67
$[Zn(H_2O)_6]SO_4·H_2O$	—	541	364	0.64
$[Zn(D_2O)_6]SO_4·D_2O$	467	392	358	0.64
$[Mg(H_2O)_6]SO_4·H_2O$	—	460	310	0.32
$[Mg(D_2O)_6]SO_4·D_2O$	474	391	—	0.32

^{a}UBF field (mdyn/Å).
^{b}Ni(H_2O)_4Cl_2.
^{c}Mn(H_2O)_4Cl_2.

(ρ_r), twisting (ρ_t), and wagging (ρ_w) modes of the coordinate water molecule at 825 (629), 575 (420), and 500 (390) cm^{-1}, respectively, in IR spectrum (the number in the brackets indicates the frequency of the D$_2$O complex). In Cs$_2$[InBr$_5$(H$_2$O)], the ρ_r, ρ_w, and ν(In–O) were assigned at 520, 380, and 260 cm^{-1}, respectively.[294] In solid K$_2$[FeCl$_5$(H$_2$O)], however, the frequency order

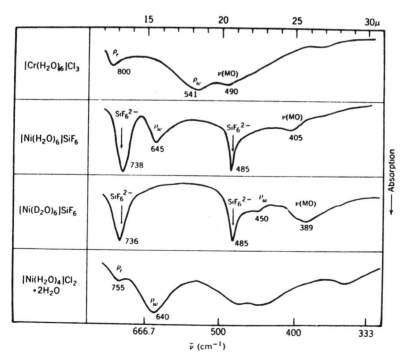

Fig. III-32. Infrared spectra of aquo complexes in the low-frequency region.[292]

of these bands is different; $600(\rho_w) > 460(\rho_t) > 390$ cm^{-1} [ν(Fe–O)].[295] The ν(Co–O) of [Co(NH$_3$)$_5$(H$_2$O)]Cl$_3$ and [Co(NH$_3$)$_5$(OH)]Cl$_2$ are assigned at 502 and 531 cm^{-1}, respectively.[296] Complete vibrational analysis has been made for single crystals of Ni(H$_2$O)$_4$Cl$_2$·2H$_2$O and related Co(II) complex.[297] α-Alums such as CsM(SO$_4$)$_2$·12H$_2$O (M = Co or Ir) contain the [M(H$_2$O)$_6$]$^{3+}$ ions, and their single crystal Raman spectra have been assigned by Best et al.[297a] An ab initio method has been employed to calculate vibrational frequencies of the [Na(H$_2$O)$_n$]$^+$ ion (n = 1~4).[298]

Vibrational spectroscopy is very useful in elucidating the structures of aquo complexes. For example, TiCl$_3$·6H$_2$O should be formulated as *trans*-[Ti(H$_2$O)$_4$Cl$_2$]Cl·2H$_2$O since it exhibits one TiO stretching (500 cm^{-1}, E_u) and one TiCl stretching (336 cm^{-1}, A_{2u}) mode.[299] Chang and Irish[300] showed from infrared and Raman studies that the structures of the tetrahydrates and dihydrates resulting from the dehydration of Mg(NO$_3$)$_2$·6H$_2$O are as follows:

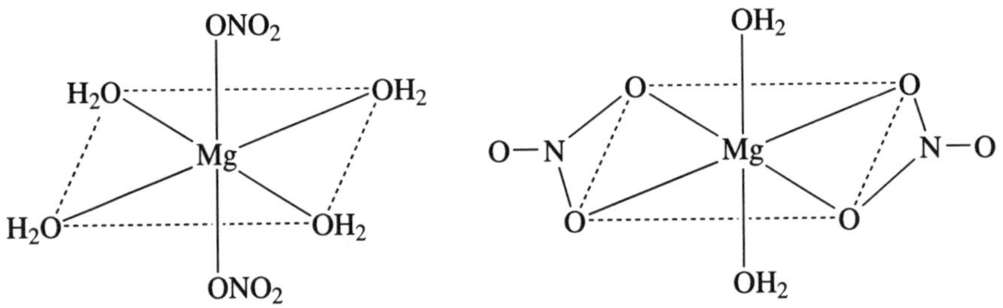

Raman spectra of aqueous solutions of inorganic salts have been studied extensively. For example, Hester and Plane[301] observed polarized Raman bands in the 400–360 cm^{-1} region for the nitrates, sulfates, and perchlorates of Zn(II), Hg(II), and Mg(II), and assigned them to the MO stretching modes of the hexacoordinated aquo complex ions.

A number of hydrated inorganic salts have also been studied by the inelastic neutron scattering (INS) technique.[302,303] Since the proton scattering cross section is quite large, the INS spectrum reflects mainly the motion of the protons in the crystal. Furthermore, INS spectroscopy has no selection rules involving dipole moments or polarizabilities. Thus, it serves as a complementary tool to vibrational spectroscopy in studying the hydrogen vibrations of hydrated salts.

(3) Aquo Complexes in Inert Gas Matrices

Cocondensation reactions of alkali halide vapors with H$_2$O in Ar matrices (14 K) produce 1 : 1 adducts of a pyramidal structure (C$_s$ symmetry):

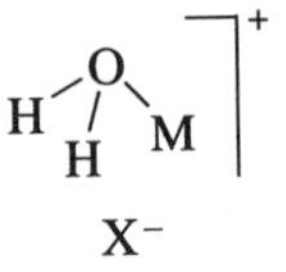

These aquo complexes exhibit two $\nu(OH)$ at 3300–3000 and two $\delta(OH)$ at 700–400 cm^{-1} in agreement with C_s symmetry.[304] Cocondensation reaction of Li vapor with H_2O in Kr matrices yields the 1:1 complex, $Li(H_2O)$ which can be characterized by three internal modes of water.[305] Similar studies with alkaline earth metal vapors show that the $\delta(H_2O)$ is down-shifted by 15 cm^{-1} for Mg and 30 cm^{-1} for Ca, Sr, and Ba upon complex formation.[306]

(4)   Hydroxo (OH) Complexes

The spectra of hydroxo complexes are expected to be similar to those of the metal hydroxides discussed in Sec. II-I of Part A. The hydroxo group can be distinguished from the aquo group since the former lacks the HOH bending mode near 1600 cm^{-1}. Furthermore, the hydroxo complex exhibits the MOH bending mode below 1200 cm^{-1}. For example, this mode is at 1150 cm^{-1} for the $[Sn(OH)_6]^{2-}$ ion[307] and at ca. 1065 cm^{-1} for the $[Pt(OH)_6]^{2-}$ ion.[308] The OH group also forms a bridge between two metals. For example,

exhibits the bridging OH bending mode at 955 cm^{-1}; this is shifted to 710 cm^{-1} upon deuteration.[309] For other bridging hydroxo complexes, the following references are given: Cu(II) complexes (309, 310), Cr(III) and Fe(III) complexes (311), Co(III) complexes (312), and Pb(II) complexes (313).

III-8. COMPLEXES OF ALKOXIDES, ALCOHOLS, ETHERS, KETONES, ALDEHYDES, ESTERS, AND CARBOXYLIC ACIDS

(1)   Complexes of Alkoxides and Alcohols

Metal alkoxides, $M(OR)_n$ (R: alkyl), exhibit $\nu(CO)$ at ca. 1000 cm^{-1} and $\nu(MO)$ at 600–300 cm^{-1},[314] Infrared spectra have been reported for various alkoxides of Er(III)[315] and isopropoxides of rare-earth metals.[316] Complete assignments of the IR and Raman spectra of $M(OCH_3)_6$ (M = Mo and W) and the $Sb(OCH_3)_6^-$ ion have been made based on normal coordinate analysis ($C_{3i} \equiv S_6$ symmetry). The $\nu(MO_6)$ and $\delta(MO_6)$ vibrations are at 600–450 and 400–200 cm^{-1}, respectively.[317]

The infrared spectra of alcohol complexes, $[M(EtOH)_6]Y_2$, where M is a divalent metal and Y is ClO_4^-, BF_4^-, and NO_3^-, have been measured by van Leeuwen.[318] As expected, the anions have considerable influence on $\nu(OH)$ and $\delta(MOH)$. In ethylene glycol complexes with MX_2 (X = Cl, Br, and I), $\nu(OH)$ are shifted to lower frequencies and $\delta(CCO)$ to higher frequencies relative to those of free ligand. It was shown that ethylene glycol serves as a bidentate chelating as well as a unidentate ligand, and that the *gauche* form prevails in the complexes.[319] Normal coordinate analyses have been carried out to assign the IR spectra of $[M(ROH)_6]X_2$ (M = Mg and Ca, R = CH_3 and C_2H_5, and X = Cl and Br). The bands at 305 and 275 cm^{-1} of $Mg(CH_3OH)_6Br_2$ and its Ca analog are primarily the $\nu(M—O)$ and the corresponding force constants are 0.42 and 0.35 mdyn/Å, respectively.[320]

(2) Complexes of Ethers

The vibrational spectra of diethyl ether complexes with $MgBr_2$ and MgI_2 have been assigned completely;[321] $\nu(MgO)$ at 390–300 cm^{-1}. The solid-state Raman spectra of 1:1 and 1:2 adducts of 1,4-dioxane with metal halides show that the ligand is bridging between metals in the chair conformation.[322]

When the oxygen atoms of the crown ether (18-crown-6) coordinate to Ba(II)[323] and Sb(III),[324] the $\nu(COC)$ band near 1100 cm^{-1} is shifted by 14 and 30 cm^{-1}, respectively, to a lower frequency. In the case of Ln(NCS)$_3$ (13-crown-4)·2H$_2$O (Ln: La, Pr, etc.), the red shift of the $\nu(COC)$ band is in the range of 76–64 cm^{-1}, and the $\nu(Ln–O)$ band appears at 390–370 cm^{-1}.[325] According to X-ray analysis, the 18-crown-6 ring takes the $\mathbf{D}_{3d}$ structure in the K$^+$ complex, and the $\mathbf{C}_i$ structure in the uncomplexed state (Fig. III-33). Takeuchi et al.[326] assigned the IR/Raman spectra of these two compounds in the solid state via normal coordinate analysis, and elucidated the ring conformations of other metal complexes in methanol solution by comparing their Raman spectra with those of the known structures.

(3) Complexes of Other Oxygen Donors

There are many coordination compounds with weakly coordinating ligands containing oxygen donors. These include ketones, aldehydes, esters, and some nitro compounds. Driessen and Groeneveld[327–329] and Driessen et al.[330] prepared metal complexes of these ligands through the reaction

$$MCl_2 + 6L + 2FeCl_3 \xrightarrow[CH_3NO_2]{} [ML_6](FeCl_4)_2$$

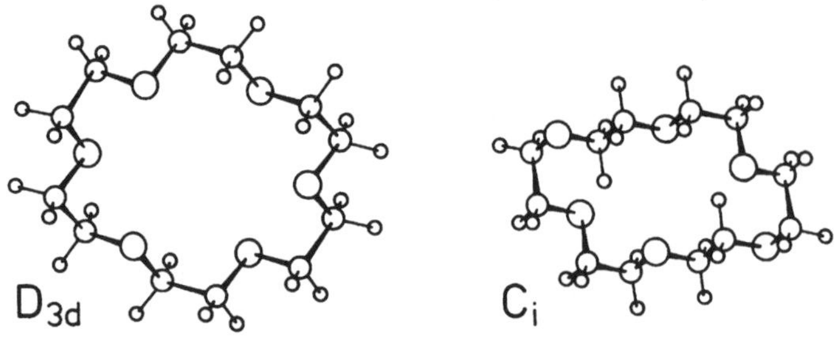

Fig. III-33. Two conformers of the 18-crown-6 ring (reproduced with permission from Ref. 326).

in a moisture-free atmosphere; CH_3NO_2 was chosen as the solvent because it is the weakest ligand available. In acetone complexes, $\nu(C{=}O)$ are lower, and $\delta(CO)$, $\pi(CO)$, and $\delta(CCC)$ are higher than those of free ligand.[327] Similar results have been obtained for complexes of acetophenone, chloracetone, and butanone.[328] In the [Li(acetone)$_4$]$^+$ ion, however, the $\nu(C{=}O)$, $\nu_a(CC)$, and $\nu_s(CC)$ all shift to higher frequencies upon coordination to the Li ion.[331]

In metal complexes of acetaldehyde, $\nu(C{=}O)$ are lower and $\delta(CCO)$ are higher than those of free ligand.[329] In ester complexes,[330] $\nu(C{=}O)$ shifts to lower and $\nu(C{-}O)$ to higher frequency by complex formation. When these shifts are dependent on the metal ions, the magnitudes of the shifts follow the well-known Irving–Williams order: Mn(II) < Fe(II) < Co(II) < Ni(II) < Cu(II) > Zn(II).

Formamide($HCONH_2$) coordinates to metal ions via the O atom, and the $\nu(M{-}O)$ vibrations appear in the 304–230 cm^{-1} range. In NiCl$_2$(NMF)$_4$ and NiCl$_2$(DMF)$_4$ (NMF: N-methylformamide; DMF: dimethylformamide), the Ni(II) ion is coordinated by the N as well as O atoms, and the $\nu(Ni{-}N)$ and $\nu(Ni{-}O)$ vibrations are observed at 500–480 and 420–380 cm^{-1}, respectively.[332]

(4) Complexes of Carboxylic Acids

Extensive infrared studies have been made on metal complexes of carboxylic acids. Table III-18 gives the infrared frequencies and band assignments for the

TABLE III-18. Infrared Frequencies and Band Assignments for Formate and Acetate Ions (cm^{-1})[333]

[HCOO]$^-$		[CH$_3$COO]$^-$			
Na Salt	Aqueous Solution	Na Salt	Aqueous Solution	C_{2v}	Band Assignment
2841	2803	2936	2935	A_1	$\nu(CH)$
—	—	—	1344		$\delta(CH_3)$
1366	1351	1414	1413		$\nu_s(COO)$
—	—	924	926		$\nu(CC)$
772	760	646	650		$\delta(OCO)$
—	—	—	—	A_2	$\rho_t(CH_3)$
—	—	2989	3010 or 2981	B_1	$\nu(CH)$
1567	1585	1578	1556		$\nu_a(COO)$
—	—	1430	1429		$\delta(CH_3)$
—	—	1009	1020		$\rho_r(CH_3)$
1377	1383	460	471		$\delta(CH)$ or $\rho_r(COO)$
—	—	2989	2981 or 3010	B_2	$\nu(CH)$
—	—	1443	1456		$\delta(CH_3)$
—	—	1042	1052		$\rho_r(CH_3)$
1073	1069	615	621		$\pi(CH)$ or $\pi(COO)$

formate and acetate ions obtained by Itoh and Bernstein.[333] The carboxylate ion may coordinate to a metal in one of the following modes:

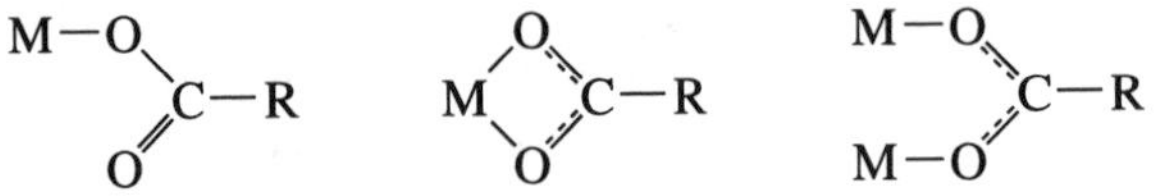

Deacon and Phillips[334] made careful examinations of IR spectra of many acetates and trifluoroacetates having known X-ray crystal structures, and arrived at the following conclusions:

1. Unidentate complexes (structure I) exhibit the Δ values $[\nu_a(CO_2^-)-\nu_s-(CO_2^-)]$ which are much greater than the ionic complexes.

2. Chelating (bidentate) complexes (structure II) exhibit Δ values which are significantly less than the ionic values.

3. The Δ values for bridging complexes (structure III) are greater than those of chelating (bidentate) complexes, and close to the ionic values.

TABLE III-19. Carboxyl Stretching Frequencies and Structures of Carboxylate Complexes (cm^{-1})

Compound	$\nu_a(COO)^a$	$\nu_s(COO)^a$	Δ	Structure	Ref.
HCOO$^-$	1567	1366	201	Ionic	333
CH$_3$COO$^-$ (OAc$^-$)	1578	1414	164	Ionic	333
Rh(OAc)(CO)(PPh$_3$)$_2$	1604	1376	228	Unidentate	335
Ru(OAc)(CO)$_2$(PPh$_3$)	1613	1315	298	Unidentate	335
Si(OAc)$_4$	1745^b	1290^b	455	Unidentate	336
Ge(OAc)$_4$	1710^b	1280^b	430	Unidentate	336
RuCl(OAc)(CO)(PPh$_3$)$_2$	1507	1465	42	Bidentate	335
RuH(OAc)(PPh$_3$)$_2$	1526	1449	77	Bidentate	335
Ph$_2$Sn(CH$_3$—COO)$_2$	1610	1335	265	Asym. bidentate	337
Ph$_2$Sn(CH$_2$Cl—COO)$_2$	1620	1240	380	Asym. bidentate	337
Ph$_2$Te(CCl$_3$—COO)$_2$	1705	1270	435	Asym. bidentate	337
Rh$_2$(OAc)$_2$(CO)$_3$(PPh$_3$)	1580	1440	140	Bridging	338
[Ru(CO)$_2$(C$_2$H$_5$COO)]$_n$	1548	1410	138	Bridging	339
[Cr$_3$O(OAc)$_6$(H$_2$O$_3$)]$^+$	1621	1432	189	Bridging	340
[Mn$_2$O$_2$(OAc)]$^{2+}$	1548	1387	171	Bridging	341
[Pd(OAc)$_2$(PPh$_3$)]$_2$	1629	1314	315	Unidentate	342
	1580	1411	169	Bridging	
CrO$_2$(OAc)$_2$	1710	1240	470	Unidentate	343
	1610	1420	190	Bidentate	
Cp$_2$Zr[Cr(CO)$_3$(RCOO)]$_2$ ^{c}	1641	1329	312	Unidentate	344
	1542	1377	165	Bidentate	

^{a}These correspond to the $\nu(C{=}O)$ (free) and $\nu(C{-}O)$ (coordinated) of the unidentate carboxylates, respectively.
^{b}IR frequency.
^{c}R = C$_6$H$_5$.

As seen in Table III-19, these criteria hold except for asymmetric bidentates such as $Ph_2Sn(CH_3COO)_2$ where the two Sn–O bond distances are markedly different:

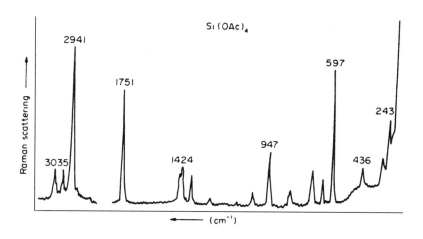

In these cases, Δ values are comparable to those of unidentate complexes.[337] Table III-19 also shows three carboxylate complexes in which two modes of coordination are mixed. Figure III-34 shows the Raman spectra of $Si(OAc)_4$ and $Ge(OAc)_4$ which contain only unidentate acetato ligands.[336] According to

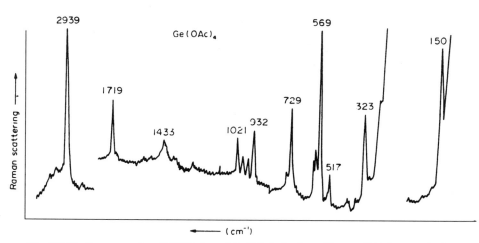

Fig. III-34. Raman spectra of $Si(OAc)_4$ and $Ge(OAc)_4$ in the solid state (514.5 nm excitation) (reproduced with permission from Ref. 336).

Stoilova et al.,[345] unidentate acetates exhibit three bands (COO deformation) at 920–720 cm^{-1} and a strong band [$\pi(CO_2)$] at 540 cm^{-1} which are absent in bridging complexes and reduced in number in bidentate complexes. Infrared spectra of formates have been reviewed by Busca and Lorenzelli.[346]

The linkage isomerism involving the acetate group has been reported by Baba and Kawaguchi:[347]

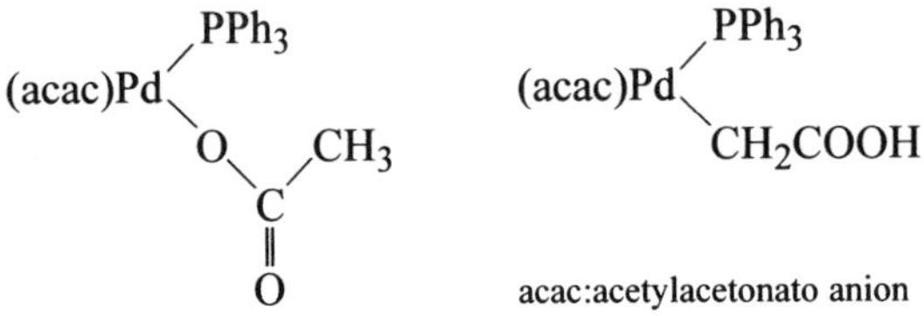

acac:acetylacetonato anion

The O-isomer exhibits $\nu(C{=}O)$ at 1640 cm^{-1}, whereas the C-isomer shows $\nu(C{=}O)$ at 1670 and 1650 and $\nu(OH)$ at 2700–2500 cm^{-1}. It is also possible to distinguish two acetato groups having different *trans* ligands by their frequencies:

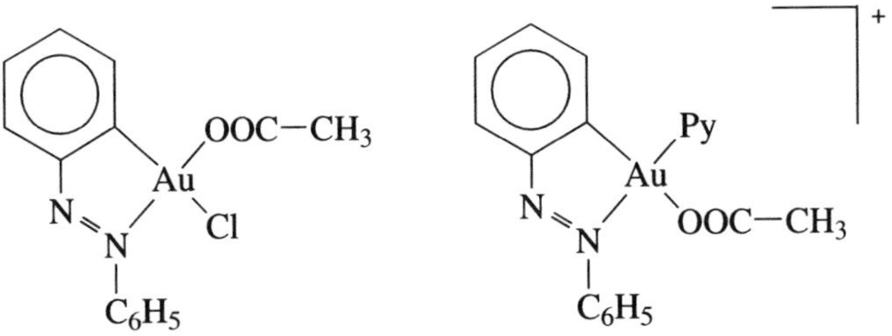

(A)   (B)

Complex (A) exhibits the $\nu_a(COO)$ and $\nu_s(COO)$ at 1665 and 1360 cm^{-1}, respectively, whereas complex (B) exhibits these vibrations at 1620 and 1300 cm^{-1}, respectively.[348] The IR spectra of metal glycolato ($CH_2(OH){-}COO^-$) complexes have been assigned based on normal coordinate calculations.[349]

III-9. COMPLEXES OF AMINO ACIDS, EDTA, AND RELATED LIGANDS

(1)   Complexes of Amino Acids

Amino acids exist as zwitterions in the crystalline state. Table III-20 gives band assignments made for the zwitterions of glycine[350] and α-alanine.[351] According to X-ray analysis, two glycino anions (gly) in [Ni(gly)$_2$]·2H$_2$O,[352] for example, coordinate to the metal by forming a *trans*-planar structure, and the noncoor-

TABLE III-20. Infrared Frequencies and Band Assignments of Glycine and α-Alanine in the Crystalline State (cm^{-1})[350,351]

Glycine	α-Alanine	Band Assignment
1610	1597	$\nu_a(COO^-)$
1585	1623	$\delta_d(NH_3^+)$
1492	1534	$\delta_s(NH_3^+)$
—	1455	$\delta_d(CH_3)$
1445	—	$\delta(CH_2)$
1413	1412	$\nu_s(COO^-)$
—	1355	$\delta_s(CH_3)$
1333	—	$\rho_w(CH_2)$
—	1308	$\delta(CH)$
1240 (R)	—	$\rho_t(CH_2)$
1131 ⎱ 1110 ⎰	1237 ⎱ 1113 ⎰	$\rho_r(NH_3^+)^a$
1003	1148	$\nu_a(CCN)^a$
—	1026 ⎱ 1015 ⎰	$\rho_r(CH_3)^a$
910	—	$\rho_r(CH_2)$
893	918 ⎱ 852 ⎰	$\nu_s(CCN)^a$
694	648	$\rho_w(COO^-)$
607	771	$\delta(COO^-)$
516	492	$\rho_t(NH_3^+)$
504	540	$\rho_r(COO^-)$

^{a}These bands are coupled with other modes in α-alanine.

dinating C=O groups are hydrogen-bonded to the neighboring

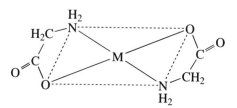

molecule or water of crystallization, or weakly bonded to the metal of the neighboring complex. Thus $\nu(CO_2)$ of amino acid complexes are affected by coordination as well as by intermolecular interactions.

To examine the effects of coordination and hydrogen bonding, Nakamoto et al.[353] made extensive IR measurements of the COO stretching frequencies of various metal complexes of amino acids in D_2O solution, in the hydrated crystalline state, and in the anhydrous crystalline state. The results showed that, in any one physical state, the same frequency order is found for a series of metals, regardless of the nature of the ligand. The antisymmetric frequencies increase, the symmetric frequencies decrease, and the separation between the two frequencies increases in the following order of metals:

$$Ni(II) < Zn(II) < Cu(II) < Co(II) < Pd(II) \approx Pt(II) < Cr(III)$$

Although there are several exceptions to this order, these results generally indicate that the effect of coordination is still the major factor in determining the frequency order in a given physical state. The above frequency order indicates the increasing order of the metal–oxygen interaction since the COO group becomes more asymmetrical as the metal–oxygen interaction becomes stronger.

To give theoretical band assignments on metal glycino complexes, Condrate and Nakamoto[354] carried out a normal coordinate analysis on the metal–glycino chelate ring. Figure III-35 shows the infrared spectra of bis(glycino) complexes of Pt(II), Pd(II), Cu(II), and Ni(II). Table III-21 lists the observed frequencies and theoretical band assignments. The CH$_2$ group frequencies are not listed, since they are not metal-sensitive. It is seen that the C=O stretching, NH$_2$

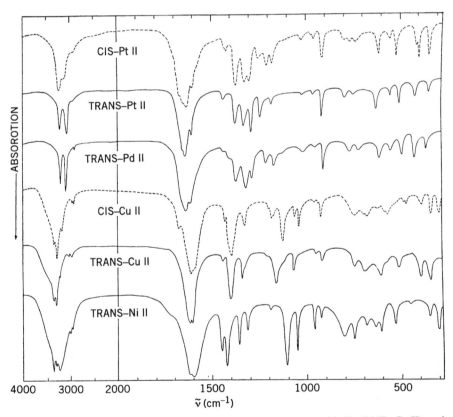

Fig. III-35. Infrared spectra of *cis*- or *trans*-bis(glycino) complexes of Pt(II), Pd(II), Cu(II), and Ni(II).[354]

TABLE III-21. Observed Frequencies and Band Assignments of Bis(glycino) Complexes (cm^{-1})^{354}

trans-[Pt(gly)$_2$]	trans-[Pd(gly)^2]	trans-[Cu(gly)$_2$]	trans-[Ni(gly)$_2$]	Band Assignment
3230 ⎱	3230 ⎱	3320 ⎱	3340 ⎱	
3090 ⎰	3120 ⎰	3260 ⎰	3280 ⎰	$\nu(NH_2)$
1643	1642	1593	1589	$\nu(C{=}O)$
1610	1616	1608	1610	$\delta(NH_2)$
1374	1374	1392	1411	$\nu(C{-}O)$
1245	1218	1151	1095	$\rho_t(NH_2)$
1023	1025	1058	1038	$\rho_w(NH_2)$
792	771	644	630	$\rho_r(NH_2)$
745	727	736	737	$\delta(C{=}O)$
620	610	592	596	$\pi(C{=}O)$
549	550	481	439	$\nu(MN)$
415	420	333	290	$\nu(MO)$
2.10	2.00	0.90	0.70	$K(M{-}N)$ (mdyn/Å)^a
2.10	2.00	0.90	0.70	$K(M{-}O)$ (mdyn/Å)^a

^{a}UBF.

rocking, and MN and MO stretching bands are metal sensitive and are shifted progressively to higher frequencies as the metal is changed in the order Ni(II) < Cu(II) < Pd(II) < Pt(II). Table III-21 shows that both the MN and MO stretching force constants also increase in the same order of the metals. These results provide further support to the preceding discussion of the M–O bonds of glycino complexes.

To give definitive band assignments in the low-frequency region of *bis*(glycino) complexes of Ni(II), Cu(II), and Co(II), Kincaid and Nakamoto[355] carried out H–D, ^{14}N–^{15}N, ^{58}Ni–^{62}Ni, and ^{63}Cu–^{65}Cu substitutions, and performed normal coordinate analyses on the skeletal modes of bis(glycino) complexes. Their results show that, in *trans*-[M(gly)$_2$]2H$_2$O, the infrared-active ν(MN) and ν(MO) are at 483 and 337 cm^{-1}, respectively, for the Cu(II) complex, and at 442 and 289 cm^{-1}, respectively, for the Ni(II) complex. Both modes are coupled strongly with other skeletal modes, however. Use of multiple isotope labeling techniques in assigning IR spectra of amino acid complexes has been extended to [Cd(gly)$_2$]·H$_2$O,[356] *cis*-[Ni(gly)$_2$(ImH)$_2$],[357] and [M(L–Ala)$_2$] [M = Ni(II) and Cu(II)].[358]

Square-planar bis(glycino) complexes can take the *cis* or the *trans* configuration. As expected from symmetry consideration, the *cis*-isomer exhibits more bands in infrared spectra than does the *trans*-isomer (see Fig. III-35). In the low-frequency region, the *cis*-isomer exhibits two ν(MN) and two ν(MO), whereas the *trans*-isomer exhibits only one for each of these modes.[354] This criterion has been used by Herlinger et al. to assign the geometry of a series of bis(amino acidato)Cu(II) complexes.[359,360] Octahedral tris(glycino) complexes may take the *fac* and *mer* configurations shown in Fig. III-36. For example, [Co(gly)$_3$]

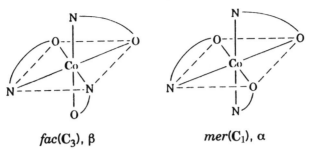

fac(C₃), β   *mer*(C₁), α

Fig. III-36. Structures of *fac*- and *mer*-tris(glycino) complexes.

exists in two forms: purple crystals (dihydrate, α-form) and red crystals (mono-hydrate, β-form). The α-form is assigned to the *mer* configuration since it exhibits more infrared bands than does the β-form (*fac* configuration).[361]

Glycine also coordinates to the Pt(II) atom as a unidentate ligand:

$$-\overset{|}{\underset{|}{Pt}}-NH_2-CH_2-C\overset{O}{\underset{OH}{\diagup}} \qquad -\overset{|}{\underset{|}{Pt}}-NH_2-CH_2-C\overset{O^{-1/2}}{\underset{O^{-1/2}}{\diagup}}$$

The carboxyl group is not ionized in *trans*-[Pt(glyH)₂X₂] (X: a halogen), whereas it is ionized in *trans*-[Pt(gly)₂(NH₃)₂]. The former exhibits the union-ized COO stretching band near 1710 cm⁻¹, while the latter shows the ionized COO stretching band near 1610 cm⁻¹.[362]

The distinction between unidentate and bidentate glycino complexes of Pt(II) can be made readily from their infrared spectra. Figure III-37 illustrates the infrared spectra of *trans*-[Pt(glyH)₂Cl₂] and K[Pt(gly)Cl₂] in the COO stretching and PtO stretching regions. The bidentate (chelated) glycino group absorbs at 1643 cm⁻¹, unlike either the ionized unidentate group (1610 cm⁻¹) or the union-ized unidentate group (1710 cm⁻¹). Furthermore, the bidentate glycino group exhibits the PtO stretching band at 388 cm⁻¹, whereas the unidentate glycino group has no absorption between 470 and 350 cm⁻¹. Figure III-37 also shows the spectrum of [Pt(gly)(glyH)Cl], in which both the unidentate and bidentate glycino groups are present. It is seen that the spectrum of this compound can be interpreted as a superposition of the spectra of the former two compounds.[362]

The Cu(III) complexes of tetraglycine and tetraglycineamide exhibit the N(amide)–Cu(III) CT absorption at 365 nm. Using the 363.8 nm excitation, Kincaid et al.[363] were able to resonance-enhance the ν(Cu–N) vibrations at 420 and 417 cm⁻¹, respectively.

Metal complexes with *N*-methylglycine (sarcosine) and *N*-phenylglycine, of the ML₂·nH₂O type, take the chelate ring structures similar to that of the glycine complexes, and their IR spectra have been assigned by Inomata et al.[364] based on normal coordinate calculations. These ligands also form metal complexes of the type CoCl₂(HL)·2H₂O and MCl₂(HL) (M = Zn and Cd) in which the

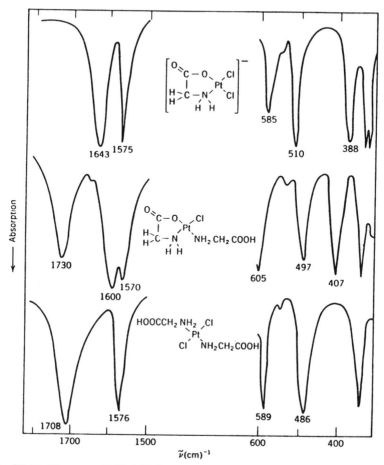

Fig. III-37. IR spectra of K[Pt(gly)Cl$_2$], [Pt(gly)(glyH)Cl], and *trans*-[Pt(glyH)$_2$Cl$_2$].[362]

zwitterion of the amino acid is coordinated to the metal via the carboxyl oxygen atom.[364]

(2)   Complexes of EDTA and Related Ligands

From the infrared spectra observed in the solid state, Busch and co-workers[365] determined the coordination numbers of the metals in metal chelate compounds of EDTA and its derivatives:

$$\begin{array}{c} \text{HOOCH}_2\text{C} \\ \text{HOOCH}_2\text{C} \end{array} \!\! N-\text{CH}_2-\text{CH}_2-N \!\! \begin{array}{c} \text{CH}_2\text{COOH} \\ \text{CH}_2\text{COOH} \end{array}$$

Ethylenediaminetetraacetic acid
(EDTA) or (H$_4$Y)

The method is based on the simple rule that the unionized and uncoordinated COO stretching band occurs at 1750–1700 cm^{-1}, whereas the ionized and coordinated COO stretching band is at 1650–1590 cm^{-1}. The latter frequency depends on the nature of the metal: 1650–1620 cm^{-1} for metals such as Cr(III) and Co(III), and 1610–1590 cm^{-1} for metals such as Cu(II) and Zn(II). Since the free ionized COO$^-$ stretching band is at 1630–1575 cm^{-1}, it is also possible to distinguish the coordinated and free COO$^-$ stretching bands if a metal such as Co(III) is chosen for complex formation. Table III-22 shows the results obtained by Busch et al.

Tomita and Ueno[367] studied the infrared spectra of metal complexes of NTA, using the method described above. They concluded that NTA

$$\begin{matrix} & \diagup CH_2COOH \\ N & \!\!\!-CH_2COOH \\ & \diagdown CH_2COOH \end{matrix}$$ Nitrilotriacetic acid (NTA)

acts as a quadridentate ligand in complexes of Cu(II), Ni(II), Co(II), Zn(II), Cd(II), and Pb(II), and as a tridentate in complexes of Ca(II), Mg(II), Sr(II), and Ba(II).

Krishnan and Plane[368] studied the Raman spectra of EDTA and its metal complexes in aqueous solution. They noted that $\nu(MN)$ appears strongly in the 500–400 cm^{-1} region for Cu(II), Zn(II), Cd(II), Hg(II), and so on, and that its frequency decreases with an increasing radius of the metal ion, independently

TABLE III-22. Antisymmetric COO Stretching Frequencies and Number of Functional Groups Used for Coordination in EDTA Complexes (cm^{-1})[365]

Compound[a]	Un-ionized COOH	Coordinated COO$^-$ $\cdots$M	Free COO$^-$	Number of Coordinated Groups
H$_4$[Y]	1698[b]	—	—	
Na$_2$[H$_2$Y]	1668[b]	—	1637[b]	
Na$_4$[Y]	—	—	1597[b]	
Ba[Co(Y)]$_2$·4H$_2$O	—	1638	—	6
Na$_2$[Co(Y)Cl]	—	1648	1600	5
Na$_2$[Co(Y)NO$_2$]	—	1650	1604	5
Na[Co(HY)Cl]·$\frac{1}{2}$H$_2$O	1750	1650	—	5
Na[Co(HY)NO$_2$]·H$_2$O	1745	1650	—	5
Ba[Co(HY)Br]·9H$_2$O	1723	1628	—	5
Na[Co(YOH)Cl]·$\frac{3}{2}$H$_2$O	—	1658	—	5
Na[Co(YOH)Br]·H$_2$O	—	1654	—	5
Na[Co(YOH)NO$_2$]	—	1652	—	5
[Pd(H$_2$Y)]·3H$_2$O	1740	1625	—	4
[Pt(H$_2$Y)]·3H$_2$O	1730	1635	—	4
[Pd(H$_4$Y)Cl$_2$]·5H$_2$O	1707, 1730	—	—	2
[Pt(H$_4$Y)Cl$_2$]·5H$_2$O	1715, 1530	—	—	2

[a]Y = tetranegative ion; HY = trinegative ion; H$_2$Y = dinegative ion; H$_4$Y = neutral species of EDTA; YOH = trinegative ion of HEDTA (hydroxyethylenediaminetriacetic acid).
[b]Reference 366.

of the stability of the metal complex. McConnell and Nuttall[369] assigned the $\nu(MN)$ and $\nu(MO)$ of $Na_2[M(EDTA)]2H_2O$ (M = Sn and Pb) in their Raman and infrared spectra.

III-10. INFRARED SPECTRA OF AQUEOUS SOLUTIONS

Since water is a weak Raman scatterer, Raman spectra of samples in aqueous solution can be measured without major interference from water vibrations. On the other hand, infrared spectroscopy of aqueous solution suffers from strong absorption of bulk water which interferes with IR absorption of the sample. Even so, it is sometimes necessary to measure aqueous IR spectra because some vibrations are inherently weak in Raman spectra.

To measure IR spectra of aqueous solution, it is common to use very thin layers (0.01–0.05 mm thick) of solutions of relatively high concentrations (5–20%) which are sandwiched between two plates of water-insoluble crystals such as CaF_2 and KRS-5 (TlBr/TlI). Figure III-38 displays the IR spectra of H_2O and D_2O obtained by using a CaF_2 cell (4000–1000 cm^{-1}, 0.03 mm thick) and a KRS-5 cell (1200–250 cm^{-1}, 0.015 mm thick) which show that at least two regions, 2800–1800 and 1500–950 cm^{-1}, are relatively free from H_2O absorption. These spectral "window" regions can be shifted to 2150–1250 and 1100–750 cm^{-1}, respectively, in D_2O.[369a] The combination of a recently developed cylindrical internal reflection (CIR) cell[369b] with a FTIR spectrometer may be best suited to IR studies of aqueous solution.[370] The following examples demonstrate the utility of aqueous IR spectroscopy in elucidating the structures of complex ions in solution equilibria.

The $C{\equiv}N$ stretching band (2200–2000 cm^{-1}) can be measured in aqueous solution since it is in the "window" region. Thus, the solution equilibria of cyano complexes have been studied extensively by using aqueous infrared

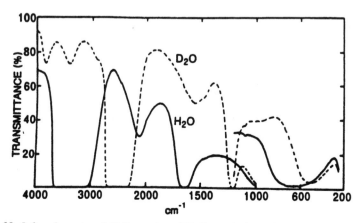

Fig. III-38. Infrared spectra of H_2O vs. air and D_2O vs. air (reproduced with permission from Ref. 369a).

spectroscopy (Sec. III-15). Fronaeus and Larsson[371] extended similar studies to thiocyanato complexes that exhibit the C≡N stretching bands in the same region. They[372] also studied the solution equilibria of oxalato complexes in the 1500–1200 cm^{-1} region, where the CO stretching bands of the coordinated oxalato group appear. Larsson[373] studied the infrared spectra of metal glycolato complexes in aqueous solution. In this case, the C—OH stretching band near 1060 cm^{-1} was used to elucidate the structures of the complex ions in equilibria.

The COO stretching bands of NTA, EDTA, and their metal complexes appear between 1750 and 1550 cm^{-1} (Sec. III-9). As stated above, this region is free from D$_2$O absorption. Nakamoto et al.,[374] therefore, studied the solution (D$_2$O) equilibria of NTA, EDTA, and related ligands in this frequency region. By combining the results of potentiometric studies with the spectra obtained as a function of the pH (pD) of the solution, it was possible to establish the following COO stretching frequencies:

Type A, un-ionized carboxyl (R$_2$N—CH$_2$COOH), 1730–1700 cm^{-1}

Type B, α-ammonium carboxylate (R$_2$N$^+$H—CH$_2$COO$^-$), 1630–1620 cm^{-1}

Type C, α-aminocarboxylate (R$_2$N—CH$_2$COO$^-$), 1585–1575 cm^{-1}

As stated in Sec. III-9, the coordinated (ionized) COO group absorbs at 1650–1620 cm^{-1} for Cr(III) and Co(III), and at 1610–1590 cm^{-1} for Cu(II) and Zn(II). Thus it is possible to distinguish the coordinated COO group from those of Types B and C if a proper metal ion is selected.

Tomita et al.[375] studied the complex formation of NTA with Mg(II) by aqueous infrared spectroscopy. Figure III-39 shows the infrared spectra of equimolar mixtures of NTA and MgCl$_2$ at concentrations about 5–10% by weight. The spectra of the mixture from pD 3.2 to 4.2 exhibit a single band at 1625 cm^{-1}, which is identical to that of the free H(NTA)$^{2-}$ ion in the same pD range.[376] This result indicates that no complex formation occurs in this pD range, and that the 1625-cm^{-1} band is due to the H(NTA)$^{2-}$ ion (Type B). If the pD is raised to 4.2, a new band appears at 1610 cm^{-1}, which is not observed for the free NTA solution over the entire pD range investigated. Figure III-39 shows that this 1610-cm^{-1} band becomes stronger, and the 1625-cm^{-1} band becomes weaker, as the pD increases. It was concluded that this change is due mainly to a shift of the following equilibrium in the direction of complex formation:

$$
\underset{\substack{\text{1625 cm}^{-1}\\ \text{(Type B)}}}{\text{HN}^+\!\!\begin{array}{l}\diagup\text{CH}_2\text{COO}^-\\-\text{CH}_2\text{COO}^-\\\diagdown\text{CH}_2\text{COO}^-\end{array}} + \text{Mg}^{2+} \;\rightleftharpoons\; \underset{\substack{\text{1610 cm}^{-1}}}{\left[\text{N}\begin{array}{l}\diagup\text{CH}_2\text{COO}^-\\-\text{CH}_2\text{COO}^-\!\!-\text{Mg}\\\diagdown\text{CH}_2\text{COO}^-\end{array}\right]^-} + \text{H}^+
$$

By plotting the intensity of these two bands as a function of pD, the stability constant of the complex ion was calculated to be 5.24. This value is in good agreement with that obtained from potentiometric titration (5.41).

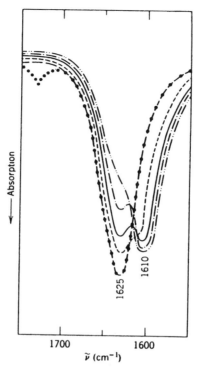

Fig. III-39. Infrared spectra of Mg–NTA complex in D_2O solutions: (●●●●) pD 3.2; (– – – –) pD 4.2; (– – –) pD 5.5; (—) pD 6.8; (–·–·–), pD 10.0; and (–··–··–), pD 11.6.[375]

Martell and Kim[377-380] carried out an extensive study on solution equilibria involving the formation of Cu(II) complexes with various polypeptides. As an example, the glycylglycino–Cu(II) system is discussed below.[378] Figure III-40 illustrates the infrared spectra of free glycylglycine in D_2O solution as a function of pD. The observed spectral changes were interpreted in terms of the solution equilibria shown below:

$$\overset{+}{H_3N}-CH_2-\overset{\overset{\displaystyle O}{\|}}{C}-\overset{\overset{\displaystyle}{|}}{\underset{\displaystyle H}{N}}-CH_2COO^-$$

O 1665 cm^{-1} (Type D)

1595 cm^{-1} (Type C)

II

$pK_1 = 3.21$ $pK_2 = 8.12$

$$\overset{+}{H_3N}-CH_2-\overset{\overset{\displaystyle O}{\|}}{C}-\overset{\overset{\displaystyle}{|}}{\underset{\displaystyle H}{N}}-CH_2COOH$$

O 1675 cm^{-1} (Type D)

1720 cm^{-1}

(Type A)

I

$$H_2N-CH_2-\overset{\overset{\displaystyle O}{\|}}{C}-\overset{\overset{\displaystyle}{|}}{\underset{\displaystyle H}{N}}-CH_2COO^-$$

O 1630 cm^{-1} (Type D)

1595 cm^{-1}

(Type C)

III

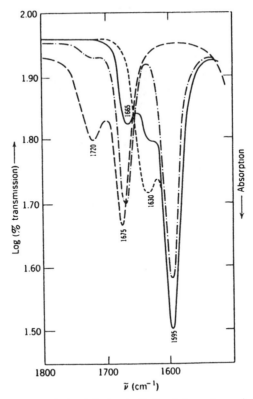

Fig. III-40. Infrared spectra of glycylglycine in D_2O solutions: (----) pD 1.75; (-·--·-) pD 4.31; (—) pD 8.77; and (-----) pD 10.29.[378]

Band assignments have been made by using the criteria given previously. In addition, Type-D frequency (1680–1610 cm^{-1}) was introduced to denote the peptide carbonyl group. The exact frequency of this group depends on the nature of the neighboring groups.

Figure III-41 shows the infrared spectra of glycylglycine mixed with copper chloride at equimolar ratio in D_2O solution.[379] At pD = 3.58, the ligand exhibits three bands at 1720, 1675, and 1595 cm^{-1} (Fig. III-40). This result indicates that I and II are in equilibrium. At the same pD value, however, the mixture exhibits one extra band at 1625 cm^{-1}. This band was attributed to the metal complex (IV), which was formed by the following reaction:

$$\text{I, II} + \text{Cu}^{2+} \longrightarrow \left[\begin{array}{c} \text{NH—CH}_2\text{COO}^- \\ \text{1598 cm}^{-1} \\ \text{H}_2\text{C—C} \\ \text{H}_2\text{N} \quad \text{O 1625 cm}^{-1} \\ \text{Cu} \\ \text{H}_2\text{O} \quad \text{OH}_2 \end{array} \right]^+ + \text{xH}^+$$

$$\text{IV}$$

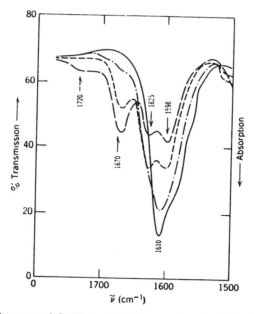

Fig. III-41. Infrared spectra of Cu(II)-glycylglycino complexes in D_2O solutions: (——) pD 3.58; (- - - - -) pD 4.24; (-·-·-) pD 5.18; and (—) pD 10.65.

At pD = 5.18, the solution exhibits one broad band at about 1610 cm^{-1}. This result was interpreted as an indication that the following equilibrium was shifted almost completely to the right-hand side, and that the 1610-cm^{-1} band is an overlap of two bands at 1610 and 1598 cm^{-1}:

$$\text{II} + \text{Cu}^{2+} \longrightarrow \left[\begin{array}{c} \text{structure V} \end{array} \right] + 2\text{H}^+$$

V

The shift of the peptide carbonyl stretching band from 1625 (IV) to 1610 (V) cm^{-1} may indicate the ionization of the peptide NH hydrogen, since such an ionization results in the resonance of the O—C—N system, as indicated by the dotted line in structure V. Kim and Martell[380] also studied the triglycine and tetraglycine Cu(II) systems. Later, Tasumi et al.[381] carried out similar studies in a wider frequency range (1800–1200 cm^{-1}). Kruck and Sarker[382] studied the equilibria of the Cu(II)-L-histidine system in D_2O.

III-11. COMPLEXES OF OXALATO AND RELATED LIGANDS

(1) Oxalato Complexes

The oxalato anion (ox^{2-}) coordinates to a metal as a unidentate (I) or bidentate (II) ligand:

$$
\left[(NH_3)_5Co-O\diagdown_{C}\diagup^{O}_{\diagdown C\diagdown OH} \right]^{2+}
\qquad
\left[(NH_3)_4Co\diagup^{O-C}_{\diagdown O-C\diagdown O}^{\diagup O} \right]^{+}
$$

I   II

The bidentate chelate structure (II) is most common. Fujita et al.[383] carried out normal coordinate analyses on the 1 : 1 (metal–ligand) model of the $[M(ox)_2]^{2-}$ and $[M(ox)_3]^{3-}$ series, and obtained the band assignments listed in Table III-23. In the divalent metal series, $\nu(C{=}O)$ (average of ν_1 and ν_7) becomes higher, and $\nu(C{-}O)$ (ν_2 and ν_8) becomes lower, as $\nu_4(MO)$ becomes higher in the order Zn(II) < Cu(II) < Pd(II) < Pt(II) (see Fig. III-42). This relation holds in spite of the fact that ν_2, ν_4, and ν_8 are all coupled with other vibrations.

In the trivalent metal series, Hancock and Thornton[384] found that ν_{11} (MO stretching) follows the same trend as the crystal field stabilization energies

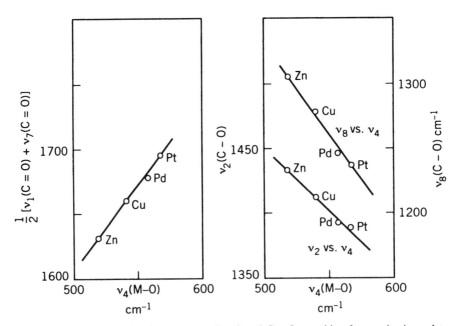

Fig. III-42. M–O stretching frequency vs. C=O and C—O stretching frequencies in oxalato complexes of divalent metals.[383]

TABLE III-23. Frequencies and Band Assignments of Chelating Oxalato Complexes (cm^{-1})[383]

K₂[Zn(ox)₂]·2H₂O	K₂[Cu(ox)₂]·2H₂O	K₂[Pd(ox)₂]·2H₂O	K₂[Pt(ox)₂]·3H₂O	K₃[Fe(ox)₃]·3H₂O	K₃[V(ox)₃]·3H₂O	K₃[Cr(ox)₃]·3H₂O	K₃[Co(ox)₃]·3H₃O	K₃[Al(ox)₃]·3H₂O	[Cr(NH₃)₄(ox)]·Cl	Band Assignment	
1632	(1720) 1672	1698	1709	1712	1708	1708	1707	1722	1704	ν_a(C=O)	ν_7
—	1645	1675, 1657	1674	1677, 1649	1675, 1642	1684, 1660	1670	1700, 1683	1668	ν_a(C=O)	ν_1
1433	1411	1394	1388	1390	1390	1387	1398	1405	1393	ν_s(CO) + ν(CC)	ν_2
1302	1277	1245 (1228)	1236	1270, 1255	1261	1253	1254	1292, 1269	1258	ν_s(CO) + δ(O—C=O)	ν_8
890	886	893	900	885	893	893	900	904	914, 890	ν_s(CO)	ν_3
785	795	818	825	797, 785	807, 797	810, 798	822, 803	820, 803	804	δ(O—C=O) + ν(MO)	ν_9
622	593	610	—	580	581	595	—	—	—	Crystal water?	
519	541	556	575, 559	528	531	543	565	587	545	ν(MO) + ν(CC)	ν_4
519	481	469	469	498	497	485	472	436	486, 469	Ring. def. + δ(O—C=O)	ν_{10}
428, 419	420	417	405	366	368	415	446	485	366	ν(MO) + ring def.	ν_{11}
377, 364	382, 370	368	370	340	336	358	364	364	347	δ(O—C=O) + ν(CC)	ν_5
291	339	350	328	—	—	313	332	—	328	π	

(CFSE) of these metals, namely:

	Sc d^0		V d^2		Cr d^3		Mn d^4		Fe d^5		Co d^6		Ga d^{10}
ν(MO)(cm)$^{-1}$	340	<	367	<	416	>	372	>	354	<	446	>	368
CFSE (10^3 cm^{-1})	0	<	10.2	<	21.2	>	10.2	>	0	<	27.0	>	0

Both quantities are maximized at the d^3 and d^6 configurations (d^4 and d^5 ions are in high-spin states). The IR spectra of [Ir(ox)Cl$_4$]$^{3-}$ (**C**$_{2v}$), [Ir(ox)$_2$Cl$_2$]$^{3-}$ (*trans*, **D**$_{2h}$; *cis*, **C**$_2$), and [Ir(ox)$_3$]$^{3-}$ (**D**$_3$) have been assigned by Gouteron.[385] More recently, the IR and Raman spectra of the [Co(ox)$_2$]$^{2-}$ (**D**$_{2h}$)[386] and [Os(ox)X$_4$]$^{2-}$ (X = Cl, Br, and I) (**C**$_{2v}$) ions have been assigned.[387]

The oxalato anion may act as a bridging group between metal atoms. According to Scott et al.,[388] the oxalato anion can take the following four bridging structures:

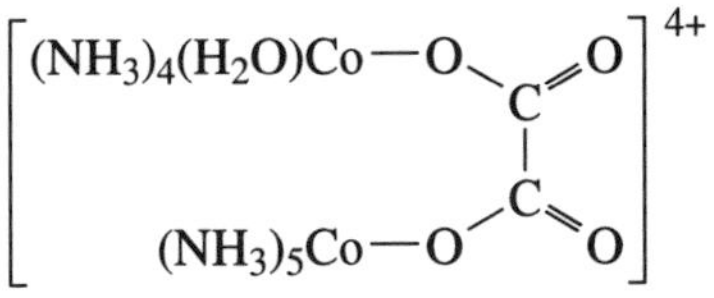

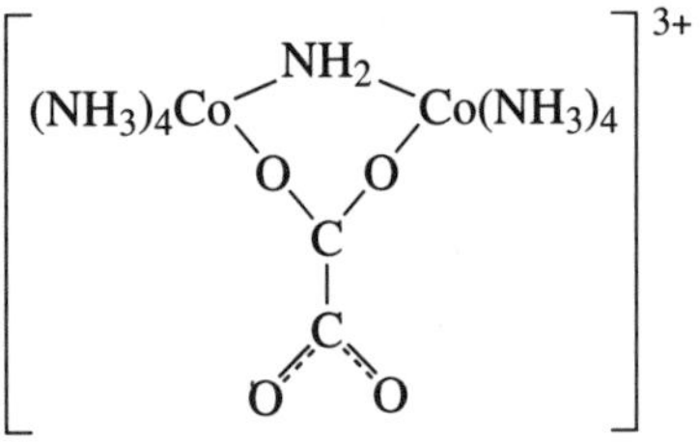

Table III-24 lists the ν(CO) of each type. The spectrum of the tetradentate complex (VI) is the most simple. Because of its high symmetry [**D**$_{2h}$ (planar) or **D**$_{2d}$ (twisted)], it exhibits only two ν(CO). The spectra of bidentate complexes (III and IV) show four ν(CO), as expected from the **C**$_{2v}$ symmetry. The spectrum of the tridentate complex (V) should show four ν(CO), although only three are observed. The tetradentate bridging structure (VI) is also found in

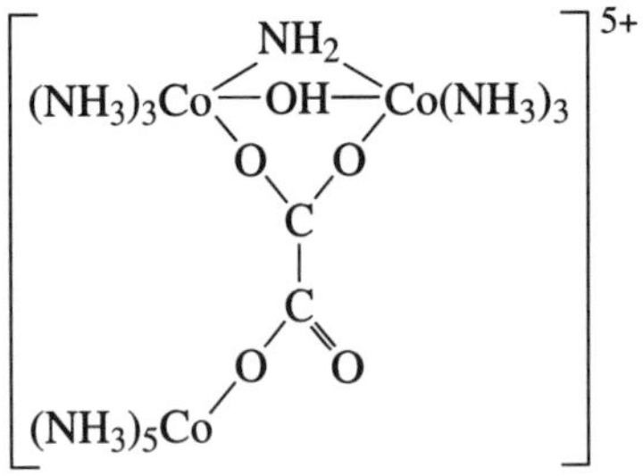

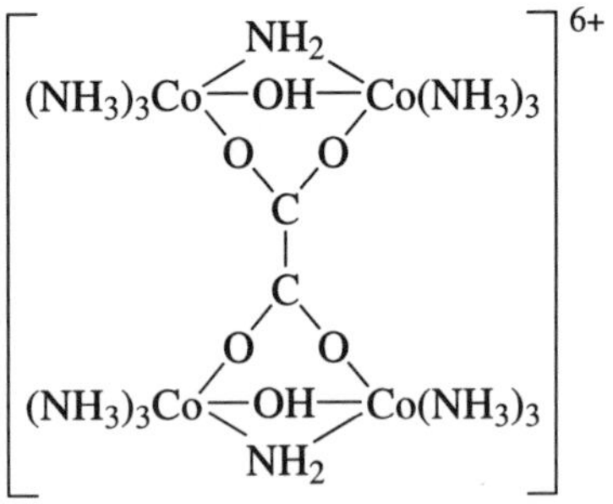

TABLE III.24. CO Stretching Vibrations of Co(III) Oxalato Complexes (cm^{-1})

Compound	Symmetry[a]	$\nu(CO)$			
I	C_s/C_1	1761	1682 1665	1400	1260
II	C_{2v}	1696	1667	1410	1268
III	C_{2v}/C_2	1721 } 1701 }	1629 } 1670 }	1439 } 1430 }	1276 } 1250 }
IV	C_{2v}/C_2	1755	1626	1318	1284
V	C_s/C_1	1650	1610	1322	—
VI	D_{2h}/D_{2d}	—	1628	1345	—

$[(MoFe_3S_4Cl_4)_2(ox)]^{4-}$ which exhibits the $\nu(CO)$ at 1630 cm^{-1}. This

frequency is much lower than that of the chelating oxalato group in $[MoFe_3S_4Cl_4(ox)]^{3-}$ (1670 cm^{-1}).[389]

The Raman spectra of metal oxalato complexes have also been examined to investigate the solution equilibria and the nature of the M—O bond.[390]

(2) Complexes of Related Ligands

Vibrational assignments have been made on metal oxamido complexes of V_h symmetry.[391]

(M = Ni or Cu)

and the *cis*- and *trans*-dimethyloxamido complexes of dimethylgallium.[392]

Cis(C_{2v})   Trans(C_{2h})

The IR and Raman spectra of the Ni(II) and Cu(II) complexes of oxamic hydrazine have been assigned using ^{58}Ni and ^{62}Ni isotopes.[393]

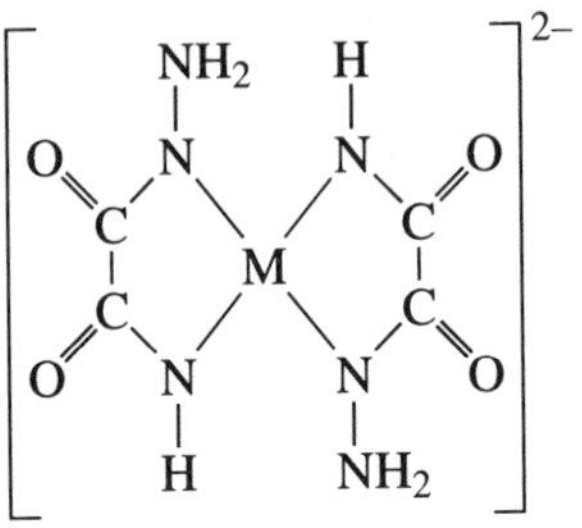

$(\mathbf{D}_{2h})$

The ν(Ni–NH) and ν[Ni–N(NH$_2$)] of the ^{58}Ni complex are at 439 and 428 cm^{-1}, respectively, in the IR spectrum.

Biuret (NH$_2$CONHCONH$_2$) is known to form the following two types of chelate rings:

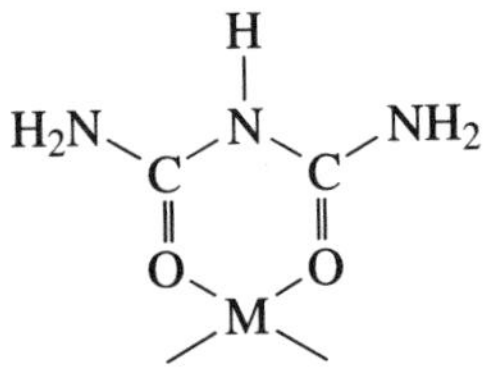

VII 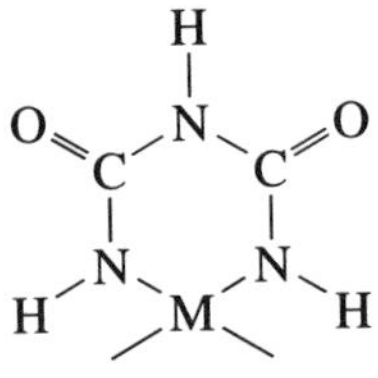   VIII

Violet crystals of composition K$_2$[Cu(biureto)$_2$]·4H$_2$O are obtained when the Cu(II) ion is added to an alkaline solution of biuret, whereas pale blue–green crystals of composition [Cu(biuret)$_2$]Cl$_2$ result when the Cu(II) ion is mixed with biuret in neutral (alcoholic) solution. The former contains the N-bonded chelate ring structure (VIII), while the latter consists of the O-bonded chelate rings (VII). Kedzia et al.[394] carried out normal coordinate analyses of both compounds. The Co(II) complex forms the N-bonded chelate ring, whereas the Zn complex forms the O-bonded ring structure.[394] In [Cd(biuret)$_2$]Cl$_2$, the biuret molecules are bonded to the metal as follows:[395]

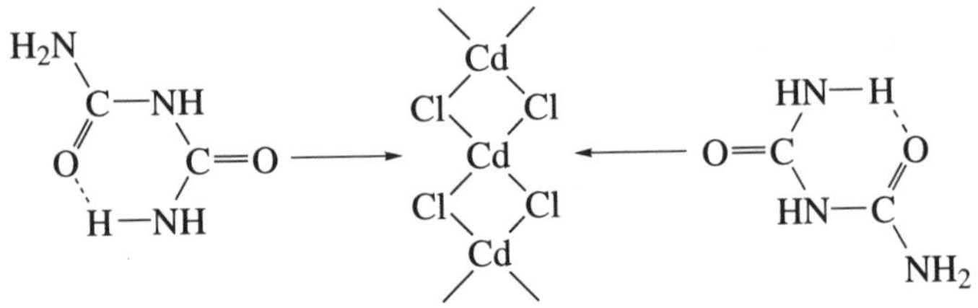

IX

Saito et al.[396] carried out normal coordinate analysis on the ligand portion of the Cd complex. Thamann and Loehr[397] assigned the Raman spectra of N-bonded Cu(II) and Cu(III) complexes of biuret and oxamide based on normal coordinate calculations. The vibrations which are predominantly ν(Cu–N) appear at 320–291 cm^{-1} for the Cu(II) and at 344–320 cm^{-1} for the Cu(III) complexes. The corresponding force constants were 1.04–0.96 mdyn/Å for the former and 1.46–1.35 mdyn/Å for the latter.

III-12. COMPLEXES OF SULFATE, CARBONATE, AND RELATED LIGANDS

When a ligand of relatively high symmetry coordinates to a metal, its symmetry is lowered and marked changes in the spectrum are expected because of changes in the selection rules. This principle has been used extensively to determine whether acido anions such as SO_4^{2-} and CO_3^{2-} coordinate to metals as unidentate, chelating bidentate, or bridging bidentate ligands. Although symmetry lowering is also caused by the crystalline environment, this effect is generally much smaller than the effect of coordination.

(1) Sulfato (SO_4) Complexes

The free sulfate ion belongs to the high-symmetry point group $\mathbf{T}_d$. Of the four fundamentals, only ν_3 and ν_4 are infrared-active. If the symmetry of the ion is lowered by complex formation, the degenerate vibrations split and Raman-active modes appear in the infrared spectrum. The lowering of symmetry caused by coordination is different for the unidentate and bidentate complexes, as shown below:

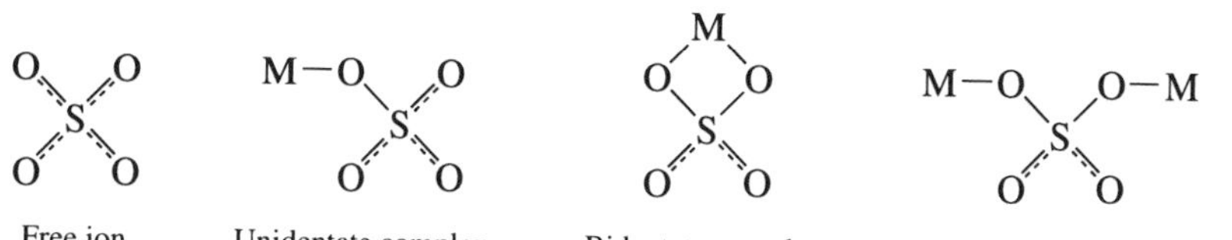

| Free ion ($\mathbf{T}_d$) | Unidentate complex ($\mathbf{C}_{3v}$) | Bidentate complex ($\mathbf{C}_{2v}$) | Bridged bidentate complex ($\mathbf{C}_{2v}$) |

The change in the selection rules caused by the lowering of symmetry was shown in Table II-6f of Part A. Table III-25 and Fig. III-43 give the frequencies and the spectra of typical Co(III) sulfato complexes obtained by Nakamoto et al.[398] In [Co(NH$_3$)$_6$]$_2$(SO$_4$)$_3\cdot$5H$_2$O, ν_3 and ν_4 do not split and ν_2 does not appear, although ν_1 is observed, it is very weak. We conclude, therefore, that the symmetry of the SO_4^{2-} ion is approximately $\mathbf{T}_d$. In [Co(NH$_3$)$_5$SO$_4$]Br, both ν_1 and ν_2 appear with medium intensity; moreover, ν_3 and ν_4 each splits into two bands. This result can be explained by assuming a lowering of symmetry

TABLE III-25. Vibrational Frequencies of Co(III) Sulfato Complexes (cm^{-1})[398]

Compound	Symmetry	ν_1	ν_2	ν_3	ν_4
Free SO$_4^{2-}$ ion	T$_d$	—	—	1104 (vs)[a]	613 (s)
[Co(NH$_3$)$_6$]$_2$(SO$_4$)$_3$·5H$_2$O	T$_d$	973 (vw)	—	1130–1140 (vs)	613 (s)
[Co(NH$_3$)$_5$SO$_4$]Br	C$_{3v}$	970 (m)	438 (m)	$\left\{\begin{array}{l}1032\text{–}1044\ (s)\\1117\text{–}1143\ (s)\end{array}\right.$	$\left\{\begin{array}{l}645\ (s)\\604\ (s)\end{array}\right.$
$\left[\begin{array}{c}\text{NH}_2\\(\text{NH}_3)_4\text{Co}\diagdown\ \diagup\text{Co(NH}_3)_4\\ \text{SO}_4\end{array}\right][\text{NO}_3]_3$	C$_{2v}$	995 (m)	462 (m)	$\left\{\begin{array}{l}1050\text{–}1060\ (s)\\1170\ (s)\\1105\ (s)\end{array}\right.$	$\left\{\begin{array}{l}641\ (s)\\610\ (s)\\571\ (m)\end{array}\right.$

[a] vs = very strong; s = strong; m = medium; vw = very weak.

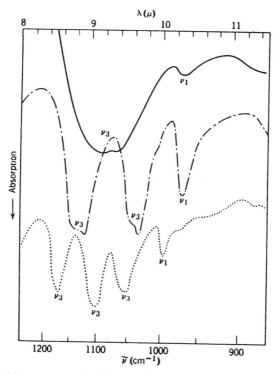

Fig. III-43. Infrared spectra of $[Co(NH_3)_6]$ $(SO_4)_3 \cdot 5H_2O$ (solid line); $[Co(NH_3)_5SO_4]Br$ (dot–dash line); and

$$\left[(NH_3)_4Co \overset{\overset{\displaystyle NH_2}{\diagup}}{\underset{\underset{\displaystyle SO_4}{\diagdown}}{}} Co(NH_3)_4 \right] (NO_3)_3 \text{ (dotted line).}^{398}$$

from $\mathbf{T}_d$ to $\mathbf{C}_{3v}$ (unidentate coordination). In

$$\left[(NH_3)_4Co \overset{\overset{\displaystyle NH_2}{\diagup}}{\underset{\underset{\displaystyle SO_4}{\diagdown}}{}} Co(NH_3)_4 \right] (NO_3)_3$$

both ν_1 and ν_2 appear with medium intensity, and ν_3 and ν_4 each splits into three bands. These results suggest that the symmetry is further lowered and probably reduced to $\mathbf{C}_{2v}$, as indicated in Table II-6f. Thus, the SO_4^{2-} group in this complex is concluded to be a bridging bidentate as depicted in the foregoing diagram.

The chelating bidentate SO_4^{2-} group was discovered by Barraclough and Tobe,[399] who observed three bands (1211, 1176, and 1075 cm^{-1}) in the ν_3 region of $[Co(en)_2SO_4]Br$. These frequencies are higher than those of the bridging bidentate complex listed in Table III-25. Eskenazi et al.[400] also found the same trend in Pd(II) sulfato complexes. Thus the distinction between bridging and chelating sulfato complexes can be made on this basis. Table III-26 lists

TABLE III-26. Vibrational Frequencies and Modes of Coordination of Various Sulfato Complexes (cm^{-1})

Compound	Mode of Coordination	ν_1	ν_2	ν_3	ν_4	Ref.
[Cr(H$_2$O)$_5$SO$_4$]Cl·$\frac{1}{2}$H$_2$O	Unidentate	1002	—	1118 1068	—	401
[VO(SO$_4$)$_2$ (H$_2$O)$_3$]$^{2-}$	Unidentate	—	483	1140 1046	640 619	402
[Cu(bipy)SO$_4$]·2H$_2$O (polymeric)	Bridging bidentate	971	—	1163 1096 1053–1035	—	403
Ni(morpholine)$_2$SO$_4$ (polymeric)	Bridging bidentate	973	493	1177 1094 1042	628 612 593	404
[Co$_2${(SO$_4$)$_2$OH}(NH$_3$)$_6$]Cl	Bridging bidentate	966	—	1180 1101 1048	645 598	405
Pd(NH$_3$)$_2$SO$_4$	Bridging bidentate	960	—	1195 1110 1035	—	400
Pd(phen)SO$_4$	Chelating bidentate	955	—	1240 1125 1040–1015	—	400
Pd(PPh$_3$)$_2$SO$_4$	Chelating bidentate	920	—	1265 1155 1110	—	406
Ir(PPh$_3$)$_2$(CO)I(SO$_4$)	Chelating bidentate	856	549	1296 1172 880	662 610	407
K$_3$[Fe(SO$_4$)F]	Chelating bidentate	—	—	1225 1130 1020	—	408
Tl[VO$_2$SO$_4$]	Chelating bidentate	1000	455 400	1255 1160 1125	720 570	409

the observed frequencies of the sulfato groups and the modes of coordination as determined from the spectra.

The symmetries of the sulfate ions in metal salts at various stages of hydration have been discussed using IR spectra.[410] Normal coordinate analyses have been made on Co(III) ammine complexes containing sulfato groups.[411,412]

(2) Perchlorato (ClO$_4$) Complexes

In general, the perchlorate (ClO$_4^-$) ion coordinates to a metal when its complexes are prepared in nonaqueous solvents. The structure and bonding of metal complexes containing these weakly coordinating ligands have been reviewed briefly by Rosenthal.[413] Infrared and Raman spectroscopy has been used extensively to determine the mode of coordination of the ClO$_4^-$ ligand. The structures listed in

TABLE III-27. ClO Stretching Frequencies of Perchlorato Complexes (cm^{-1})

Complex	Structure	ν_3	ν_4	Ref.
K[ClO$_4$]	Ionic	1170–1050	(935)^a	
Cu(ClO$_4$)$_2$·6H$_2$O	Ionic	1160–1085	(947)^a	414
Cu(ClO$_4$)$_2$·2H$_2$O	Unidentate	$\begin{cases} 1158 \\ 1030 \end{cases}$	920	414
Cu(ClO$_4$)$_2$	Bidentate	$\begin{cases} 1270\text{–}1245 \\ 1130 \\ 948\text{–}920 \end{cases}$	1030	414
Mn(ClO$_4$)$_2$·2H$_2$O	Bidentate	$\begin{cases} 1210 \\ 1138 \\ 945 \end{cases}$	1030	415
Co(ClO$_4$)$_2$·2H$_2$O	Bidentate	$\begin{cases} 1208 \\ 1125 \\ 935 \end{cases}$	1025	415
[Ni(en)$_2$(ClO$_4$)$_2$]^b	Bidentate	$\begin{cases} 1130 \\ 1093 \\ 1058 \end{cases}$	962	416
Ni(CH$_3$CN)$_4$(ClO$_4$)$_2$	Unidentate	$\begin{cases} 1135 \\ 1012 \end{cases}$	912	417
Ni(CH$_3$CN)$_2$(ClO$_4$)$_2$	Bidentate	$\begin{cases} 1195 \\ 1106 \\ 1000 \end{cases}$	920	417
[Ni(4-Me-py)$_4$](ClO$_4$)$_2$	Ionic	1040–1130	(931)^a	418
Ni(3-Br-py)$_4$(ClO$_4$)$_2$	Unidentate	$\begin{cases} 1165\text{–}1140 \\ 1025 \end{cases}$	920	418
GeCl$_3$ (ClO$_4$)	Unidentate	1265, 1240	1030	419

^{a} Weak.
^{b} Blue form.

Table III-27 were determined on the basis of the same symmetry selection rules as are used for sulfato complexes. Chausse et al.[420] concluded from their IR and Raman study that [Al(ClO$_4$)$_n$]$^{-(n-3)}$ contain two unidentate and two bidentate for $n = 4$, four unidentate and one bidentate for $n = 5$, and six unidentate ligands for $n = 6$. In polymeric M(ClO$_4$)$_3$ (M = In and Tl), the ClO$_4^-$ ion acts as a bridging bidentate ligand.[421]

According to Pascal et al., the ClO$_4^-$ ligands in M(ClO$_4$)$_2$ (M = Ni and Co are bridging tridentate.[422]

In Ni(ClO$_4$)$_2$ (IR), the ν(ClO$_t$), ν_a(ClO$_b$), and ν_s(ClO$_b$) are at 1300, 1030, and 960 cm^{-1}, respectively. This type of coordination has also been proposed for M(ClO$_4$)$_3$ (M = Y, La, Nd, Sm, etc.),[423] and for Ce(ClO$_4$)$_3$[424] and Mn(ClO$_4$)$_2$.[425]

(3) Complexes of Other Tetrahedral Ligands

Many tetrahedral anions coordinate to a metal as unidentate and bidentate ligands, and their modes of coordination have been determined by the same method as is used for the SO_4^{2-} and ClO_4^- ions. Thus, the PO_4^{3-} ion is a unidentate in $[Co(NH_3)_5PO_4]$ and a bidentate in $[Co(NH_3)_4PO_4]$.[426] Vibrational spectra have been reported for unidentate and bidentate complexes of the AsO_4^{3-},[427] CrO_4^{2-}, and MoO_4^{2-} ions.[428] The SeO_4^{2-} ion in $[Co(NH_3)_5SeO_4]Cl$ is a unidentate,[429] whereas it is a bridging bidentate ligand in $[Co_2\{SeO_4)_2OH\}-(NH_3)_6]Cl$.[405] The latter structure is also reported for $(NH_4)_2UO_2(SeO_4)_2\cdot4H_2O$.[430]

The Raman spectrum of solid $Ni(H_2PO_2)_2$ is interpreted as that of the two ions, $[Ni(H_2PO_2)]^+$ and $H_2PO_2^-$, the hypophosphite ion in the former being a chelating bidentate.[431] In polymeric $UCl(H_2PO_2)_3\cdot2H_2O$, however, the $H_2PO_2^-$ ion serves as a bridging bidentate with $\nu_a(PO_2)$ and $\nu_s(PO_2)$ at 1234 and 1058 cm^{-1}, respectively.[432] Bridging bidentate phosphinates ($R_2PO_2^-$, where R is a phenyl) of Ru(II) exhibit the $\nu_a(PO_2)$ and $\nu_s(PO_2)$ at approximately 1145 and 1035 cm^{-1}, respectively.[433]

The $S_2O_3^{2-}$ ion can coordinate to a metal in a variety of ways. According to Freedman and Straughan,[434] $\nu_a(SO_3)$ near 1130 cm^{-1} is most useful as a structural diagnosis: >1175 (S-bridging); 1175–1130 (S-coordination); ~1130 (ionic $S_2O_3^{2-}$); < 1130 cm^{-1} (O-coordination). On the basis of this criterion, they proposed polymeric structures linked by O-bridges for thiosulfates of UO_2^{2+} and ZrO_2^{2+}. In the $[OsO_2(S_2O_3)_2]^{2-}$ ion, the thiosulfate ion is a S-bonded unidentate with the $\nu(S-S)$ at 409 cm^{-1}, which is much lower than that of the free ligand (434 cm^{-1}).[435]

The fluorosulfate (SO_3F^-) ion is a unidentate in $[Sn(SO_3F)_6]^{2-}$,[436], but is a unidentate as well as a bidentate in $VO(SO_3F)_3$.[437] Similarly, only unidentate coordination is seen in $[Ru(SO_3F)_6]^{2-}$, whereas $[Ru(SO_3F)_5]^-$ may contain both unidentate and bidentate ligands.[438]

(4) Carbonato(CO_3) Complexes

The unidentate and bidentate (chelating) coordinations shown below are found in the majority of carbonato complexes.

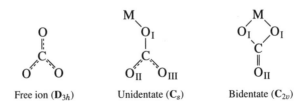

Free ion (D_{3h})   Unidentate (C_s)   Bidentate (C_{2v})

The selection rule changes as shown in Table I-18 of Part A. In C_{2v} and

C_s,* the ν_1 vibration, which is forbidden in the free ion, becomes infrared-active and each of the doubly degenerate vibrations, ν_3 and ν_4, splits into two bands. Although the number of infrared-active fundamentals is the same for C_{2v} and C_s, the splitting of the degenerate vibrations is larger in the bidentate than in the unidentate complex.[398] For example, $[Co(NH_3)_5CO_3]Br$ exhibits two CO stretchings at 1453 and 1373 cm^{-1}, whereas $[Co(NH_3)_4CO_3]Cl$ shows them at 1593 and 1265 cm^{-1}. In organic carbonates such as dimethyl carbonate, $(CH_3O_I)_2CO_{II}$, this effect is more striking because the CH_3—O_I bond is strongly covalent. Thus, the CO_{II} stretching is observed at 1870 cm^{-1}, whereas the CO_I stretching is at 1260 cm^{-1}. Gatehouse and co-workers[439] showed that the separation of the CO stretching bands increases along the following series:

basic salt < carbonato complex < acid < organic carbonate

Fujita et al.[440] carried out normal coordinate analysis on unidentate and bidentate carbonato complexes of Co(III). According to their results the CO stretching force constant, which is 5.46 for the free ion, becomes 6.0 for the C–O_{II} bonds and 5.0 for the C–O_I bond of the unidentate complex, whereas it becomes 8.5 for the C–O_{II} bond and 4.1 for the C–O_I bonds of the bidentate complex (all are UBF force constants in units of mdyn/Å). The observed and calculated frequencies and theoretical band assignments are shown in Table III-28. Normal coordinate analyses on carbonato complexes have also been carried out by Hester and Grossman[441] and Goldsmith and Ross.[442]

As is shown in Table III-28, normal coordinate analysis predicts that the highest-frequency CO stretching band belongs to the B_2 species in the unidentate and the A_1 species in the bidentate complex. Elliott and Hathaway[443] studied the polarized infrared spectra of single crystals of $[Co(NH_3)_4CO_3]Br$ and confirmed these symmetry properties. As will be shown later for nitrato complexes, Raman polarization studies are also useful for this purpose.

According to X-ray analysis, the carbonato groups in $[(NH_3)_3Co(\mu$-$OH)_2(\mu$-$CO_3)Co(NH_3)_3]SO_4 \cdot 5H_2O$[444] and $[(teed)CuCl(CO_3)CuCl(teed)]$ (teed: N,N,N',N'-tetraethyl-ethylenediamine)[445] take the bridging and tridentate (bridging) structures, respectively.

No simple criteria have been established to distinguish these structures from common unidentate and bidentate (chelating) coordination based on vibrational

---

*The symmetry of the unidentate carbonato group is C_{2v} if the metal atom is ignored.

TABLE III-28. Calculated and Observed Frequencies of Unidentate and Bidentate Co(III) Carbonato Complexes (cm^{-1})[440]

Species (C_{2v})[a]	$\nu_1(A_1)$	$\nu_2(A_1)$	$\nu_3(A_1)$	$\nu_4(A_1)$	$\nu_5(B_2)$	$\nu_6(B_2)$	$\nu_7(B_2)$	$\nu_8(B_1)$
Calculated frequency	1376	1069	772	303	1482	676	92	—
Assignment	$\nu(CO_{II})$ $+ \nu(CO_I)$	$\nu(CO_I)$ $+ \nu(CO_{II})$	$\delta(O_{II}CO_{II})$	$\nu(CoO_I)$	$\nu(CO_{II})$	$\rho_r(O_{II}CO_{II})$	$\delta(CoO_IC)$	π
[Co(NH$_3$)$_5$CO$_3$]Br	1373	1070	756	362	1453	678	—	850
[Co(ND$_3$)$_5$CO$_3$]Br	1369	1072	751	351	1471	687	—	854
[Co(NH$_3$)$_5$CO$_3$]I	1366	1065	776	360	1449	679	—	850
[Co(ND$_3$)$_5$CO$_3$]I	1360	1063	742	341	1467	687	—	853

Species (C_{2v})[a]	$\nu_1(A_1)$	$\nu_2(A_1)$	$\nu_3(A_1)$	$\nu_4(A_1)$	$\nu_5(B_2)$	$\nu_6(B_2)$	$\nu_7(B_2)$	$\nu_8(B_1)$
Calculated frequency	1595	1038	771	370	1282	669	429	—
Assignment	$\nu(CO_{II})$	$\nu(CO_I)$	Ring def. $+ \nu(CoO_I)$	$\nu(CoO_I)$ $+$ ring def.	$\nu(CO_I)$ $+ \delta(O_ICO_{II})$	$\delta(O_ICO_{II})$ $+ \nu(CO_I)$ $+ \nu(CoO_I)$	$\nu(CoO_I)$	π
[Co(NH$_3$)$_4$CO$_3$]Cl	1593	1030	760	395	1265	673	430	834
[Co(ND$_3$)$_4$CO$_3$]Cl	1635 } 1607	(1031)[b]	753	378	1268	672	418	832
[Co(NH$_3$)$_4$CO$_3$]ClO$_4$	1602	—[c]	762	392	1284	672	428	836
[Co(ND$_3$)$_4$CO$_3$]ClO$_4$	1603	—[c]	765	374	1292	676	415	835

[a]Symmetry assuming a linear Co—O—C bond (see Ref. 440).
[b]Overlapped with δ_s(ND$_3$).
[c]Hidden by [ClO$_4$]$^-$ absorption.

frequencies. However, Greenaway et al.[446] have demonstrated that the bridging and bidentate carbonato ligands can be distingished if the angular distortion ($\Delta\alpha$), the difference between the largest and smallest OCO angles, is known from X-ray analysis. These workers found that the frequency separation ($\Delta\nu$) between the two highest $\nu(CO)$ bands increases linearly with $\Delta\alpha$. As an example, $Na_2[Cu(CO_3)_2]$ contains one bidentate and one bridging carbonato ligand. Using their correlation, they were able to assign the 1610 and 1328 cm^{-1} bands to the bidentate ($\Delta\alpha = 11.2°$ and $\Delta\nu = 282$ cm^{-1}) and the 1525 and 1380 cm^{-1} bands to the bridging carbonato ligands ($\Delta\alpha = 7.7°$ and $\Delta\nu = 145$ cm^{-1}).

The IR spectrum of K_2CO_3 in a N_2 matrix indicates that the CO_3 group coordinates in a bidentate fashion to one of the K atom and in a unidentate fashion to the other K atom.[447] Busca and Lorenzelli[448] reviewed the IR spectra and modes of coordination of carbonate, bicarbonate, and formate ions, and of CO_2 in metal complexes.

(5) Nitrato (NO_3) Complexes

The structures and vibrational spectra of a large number of nitrato complexes have been reviewed by Addison et al.[449] and Rosenthal.[413] X-Ray analyses show that the NO_3^- ion coordinates to a metal as a unidentate, symmetric, and asymmetric chelating bidentate, and bridging bidentate ligand of various structures. It is rather difficult to differentiate these structures by vibrational spectroscopy since the symmetry of the nitrate ion differs very little among them (C_{2v} or C_s). Even so, vibrational spectroscopy is still useful in distinguishing unidentate and bidentate ligands.

Originally, Gatehouse et al.[450] noted that the unidentate NO_3 group exhibits three NO stretching bands, as expected for its C_{2v} symmetry. For example, $[Ni(en)_2(NO_3)_2]$ (unidentate) exhibits three bands as follows:

$$\begin{array}{lll} \nu_5(B_2) & 1420 \text{ cm}^{-1} & \nu_a(NO_2) \\ \nu_1(A_1) & 1305 \text{ cm}^{-1} & \nu_s(NO_2) \\ \nu_2(A_1) & (1008) \text{ cm}^{-1} & \nu(NO) \end{array}$$

whereas $[Ni(en)_2NO_3]ClO_4$ (chelating bidentate) exhibits three bands at the following:

$$\begin{array}{lll} \nu_1(A_1) & 1476 \text{ cm}^{-1} & \nu(N{=}O) \\ \nu_5(B_2) & 1290 \text{ cm}^{-1} & \nu_a(NO_2) \\ \nu_2(A_1) & (1025) \text{ cm}^{-1} & \nu_s(NO_2) \end{array}$$

The separation of the two highest-frequency bands is 115 cm^{-1} for the unidentate complex, whereas it is 186 cm^{-1} for the bidentate complex. Thus Curtis and Curtis[451] concluded that $[Ni(dien)(NO_3)_2]$ contains both types, since it exhibits

TABLE III-29. NO Stretching Frequencies of Unidentate and Bidentate Nitrato Complexes (cm^{-1})

Compound	Mode of Coordination	ν_5	ν_1	ν_2	$\nu_5 - \nu_1$	Ref.
$Re(CO)_5NO_3$	Unidentate	1497	1271	992	226	452
cis-$[Pt(NH_3)_2(NO_3)_2]$	Unidentate	1510	1275	997	235	453
$Sn(NO_3)_4$	Chelating bidentate	1630	1255	983	375	454
$K[UO_2(NO_3)_3]$	Chelating bidentate	1555 1521	1271	1025	284 250	455
$Co(NO_3)_3$	Chelating bidentate	1619	1162	963	457	456
$Na_2[Mn(NO_3)_4]$	Chelating bidentate	1490	1280	1041 1036	210	457
$Cu(NO_3)_2MeNO_2$	Bridging bidentate	1519	1291	1008	228	458
$Zn(bt)_2(NO_3)_2$ [a]	Chelating bidentate	1485	1300	—	185	459
$Ni(dmpy)_2(NO_3)_2$ [b]	Chelating bidentate	1513	1270	1013	243	460
$Th(NO_3)_4$ $(tmu)_2^c$	Chelating bidentate	1530	1278	1023	252	461
$Ln(NO_3)_3$ $(DMSO)_n$ (Ln = La,Ce,.)	Chelating bidentate	1500	1295	1030	305	462

[a]bt = benzothiazole.
[b]dmpy = 2,6-dimethyl-4-pyrone.
[c]tmu = tetramethylurea.

bands due to unidentate (1440 and 1315 cm^{-1}) and bidentate (1480 and 1300 cm^{-1}) groups. In general, the separation of the two highest-frequency bands is larger for bidentate than for unidentate coordination if the complexes are similar. However, this rule does not hold if the complexes are markedly different. This is clearly shown in Table III-29.

Lever et al.[463] proposed using the combination band, $\nu_1 + \nu_4$, of free NO_3^- which appears in the 1800–1700 cm^{-1} region for structural diagnosis. Upon coordination, ν_4 (E', in-plane bending) near 700 cm^{-1} splits into two bands, and the magnitude of this splitting is expected to be larger for bidentate than for unidentate ligands. This should be reflected in the separation of two ($\nu_1 + \nu_4$) bands in the 1800–1700 cm^{-1} region. According to Lever et al., the NO_3^- ion is bidentate if the separation is ca. 66–20 cm^{-1} and is unidentate if it is ca. 26–5 cm^{-1}.

As stated previously, the highest-frequency CO stretching band of the carbonato complexes belongs to the A_1 species in the bidentate and to the B_2 species in the unidentate complex. The same holds true for the nitrato complex. Ferraro et al.[464] showed that all the nitrato groups in $Th(NO_3)_4(TBP)_2$ coordinate to the metal as bidentate ligands since the Raman band at 1550 cm^{-1} is polarized (TBP: tributylphosphate). This rule holds very well for other compounds.[465] According to Addison et al.,[449] the intensity pattern of the three

NO stretching bands in the Raman spectrum can also be used to distinguish unidentate and symmetrical bidentate NO_3 ligands. The middle band is very strong in the former, whereas it is rather weak in the latter.

The use of far-infared spectra to distinguish unidentate and bidentate nitrato coordination has been controversial. Nuttall and Taylor[466] suggested that unidentate and bidentate complexes exhibit one and two MO stretching bands, respectively, in the 350–250 cm^{-1} region. Bullock and Parrett[467] showed, however, that such a simple rule is not applicable to many known nitrato complexes. Ferraro and Walker[468] assigned the MO stretching bands of anhydrous metal nitrates such as $Cu(NO_3)_2$ and $Pr(NO_3)_3$.

Several workers studied the Raman spectra of metal nitrates in aqueous solution and molten states. For example, Irish and Walrafen[469] found that E' mode degeneracy is removed even in dilute solutions of $Ca(NO_3)_2$. This, combined with the appearance of the A_1' mode in the infrared, suggests C_{2v} symmetry of the NO_3^- ion. Using FTIR and Raman spectroscopy, Castro and Jagodzinski[470] have shown that the $Cu(NO_3)^+$ ion of C_{2v} symmetry is formed when copper nitrate hydrate is dissolved in H_2O and acetone at a high solute concentration. The Raman band at 335 cm^{-1} was assigned to the $\nu_s(Cu-O)$ of this chelating bidentate complex. Hester and Krishnan[471] studied the Raman spectra of $Ca(NO_3)_2$ dissolved in molten KNO_3 and $NaNO_3$. Their results suggest an asymmetric perturbation of the NO_3^- ion by the Ca^{2+} ion through ion-pair formation.

(6) Sulfito (SO_3), Selenito (SeO_3), and Sulfinato (RSO_2) Complexes

The pyramidal sulfite (SO_3^{2-}) ion may coordinate to a metal as a unidentate, bidentate, or bridging ligand. The following two structures are probable for unidentate coordination:

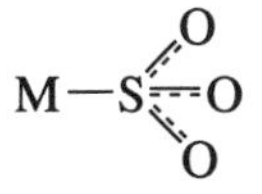

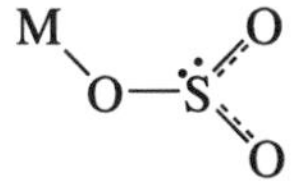

If coordination occurs through sulfur, the C_{3v} symmetry of the free ion will be preserved. If coordination occurs through oxygen, the symmetry may be lowered to C_s. In this case, the doubly degenerate vibrations of the free ion will split into two bands. It is anticipated[472] that coordination through sulfur will shift the SO stretching bands to higher frequencies, whereas coordination through oxygen will shift them to lower frequencies, than those of the free ion. On the basis of these criteria, Newman and Powell[473] showed that the sulfito groups in $K_6[Pt(SO_3)_4]\cdot 2H_2O$ and $[Co(NH_3)_5(SO_3)]Cl$ are S-bonded and those in $Tl_2[Cu(SO_3)_2]$ are O-bonded. Baldwin[474] suggested that the sulfito groups in cis- and trans-$Na[Co(en)_2(SO_3)_2]$ and $[Co(en)_2(SO_3)X]$ (X = Cl or OH) are S-

TABLE III-30. Infrared Spectra of Unidentate Sulfito Complexes (cm^{-1})

Compound	Structure	$\nu_3(E)$	$\nu_1(A_1)$	$\nu_2(A_1)$	$\nu_4(E)$	Ref.
Free SO_3^{2-}	—	933	967	620	469	
$K_6[Pt(SO_3)_4]\cdot2H_2O$	S-bonded	1082–1057	964	660	540	473
$[Co(NH_3)_5(SO_3)]Cl$	S-bonded	1110	985	633	519	473
trans-$Na[Co(en)_2(SO_3)_2]$	S-bonded	1068	939	630	—	474
$[Co(en)_2(SO_3)Cl]$	S-bonded	1117–1075	984	625	—	474
$Tl_2[Cu(SO_3)_2]$	O-bonded	902 862 }	989	673	506 460 }	473
$(NH_4)_9[Fe(SO_3)_6]$	O-bonded	943	815	638	520	476

bonded, since they show only two SO stretchings between 1120 and 930 cm^{-1}. According to Nyberg and Larsson,[475] the appearance of a strong SO stretching band above 975 and below 960 cm^{-1} is an indication of S- and O-coordination, respectively. Table III-30 lists typical results obtained for unidentate complexes.

The structures of complexes containing bidentate sulfito groups are rather difficult to deduce from their infrared spectra. Bidentate sulfito groups may be chelating or bridging through either oxygen or sulfur or both, all resulting in C_s symmetry. Baldwin[474] prepared a series of complexes of the type $[Co(en)_2(SO_3)]X$ (X = Cl, I, or SCN), which are monomeric in aqueous solution. They show four strong bands in the SO stretching region (one of them may be an overtone or a combination band). She suggests a chelating structure in which two oxygens of the sulfito group coordinate to the Co(III) atom. Newman and Powell[473] obtained the infrared spectra of $K_2[Pt(SO_3)_2]\cdot2H_2O$, $K_3[Rh(SO_3)_3]\cdot2H_2O$, and other complexes for which bidentate coordination of the sulfito group is expected. It was not possible, however, to determine their structures from infrared spectra alone.

The mode of coordination of the selenite ion (SeO_3^{2-}) is similar to that of the sulfite ion. Two types of unidentate complexes are expected. The O-coordinated complex exhibits $\nu_3(E)$ and $\nu_1(A_1)$ at 755 and 805 cm^{-1}, respectively, for $[Co(NH_3)_5(SeO_3)]Br\cdot H_2O$,[477] whereas the Se-coordinated complex, $[Co(NH_3)_5(SeO_3)]ClO_4$,[478] shows them at 823 and 860 cm^{-1}, respectively.

Four types of coordination are probable for sulfinato (RSO_2^-, R = CH_3, CF_3, Ph, etc.) groups:

The SO stretching bands at 1200–850 cm^{-1} are useful in distinguishing these structures.[479,480]

III-13. COMPLEXES OF β-DIKETONES

(1) Complexes of Acetylacetonato Ion

A number of β-diketones form metal chelate rings of Type A:

Type A

Among them, acetylacetone (acacH) is most common ($R_I = R_{III} = CH_3$ and R_{II} = H). Infrared spectra of $M(acac)_2$- and $M(acac)_3$-type complexes have been studied extensively. Theoretical band assignments were first made by Nakamoto and Martell,[481] who carried out normal coordinate analysis on the 1:1 model of $Cu(acac)_2$. Mikami et al.[482] performed normal coordinate analyses on the 1:2 (square-planar) and 1:3 (octahedral) models of various acac complexes. Figure III-44 shows the infrared spectra of six acac complexes, and Table III-31 lists the observed frequencies and band assignments for the Cu(II), Pd(II), and Fe(III) complexes obtained by Mikami et al. In this table, the 1577- and 1529-cm^{-1} bands of $Cu(acac)_2$ are assigned to $\nu(C{=}{=}{=}C)$ coupled with $\nu(C{=}{=}{=}O)$ and $\nu(C{=}{=}{=}O)$ coupled with $\nu(C{=}{=}{=}C)$, respectively. Junge and Musso[483] have measured the ^{13}C and ^{18}O isotope shifts of these bands and concluded that the above assignments must be reversed.

The $\nu(MO)$ of acac complexes are most interesting since they provide direct information about the M–O bond strength. Using the metal isotope technique, Nakamoto et al.[484] assigned the MO stretching bands of acetylacetonato complexes at the following frequencies (cm^{-1}):

$Cr(acac)_3$	$Fe(acac)_3$	$Pd(acac)_2$	$Cu(acac)_2$	$Ni(acac)_2(py)_2$
463.4	436.0	466.8	455.0	438.0
358.4	300.5	297.1	290.5	270.8
		265.9		

Both normal coordinate calculations and isotope shift studies show that the bands near 450 cm^{-1} are coupled with the C—CH_3 bending mode, whereas those in the low-frequency region are relatively pure MO stretching vibrations. Figure III-45 shows the actual tracings of the infrared spectra of $^{50}Cr(acac)_3$ and its ^{53}Cr analog. It is seen that two bands at 463.4 and 358.4 cm^{-1} of the former give negative shifts of 3.0 and 3.9 cm^{-1}, respectively, whereas other bands (ligand vibrations) produce negligible shifts by the ^{50}Cr–^{53}Cr substitution.

Recently, Schönherr et al.[485] carried out normal coordinate analysis on

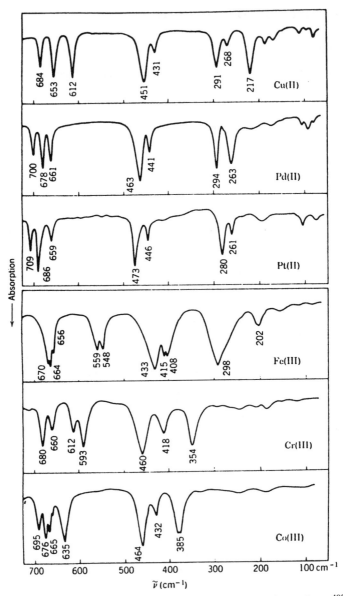

Fig. III-44. Infrared spectra of bis- and tris-(acetylacetonato) complexes.[482]

Sn(acac)Cl$_4$. The Sn–O stretching force constant (GVF) was 1.56 mdyn/Å. Handa et al.[486] observed the following trends (cm^{-1}) in the RR spectra of the Fe(III)–acac system in CH$_3$CN solution:

	Fe(acac)$^{2+}$		Fe(acac)$_2^+$		Fe(acac)$_3$
ν(CC) + ν(CO)	1554	<	1578	<	1603
ν(Fe–O)	474	>	462	>	451

TABLE III-31. Observed Frequencies* and Band Assignments of Acetylacetonato Complexes (cm^{-1})^{482}

Cu(acac)$_2$	Pd(acac)$_2$	Fe(acac)$_3$	Predominant Mode
3072	3070	3062	$\nu(CH)$
2987 ⎫	2990 ⎫	2895 ⎫	
2969 ⎬	2965 ⎬	2965 ⎬	$\nu(CH_3)$
2920 ⎭	2920 ⎭	2920 ⎭	
1577	1569	1570	$\nu(C \cdots C) + \nu(C \cdots O)$
1552	1549	—	combination
1529	1524	1525	$\nu(C \cdots O) + \nu(C \cdots C)$
1461	(1425)	1445	$\delta(CH) + \nu(C \cdots C)$
1413	1394	1425	$\delta_d(CH_3)$
1353	1358	1385 ⎫ 1360 ⎭	$\delta_s(CH_3)$
1274	1272	1274	$\nu(C-CH_3) + \nu(C \cdots C)$
1189	1199	1188	$\delta(CH) + \nu(C-CH_3)$
1019	1022	1022	$\rho_r(CH_3)$
936	937	930	$\nu(C \cdots C) + \nu(C \cdots O)$
780	786 ⎫ 779 ⎭	801 ⎫ 780 ⎬ 771 ⎭	$\pi(CH)$
684	700	670 ⎫ 664 ⎭	$\nu(C-CH_3) +$ ring deformation $+ \nu(MO)$
653	678	656	$\pi\left(CH_3-C\diagdown^C_O\right)$
612	661	559 ⎫ 548 ⎭	Ring deformation $+ \nu(MO)$
451	463	433	$\nu(MO) + \nu(C-CH_3)$
431	441	415 ⎫ 408 ⎭	Ring deformation
291	294	298	$\nu(MO)$
1.45	1.85	1.30	$K(M-O)$ (mdyn/Å) (UBF)

*IR spectra in the solid state.

These orders suggest that the Fe–O bond becomes weaker as the number of the coordinated acac ligand increases because the Lewis acidity of the metal ion decreases in the same order.

Complexes of the M(acac)$_2$X$_2$-type may take the *cis* or *trans* structure. Although steric and electrostatic considerations would favor the *trans*-isomer, the greater stability of the *cis*-isomer is expected in terms of metal–ligand π-bonding. This is the case for Ti(acac)$_2$F$_2$, which is "*cis*" with two ν(TiF) at 633 and 618 cm^{-1}.[487] In the case of Re(acac)$_2$Cl$_2$, however, both forms can be isolated; the *trans*-isomer exhibits ν(ReO) and ν(ReCl) at 464 and 309 cm^{-1}, respectively, while each of these bands splits into two in the *cis*-isomer [472 and 460 cm^{-1} for ν(ReO) and 346 and 333 cm^{-1} for ν(ReCl) in the infrared].[488] For VO(acac)$_2$L, where L is a substituted pyridine, *cis*- and *trans*-isomers are expected. According to Caira et al.,[489] these structures can be distinguished by their infrared spectra. The ν(V=O) and ν(V—O) of the *cis*-isomer are lower than those of the *trans*-isomer. For example, ν(V=O) of VO(acac)$_2$ is 999

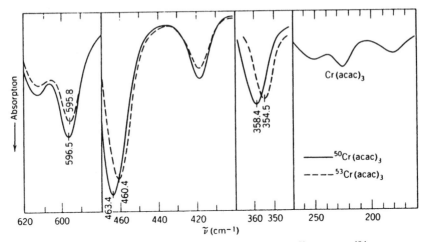

Fig. III-45. Infrared spectra of ^{50}Cr(acac)$_3$ and its ^{53}Cr analog.[484]

cm^{-1}, and this band shifts to 959 cm^{-1} for 4-Et-py (*cis*) and to 973 cm^{-1} for py (*trans*). Furthermore, the ν(V—O) of the *cis*-isomer splits into two bands.

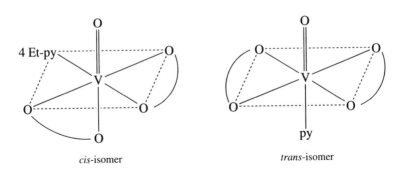

Vibrational spectra of acac complexes have been studied by many other investigators. Only references are cited for the following: Raman spectra of tris (acac) complexes,[390] IR spectra of acac complexes of rare-earth metals,[490] IR spectra of Cs[Os(acac)X$_4$] (X = Cl, Br, and I),[491] relationships between CH and CH$_3$ stretching frequencies and ^{13}C–H spin–spin coupling constants,[492] relationships between ν(C$\equiv$O), ν(C$\equiv$C), and ^{13}C NMR shifts of CO groups.[493] An extensive review on IR spectra of metal complexes of β-diketones has been made by Thornton.[494]

According to X-ray analysis,[495] the hexafluoroacetylacetonato ion (hfa) in [Cu(hfa)$_2${Me$_2$N-(CH$_2$)$_2$-NH$_2$}$_2$] coordinates to the metal as a unidentate via one of its O atoms. This compound exhibits ν(C=O) at 1675 and 1615 cm^{-1}, values slightly higher than those for Cu(hfa)$_2$, in which the hfa ion is chelated to the metal (1644 and 1614 cm^{-1}).

The following dimeric bridging structue has been proposed for [CoBr(acac)]$_2$:

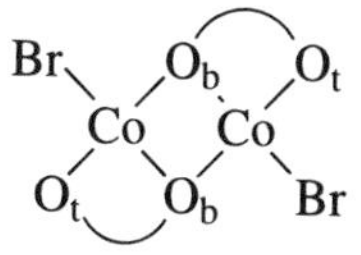

$\nu(CoO_t)$ and $\nu(CoO_b)$ were assigned at 435 and 260 cm^{-1}, respectively.[496] In [Ni(acac)$_2$]$_3$ and [Co(acac)$_4$]$_4$, the O atoms of the acac ion serve as a bridge between two metal atoms.[497] However, no band assignments are available on these polymeric species.

(2) Complexes of Neutral Acetylacetone

In some compounds, the keto form of acetylacetone forms a chelate ring of type B:

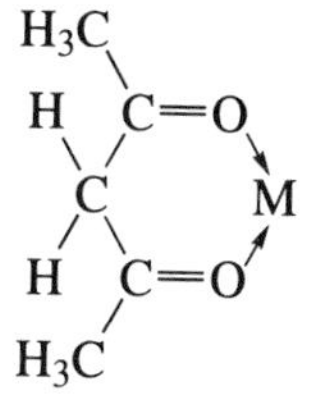

Type B

This particular type of coordination was found by van Leeuwen[498] in [Ni(acacH)$_3$](ClO$_4$)$_2$ and its derivatives, and by Nakamura and Kawaguchi[499] in Co(acacH)Br$_2$. These compounds were prepared in acidic or neutral media, and exhibit strong $\nu(C{=}O)$ bands near 1700 cm^{-1}. Similar ketonic coordination was proposed for Ni(acacH)$_2$Br$_2$[500] and M(acacH)Cl$_2$ (M = Co and Zn).[501]

According to X-ray analysis,[502] the acetylacetone molecule in Mn(acacH)$_2$Br$_2$ is in the enol form and is bonded to the metal as a unidentate via one of its O atoms:

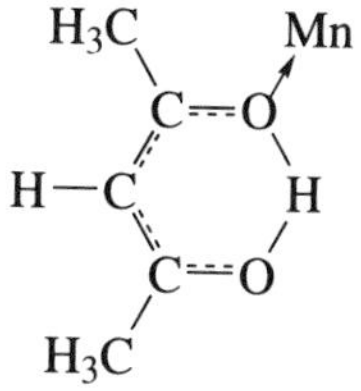

Type C

The C---O and C---C stretching bands of the enol ring were assigned at 1627 and 1564 cm^{-1}, respectively.

(3)   C-bonded Acetylacetonato Complexes

Lewis and co-workers[503] reported the infrared and NMR spectra of a number of Pt(II) complexes in which the metal is bonded to the γ-carbon atom of the acetylacetonato ion:

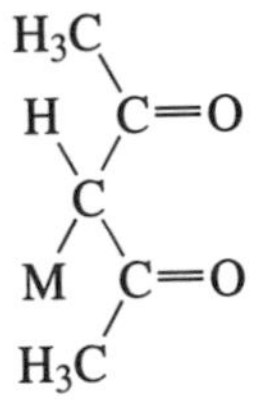

$$
\begin{array}{c}
H_3C \\
\quad\diagdown \\
H \quad C{=}O \\
\diagdown \diagup \\
C \\
\diagup \diagdown \\
M \quad C{=}O \\
\diagup \\
H_3C
\end{array}
$$

Type D

Behnke and Nakamoto carried out normal coordinate analysis on the $[Pt(acac)Cl_2]^-$ ion, in which the acac ion is chelated to the metal (Type A),[504] and on the $[Pt(acac)_2Cl_2]^{2-}$ ion, in which the acac ion is C-bonded to the metal (Type D).[505] Table III-32 lists the observed frequencies and band assignments for these two types, and Fig. III-46 shows the infrared spectra of these two compounds. The results indicate that (1) two $\nu(C{=}O)$ of Type D are higher than those of Type A, (2) two $\nu(C{-}C)$ of type D are lower than those of Type A, and (3) $\nu(PtC)$ of Type D is at 567 cm^{-1}, while $\nu(PtO)$ of Type A are at 650 and 478 cm^{-1}. Figure III-46 also shows that the structure of $K[Pt(acac)_2Cl]$ is

TABLE III-32. Observed Frequencies, Band Assignments, and Force Constants for K[Pt(acac)Cl$_2$] and Na$_2$[Pt(acac)$_2$Cl$_2$]·2H$_2$O

K[Pt(acac)Cl$_2$] (O-bonded, Type A)	Na$_2$[Pt(acac)$_2$Cl$_2$]·2H$_2$O (C-bonded, Type D)	Band Assignment
—	1652, 1626	$\nu(C{=}O)$
1563, 1380	—	$\nu(C\text{---}O)$
1538, 1288	—	$\nu(C\text{---}C)$
—	1350, 1193	$\nu(C{-}C)$
1212, 817	1193, 852	$\delta(CH)$ or $\pi(CH)$
650, 478	—	$\nu(PtO)$
—	567	$\nu(PtC)$
$K(C\text{---}O) = 6.50$	$K(C{=}O) = 8.84$	
$K(C\text{---}C) = 5.23$	$K(C{-}C) = 2.52$	UBF constant
$K(C{-}CH_3) = 3.58$	$K(C{-}CH_3) = 3.85$	(mdyn/Å)
$K(Pt{-}O) = 2.46$	$K(Pt{-}C) = 2.50$	
$K(C{-}H) = 4.68$	$K(C{-}H) = 4.48$	
$\rho = 0.43^a$		

^{a}The stretching–stretching interaction constant (ρ) was used for Type A because of the presence of resonance in the chelate ring.

as follows:

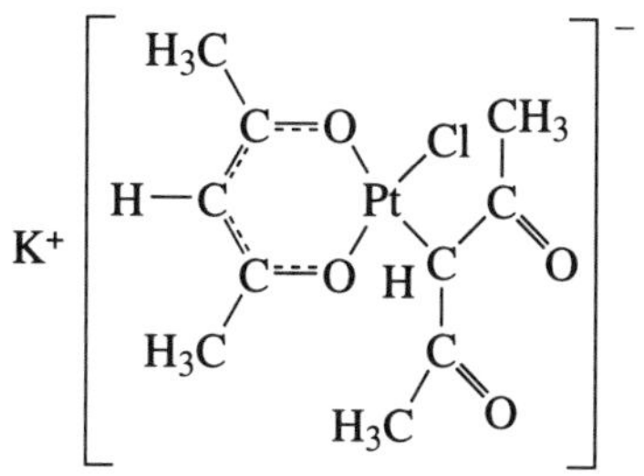

since its spectrum is roughly a superposition of those types A and D. Similarly, the infrared spectrum of K[Pt(acac)$_3$)][503] is interpreted as a superposition of spectra of Types A, D, and D′, in which two C—O bonds are transoid.[506]

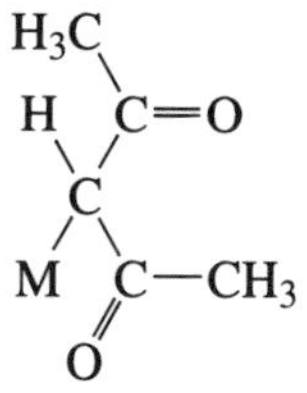

Type D′

The C-bonded acac ion was found in Hg$_2$Cl$_2$(acac),[507] Au(acac)(PPh$_3$),[508] and Pd(acac)$_2$(PPh$_3$).[509] In the last compound, one acac group is Type A and the other Type D. In all these cases, the ν(C=O) of the Type-D acac groups are at 1700–1630 cm^{-1}.

As discussed above, K[Pt(acac)$_2$Cl] contains one Type-A and one type-D acac group. If a solution of K[Pt(acac)$_2$Cl] is acidified, its Type-D acac group is converted into Type-E:

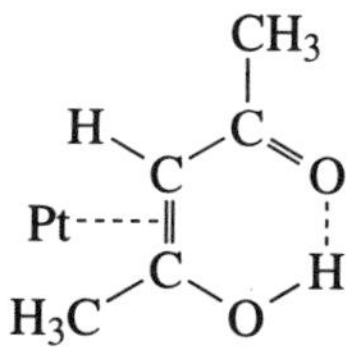

Type E

This structure was first suggested by Allen et al.,[510] based on NMR evidence. Behnke and Nakamoto[511] showed that the infrared spectrum of [Pt(acac)-

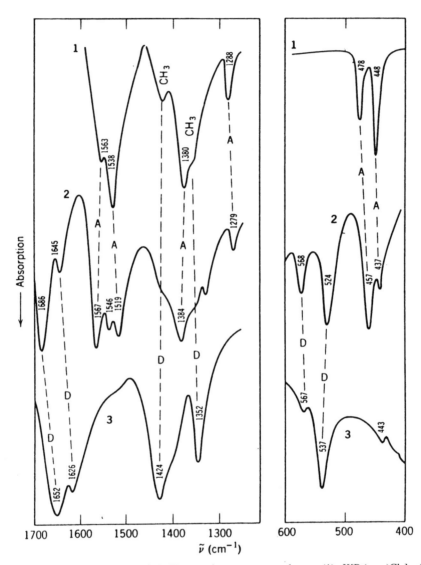

Fig. III-46. Infrared spectra of Pt(II) acetylacetonato complexes: (1) K[Pt(acac)Cl$_2$]; (2) K[Pt(acac)$_2$Cl]; and (3) Na$_2$[Pt(acac)$_2$Cl$_2$]H$_2$O. A and D denote the bands characteristic of Types A and D, respectively.[504,505]

(acacH)Cl] thus obtained can be interpreted as a superposition of spectra of Types A and E.

That the two O atoms of the C-bonded acac group (Type D) retain the ability to coordinate to a metal was first demonstrated by Lewis and Oldham,[512] who prepared neutral complexes of the following type:

Using the metal isotope technique, Nakamura and Nakamoto[513] assigned the $\nu(\text{NiO})$ of $\text{Ni[Pt(acac)}_2\text{Cl]}_2$ at 279 and 266 cm^{-1}. These values are relatively close to the $\nu(\text{NiO})$ of $\text{Ni(acacH)}_2\text{Br}_2$ (264 and 239 cm^{-1}), discussed previously. Thus, the newly formed Ni-acac ring retains its keto character and is close to Type B. Other types prepared by Kawaguchi and co-workers include the following:

Kawaguchi[516] reviewed the modes of coordination of β-diketones in a variety of metal complexes.

(4) Complexes of Other β-Diketones

In a series of metal tropolonato complexes, Hulett and Thornton[517] noted a parallel relationship between the $\nu(\text{MO})$ and the CFSE energy. These workers assigned the $\nu(\text{MO})$ of trivalent metal tropolonates in the 660–580 cm^{-1} region, based on the ^{16}O–^{18}O isotope shifts observed for the Cu(II) complex.[518] Using the metal isotope technique, Hutchinson et al.[519] assigned the $\nu(\text{MO})$ at the following frequencies (cm^{-1}):

V(III)		Cr(III)		Mn(III)		Fe(III)		Co(III)
377	~	361	>	338	>	317	<	371
319	<	334	>	268	>	260	<	360

It was found that these frequencies still follow the order predicted by the CFSE.

2,4,6-Heptanetrione forms $1:1$ and $1:2$ (metal : ligand) complexes with Cu(II):[520]

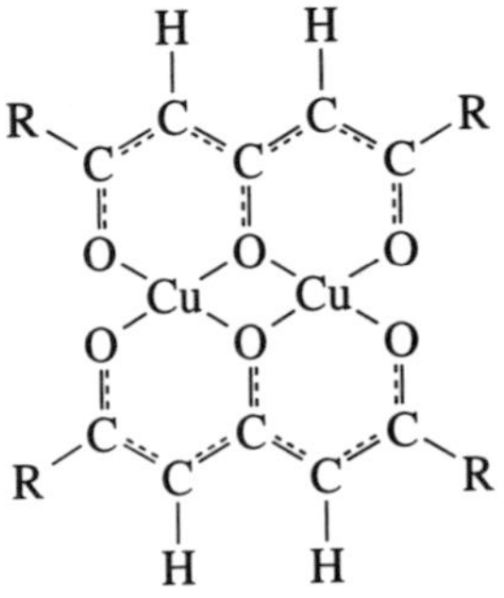

1:1 Complex 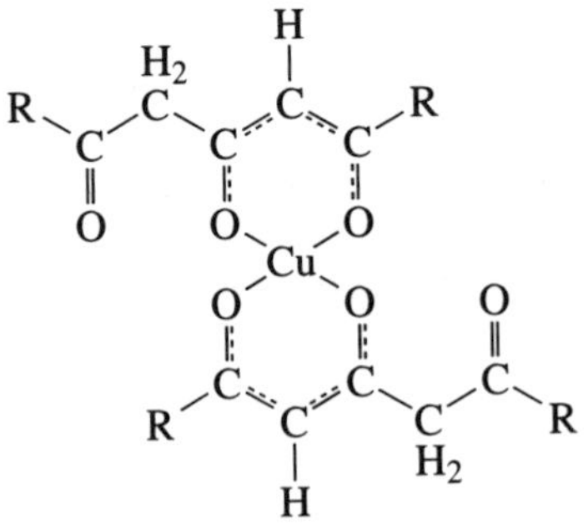   1:2 Complex

Both complexes exhibit multiple bands due to Type-A rings in the 1600–1500 cm^{-1} region. However, the $1:2$ complex exhibits $\nu(C{=}O)$ of the uncoordinated C=O groups near 1720 cm^{-1}.

III-14. COMPLEXES OF UREA, SULFOXIDES, AND RELATED LIGANDS

(1) Complexes of Urea and Related Ligands

Penland et al.[521] first studied the infrared spectra of urea complexes to determine whether coordination occurs through nitrogen or oxygen. The electronic structure of urea may be represented by a resonance hybrid of structures I, II, and III, with each contributing roughly an equal amount:

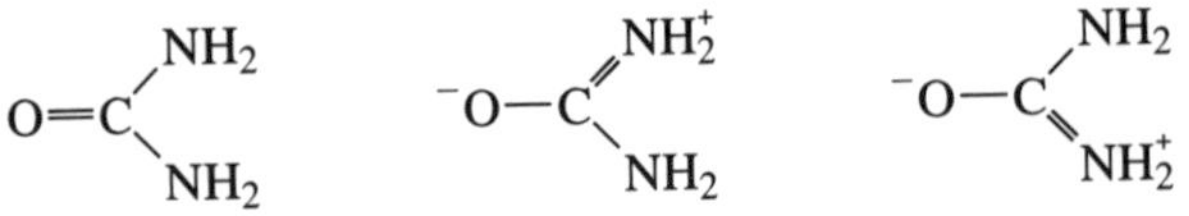

If coordination occurs through nitrogen, the contributions of structures II and III will decrease. This results in an increase of the CO stretching frequency with a decrease in the CN stretching frequency. The NH stretching frequency

TABLE III-33. **Some Vibrational Frequencies of Urea and Its Metal Complexes (cm^{-1})[521]**

[Pt(urea)$_2$Cl$_2$]	Urea	[Cr(urea)$_6$]Cl$_3$	Predominant Mode
3390 ⎫ 3290 ⎭	3500 ⎫ 3350 ⎭	3440 ⎫ 3330 ⎭	ν(NH$_2$), free
3130 ⎫ 3030 ⎭		3190	ν(NH$_2$), bonded
1725	1683	1505^a	ν(C=O)
1395	1471	1505^a	ν_a(CN)

$^a\nu$(C=O) and ν(C—N) couple in the Cr complex.

in this case may fall in the same range as the value for the amido complexes (Sec. III-1). If coordination occurs through oxygen, the contribution of structure I will decrease. This may result in a decrease of the CO stretching frequency but no appreciable change in the NH stretching frequency. Since the spectrum of urea itself has been analyzed completely,[522] band shifts caused by coordination can be checked immediately. The results shown in Table III-33 indicate that coordination occurs through nitrogen in the Pt(II) complex, and through oxygen in the Cr(III) complex. It was also found that Pd(II) coordinates to the nitrogen, whereas Fe(III), Zn(II), and Cu(II) coordinate to the oxygen of urea. The infrared spectra of tetramethylurea (tmu) complexes of lanthanide elements, [Ln(tmu)$_6$](ClO$_4$)$_3$, indicate the presence of O-coordination.[523] Similar conclusions have been obtained for ThL$_3$Cl$_4$ (L: N,N'-dialkylurea) inasmuch as the ν(CO) is downshifted by 110–60 cm^{-1} upon coordination.[524]

From infrared studies on thiourea [(NH$_2$)$_2$CS] complexes, Yamaguchi et al.[525] found that all the metals studied (Pt, Pd, Zn, and Ni) form M–S bonds, since the CN stretching frequency increases and the CS stretching frequency decreases upon coordination, without an appreciable change in the NH stretching frequency. On the basis of the same criterion, thiourea complexes of Fe(II),[526] Mn(II), Co(II), Cu(I), Hg(II), Cd(II), and Pb(II) were shown to be S-bonded.[527] Several investigators[528–530] studied the far-infrared spectra of thiourea complexes and assigned the MS stretching bands between 300 and 200 cm^{-1}. Thus far, the only metal reported to be N-bonded is Ti(IV).[531] Infrared spectra of alkylthiourea complexes have also been studied. Lane and colleagues[532] studied the infrared spectra of methylthiourea complexes and concluded that methylthiourea forms M–S bonds with Zn(II) and Cd(II) and M–N bonds with Pd(II), Pt(II), and Cu(I). For other alkylthiourea complexes, see Refs. 533 and 534. Infrared spectra of selenourea (su) complexes of Co(II), Zn(II), Cd(II), and Hg(II) exhibit ν(MSe) in the 245–167 cm^{-1} region.[535] The Raman spectra of [Pd(su)$_4$]$^{2+}$ and [Pt(su)$_4$]$^{2+}$ ions exhibit the A_{1g} ν(MSe) at 178 and 191 cm^{-1}, respectively.[536]

Linkage isomerism was found for the formamidopentamminecobalt (III): [(NH$_3$)$_5$Co(—NH$_2$CHO)]$^{3+}$ and [(NH$_3$)$_5$Co(—OCHNH$_2$)]$^{3+}$. Although little difference was found in the ν(C=O) region, the N-isomer showed the aldehyde ν(CH) at 2700 cm^{-1}, whereas such a band was not obvious in the O-isomer.[537]

(2) Complexes of Sulfoxides and Related Compounds

Cotton et al.[538] studied the infrared spectra of sulfoxide complexes to see whether coordination occurs through oxygen or sulfur. The electronic structure of sulfoxides may be represented by a resonance hybrid of these structures:

If coordination occurs through oxygen, the contribution of structure V will decrease and result in a decrease in $\nu(S{=}O)$. If coordination occurs through sulfur, contribution of structure IV will decrease and may result in an increase in $\nu(S{=}O)$. It has been concluded that coordination occurs through oxygen in the $Co(DMSO)_6^{2+}$ ion, since the $\nu(S{=}O)$ of this ion absorbs at $1100{-}1055$ cm^{-1}. On the other hand, coordination may occur through sulfur in $PdCl_2(DMSO)_2$ and $PtCl_2(DMSO)_2$, since $\nu(S{=}O)$ of these compounds ($1157{-}1116$ cm^{-1}) are higher than the value for the free ligand. Other ions such as Mn(II), Fe(II, III), Ni(II), Cu(II), Zn(II), and Cd(II) are all coordinated through oxygen, since the DMSO complexes of these metals exhibit $\nu(S{=}O)$ between 960 and 910 cm^{-1}. Drago and Meek,[539] however, assigned $\nu(S{=}O)$ of O-bonded complexes in the $1025{-}985$ cm^{-1} region, since they are metal sensitive. The bands between 960 and 930 cm^{-1}, which were previously assigned to $\nu(S{=}O)$ are not metal sensitive and are assigned to $\rho_r(CH_3)$. Even so, $\nu(S{=}O)$ of O-bonded complexes are lower than the value for free DMSO. To confirm $\nu(S{=}O)$ assignments, it is desirable to compare the spectra of the corresponding DMSO-d_6 complexes since $\rho_r(CD_3)$ is outside the $\nu(S{=}O)$ region. Table III-34 lists $\nu(S{=}O)$ of typical compounds.

Wayland and Schramm[540] found the first example of mixed coordination of DMSO in the $[Pd(DMSO)_4]^{2+}$ ion; it exhibits two S-bonded $\nu(S{=}O)$ at 1150 and 1140 cm^{-1}, and two O-bonded $\nu(S{=}O)$ at 920 and 905 cm^{-1}. Thus, the infrared spectrum is most consistent with a configuration in which two S-bonded and two O-bonded DMSO are in the *cis* position. The infrared and NMR spectra of $Ru(DMSO)_4Cl_2$ suggested a mixing of O- and S-coordination; $\nu(S{=}O)$ at 1120 and 1090 cm^{-1} for S-coordination and at 915 cm^{-1} for O-coordination.[550] X-Ray analysis[551] has since shown that two Cl atoms are in the *cis* positions of an octahedron and the remaining positions are occupied by one O-bonded and three S-bonded DMSO ligands. Infrared spectra show that all DMSO ligands in $Ru(DMSO)_3Cl_3$ are O-bonded while O- and S-bonded DMSO ligands are mixed in $M(DMSO)_3Cl_3$ (M = Os and Rh).[552] In contrast, a recent study shows that all the DMSO ligands are S-bonded in the *fac*-isomer of $RuCl_3(DMSO)_3$ but O- and S-bonded DMSO ligands are mixed in the *mer*-isomer.[552a] In *mer,cis*-$RuCl_3(DMSO)_2(NH_3)$, one DMSO which is *trans* to the NH_3 is S-bonded, and

TABLE III-34. SO Stretching Frequencies of DMSO Complexes (cm^{-1})

Compound	$\nu(S=O)$	Bonding	Ref.
Sn(DMSO)$_2$Cl$_4$	915	O	540
[Cr(DMSO)$_6$](ClO$_4$)$_3$	928	O	540
[Ni(DMSO)$_6$](ClO$_4$)$_2$	955	O	540
[Ln(DMSO)$_8$](ClO$_4$)$_3$, (Ln = La, Ce, Pr, Nd)	998–992	O	541
[Al(DMSO)$_6$]X$_3$, (X = Cl, Br, I)	1000–1008	O	542
CdAg$_6$I$_8$(DMSO)$_8$	1000	O	543
[Ru(NH$_3$)$_5$(DMSO)](PF$_6$)$_2$	1045	S	544
trans-[Pd(DMSO)$_2$Cl$_2$]	1116	S	545
cis-[Pt(DMSO)$_2$Cl$_2$]	1135 1160	S	546
cis[PtCl$_2$(quinoline) (DMSO)]	1120	S	547
[Pt(R$_2$SO) (μ-Cl)Cl]$_2^a$	1142	S	548
cis-RuCl$_2$ [CH$_3$C(CH$_2$S-Et)$_3$] (DMSO)	1080	S	549

^{a}R = C$_2$H$_5$.

the other DMSO which is *trans* to the Cl is O-bonded. These two DMSO ligands exhibit the $\nu(S=O)$ at 1088 and 910 cm^{-1}, respectively.[553]

Interaction of DMSO with lanthanide perchlorates in anhydrous CH$_3$CN has been studied by FTIR and Raman spectroscopy.[554] The magnitude of downshifts of the $\nu(S=O)$ increases with the increasing atomic number of the Ln(III) ion from −49 to −58 cm^{-1}. In free (CF$_3$)$_2$SO, the $\nu(S=O)$ is at 1242 cm^{-1}. This band is shifted to 1130 cm^{-1} (IR) in [{(CF$_3$)$_2$SO}XeF]SbF$_6$, indicating O-coordination of the sulfrane ligand.[555]

Complete assignments on infrared and Raman spectra of *trans*-Pd(DMSO)$_2$X$_2$ (X = Cl and Br) and their deuterated analogs have been made by Tranquille and Forel.[556] Berney and Weber[557] found the order of ν(MO) in the [M(DMSO)$_6$]$^{n+}$ ion to be as follows:

M =	Cr(III)		Ni(II)		Co(II)		Zn(II)		Fe(II)		Mn(II)
ν(MO) (cm^{-1})	529	>	444	>	436	>	431	>	438 415	>	418

Griffiths and Thornton[558] made band assignments of these DMSO complexes based on d_6 and ^{18}O substitution of DMSO.

Ligands such as DPSO(diphenylsulfoxide) and TMSO (tetramethylenesulfoxide) do not exhibit the CH_3 rocking bands near 950 cm^{-1}. Thus, the SO stretching bands of metal complexes containing these ligands can be assigned without difficulty. In a series of O-bonded DMSO and TMSO complexes, the S=O stretching force constant decreases linearly as the M–O stretching force constant increases.[559] Table III-35 lists the SO stretching frequencies and the magnitude of band shifts in DPSO complexes.[560] van Leeuwen and Groeneveld[560] noted that the shift becomes larger as the electronegativity of the metal increases. In Table III-35, the metals are listed in the order of increasing electronegativity.

In $[M(DTHO_2)_3]^{2+}$ [M = Co(II), Ni(II), Mn(II), etc.], the metals are O-bonded since the $\nu(S=O)$ of free ligand (1055–1015 cm^{-1}) are shifted to lower frequencies by 40–22 cm^{-1}:

2,5-Dithiahexane-2,5-dioxide (DTHO₂)

On the other hand, the metals are S-bonded in $M(DTHO_2)Cl_2$ [M = Pt(II) and Pd(II)] since $\nu(S=O)$ are shifted to higher frequencies by 108–77 cm.[1,561] Dimethylselenoxide, $(CH_3)_2Se=O$, forms complexes of the $MCl_2(DMSeO)_n$ type, where M is Hg(II), Cd(II), Cu(II), and so on, and n is 1, $1\frac{1}{2}$, or 2. The $\nu(Se=O)$ of the free ligand (800 cm^{-1}) is shifted to the 770–700 cm^{-1} region, indicating the O-bonding in these complexes.[562]

TABLE III-35. Shifts of SO Stretching Bands in DPSO and DMSO Complexes (cm^{-1})[560]

| Metal | DPSO Complex | | DMSO Complex |
	$\nu(SO)$	Shift	Shift
Ca(II)	1012–1035	0– (−23)	—
Mg(II)	1012	−23	—
Mn(II)	983–991	−45	−41
Zn(II)	987–988	−47	—
Fe(II)	987	−48	—
Ni(II)	979–982	−55	−45
Co(II)	978–980	−56	−51
Cu(II)	1012, 948	−23, −87	−58
Al(III)	942	−93	—
Fe(III)	931	−104	—

III-15. CYANO AND NITRILE COMPLEXES

(1) Cyano Complexes

The vibrational spectra of cyano complexes have been studied extensively and these investigations are reviewed by Sharp,[563] Griffith,[564] Rigo and Turco,[565] and Jones and Swanson.[566]

(a) CN Stretching Bands. Cyano complexes can be identified easily since they exhibit sharp $\nu(CN)$ at 2200–2000 cm^{-1}. The $\nu(CN)$ of free CN$^-$ is 2080 cm^{-1} (aqueous solution). Upon coordination to a metal, the $\nu(CN)$ shift to higher frequencies, as shown in Table III-36. The CN$^-$ ion acts as a σ-donor by donating electrons to the metal and also as a π-acceptor by accepting electrons from the metal. σ-Donation tends to raise the $\nu(CN)$ since electrons are removed from the 5σ orbital, which is weakly antibonding, while π-backbonding tends to decrease the $\nu(CN)$ because the electrons enter into the antibonding $2p\pi^*$ orbital. In general, CN$^-$ is a better σ-donor and a poorer π-acceptor than CO. Thus, the $\nu(CN)$ of the complexes are generally higher than the value for free CN$^-$, whereas the opposite prevails for the CO complexes (Sec. III-17).

According to El-Sayed and Sheline,[575] the $\nu(CN)$ of cyano complexes are governed by (1) the electronegativity, (2) the oxidation state, and (3) the coordination number of the metal. The effect of electronegativity is seen in the order:

$$[Ni(CN)_4]^{2-} \qquad [Pd(CN)_4]^{2-} \qquad [Pt(CN)_4]^{2-}$$
$$2128 \quad < \quad 2143 \quad < \quad 2150 \qquad cm^{-1}$$

Since the electronegativity of Ni(II) is smallest, the σ-donation will be the least, and the $\nu(CN)$ is expected to be the lowest. The effect of oxidation state is seen in the frequency order.[576]

$$[V(CN)_6]^{5-} \qquad [V(CN)_6]^{4-} \qquad [V(CN)_6]^{3-}$$
$$1910 \quad < \quad 2065 \quad < \quad 2077 \qquad cm^{-1}$$

The higher the oxidation state, the stronger the σ-bonding, and the higher the $\nu(CN)$. The effect of coordination number[577–579] is evident in the frequency order:

$$[Ag(CN)_4]^{3-} \qquad [Ag(CN)_3]^{2-} \qquad [Ag(CN)_2]^-$$
$$2092 \quad < \quad 2105 \quad < \quad 2135 \qquad cm^{-1}$$

Here an increase in the coordination number results in a decrease in the positive charge on the metal, which, in turn, weakens the σ-bonding, thus decreasing

TABLE III-36. C≡N Stretching Frequencies of Cyano Complexes (cm^{-1})

Compound	Symmetry	ν(CN)	Ref.
Tl[Au(CN)$_2$]	**D**$_{\infty h}$	2164 ($\sum_g^+$), 2141 ($\sum_u^+$)	567, 568
K[Ag(CN)$_2$]	**D**$_{\infty h}$	2146 ($\sum_g^+$), 2140 ($\sum_u^+$)	569
K$_2$[Ni(^{12}C^{14}N)$_4$]	**D**$_{4h}$	2143.5 (A_{1g}), 2134.5 (B_{1g}), 2123.5 (E_u)	570
K$_2$[Pd(^{12}C^{14}N)$_4$]	**D**$_{4h}$	2160.5 (A_{1g}), 2146.4 (B_{1g}), 2135.8 (E_u)	570
K$_2$[Pt(^{12}C^{14}N)$_4$]	**D**$_{4h}$	2168.0 (A_{1g}), 2148.8 (B_{1g}), 2133.4 (E_u)	570
Na$_3$[Ni(CN)$_5$]	**C**$_{4v}$	2130 (A_1), 2117 (B_1), 2106 (E), 2090 (A_1)	571
Na$_3$[Co(CN)$_5$]	**C**$_{4v}$	2115 (A_1), 2110 (B_1), 2096 (E), 2080 (A_1)	571
K$_3$[Mn(CN)$_6$]	**O**$_h$	2129 (A_{1g}), 2129 (E_g), 2112 (F_{1u})	572, 573
K$_4$[Mn(CN)$_6$]	**O**$_h$	2082 (A_{1g}), 2066 (E_g), 2060 (F_{1u})	572
K$_3$[Fe(CN)$_6$]	**O**$_h$	2135 (A_{1g}), 2130 (E_g), 2118 (F_{1u})	572
K$_4$[Fe(CN)$_6$]·3H$_2$O	**O**$_h$	2098 (A_{1g}), 2062 (E_g), 2044 (F_{1u})	572
K$_3$[Co(CN)$_6$]	**O**$_h$	2150 (A_{1g}), 2137 (E_g), 2129 (F_{1u})	572
K$_4$[Ru(CN)$_6$]·3H$_2$O	**O**$_h$	2111 (A_{1g}), 2071 (E_g), 2048 (F_{1u})	572
K$_3$[Rh(CN)$_6$]	**O**$_h$	2166 (A_{1g}), 2147 (E_g), 2133 (F_{1u})	572
K$_2$[Pd(CN)$_6$]	**O**$_h$	2185 (F_{1u})	574
K$_4$[Os(CN)$_6$]·3H$_2$O	**O**$_h$	2109 (A_{1g}), 2062 (E_g), 2036 (F_{1u})	572
K$_3$[Ir(CN)$_6$]	**O**$_h$	2167 (A_{1g}), 2143 (E_g), 2130 (F_{1u})	572

the ν(CN). The ν(CN) of A$_3$[M(CN)$_6$]-type salts (M = Fe and Co) are sensitive to the nature of the counterion (A). Thus, Fernandez-Beltran et al.[580] used this fact to examine the CN ligand–counterion interaction quantitatively.

Other cyano complexes which are not included in Table III-36 are Na[Cu(CN)$_2$]2H$_2$O (polymeric chain),[581] Na$_2$[Cu(CN)$_3$]·3H$_2$O (**D**$_{3h}$),[582] Cs[Hg(CN)$_3$] (**D**$_{3h}$),[583] and K$_2$[Zn(CN)$_4$] (**T**$_d$).[584] The symmetry of the [Mo(CN)$_7$]$^{4-}$ ion may be **D**$_{5h}$[585] or **C**$_{2v}$.[586] The pentagonal–bipyramidal structure (**D**$_{5h}$) has been proposed for [Re(CN)$_7$]$^{4-}$,[587] [Tc(CN)$_7$]$^{4-}$,[588] and [W(CN)$_7$]$^{5-}$,[589] based on their IR and Raman spectra either in the solid state or in solution or both. According to X-ray analysis,[590] the [Mo(CN)$_8$]$^{4-}$ ion in K$_4$[Mo(CN)$_8$]·2H$_2$O is definitely **D**$_{2d}$ (dodecahedron). On the other hand, a Raman study[591] supported the **D**$_{4d}$ (archimedean–antiprism) structure of the [Mo(CN)$_8$]$^{4-}$ ion in aqueous solution. The stereochemical conversion of the [Mo(CN)$_8$]$^{4-}$ ion from **D**$_{2d}$ (solid) to **D**$_{4d}$ (solution) symmetry was confirmed by Hartman and Miller[592] and Parish et al.[593] Similar conversions were proposed for the [W(CN)$_8$]$^{4-}$,[592,593] and [Nb(CN)$_8$]$^{4-}$[594] ions. However, Long and Vernon[595] claim that the **D**$_{2d}$ geometry is maintained even in aqueous solution. Both X-ray and Raman studies confirm the **D**$_{2d}$ structure for K$_5$[Nb(CN)$_8$] in the solid state although the **D**$_{4d}$ structure prevails in solution.[596]

According to the results of X-ray analysis,[597] the unit cell of [Cr(en)$_3$]-[Ni(CN)$_5$]·1$\frac{1}{2}$H$_2$O contains both square-pyramidal (**C**$_{4v}$) and trigonal-bipyramidal (**D**$_{3h}$) structures of the [Ni(CN)$_5$]$^{3-}$ ion. Terzis et al.[598] showed that the complicated vibrational spectrum of this crystal in the ν(CN) region is simplified dramatically when it is dehydrated. These spectral changes suggest that the **D**$_{3h}$ (somewhat distorted) units have been converted to **C**$_{4v}$ geometry upon dehydration. Basile et al.[599] showed that such conversion from D_{3h} to C_{4v} also occurs when the crystal is subjected to high pressure. Hellner et al.[600] observed the splitting of the degenerate $\nu(CN)$ of $K_2[Zn(CN)_4]$ and a partial reduction of the central metal in $K_3[M(CN)_6]$ [M = Fe(III) and Mn(III)] when these crystals are under high external pressure.

Jones and Penneman[577–579] made an extensive infrared study of the equilibria of cyano complexes in aqueous solution. (For aqueous infrared spectroscopy, see Sec. III-10.) Figure III-47 shows the infrared spectra of aqueous silver cyano complexes obtained by changing the ratio of Ag^+ to CN^- ions. Table III-37 lists the frequencies and extinction coefficients from which equilibrium constants can be calculated. Chantry and Plane[602] studied the same equilibria using Raman spectroscopy.

*(b) **Lower-Frequency Bands.*** In addition to $\nu(CN)$, the cyano complexes exhibit $\nu(MC)$, $\delta(MCN)$, and $\delta(CMC)$ bands in the low-frequency region. Figure III-48 shows the infrared spectra of $K_3[Co(CN)_6]$ and $K_2[Pt(CN)_4]\cdot3H_2O$. Normal coordinate analyses have been carried out on various hexacyano complexes to assign these low-frequency bands (Table III-38). The results of these calculations indicate that the $\nu(MC)$, $\delta(MCN)$, and $\delta(CMC)$ vibrations appear in the regions 600–350, 500–350, and 130–60 cm^{-1}, respectively. The MC and C≡N stretching force constants obtained are also given in Table III-38.

Nakagawa and Shimanouchi[607] noted that the MC stretching force constant increases in the order Fe(III) < Co(III) < Fe(II) < Ru(II) < Os(II), and the C≡N stretching force constant decreases in the same order of metals. This result was interpreted as indicating that the M–C π-bonding is increasing in the above order. The degree of M–C π-bonding may be proportional to the number of

TABLE III-37. Frequencies and Molecular Extinction Coefficients of Cyano Complexes in Aqueous Solutions

Ion	Frequency (cm^{-1})	Molecular Extinction Coefficient	Ref.
Free [CN]$^-$	2080 ± 1	29 ± 1	577
[Ag(CN)$_2$]$^-$	2135 ± 1	264 ± 12	577
[Ag(CN)$_3$]$^{2-}$	2105 ± 1	379 ± 23	577
[Ag(CN)$_4$]$^{3-}$	2092 ± 1	556 ± 83	577
[Cu(CN)$_2$]	2125 ± 3	165 ± 25	578
[Cu(CN)$_3$]$^{2-}$	2094 ± 1	1090 ± 10	578
[Cu(CN)$_4$]$^{3-}$	2076 ± 1	1657 ± 15	578
[Zn(CN)$_4$]$^{2-}$	2149	113	579
[Cd(CN)$_4$]$^{2-}$	2140	75	579
Hg(CN)$_2$	2194	3	579
[Hg(CN)$_3$]$^-$	2161	26	579
[Hg(CN)$_4$]$^{2-}$	2143	113	579
[Ni(CN)$_4$]$^{2-}$	2124 ± 1	1068 ± 95	601
[Ni(CN)$_5$]$^{3-}$	2102 ± 2	1730 ± 230	601

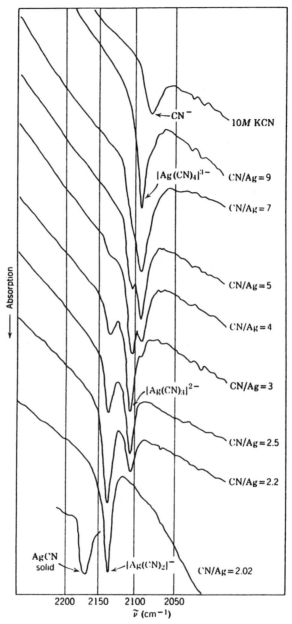

Fig. III-47. Infrared spectra of silver cyano complexes in aqueous solutions.[577]

d-electrons in the t_{2g} electronic level. According to Jones,[608] the integrated absorption coefficient of the C≡N stretching band (F_{1u}) becomes larger as the number of d-electrons in the t_{2g} level increases. Thus the results shown in Table III-39 suggest that the M–C π-bonding increases in the order Cr(III) < Mn(III) < Fe(III) < Co(III). The order of ν(MC) shown in the same table con-

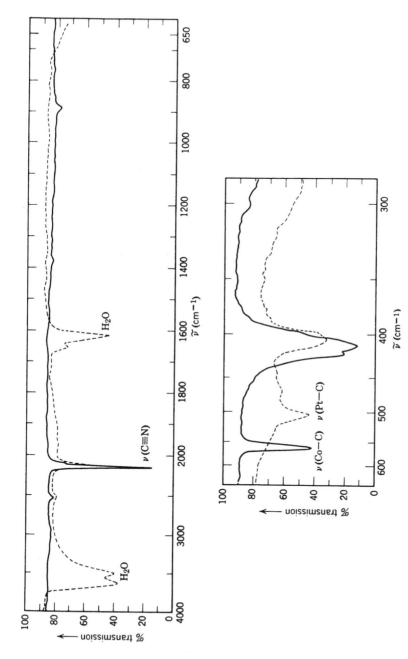

Fig. III-48. Infrared spectra of $K_3[Co(CN)_6]$ (solid line) and $K_2[Pt(CN)_4] \cdot 3H_2O$ (broken line).

TABLE III-38. Vibrational Frequencies and Band Assignments of Hexacyano Complexes (cm⁻¹)

		$[Cr(CN)_6]^{3-}$	$[Co(CN)_6]^{3-}$	$[Ir(CN)_6]^{3-}$	$[Rh(CN)_6]^{3-}$	$[Co(CN)_6]^{3-}$	$[Fe(CN)_6]^{4-}$	$[Fe(CN)_6]^{3-}$	$[Ru(CN)_6]^{4-}$	$[Os(CN)_6]^{4-}$
A_{1g}	ν(MC)	374	408	469	445	406	(410)	(390)	(460)	(480)
E_g	ν(MC)	336	(391)	450	435	(375)	(390)	—	(410)	(450)
F_{1g}	δ(MCN)	536	(358)	(415)	(380)	(380)	(350)	—	(340)	(360)
F_{1u}	ν(MC)	457	564	520	520	565	585	511	550	554
	δ(MCN)	694	416	398	386	414	414	387	376	392
	δ(CMC)	124	(84)	(82)	(88)	—	—	89	—	—
F_{2g}	δ(MCN)	536	(480)	483	(475)	—	(420)	—	(400)	(430)
	δ(CMC)	106	98	95	94	98	—	99	—	—
F_{2u}	δ(MCN)	496	(440)	445	—	(395)	—	—	(365)	(390)
	δ(CMC)	95	(72)	(69)	—	—	—	70	—	—
Force field		GVF	GVF	GVF	GVF	UBF	UBF	UBF	UBF	UBF
K(M—C) (mdyn/Å)		1.928	2.063	2.704	2.366	2.308	2.428	1.728	2.793	3.343
K(C≡N) (mdyn/Å)		16.422	16.767	16.678	16.831	16.5	15.1	17.0	15.3	14.9
References		603	604	605	605	607	607	607	607	607
			605					606		
			606							

TABLE III-39. Relation between Infrared Spectrum and Electronic Structure in Hexacyano Complexes[608]

Compound	Number of d-Electrons in t_{2g} Level	ν(CN) (cm^{-1})	ν(MC) (cm^{-1})	Integrated Absorption Coefficient (mole^{-1} cm^{-2})
K$_3$[Cr(CN)$_6$]	3	2128	339	2,100
K$_3$[Mn(CN)$_6$]	4	2112	361	8,200
K$_3$[Fe(CN)$_6$]	5	2118	389	12,300
K$_3$[Co(CN)$_6$]	6	2129	416	18,300

firms this conclusion. Griffith and Turner[572] found a similar trend in the Fe(II) < Ru(II) < Os(II) series. Nakagawa and Shimanouchi[609] carried out complete normal coordinate analyses on K$_3$[M(CN)$_6$] [M = Fe(III) and Cr(III)] crystals, including all lattice modes. Jones et al.[610] also performed complete normal coordinate analyses on crystalline Cs$_2$Li[Fe(CN)$_6$], including its ^{13}C, ^{15}N, and ^{6}Li analogs.

Normal coordinate analyses have been made on tetrahedral, square-planar, and linear cyano complexes of various metals; Table III-40 gives the results of these studies. Far-infrared spectra of various cyano complexes have been measured.[622] An ultraviolet and infrared study[623] showed that the [Ni(CN)$_4$]$^{2-}$ and [Ni(CN)$_5$]$^{3-}$ ions are in equilibrium in a solution containing Na$_2$[Ni(CN)$_4$], KCN, and KF. The integrated absorption coefficient of the C≡N stretching band increases in the order Hg(II) < Ag(I) < Au(I) in linear dicyano complexes, indicating that the M–C π-bonding increases in the same order.[608] From the measurements of infrared dichroism, Jones determined the orientation of [Ag(CN)$_2$]$^-$ and [Au(CN)$_2$]$^-$ ions in their potassium salts.[620,621] His results are in good agreement with those of X-ray analysis.

Jones and co-workers[624] carried out an extensive study on mixed cyano–halide complexes such as [Au(CN)$_2$X$_2$]$^-$, where X is Cl$^-$, Br$^-$, or I$^-$. The infrared spectra of K$_2$[Pt(CN)$_4$X$_2$][625] and K$_2$[Pt(CN)$_5$X][626] (X = Cl$^-$, Br$^-$, I$^-$, etc.) have been reported. Vibrational spectra of cyano complexes containing NO groups are discussed in Sec. III-18.

(c) Bridged Cyano Complexes. If the M—C≡N group forms a M—C≡N—M′ type bridge, ν(C≡N) shifts to a higher, and ν(MC) to a lower, frequency. The higher-frequency shift of ν(C≡N) should be noted since the opposite trends are observed for bridging carbonyl and halogeno complexes. Shriver[627] observed that ν(C≡N) of K$_2$[Ni(CN)$_4$] at 2130 cm^{-1} shifts to 2250 cm^{-1} in K$_2$[Ni(CN)$_4$]·4BF$_3$ because of the formation of the Ni—C≡N—BF$_3$ type bridge. They[628] also found that, for KFeCr(CN)$_6$, the green isomer containing the Fe(II)—C≡N—Cr(III) bridges exhibits ν(C≡N) at 2092 cm^{-1}, while the red isomer containing the Cr(III)—C≡N—Fe(II) bridges shows ν(C≡N) at 2168 and 2114 cm^{-1}. Brown et al.[629] studied the mechanism of conversion from green to red isomer by combining infrared and

TABLE III-40. Frequencies and Band Assignments of the Lower-Frequency Bands of Cyano Complexes (cm^{-1})

Ion	Symmetry	ν(MC)	δ(MCN)	δ(CMC)	Force Constant[a]		Ref.
					K(M—C)	K(C≡N)	
$[Cu(CN)_4]^{3-}$	T_d	364(IR)⎫ 288(R)⎬	324(R) 306(IR)	(74) (63)	1.25–⎫ 1.30⎬	16.10–⎫ 16.31⎬	611 612
$[Zn(CN)_4]^{2-}$	T_d	359(IR)[b] 342(R)	315(IR)[b] 230(R)	71(R)	1.30	17.22	613 584
$[Cd(CN)_4]^{2-}$	T_d	316(IR)[b] 324(R)	250(R)[b] 194(R)	61(R)	1.28	17.13	613
$[Hg(CN)_4]^{2-}$	T_d	330(IR)[b] 335(R)	235(R)[b] 180(R)	54(R)	1.53	17.08	613
$[Pt(CN)_4]^{2-}$	D_{4h}	505(IR) 465(R) 455(R)	318(R) 300(IR)	95(R)	3.425	16.823	614 615
$[Ni(CN)_4]^{2-}$	D_{4h}	543(IR) (419) (405)	433(IR) 421(IR) 488(IR) (325)	(54)	2.6	16.67	616
$[Au(CN)_4]^{-}$	D_{4h}	462(IR) 459(R) 450(R)	415(IR)	110(R)	3.28–⎫ 3.42⎬	17.40–⎫ 17.44⎬	617
$[Hg(CN)_2]$	$D_{\infty h}$	442(IR) 412(R)	341(IR) 275(R)	(100)	2.607	17.62	618 619
$[Ag(CN)_2]^{-}$	$D_{\infty h}$	390(IR) (360)	(310) 250(R)	(107)	1.826	17.04	620
$[Au(CN)_2]^{-}$	$D_{\infty h}$	427(IR) 445(R)	(368) 305(R)	(100)	2.745	17.17	621

[a]Force constants (mdyn/Å) were obtained by using the GVF field for all ions except the $[Pt(CN)_4]^{2-}$ ion, for which the UBF field was used.
[b]Coupled vibrations between ν(MC) and δ(MCN).

Mossbauer spectroscopy with other techniques. The $\nu(C\equiv N)$ and $\nu(Fe-C)$ of crystalline $Cs_2Mg[Fe(CN)_6]$ are higher by 40 cm^{-1} than those of the $[Fe(CN)_6]^{4-}$ ion in aqueous solution.[630] The same trend is seen for crystalline $Mn_3[Co(CN)_6]\cdot xH_2O$ and the $[Co(CN)_6]^{3-}$ ion in aqueous solution.[631] These observations suggest the presence of strong interaction of the $Fe-C\equiv N\cdots Mg$ or $Co-C\equiv N\cdots Mn$ type in the solid state. The bridging $\nu(C\equiv N)$ of the $[(NC)_5Fe^{II}-CN-Co^{III}(CN)_5]^{6-}$ and $[(NC)_5Fe^{III}-CN-Co^{III}(CN)_5]^{5-}$ ions are at 2130 and 2185 cm^{-1}, respectively.[632] The infrared and Mossbauer spectra of $K_4[Fe(CN-SbX_3)_6]$ (X = F and Cl) and $K_4[Fe(CN-SbX_3)_4(CN)_2]$ (X = Cl and Br) have been studied.[633] As expected, the infrared spectrum of Prussian blue is identical to that of Turnbull's blue.[634]

Finally, partially oxidized tetracyanoplatinates such as $K_2[Pt(CN)_4]Br_{0.3}\cdot$ $3H_2O$ are known as one-dimensional (linear chain) conductors.[635] In these compounds, the planar $Pt(CN)_4^{2-}$ ions are stacked in one direction, and the $Pt\cdots Pt$ distances (2.88 Å) are much shorter than that of the parent compound, $K_2[Pt(CN)_4]\cdot 3H_2O$ (3.478 Å). The oxidation state of the Pt atom in $K_2[Pt(CN)_4]Br_{0.33}\cdot 3H_2O$ is +2.33. As a result, its $\nu(CN)$ [2182(A_{1g}), 2165(B_{1g}) cm^{-1}] are between those of $K_2[Pt^{II}(CN)_4]\cdot 3H_2O$ (2168, 2149 cm^{-1}) and $K_2[Pt^{IV}(CN)_4Cl_2]$ (2196 and 2186 cm^{-1}).[636]

(2) Nitrile and Isonitrile Complexes

Nitriles ($R-C\equiv N$, R = alkyl or phenyl) form a number of metal complexes by coordination through their N atoms. Again, $\nu(CN)$ becomes higher upon complex formation. For example, Walton[637] measured the infrared spectra of $MX_2(RCN)_2$-type compounds, where M is Pt(II) and Pd(II) and X is Cl^- and Br^-. When R is phenyl, the $\nu(CN)$ are near 2285 cm^{-1}, which is higher than the value for the $\nu(CN)$ of free benzonitrile (2231 cm^{-1}). It was noted that the $\nu(CN)$ of benzonitrile (2231 cm^{-1}) shifts to a higher frequency (2267 cm^{-1}) when it coordinates to the pentammine Ru(III) species but to a lower frequency (2188 cm^{-1}) when coordinated to the pentammine Ru(II) species. This result may indicate that the latter species has unusually strong π-back-bonding ability.[638] A strong band at 174 cm^{-1} of $ZnCl_2(CH_3CN)_2$ was suggested to be $\nu(ZnN)$.[639] The $\nu(MN)$ bands of other acetonitrile complexes have been assigned in the 450–160 cm^{-1} region.[640]

In solution, $Fe(PEt_3)_2(CO)_2(Et-C\equiv N)$ exists as a mixture of the following isomers:

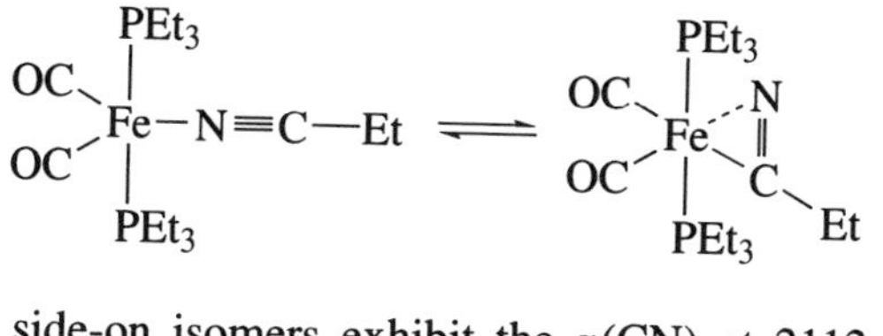

The end-on and side-on isomers exhibit the $\nu(CN)$ at 2112 and 1625 cm^{-1}, respectively.[641] The latter frequency is extremely low because of its η^2-

bonding. This type of bonding is also found in $Mo(Cp)_2(CH_3CN)$, which exhibits the $\nu(CN)$ at 1725 cm^{-1}.[642] Farona and Kraus[643] observed $\nu(CN)$ of $Mn(CO)_3(NC{-}CH_2{-}CH_2{-}CN)Cl$ at 2068 cm^{-1}, although $\nu(CN)$ of free succinonitrile (sn) is at 2257 cm^{-1}. This large shift to a lower frequency was attributed to the chelating bidentate coordination through its CN triple bonds:

According to X-ray analysis,[644] the complex ion in $[Cu(sn)_2]NO_3$ takes a polymeric chain structure in which the ligand is in the *gauche* conformation:

In these dinitrile complexes, $\nu(CN)$ are shifted to higher frequencies upon coordination. As in the case of ethylenediamine complexes (Sec. III-2), infrared spectroscopy has been used to determine the conformation of the ligand in metal complexes. The Cu(I) complex, which is known to contain the *gauche* conformation, exhibits two CH$_2$ rocking modes at 966 and 835 cm^{-1}, whereas the Ag(I) complex, $Ag(sn)_2BF_4$, shows a single CH$_2$ rocking mode at 770 cm^{-1}, which is characteristic of the *trans* conformation.[645]

There are four rotational isomers for glutaronitrile (gn), $NC{-}CH_2{-}CH_2{-}CH_2{-}CN$, which are spectroscopically distinguishable. Figure III-49 shows

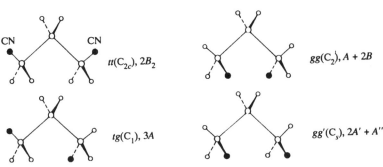

Fig. III-49. Rotational isomers of glutaronitrile.

the conformation, the symmetry, and the number of infrared-active CH_2 rocking vibrations for each isomer. According to X-ray analysis on $Cu(gn)_2NO_3$,[646] the ligand in this complex is in the *gg* conformation. The infrared spectrum of this complex is very similar to that of solid glutaronitrile in the stable form. Matsubara[647] therefore concluded that the latter also takes the *gg* conformation. However, the spectrum of solid glutaronitrile in the metastable form (produced by rapid cooling) is different from that of the *gg* conformation and it could have been *tt*, *tg*, or *gg'*. The *tt* conformation was excluded because of the absence of the 730 cm^{-1} band characteristic of the *trans*-planar methylene chain,[648] and the *gg'* conformation was considered to be improbable because of steric repulsion between two CN groups. This left only the *tg* conformation for the metastable solid. The complicated spectrum of liquid glutaronitrile was accounted for by assuming that it is a mixture of the *tg*, *gg*, and *tt* conformations. Kubota and Johnston,[649] using these results, have been able to show that the glutaronitrile molecules in $Ag(gn)_2ClO_4$ and $Cu(gn)_2ClO_4$ are in the *gg* conformation, while those in $TiCl_4gn$ and $SnCl_4gn$ have the *tt* conformation. Table III-41 summarizes the CH_2 rocking frequencies of glutaronitrile and its metal complexes. An infrared study similar to the above has been extended to adiponitrile [$NC-(CH_2)_4-CN$] and its Cu(I) complex.[650]

Cotton and Zingales[651] studied the $N{\equiv}C$ stretching bands of isonitrile complexes. When isonitriles are coordinated to zero-valent metals such as Cr(O), back donation of electrons from the metal to the ligand is extensive and the $N{\equiv}C$ stretching band is shifted to a lower frequency. For monopositive and dipositive metal ions, little or no back-donation occurs and the $N{\equiv}C$ stretching band is shifted to a higher frequency as a result of the inductive effect of the metal ion. Sacco and Cotton[652] obtained the infrared spectra of $Co(CH_3NC)_4X_2$ and $[Co(CH_3NC)_4][CoX_4]$-type compounds (X = Cl, Br, etc.). Dart et al.[653] report the $\nu(NC)$ of bis(phosphine) tris(isonitrile) complexes of Co(I). Boorman et al.[654] made rather complete assignments of vibrational spectra of some isonitrile complexes of Co(I) and Co(II) in the 4000–33 cm^{-1} region.

TABLE III-41. Infrared-Active CH_2 Rocking Frequencies of Glutaronitrile and Its Metal Complexes (cm^{-1})

Liquid[a]	945 (*tg*)	904 (*gg*)	835 (*tg*, *gg*)	757 (*tg*, *gg*)	737 (*tt*)[b]
Solid[a]					
(metastable)	943 (*tg*)	—	839 (*tg*)	757 (*tg*)	—
Solid[a] (stable)	—	903 (*gg*)	837 (*gg*)	768 (*gg*)	—
$Cu(gn)_2NO_3$[a]	—	913 (*gg*)	830 (*gg*)[c]	778 (*gg*)	—
$Cu(gn)_2ClO_4$[d]	—	908 (*gg*)	875 (*gg*)	767 (*gg*)	—
$Ag(gn)_2ClO_4$[d]	—	904 (*gg*)	872 (*gg*)	772 (*gg*)	—
$SnCl_4(gn)$[d]	—	—	—	—	733 (*tt*)
$TiCl_4(gn)$[d]	—	—	—	—	730 (*tt*)

[a]Reference 647.
[b]The *tt* form should exhibit two infrared-active CH_2 rocking vibrations. The other one is not known, however.
[c]Overlapped with a NO_3^- absorption.
[d]Reference 649.

III-16. THIOCYANATO AND OTHER PSEUDOHALOGENO COMPLEXES

The CN^-, OCN^-, SCN^-, $SeCN^-$, CNO^-, and N_3^- ions are called pseudohalide ions, since they resemble halide ions in their chemical properties. These ions may coordinate to a metal through either one of the end atoms. As a result, the following linkage isomers are possible:

M—CN, cyano complex   M—NC, isocyano complex
M—OCN, cyanato complex   M—NCO, isocyanato complex
M—SCN, thiocyanato complex   M—NCS, isothiocyanato complex
M—SeCN, selenocyanato complex   M—NCSe, isoselenocyanato complex
M—CNO, fulminato complex   M—ONC, isofulminato complex

Two compounds are called true linkage isomers if they have exactly the same composition and two of the different linkages mentioned above. A well-known example is nitro (and nitrito) pentammine Co(III) chloride, discussed in Sec. III-6. A pair of true linkage isomers is difficult to obtain since, in general, one form is much more stable than the other. As will be shown later, a number of new linkage isomers have been isolated, and infrared spectroscopy has proved to be very useful in distinguishing them. Burmeister[655] reviewed linkage isomerism in metal complexes. Bailey et al.[656] and Norbury[657] reviewed the infrared spectra of SCN, SeCN, NCO, and CNO complexes and their linkage isomers in detail.

(1) Thiocyanato (SCN) Complexes

The SCN group may coordinate to a metal through the nitrogen or the sulfur or both (M—NCS—M′). In general, Class A metals (first transition series, such as Cr, Mn, Fe, Co, Ni, Cu, and Zn) form M–N bonds, whereas Class B metals (second half of the second and third transition series, such as Rh, Pd, Ag, Cd, Ir, Pt, Au, and Hg) form M–S bonds.[658] However, other factors, such as the oxidation state of the metal, the nature of other ligands in a complex, and steric consideration, also influence the mode of coordination.

Several empirical criteria have been developed to determine the bonding type of the NCS group in metal complexes.

1. The CN stretching frequencies are generally lower in N-bonded complexes (near and below 2050 cm^{-1}) than in S-bonded complexes (near 2100 cm^{-1}).[659] The bridging (M—NCS—M′) complexes exhibit $\nu(CN)$ well above 2100 cm^{-1}. However, this rule must be applied with caution since $\nu(CN)$ are affected by many other factors.[656]

2. Several workers[660–662] considered $\nu(CS)$ as a structural diagnosis: 860–780 cm^{-1} for N-bonded, and 720–690 cm^{-1} for S-bonded, complexes. However, this band is rather weak and is often obscured by the presence of other bands in the same region.

3. It was suggested[661,662] that the N-bonded complex exhibits a single sharp δ(NCS) near 480 cm^{-1}, whereas the S-bonded complex shows several bands of low intensity near 420 cm^{-1}. However, these bands are also weak and tend to be obscured by other bands.

4. Several workers[663–666] used the integrated intensity of ν(CN) as a criterion; it is larger than 9×10^4 M^{-1} cm^{-2} per NCS$^-$ for N-bonded complexes, and close to or smaller than 2×10^4 M^{-1} cm^{-2} for S-bonded complexes. However, this rule is also difficult to apply when the spectrum consists of multiple components or when the dissociation occurs in solution.

5. Some workers[667,668] proposed using ν(MN) and ν(MS) in the far-infrared region as a criterion; in general, ν(MN) is higher than ν(MS). However, these frequencies are very sensitive to the overall structure of the complex and the nature of the central metal. Thus extreme caution must be taken in applying this criterion.

It is clear that only a combination of these five criteria would provide reliable structural diagnosis. Table III-42 lists the vibrational frequencies of typical isothiocyanato and thiocyanato complexes. The ν(MN) and ν(MS) vibrations of some of these and other complexes were assigned by using metal isotopes[672] and ^{15}N-substituted ligands.[673] Figure III-50 shows the IR and Raman spectra of (TBA)$_3$ [Ru(NCS)$_6$] (TBA: tetrabutylammonium ion) obtained by Fricke and Preetz.[674] Bütje and Preetz[675] obtained the IR and Raman spectra of all 10 isomeric complexes of [Os(NCS)$_n$(SCN)$_{6-n}$]$^{3-/2-}$ (n = 0~6). Preetz and co-workers also studied the vibrational spectra of Re(IV) complexes involving NCS/SCN linkage isomerism.[676] In the [M(NCS)$_4$]$^{2-}$ (M = Zn, Cd, and Hg) series, aqueous Raman studies by Yamaguchi et al.[677] show that all the ligands are N-bonded in the Zn and S-bonded in the Hg complexes, but both types coexist in the Cd complex. The Raman spectra of Cd(NCS)$_2$ dissolved in DMSO suggests that the Cd atom is S-bonded in this case.[678]

Clark and Williams[667] measured the infrared spectra of tetrahedral M(NCS)$_2$L$_2$, monomeric octahedral M(NCS)$_2$L$_4$, and polymeric octahedral M(NCS)$_2$L$_2$-type complexes (M = Fe, Co, Ni, etc.; L = py, α-pic, etc.), and studied the relationship between the spectra and stereochemistry. They found that ν(CS) are higher by 40–50 cm^{-1} for tetrahedral than for octahedral complexes for the same metal, although ν(CN) are very similar for both.

The *cis*- and *trans*-isomers of [Co(en)$_2$(NCS)$_2$]Cl·H$_2$O, for example, can be distinguished by infrared spectra in the ν(CN) region: *trans*, 2136 cm^{-1}; *cis*, 2122 and 2110 cm^{-1}.[679] Lever et al.[680] have found, however, that no splittings of ν(CN) are observed at room temperature for *cis*-octahedral ML$_2$(NCS)$_2$, where M is Co(II) and Ni(II) and L is 1,2-bis-(2′-imidazolin-2′-yl)benzene. The splitting of ν(CN) of this complex was observed only at liquid-nitrogen temperature.

Turco and Pecile[660] noted that the presence of other ligands in a complex influences the mode of the NCS bonding. For example, in Pt(NCS)$_2$L$_2$, the NCS ligand is N-bonded if L is a phosphine (π-acceptor), and is S-bonded if L is an amine. In the solid state, Ni(NCS)$_2$(PMePh$_2$)$_2$ is N-bonded (*trans*) but

TABLE III-42. Vibrational Frequencies of Isothiocyanato and Thiocyanato Complexes (cm^{-1})^a

Compound	ν(CN)	ν(CS)	δ(NCS)	Ref.
K[NCS]	2053	748	486, 471	385 (Sec. II, Part A)
(NEt$_4$)$_2$[Co(—NCS)$_4$]	2062 (s)	837 (w)	481 (m)	656
K$_3$[Cr(—NCS)$_6$]	2098 (vs)	820 (vw)	474 (s)	669
	2058 (vs)			
(NEt$_4$)$_2$[Cu(—(NCS)$_4$]	2074 (s)	835 (w)	—	670
(NEt$_4$)$_3$[Fe(—NCS)$_6$]	2098 (sh)	822 (w)	479 (m)	656
	2052 (s)			
(NEt$_4$)$_4$[Ni(—NCS)$_6$]	2109 (sh)	818 (w)	469 (m)	656
	2102 (s)			
(NEt$_4$)$_2$[Zn(—NCS)$_4$]	2074 (s)	832 (w)	480 (m)	656
(NH$_4$)[Ag(—SCN)$_2$]	2101 (s)	718 (w)	453 (m)	656
	2086 (s)			
K[Au(—SCN)$_4$]	2130 (s)	700 (w)	458 (w)	656
			413 (s)	
K$_2$[Hg(—SCN)$_4$]	2134 (m)	716 (m)	461 (m)	656
	2122 (sh)	709 (sh)	448 (m)	
	2109 (s)	703 (sh)	432 (sh)	
			419 (m)	
(NBu$_4$)$_3$[Ir(—SCN)$_6$]	2127 (m)	822 (m)	430 (w)	671
	2098 (s)	693 (w)		
K$_2$[Pd(—SCN)$_4$]	2125 (s)	703 (w)	474 (w)	656
	2095 (s)	697 (sh)	467 (w)	
			442 (m)	
			432 (m)	
K$_2$[Pt(—SCN)$_4$]	2128 (s)	696 (w)	477 (w)	656
	2099 (s)		469 (w)	
	2077 (sh)		437 (m)	
			426 (m)	

^{a}vs = very strong; s = strong; m = medium; w = weak; sh = shoulder.

its Pd analog is S-bonded (*trans*), and the Pt analog is N-bonded(*cis*).[681] For [Cr(NCS)$_4$L$_2$]$^{n-}$ ions, Contreras and Schmidt[682] proposed, based on the ν(CN) and ν(CS) of these ions, N-bonding for L = urea, glycinate ion, and so on, and S-bonding for L = thiourea, acetamide, and so on. These results have been explained in terms of the steric and electronic effects of L.

Since 1963, a variety of true linkage isomers involving the NCS group have been prepared. Table III-43 lists the ν(CN) and ν(CS) of typical pairs of these linkage isomers. Epps and Marzilli[690] isolated three linkage isomers of AsPh$_4$[Co(DMG)$_2$(NCS)$_2$]:

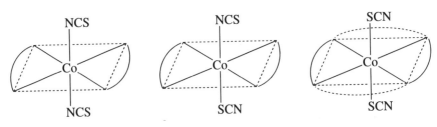

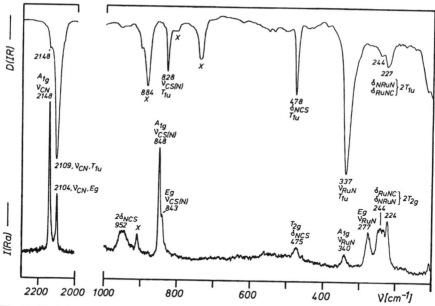

Fig. III-50. Infrared and Raman spectra of $(TBA)_3[Ru(NCS)_6]$. The symbol $\times$ indicates the TBA (tetrabutylammonium ion) band (reproduced with permission from Ref. 674).

Although all these isomers exhibit $\nu(CN)$ at 2110 cm^{-1}, they can be distinguished by the differences in the intensity of the $\nu(CN)$ band; the (NCS, NCS) isomer is the strongest, the (SCN, SCN) isomer is the weakest, and the (NCS, SCN) isomer is in between.

Both N-bonded and S-bonded NCS groups have been found in [Pd(4,4′-dimethyl-bipy)(NCS)(SCN)][691], [Pd{Ph_2P(CH_2)_3NMe_2}(NCS)(SCN)],[692] and similar Pd complexes.[693,694] Similar mixed NCS–SCN bonding was found for [PdL(NCS)(SCN)], where L is $Ph_2P(o-C_6H_4)AsPh_2$ and $Ph_2P(CH_2)_2NMe_2$.[695] These bidentate ligands contain two different donor atoms which give differ-

TABLE III-43. Vibrational Frequencies of True Linkage Isomers Involving the NCS Group (cm^{-1})

Compound	Type	$\nu(CN)$	$\nu(CS)$	Ref.
trans-[Pd(AsPh_3)_2(NCS)_2]	N-bonded	2089	854	683, 684, 685
	S-bonded	2119	—	
Pd(bipy)(NCS)_2	N-bonded	2100	842	686, 684
	S-bonded	2117	700	
		2108		
$(\pi$-Cp)Mo(CO)_3(NCS)	N-bonded	2099	—	687
	S-bonded	2114	699	
K_3[Co(CN)_5(NCS)]	N-bonded	2065	810	688
	S-bonded	2110	718	
trans-[Co(DMG)_2(py)(NCS)]	N-bonded	2128	—	689
	S-bonded	2118	—	

ent electronic effects on the NCS groups *trans* to them. Thus, the *trans* effect, together with the steric effect of these ligands, may be responsible for the mixing of the N- and S-bonding. Using the $\nu(CS)$ as a marker, Coyer et al.[696] have shown that the yellow isomer of Pt(bipy)(SCN)$_2$ containing two, *cis*, S-bonded ligands is converted into the red isomer with two, *cis*, N-bonded ligands. This "flip" can occur by heating in solution or in the solid state. In the case of [Ni(DPEA)(NCS)$_2$]$_2$[DPEA: di(2-pyridyl-β-ethyl)amine], IR spectra suggest that terminal N-bonded and bridging NCS groups are mixed [$\nu(CN)$: 2094 and 2128 cm^{-1}, respectively].[697]

Burmeister et al.[698] found that in the ML$_2$X$_2$-type complexes [M = Pd(II), Pt(II); L = a neutral ligand; X = SCN, SeCN, NCO, etc.], the mode of bonding of X to the metal is determined by the nature of the solvent. For example, Pd(AsPh$_3$)$_2$(NCS)$_2$ is N-bonded in pyridine and acetone solution, whereas it is S-bonded in DMF and DMSO solution. However, the bonding of the NCO group is insensitive to the nature of the solvent.

The NCS group also forms a bridge between two metal atoms. The CN stretching frequency of a bridging group is generally higher than that of a terminal group. For example, HgCo(NCS)$_4$(Co—NCS—Hg) absorbs at 2137 cm^{-1}, whereas (NEt$_4$)$_2$[Co(—NCS)$_4$] absorbs at 2065 cm^{-1}. According to Chatt and Duncanson,[699] the CN stretching frequencies of Pt(II) complexes are 2182–2150 cm^{-1} for the bridging and 2120–2100 cm^{-1} for the terminal NCS group. [(P(n-Pr)$_3$)$_2$Pt$_2$(SCN)$_2$Cl$_2$] (compound I) exhibits one bridging CN stretching, whereas [(P(n-Pr)$_3$)$_2$Pt$_2$(SCN)$_4$] (compound II) exhibits both bridging and terminal CN stretching bands. Thus the IR spectra suggest that the structure of each compound is as follows:

I　　　　　　　　　　　II

n-Pr: *n*-Propyl

Compound I, however, exists as two isomers, α and β, which absorb at 2162 and 2169 cm^{-1}, respectively. Chatt and Duncanson[699] originally suggested a geometrical isomerism in which two SCN groups were in a *cis* or *trans* position with respect to the central ring. Later,[700-702] "bridge isomerism" such as the following was demonstrated by X-ray analysis:

α　　　　　　　　　　　β

The IR spectra of metal complexes containing bridging NCS groups have been reported for $Sn(NCS)_2$,[703] $M(py)_2(NCS)_2[M = Mn(II), Co(II), and Ni(II)]$,[704] $[Me_3Pt(NCS)]_4$,[705] and $M[Pt(SCN)_6]$ [M = Co(II), Ni(II), Fe(II), etc.].[706]

According to X-ray analysis,[707] the $[Re_2(NCS)_{10}]^{3-}$ ion contains solely N-bonded bridging thiocyanate groups which exhibit $\nu(CN)$ near 1900 cm^{-1}.

(All terminal NCS groups are also N-bonded.)

(2) Selenocyanato (SeCN) Complexes

The SeCN group also coordinates to a metal through the nitrogen (M—NCSe) or the selenium (M—SeCN) or both (M—NCSe—M′). Again, Class A metals tend to form M–N bonds, while Class B metals prefer to form M–Se bonds. Although the number of SeCN complexes studied is much smaller than that of SCN complexes, these studies suggest the following trends:

1. $\nu(CN)$ is below 2080 cm^{-1} for N-bonded, but higher for Se-bonded, complexes. The $\nu(CN)$ of a bridged complex $[HgCo(NCSe)_4]$ is at 2146 cm^{-1}.[708]

2. The $\nu(CSe)$ is at 700–620 cm^{-1} for N-bonded and 550–500 cm^{-1} for Se-bonded complexes.

3. The $\delta(NCSe)$ of N-bonded complexes are above 400 cm^{-1}, whereas Se-bonded complexes show at least one component of $\delta(NCSe)$ below 400 cm^{-1}.

4. The integrated intensity of $\nu(CN)$ is larger for the N-bonded than for the Se-bonded group.[709]

Table III-44 lists the observed frequencies of typical N-bonded and Se-bonded complexes.

Burmeister and Gysling[714] observed that in $[PdL_2(SeCN)_2]$-type compounds the effect of changing the π-bonding ability and basicity of L on the Pd–SeCN bonding is negligible in contrast to the analogous SCN complexes. A pair of true linkage isomers has been isolated and characterized by infrared spectra for $[(\pi\text{-Cp})Fe(CO)(PPh_3)(SeCN)]$[715] and $[Pd(Et_4dien)(SeCN)]BPh_4$,[716] where Et_4dien is 1,1,7,7-tetraethyldiethylenetriamine. Vibrational spectra of osmium complexes containing —NCSe and —SeCN ligands have been reported by Preetz and co-workers.[717,718] The IR spectra of $[Ru(NH_3)_5(NCSe)]I_2 \cdot 2H_2O$ and its true linkage isomer, $[Ru(NH_3)_5(SeCN)]I_2 \cdot 2H_2O$, are reported.[719]

(3) Cyanato (OCN) Complexes

The OCN group may coordinate to a metal through the nitrogen (M—NCO) or the oxygen (M—OCN) or both. Thus far, the majority of complexes are

TABLE III-44. Vibrational Frequencies of Isoselenocyanato and Selenocyanato Complexes (cm^{-1})

Compound^a	ν(CN)	ν(CSe)	δ(NCSe)	Ref.
K[NCSe]	2070	558	424, 416	386 (Sec. II, Part A)
R$_4$[Mn($-$NCSe)$_6$]	2079, 2082 ⎱ 2070 ⎰	640 ⎱ 617 ⎰	424	709
R$_2$[Fe($-$NCSe)$_4$]	2067, 2055	673, 666	432	709
R$_4$[Ni($-$NCSe)$_6$]	2118, 2102	625	430	709
[Ni(pn)$_2$($-$NCSe)$_2$]	2096, 2083	692	—	710
R$_2'$[Co($-$NCSe)$_4$]	2053	672	433, 417	711
[Co(NH$_3$)$_5$($-$NCSe)](NO$_3$)$_2$	2116	624	—	712
R$_2$[Zn($-$NCSe)$_4$]	2087	661	429	709
[Cu(pn)$_2$($-$SeCN)$_2$]	2053, 2028	—	—	710
R$_3$[Rh($-$SeCN)$_6$]	2104, 2071	515	—	709
R$_2''$[Pd($-$SeCN)$_4$]	2114, 2105	521	410, 374	713
R$_2$[Pt($-$SeCN)$_4$]	2105, 2060	516	—	709
[Pt(bipy)($-$SeCN)$_2$]	2135, 2125	532, 527	—	712
K$_2$[Pt($-$SeCN)$_6$]	2130	519	390, 379 ⎱ 367 ⎰	711

^{a}R = [N(n-C$_4$H$_9$)$_4$]'; R' = [N(C$_2$H$_5$)$_4$]'; R'' = [N(CH$_3$)$_4$]'; pn = propylenediamine; bipy = 2,2'-bipyridine.

reported to be N-bonded. Table III-45 lists the observed frequencies of N-bonded NCO groups in typical complexes; ν_a(NCO) and ν_s(NCO) denote vibrations consisting mainly of ν(CN) and ν(CO), respectively. For ML$_2$(NCO)$_2$ (M = Pd or Pt; L = NH$_3$, py, etc.) and In(NCO)$_3$L$_3$ (L = py, DMSO, etc.), see Refs. 728 and 729, respectively. Forster and Horrocks[721] carried out normal coordinate analyses on Zn(NCX)$_4^{2-}$ (X = O, S, or Se). Complete vibrational assignments are available for the IR spectra of Zn(NCO)$_2$L$_2$, where L is NH$_3$ or pyridine.[730]

Thus far, O-bonded structures have been suggested for [M(OCN)$_6$]$^{n-}$ [M = Mo(III), Re(IV), and Re(V)].[726] Anderson and Norbury[731] prepared the first example of linkage isomers: yellow Rh(PPH$_3$)$_3$(NCO) and orange Rh(PPh$_3$)$_3$(OCN). The integrated ν(CN) intensity of the former is smaller than that of the free ion, whereas the intensity of the latter is larger than that of the free ion. Also, the latter exhibits two δ(OCN) at 607 and 590 cm^{-1}, whereas the former shows only one band at 592 cm^{-1}. An electron diffraction study shows that the previously reported structure of F$_5$Se$-$NCO is not correct; it is F$_5$Se$-$OCN with the ν_2(NCO) and ν_s(NCO) occurring at 2290 and 1104 cm^{-1}, respectively.[732]

Bridging NCO groups may take one of the following structures:

I   II   III

TABLE III-45. Vibrational Frequencies of Isocyanato Complexes (cm^{-1})

Compound	$\nu_a(NCO)^b$	$\nu_s(NCO)^b$	$\delta(NCO)$	Refs.
K[NCO]	2155	1282, 1202	630	382 (Sec. II, Part A)
Si(NCO)$_4$	2284	1482	608, 546	720
Ge(NCO)$_4$	2247	1426	608, 528	720
[Zn(NCO)$_4$]$^{2-}$	2208	1326	624	721
[Mn(NCO)$_4$]$^{2-}$	2222	1335	623	722, 723
[Fe(NCO)$_4$]$^{2-}$	2182	1337	619	722, 723
[Co(NCO)$_4$]$^{2-}$	2217 ⎱ 2179 ⎰	1325	620, 617	722, 723
[Ni(NCO)$_4$]$^{2-}$	2237 ⎱ 2186 ⎰	1330	619, 617	722
[Fe(NCO)$_4$]$^{-}$	2208 ⎱ 2171 ⎰	1370	626, 619	722
[Pd(NCO)$_4$]$^{2-}$	2200– ⎱ 2190 ⎰	1319	613, 604 ⎱ 594 ⎰	724
[Sn(NCO)$_6$]$^{2-}$	2270 ⎱ 2188 ⎰	1307	667, 622	724
[Zr(NCO)$_6$]$^{2-}$	2205	1340	628	725
[Mo(OCN)$_6$]$^{3-}$	2205	1296 ⎱ 1140 ⎰	595	726
[Ln(NCO)$_6$]$^{3-,a}$	2190	1333	633	727

^{a}Ln = Yb or Lu.
^{b}The notations ν_a and ν_b are used because the separation of the $\nu(NC)$ and $\nu(CO)$ is not distinct.

Structure I has been proposed for ML$_2$(NCO)$_2$ (M = Mn, Fe, Co, and Ni, L = 3- or 4-CN-py)[733] and Re$_2$(CO)$_8$(NCO)$_2$.[734] Thus far, II is not known. Structure III has been proposed for [(NH$_3$)$_5$Cr(NCO)Cr(NH$_3$)$_5$]Cl$_5$. It exhibits the $\nu_a(NCO)$, $\nu_s(NCO)$, $\delta(NCO)$, $\nu(Cr—NCO)$, and $\nu(Cr—OCN)$ at 2248, 1315, 605, 350, and 303 cm^{-1}, respectively.[735]

(4) Fulminato (CNO) Complexes

The fulminato (CNO$^-$) ion may coordinate to a metal through the carbon (M—CNO), the oxygen (M—ONC), or both as a bridging ligand. As stated in Sec. II-5(2) of Part A, fulminic acid (HCNO) is linear, whereas isofulminic acid (HONC) is bent. The same trend may hold for their metal complexes. Thus far, all the complexes containing the CNO group are presumed to be C-bonded. Beck and co-workers have carried out an extensive vibrational study on metal fulminato complexes. Table III-46 lists the vibrational frequencies of typical complexes obtained by these workers. A more complete listing is found in a review by Beck.[742]

Beck and Fehlhammer observed rapid isomerization:

$$\text{NEt}_4[\text{W(CO)}_5(\text{CNO})] \xrightarrow[\text{in CH}_2\text{Cl}_2]{} \text{NEt}_4[\text{W(CO)}_5(\text{NCO})]$$

TABLE III-46. Observed Frequencies of Typical Fulminato Complexes (cm^{-1})

Ion	ν(CN)	ν(NO)	δ(CNO)	Ref.
[CNO]$^-$	2052	1057	471	736
[Ag(CNO)$_2$]$^-$	2119	1144	—	737
[Au(CNO)$_2$]$^-$	2173	1180	—	737
[Fe(CNO)$_6$]$^{4-}$	2187	1040	514	737
			466	
[Hg(CNO)$_4$]$^{2-}$	2130	1143	—	738
[Ni(CNO)$_4$]$^{2-}$	2184	1122	479	739
			470	
[Zn(CNO)$_4$]$^{2-}$,^{a}	2146 (A_1)	1177 (A_1)	498 (E)	740
	2130 (F_2)	1154 (F_2)	475 (F_2)	
[Pt(CNO)$_4$]$^{2-}$,^{b}	2194 (A_{1g})	1174 (A_{1g})	476 (B_{2g})	740
	2189 (B_{1g})	1140 (B_{1g})	453 (E_g)	
Pt(PPh$_3$)$_2$ (CNO)$_2$	2183	1171	—	741

^{a}**T**$_d$ symmetry.
^{b}**D**$_{4h}$ symmetry.

The fulminato complex shows the 2ν(NO), ν(CN), and ν(NO) at 2190, 2110, and 1087 cm^{-1}, respectively, while its isocyanato isomer exhibits the ν_a and ν_s at 2235 and 1318 cm^{-1}, respectively.[743]

(5)   Azido (N$_3$) Complexes

Table III-47 lists the observed frequencies of typical azido complexes. The two N$_3$ groups around the Hg atom in Hg$_2$(N$_3$)$_2$ are in the *trans* position (**C**$_{2h}$), whereas they are in a twisted configuration (**C**$_2$) in Hg(N$_3$)$_2$. The former exhibits one ν_a(N$_3$) at 2080 cm^{-1}, whereas the latter shows two ν_a(N$_3$) at 2090 and 2045 cm^{-1}.[749] For Co(III) azido ammine complexes and [M(N$_3$)$_2$(py)$_2$] (M = Cu, Zn, and Cd), see Refs. 750 and 751, respectively. Forster and Horrocks[746] made complete assignments of vibrational spectra of the [Co(N$_3$)$_4$]$^{2-}$ and [Zn(N$_3$)$_4$]$^{2-}$ (**D**$_{2d}$) and [Sn(N$_3$)$_6$]$^{2-}$ (**D**$_{3d}$) ions. The spectra suggest that the M–NNN bonds in these anions are not linear.

The bridging azido groups are found in

$$[(PPh_3)(N_3)Pd\underset{N_3}{\overset{N_3}{\diagup\diagdown}}Pd(N_3)(PPh_3)]^{752}$$

and

$$[(acac)_2Co\underset{N_3}{\overset{N_3}{\diagup\diagdown}}Co(acac)_2]^{753}$$

TABLE III-47. Vibrational Frequencies of Azido Complexes $(cm^{-1})^a$

Compound[a]	ν_a(NNN)	ν_s(NNN)	δ(NNN)	ν(MN)	Refs.
K[N$_3$]	2041	1344	645	—	383 (Sec. II, Part A)
R$_2$[Pt(N$_3$)$_4$]	2075, 2060 2024, 2029	1276	582	394	744, 745
R[Au(N$_3$)$_4$]	2030, 2034	1261 1251	578	432	744, 745
R″$_2$[Zn(N$_3$)$_4$]	2097, 2058	1330 1282	—	—	744, 745
R$_2$[VO(N$_3$)$_4$]	2088, 2051 2092, 2060 2005	1340	652	442 405	744, 745
R$_2$[Pd(N$_3$)$_6$]	2045, 2056 2037	1262 1253	640 597	327 313	744, 745
R$_2$[Pt(N$_3$)$_6$]	2022, 2028	1275 1262 1253	578	402 397 320	744, 745
R′$_2$[Co(N$_3$)$_4$]	2089, 2050	1338 1280	642 610	368	746
R″$_2$[Mn(N$_3$)$_4$]	2058	1330 1267	650 630	317 288	747
R′$_2$[Sn(N$_3$)$_4$]	2115, 2080	1340	659 601	390 330	746
trans-R$_2$[TiCl$_4$(N$_3$)$_2$]	2072, 2060	1344	610	—	748

^{a}R = [As(Ph)$_4$]$^+$; R′ = [N(C$_2$H$_5$)$_4$]$^+$; R″ = [P(Ph)$_4$]$^+$.

The azido bridge can be either the end-on or end-to-end type shown below:

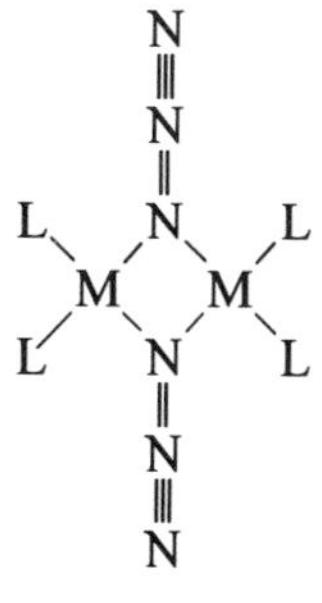

End-on (μ-1,1) 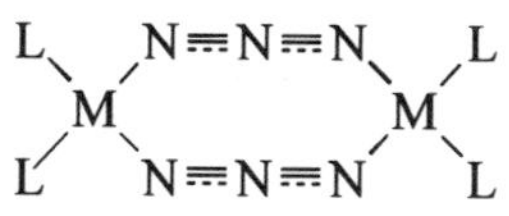   End-to-end (μ-1,3)

Distinction of these two types may be made by using an isotopically scrambled ligand such as ^{14}N—^{14}N—^{15}N since one expects three and two isotopomers for the end-on and end-to-end complexes, respectively.

X-Ray analysis[754] shows that the azido bridges in $(\mu$-N$_3)_2$NiL$_2$(ClO$_4)_2$ take the end-on structure with the ν_a(N$_3$) at ~2050 cm^{-1}. Here, L is 2,4,4-

trimethyl-1,5,9-triazacyclododeca-1-ene. Both terminal and bridging (end-on) azido groups are present in $[Rh_3(\mu\text{-dpmp})_2(CO)_3(\mu\text{-}N_3)N_3]$ (BPh$_4$). Here, dpmp is bis((diphenylphosphino)-methyl)phenylphosphine. The $\nu_a(N_3)$ of the former is at 2035 cm^{-1}, whereas that of the latter is at 2085 cm$^{-1,755}$.

III-17. COMPLEXES OF CARBON MONOXIDE AND CARBON DIOXIDE

In the last few decades, a large number of carbonyl complexes have been synthesized, and their spectra and structures have been studied exhaustively. This section describes only typical results obtained from these investigations. For more comprehensive information, several review articles[756–762] should be consulted.

Most carbonyl complexes exhibit strong and sharp $\nu(CO)$ bands at ca. 2100–1800 cm^{-1}. Since $\nu(CO)$ is generally free from coupling with other modes and is not obscured by the presence of other vibrations, studies of $\nu(CO)$ alone often provide valuable information about the structure and bonding of carbonyl complexes. In the majority of compounds, $\nu(CO)$ of free CO (2155 cm^{-1}) is shifted to lower frequencies. In terms of simple MO theory, this observation has been explained as follows. First, the σ-bond is formed by donating 5σ electrons of CO to the empty orbital of the metal (see Fig. III-51). This tends to raise $\nu(CO)$, since the 5σ orbital is slightly antibonding. Second, the π-bond is formed by back-donating the $d\pi$-electrons of the metal to an empty antibonding orbital, the $2p\pi^*$ orbital of CO. This tends to lower $\nu(CO)$. Although these two components of bonding are synergic, the net result is a drift of electrons from the metal to CO when the metal is in a relatively low oxidation state. Thus the $\nu(CO)$ of metal carbonyl complexes are generally lower than the value for free CO. The opposite trend is observed, however, when CO is complexed with metal halides in which the metals are in a relatively higher oxidation state [see Sec. III-17(6)].

If CO forms a bridge between two metals, its $\nu(CO)$ (1900–1800 cm^{-1})

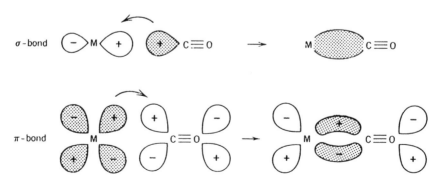

Fig. III-51. The σ- and π-bonding in metal carbonyls.

is much lower than that of the terminal CO group (2100–2000 cm^{-1}). An extremely low $\nu(CO)$ (ca. 1300 cm^{-1}) is observed when the bridging CO group forms an adduct via its O atom [see Sec. III-17(2)].[763]

(1) Mononuclear Carbonyls

Table III-48 lists the observed frequencies and band assignments of mononuclear carbonyls of tetrahedral (T_d), trigonal-bipyramidal (D_{3h}), and octahedral (O_h) structures. Complete normal coordinate analyses have been made on most of these carbonyls. Jones and co-workers[768,769] carried out extensive vibrational studies on Fe(CO)$_5$ and M(CO)$_6$ (M = Cr, Mo, and W), including their ^{13}C and ^{18}O species. They obtained the following F_{1u} force constants (mdyn/Å) from gas-phase spectra:

	Cr(CO)$_6$	Mo(CO)$_6$	W(CO)$_6$	
$F(C\equiv O)$	17.22	17.39	17.21	(GVF)
$F(M-C)$	1.64	1.43	1.80	

This result indicates that the M–C bond strength increases in the order Mo < Cr < W, an order also supported by a Raman intensity study of these compounds.[772] On the other hand, Hendra and Qurashi[773] related the Raman intensity ratio of two A_{1g} modes, $I(\nu_1$, CO stretching)$/I(\nu_2$, MC stretching), to the π-character of the M–C bond, and concluded that the M–C bond strength increases in the order Cr < W < Mo < Re(I). Kettle et al.[774] noted that the relative intensities of Raman bands of these metal carbonyls are anomalous. For example, the intensity ratios, $I(A_{1g})/I(E_g)$, are only about 0.15 for the M(CO)$_6$ series mentioned above. The origins of these anomalies have been discussed by these workers.[775,776] Raman spectra of hexacarbonyls in the vapor phase[777] and in Ar matrices[778] have been reported. Recently, Adelman and Gerrity[779] measured the UVRR spectra of Cr(CO)$_6$ and W(CO)$_6$ (cyclohexane solution) in the regions of the two lowest allowed ($A_{1g} \rightarrow F_{1u}$) CT transitions (excitation at 355 to 253 nm), and obtained evidence for Jahn–Teller distortion in their excited states.

In general, the $\nu(CO)$ is lowered as the negative charge on the metal carbonyl increases. For example, the $\nu(CO)$ (average value, see Table III-48) is in the order:

$$
\begin{array}{ccccc}
\text{Ni(CO)}_4 & & [\text{Co(CO)}_4]^- & & [\text{Fe(CO)}_4]^{2-} \\
2094 & > & 1946 & > & 1788 \text{ (cm}^{-1})
\end{array}
$$

In terms of the bonding scheme mentioned earlier, this result indicates that the metal-CO π-back bonding increases as the negative charge increases. The $\nu(CO)$ of the [Nb(CO)$_6$]$^-$ ion are at 2019 (A_{1g}), 1878–1887 (E_g), and 1860 cm^{-1} (F_{1u}).[780] The $\nu(CO)$ of the [Co(CO)$_3$]$^-$ ion is reported to be as low as

TABLE III-48. Vibrational Frequencies and Band Assignments of Mononuclear Metal Carbonyls (cm^{-1})[a]

Compound	Symmetry	State	$\nu(CO)$	$\nu(MC)$	$\delta(MCO)$	$\delta(CMC)$	Ref.
$Ni(CO)_4$	T_d	Gas	2131 (A_1) 2057.6 (F_2)	367.5 (A_1) 421 (F_2)	380 (E) 458.8 (F_2) 300 (F_1)	64 (E) 80 (F_2)	764
$[Co(CO)_4]^-$	T_d	DMF sol'n	2002 (A_1) 2890 (F_2)	431 (A_1) 556 (F_2)	523 (F_2)	91 (E)	765
$[Fe(CO)_4]^{2-}$	T_d	Aqueous sol'n	1788 (A_1) 1788 (F_2)	464 (A_1) 644 (F_2)	550 (F_2) 785 (E)	100–85 (E, F_2)	766
$Fe(CO)_5$	D_{3h}	Liquid	2116 (A_1') 2030 (A_1') 1989 (E')	418 (A_1') 381 (A_1') 482 (E')	278 (A_2') 653 (E') 559 (E') 491 (E'') 448 (E'')	107 (E') 64 (E')	767 768
$Cr(CO)_6$	O_h	Gas	2118.7 (A_{1g}) 2026.7 (E_g) 2000.4 (F_{1u})	379.2 (A_{1g}) 390.6 (E_g) 440.5 (F_{1u})	364.1 (F_{1g}) 668.1 (F_{1u}) 532.1 (F_{2g}) 510.9 (F_{2u})	97.2 (F_{1u}) 89.7 (F_{2g}) 67.9 (F_{2u})	769
$Mo(CO)_6$	O_h	Gas	2120.7 (A_{1g}) 2024.8 (E_g) 2000.3 (F_{1u})	391.2 (A_{1g}) 381 (E_g) 367.2 (F_{1u})	341.6 (F_{1g}) 595.6 (F_{1u}) 477.4 (F_{2g}) 507.2 (F_{2u})	81.6 (F_{1u}) 79.2 (F_{2g}) 60 (F_{2u})	769

Compound	Point group	Phase	CO stretch	M–C stretch	Deformation	Low frequency	Ref.
W(CO)$_6$	O_h	Gas	2126.2 (A_{1g}) 2021.1 (E_g) 1997.6 (F_{1u})	426 (A_{1g}) 410 (E_g) 374.4 (F_{1u})	361.6 (F_{1g}) 586.6 (F_{1u}) 482.0 (F_{2g}) 521.3 (F_{2u})	82.0 (F_{1u}) 81.4 (F_{2g}) 61.4 (F_{2u})	769
[V(CO)$_6$]$^-$	O_h	CH$_3$CN sol'n	2020 (A_{1g}) 1894 (E_g) 1858 (F_{1u})	374 (A_{1g}) 393 (E_g) 460 (F_{1u})	356 (F_{1g}) 650 (F_{1u}) 517 (F_{2g}) 506 (F_{2u})	92 (F_{1u}) 84 (F_{2g})	770
[Re(CO)$_6$]$^+$	O_h	CH$_3$CN sol'n	2197 (A_{1g}) 2122 (E_g) 2085 (F_{1u})	441 (A_{1g}) 426 (E_g) 356 (F_{1u})	354 (F_{1g}) 584 (F_{1u}) 486 (F_{2g}) 522 (F_{2u})	82 (F_{1u}) 82 (F_{2g})	770
[Mn(CO)$_6$]$^+$	O_h	CH$_3$CN Sol'n (IR) Solid (R)	2192 (A_{1g}) 2125 (E_g) 2095 (F_{1u})	384 (A_{1g}) 390 (E_g) 412 (F_{1u})	347 (F_{1g}) 636 (F_{1u}) 500 (F_{2g}) 500 (F_{2u})	101 (F_{1u}) 101 (F_{2g})	771

[a]The three low-frequency vibrations may be coupled.

~1610 cm^{-1} (a shoulder at 1740 cm^{-1}).[781] Highly reduced species such as Na$_4$[M(CO)$_4$] (M = Cr, Mo, and W of formal oxidation state, −IV) exhibit very low ν(CO) at 1530–1460 cm^{-1}.[782]

Conversely, as the positive charge of the complex increases, the ν(CO) is shifted to higher frequency because the σ-bonding becomes more predominant. Thus, metal carbonyl cations such as [Hg(CO)$_2$]$^{2+}$ (ν_s, 2281 and ν_a, 2278 cm^{-1})[783] and [Au(CO)$_2$]$^+$ (ν_s, 2246.0 and ν_a, 2210.5 cm^{-1})[784] show extremely high ν(CO). The [Pt(CO)$_4$]$^{2+}$ ion exhibits three ν(CO) at 2281 (A_{1g}), 2257 (B_{2g}), and 2235 cm^{-1} (E_u) which are much higher than those of neutral Pt(CO)$_4$ (2067 cm^{-1}).[784a]

Edgell and co-workers[785] attributed the band at 413 cm^{-1} of Li[Co(CO)$_4$] in a THF solution to the vibrations of the alkali ions, which form ion pairs with [Co(CO)$_4$]$^-$: For sodium and potassium salts, the corresponding bands are observed at 192 and 142 cm^{-1}, respectively. Based on computer-aided curve analysis of IR spectra of Na[Co(CO)$_4$] in the ν(CO) region, they[786] also demonstrated that there are three kinds of ion sites in THF solution each of which exhibits different spectra. Their structures and ν(CO) are shown in Fig. III-52.

(2) Polynuclear Carbonyls

Since polynuclear carbonyls take a variety of structures, elucidation of their structures by vibrational spectroscopy has been a subject of considerable interest in the past. The principles involved in these structure determinations were described in Sec. I-11 of Part A. However, the structures of some polynuclear complexes are too complicated to allow elucidation by simple application of selection rules based on symmetry. Thus the results are often ambiguous. In these cases, one must resort to X-ray analysis to obtain definitive and accurate structural information. However, vibrational spectroscopy is still useful in elucidating the structures of metal carbonyls in solution.

According to X-ray analysis,[787] Co$_2$(CO)$_8$ takes structure III of Fig. III-53. For this C_{2v} structure, five terminal and two bridging ν(CO) are expected to be infrared active; the former are observed at 2075, 2064, 2047, 2035, and 2028 cm^{-1}, and the latter are located at 1867 and 1859 cm^{-1}.[788] In addition to the usual two isomers (III and IV of Fig. III-53), the IR spectra of CO$_2$(CO)$_8$ in

	S	S	S			S	S		S	S	S

S Na$^+$ S A$^-$ S	S Na$^+$ A$^-$ S	S A^{-1} Na$^+$ A$^-$ S

S   S   S   S   S   S   S   S
S Na$^+$ S A$^-$ S   S Na$^+$ A$^-$ S   S A^{-1} Na$^+$ A$^-$ S
S   S   S   S   S   S   S   S

Solvent separated   Contact ion   Triple ion
ion pair   pair   pair

1887 cm^{-1} (T$_d$)   1899, 1856 cm^{-1} (C$_{3v}$)   1906, 1846 cm^{-1} (C$_{3v}$)

Fig. III-52. Structures and ν(CO) of Na[Co(CO)$_4$] at three ion sites in THF solution; A$^-$ and S indicate the anion and solvent, respectively.[786]

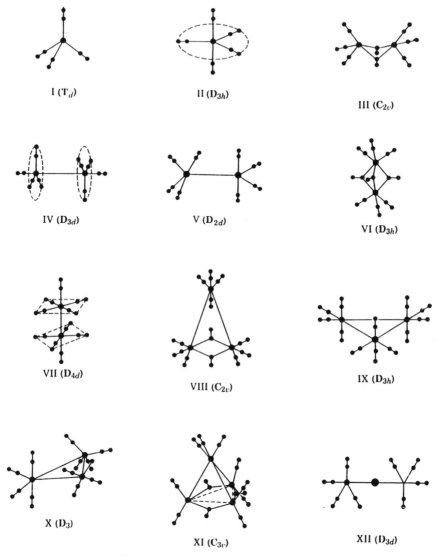

Fig. III-53. Structures of metal carbonyls.

various matrices show the presence of the third isomer (structure V) after the matrices are photolyzed.[789]

The infrared spectrum of $Fe_2(CO)_9$ was first obtained by Sheline and Pitzer,[790] who observed two terminal (2080 and 2034 cm^{-1}) and one bridging (1828 cm^{-1}) CO stretching bands. This result agrees with that expected from structure VI, determined by X-ray analysis.[791] Again according to X-ray analysis,[792] the structure of $Mn_2(CO)_{10}$ is that shown by VII of Fig. III-53. This D_{4d} structure predicts four Raman- and three infrared-active CO stretching bands. Adams et al.[793] observed the former at 2116 (A_1), 1997 (A_1), 2024 (E_2), and 1981 (E_3) cm^{-1}, and Bor[794] observed the latter at 2046 (B_2), 1984 (B_2), and 2015 (E_1) cm^{-1} in solution. Levenson et al.[795] confirmed these infrared assignments by studying the polarization properties of the three bands in a nematic liquid crystal. The structures of $Re_2(CO)_{10}$ and $Tc_2(CO)_{10}$ are similar to that of $Mn_2(CO)_{10}$. The vibrational spectra of $Re_2(CO)_{10}$, $Tc_2(CO)_{10}$, $MnTc(CO)_{10}$, and $TcRe(CO)_{10}$ are reported.[793,796–799] High-pressure IR and Raman studies show that the symmetries of $Mn_2(CO)_{10}$ and $Re_2(CO)_{10}$ change from $\mathbf{D}_{4d}$ to $\mathbf{D}_{4h}$ at pressures of 8 and 5 Kbar, respectively, owing to phase transitions.[800]

Figure III-53 (VIII) shows the structure of $Fe_3(CO)_{12}$ determined by X-ray analysis.[801] This structure can account for Mossbauer[802] and solid-state infrared spectra. In solution, the infrared spectrum does not agree with that expected for structure VIII; the bridging $\nu(CO)$ is very weak and the terminal $\nu(CO)$ region is broad without resolution. Cotton and Hunter[803] suggest that a whole range of structures varying from $\mathbf{D}_{3h}$ (structure IX) to $\mathbf{C}_{2v}$ (structure VIII) are in equilibrium, the majority being close to $\mathbf{D}_{3h}$. Johnson suggests the presence of a new isomer of $\mathbf{D}_3$ symmetry, shown by structure X.[804] According to X-ray analysis,[805] $Os_3(CO)_{12}$ takes the $\mathbf{D}_{3h}$ structure (IX), for which four terminal $\nu(CO)$ should be infrared active. Huggins et al.[806] assigned them at 2068 (E'), 2035 (A_2''), 2014 (E'), and 2002 (E') cm^{-1}. Quicksall and Spiro[807] assigned the Raman spectra of $Os_3(CO)_{12}$ and analogous $Ru_3(CO)_{12}$, for which six $\nu(CO)$ are expected in the Raman spectrum. For $Os_3(CO)_{12}$, they are observed at 2130 (A_1'), 2028 (E''), 2019 (E'), 2006 (A_1'), 2000 (E'), and 1989 (E') cm^{-1}. Vibrational spectra of solid $M_3(CO)_{12}$ (M = Ru and Os) in the $\nu(CO)$ region have been assigned based on factor group analysis.[808]

According to X-ray analysis,[809] $Co_4(CO)_{12}$ takes the $\mathbf{C}_{3v}$ structure (XI of Fig. III-53), for which six terminal and two bridging $\nu(CO)$ are infrared-active. Vibrational analyses have been made on $Co_4(CO)_{12}$ and $Rh_4(CO)_{12}$.[810] The $\nu(CO)$ frequencies are reported for $[CoRu_3(CO)_{13}]$.[811] Stammreich et al.[812] proposed structure XII of $\mathbf{D}_{3d}$ symmetry from a Raman study of $M[Co(CO)_4]_2$ (M = Cd or Hg). For this structure, three $\nu(CO)$ are Raman-active and the other three are infrared-active. The former were observed at 2107 (A_{1g}), 2030 (A_{1g}), and 1990 (E_g) cm^{-1},[812] and the latter were located at 2072 (A_{2u}), 2022 (A_{2u}), and 2007 (E_u) cm^{-1}.[813] Ziegler et al.[814] made complete vibrational assignments of the $M[Co(CO)_4]_2$ series, where M is Zn, Cd, and Hg.

Figure III-54 shows the structures of unusual bridging carbonyl compounds found by X-ray analysis. "Semibridging" carbonyls (structure I) are present in $(Cp)_2V_2(CO)_5$, and their $\nu(CO)$ have been assigned to the bands at 1871 and 1832 cm^{-1}.[815] A "semitriple bridging" carbonyl group (structure II) was found in $[(Cp)_2Rh_3(CO)_4]^-$ [816] and the band at 1693 cm^{-1} is probably due to this carbonyl. The IR band at 1662 cm^{-1} of $PtCo_2(CO)_5(\mu\text{-dppm})$ (dppm: $Ph_2P—CH_2—PPh_2$) has been assigned to a "semitriple bridging" carbonyl stretching vibration.[817] Another "semitriple bridging" carbonyl group (structure III) in $(Cp)_3Nb_3(CO)_7$ exhibits an extremely low $\nu(CO)$ at 1330 cm^{-1}.[818]

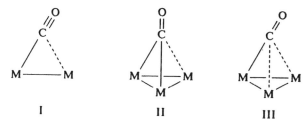

Fig. III-54. Structures of unusual bridging carbonyls.

Recently, the structures of large metal carbonyl clusters such as $[Rh_{23}N_4$-$(CO)_{38}]^{3-}$ have been determined by X-ray analysis, and characterized by IR spectroscopy.[819] According to Roth et al.,[820,821] $[Pt_{24}(CO)_{30}]^n$ exhibits six reversible one-electron redox steps ($n = 0$ to -6) in organic solvents, and each redox form gives characteristic $\nu(CO)$ (terminal and bridged) bands which are shifted to lower frequencies in a near-linear fashion as n becomes more negative. Such IR spectroelectrochemistry was found to be highly important in comparing the electronic and bonding properties of large ionizable metal clusters with those of chargeable metal surfaces.

Shriver et al.[822] found that the O atom of the bridging CO group can form a bond with a Lewis acid such as $AlEt_3$. Kristoff and Shriver[763] observed that $Co_2(CO)_8$ forms an adduct of the following type:

As expected from this structure, the adduct exhibits two bridging $\nu(CO)$ in the infrared: one at 1867 cm^{-1}, which is 15 cm^{-1} higher, and the other at 1600 cm^{-1}, which is 232 cm^{-1} lower, than that of the parent compound. In the case of $Fe_2(CO)_9AlBr_3$, only one bridging $\nu(CO)$ is observed at 1557 cm^{-1}. This suggests the following structure, which resulted from rearrangement of the CO groups of the parent compound.

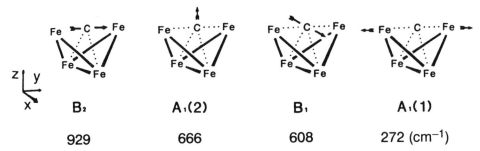

B$_2$	**A$_1$(2)**	**B$_1$**	**A$_1$(1)**
929	666	608	272 (cm^{-1})

Fig. III-55. Normal modes, symmetries, and vibrational frequencies of the iron butterfly carbide in the $[Fe_4C(CO)_{12}]^{2-}$ ion (reproduced with permission from Ref. 824).

Extensive IR studies have been made for CO adsorbed on metals, metal oxides, and catalysts, and their ν(CO) were found to be useful in elucidating the modes of CO binding on these surfaces.[823]

Metal–carbon vibrations in carbonyl carbide clusters have been assigned by several investigators. The iron butterfly carbide in the $[Fe_4C(CO)_{12}]^{2-}$ ion has an idealized symmetry of C_{2v} as shown in Fig. III-55. Using $^{12}C/^{13}C$ isotope shift data, Stanghellini et al.[824] carried out normal coordinate calculations on this skeleton. Figure III-55 illustrates the four normal modes together with their frequencies. It is seen that the first three normal modes involve the motions of the carbide carbon atom and their frequencies are between 930 and 600 cm^{-1}. The frequency of the last mode is low because it involves the motions of the Fe atoms.

For other metal carbide clusters, the metal–carbide stretching vibrations have been assigned empirically. In $[M_5C(CO)_{15}]$ (M = Ru and Os), for example, these vibrations are located in the 800–730 cm^{-1} region.[825] The C atom in $Co_6C(CO)_{12}S_2$ is located at the center of the trigonal prism formed by six Co atoms and its Co–C (carbide) vibrations were assigned at 819 and 548 cm^{-1} based on $^{12}C/^{13}C$ isotope shift data.[826] In the $[Os_{10}C(CO)_{24}]^-$ ion, the carbide C atom is at the center of the Os_{10} skeleton.[827] The $^{12}C/^{13}C$ isotope experiments show the presence of three Os–C (carbide) stretching vibrations at 772.8, 760.3, and 735.4 cm^{-1}.[828]

A carbon atom capping a three-metal array may form a CCO ligand owing to its strong affinity for CO. Sailor and Shriver[829] have demonstrated the formation of such a ligand in solid $(PPN)_2[Ru_3(CO)_6(\mu\text{-}CO)_3(\mu_3\text{-}CCO)]$ (PPN: $(PPh_3)_2N^+$ ion).

$$
\begin{matrix}
\text{O} \\
\| \\
\text{C} \quad (C_{3v}) \\
\| \\
\text{C} \\
\diagup \; | \; \diagdown \\
\text{Ru} \quad | \quad \text{Ru} \\
\text{Ru}
\end{matrix}
$$

The Raman spectra exhibits four polarized bands (A_1) which are sensitive to the $^{12}C/^{13}C$ substitution of the CCO moiety; the $\nu(CRu_3)$ (319 cm^{-1}), $\nu(C{=}C)$ (1309 cm^{-1}), and $\nu(C{=}O)$ (2024 and 1980 cm^{-1}). The last two bands are presumably due to vibrationally coupled modes between the $\nu(CCO)$ and the $\nu(CO)$ of terminal CO groups directly bonded to the Ru atoms.

For metal–metal stretching vibrations of polynuclear carbonyls, see Sec. III-24.

(3) Metal Carbonyls Containing Other Ligands

There are many metal carbonyls in which some of the CO groups are replaced by other ligands such as halogens, phosphorus derivatives, and cyclopentadienyl groups. Vibrational spectroscopy has been utilized to study the effects of these substitutions on the metal–CO bonding.

If one of the CO groups is substituted by a halogen (X), the $\nu(CO)$ tends to shift to a higher frequency since the metal–CO π-back bonding decreases as the metal becomes more electropositive by forming a M–X bond. Thus, we obtain a series such as

$Pt(CO)_4$		$[Pt(CO)Cl_3]^-$		cis-$[Pt(CO)_2Cl_2]$		$[Pt(CO)Cl_5]^-$
2067 (av)[830]	<	2097[831]	<	2163 (av)[832]	<	2184[833] (cm^{-1})

It should be noted that the oxidation state of the Pt atom has been changed from Pt(0) to Pt(II) to Pt(IV) in the series above. As discussed earlier, the $\nu(CO)$ is lowered as the negative charge on the metal carbonyl increases. Thus, the $\nu(CO)$ of $[Os(CO)Cl_5]^-$ (2121 cm^{-1}) is 170 cm^{-1} higher than that of $[Os(CO)Cl_5]^{2-}$,[834].

Table III-49 lists the observed frequencies of typical compounds for which complete band assignments have been made. Figure III-56 shows the RR spectrum of solid (TBA) [$trans$-$OsBr_4(CO)_2$] (TBA: tetra-n-butylammonium ion) obtained by Johannsen and Preetz.[839] It shows a series of overtones of the totally symmetric $\nu(Os{-}Br)$ (ν_3, 209.3 cm^{-1}) up to 10 ν_3 together with many combination bands involving other totally symmetric fundamentals ($\nu_1(CO)$, 2122.1 and $\nu_2(OsC)$, 460.4 cm^{-1}).

El-Sayed and Kaesz[840] studied the $\nu(CO)$ of $M_2(CO)_8X_2$ (M = Mn, Tc, and Re; X = Cl, Br, and I), and proposed the halogen-bridging structure I shown in Fig. III-57. Four infrared-active $\nu(CO)$ have been observed in accordance with this structure. Garland and Wilt[841] interpreted the infrared spectrum of $Rh_2(CO)_4X_2$ (X = Cl and Br) on the basis of the C_{2v} structure II (Fig. III-57) found by X-ray analysis.[842] As predicted, three infrared-active $\nu(CO)$ have been observed for this compound. Johnson et al.[843] studied the exchange of $C^{18}O$ with CO groups of $Rh_2(CO)_4X_2$ (X = Cl, Br, I, etc.) with time by following the variation of infrared spectra in the $\nu(CO)$ region. Cotton and Johnson[844] proposed the staggered structure (III) for $Fe_2(CO)_8I_2$, since only two $\nu(CO)$ were observed in the infrared.

TABLE III-49. Vibrational Frequencies of Metal Carbonyl Halides (cm^{-1})

Compound	IR or Raman and Symmetry		ν(CO)	ν(MX)	Ref.
Mn(CO)$_5$Cl	IR	(C_{4v})	2138 (A_1) 2056 (E) 2000 (A_1)	291 (A_1)	835
Mn(CO)$_5$Br	IR	(C_{4v})	2138 (A_1) 2052 (E) 2007 (A_1)	222 (A_1)^a	836
fac-[Os(CO)$_3$Cl$_3$]$^-$	Raman	(C_{3v})	2125 (A_1) 2022 (E) 2033 (E)	321 (A_1) 287 (E)	837
cis-[Os(CO)$_2$Cl$_4$]$^{2-}$	Raman	(C_{2v})	2016 (A_1) 1910 (B_2)	316 (A_1) 281 (A_1) 308 (B_2)	837 838
[Os(CO)Cl$_5$]$^{3-}$	IR	(C_{4v})	1968 (A_1)	332 (A_1)^a 316 (A_1)^a 306 (E)	837
[Pt(CO)Cl$_3$]$^-$	IR	(C_{2v})	2120 (A_1)	331 (A_1) 310 (A_1)	837 831, 832

^{a}Raman frequency.

If CO is replaced by a phosphine, ν(CO) decreases since the latter is a strong σ-donor but a weak π-acceptor. In the [Ni$_{12-n}$(PMe$_3$)$_n$(CO)$_{24-3n}$]$^{2-}$ series (n = 2, 3, and 4), both terminal and bridging ν(CO) are downshifted as n increases.[845] Ligands such as arsines, amines, and isonitriles give similar results. Table III-50 lists the ν(CO) of typical compounds. Vibrational assignments have been reported for cis-[M(CO)$_4$(L–L)] [M = Cr, Mo, and W, and L–L is Ph$_2$P—(CH$_2$)$_n$—PPh$_2$, n = 1~3],[851] [CpW(CO)$_2$(PMe$_3$)(SiH$_2$Me)],[852] and [M(CO)$_5$(CS)] (M = Cr and W) and their CSe analogs.[853] The IR spectra of M(CO)$_5$L, where M is Mo or W and L is Kr or Xe, have been measured. The lifetime of the Kr complex is ca. 0.1 sec in liquid Kr at 150 K.[854]

Low-frequency infrared species have been reported for M(CO)$_{6-n}$(PR$_3$)$_n$ (M = Cr, Mo, and W),[855] M(CO)$_{6-n}$(CH$_3$CN)$_n$ (M = Cr and W: n = 1 and 2),[856] and Fe(CO)$_4$L (L = PPh$_3$, AsPh$_3$, and SbPh$_3$).[857] References on vibrational spectra of metal carbonyls containing other ligands are cited in respective sections where vibrations of these ligands are discussed.

(4) Normal Coordinate Calculations

Normal coordinate analyses on metal carbonyl compounds have been carried out by many investigators. Among them, Jones and co-workers have made the most extensive study in this field. For example, they performed rigorous calculations on the M(CO)$_6$ (M = Cr, Mo, and W) series,[769] Fe(CO)$_5$,[768] and Mn(CO)$_5$Br,[836] including their ^{13}C and ^{18}O analogs. For the last compound, 5 stretching, 16 stretching–stretching interaction, and 33 bending–bending interaction constants (GVF) were used to calculate its 30 normal vibrations.

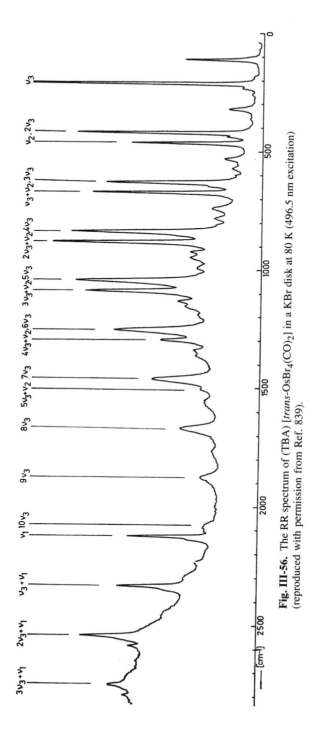

Fig. III-56. The RR spectrum of (TBA) [*trans*-OsBr$_4$(CO)$_2$] in a KBr disk at 80 K (496.5 nm excitation) (reproduced with permission from Ref. 839).

137

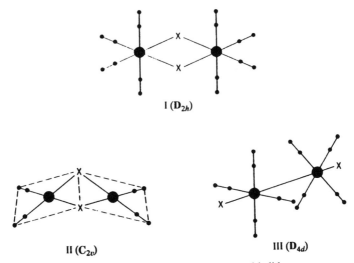

I (D_{2h})

II (C_{2v})   III (D_{4d})

Fig. III-57. Structures of metal carbonyl halides.

On the other hand, Cotton and Kraihanzel[858] developed an approxima-tion (C–K) method for calculating the CO stretching and CO—CO stretch-ing interaction constants, while neglecting all other low-frequency modes. For Mn(CO)$_5$Br, they used only the five force constants[859] shown in Fig. III-58. Since only four CO stretching bands are observed for this type of compound, it was assumed that $\frac{1}{2}k_t = k_c = k_d$ holds. This was justified on the basis of the symmetry properties of the metal $d\pi$ orbitals involved. This C–K method has since applied to many other carbonyls in making band assignments, in interpreting intensity data, and in discussing the bonding schemes of metal carbonyls.[758] It is clear that the choice of a rigorous approach (Jones) or a sim-

TABLE III-50. CO Stretching Frequencies of Metal Carbonyls Containing Other Ligands (cm^{-1})

Compound	IR or Raman and Symmetry		ν(CO)	Ref.
Ni(CO)$_3$(PMe$_3$)	Raman	(C_{3v})	2069 (A_1), 1980 (E)	846
Fe(CO)$_4$(PMe$_3$)	Raman	(C_{3v})	2051 (A_1), 1967 (A_1) 1911 (E)	847
Fe(CO)$_4$(AsMe$_3$)	Raman	(C_{3v})	2050 (A_1), 1964 (A_1) 1911 (E)	847
Co(CO)$_5$(PEt$_3$)	IR	(C_{4v})	2060 (A_1), 1973 (B_1) 1943 (A_1), 1935 (E)	848
W(CO)$_5$(NMe$_3$)	IR	(C_{4v})	2073 (A_1), 1932 (E) 1920 (A_1)	849
W(CO)$_4$(bipy)	IR	(C_{2v})	2010 (A_1), 1900 (B_1) 1874 (A_1), 1832 (B_2)	850
W(CO)$_2$(bipy)$_2$	IR	(C_2)	1778 (A_1), 1719 (B_2)	850

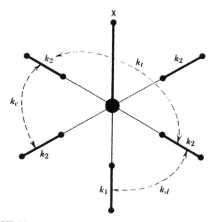

Fig. III-58. Definition of force constants for $M(CO)_5X$.

plified method (C—K) depends on the availability of observed data and the purpose of the investigation. Jones[860] and Cotton[861] discuss the merits of their respective approaches relative to the alternative.

As mentioned earlier, $\nu(CO)$ of metal carbonyls are determined by two factors: (1) donation of the 5σ-electrons to the empty metal orbital tends to raise the $\nu(CO)$ since the 5σ orbital is slightly antibonding; and (2) back donation of metal $d\pi$-electrons to the $2p\pi$ orbitals of CO tends to lower $\nu(CO)$ since the $2p\pi$ orbitals are antibonding. Vibrational spectroscopy does not allow observation of these two effects separately since the observed $\nu(CO)$ and the corresponding force constant reflect only the net result of the two counteracting components. It is possible, however, to correlate the CO stretching force constants (C–K) with the occupancies of the 5σ and $2p\pi$ orbitals as calculated by MO theory. Table III-51 lists the results obtained for d^6 carbonyl halides and dihalides by Hall and Fenske.[862] It is interesting that the *trans*-CO in $Fe(CO)_4I_2$ and the *cis*-CO in $Mn(CO)_5Cl$ have almost the same force constants since the 5σ occupancy of the former is smaller by 0.102 than that of the latter, while the $2p\pi$ occupancy of the former is larger by 0.108 than that of the latter. It is also noteworthy that the *trans*-CO in $Fe(CO)_4I_2$ and the *cis*-CO in $Cr(CO)_5Cl^-$ have identical $2p\pi$ occupancies (0.537) but substantially different force constants (17.43 and 15.58 mdyn/Å, respectively). In this case, the difference in force constants originates in the difference in the 5σ occupancies (1.293 vs. 1.457). Hall and Fenske[862] found a linear relationship between the C–K CO stretching force constants and the occupancies of the 5σ and $2p\pi$ levels:

$$k = -11.73[2\pi_x + 2\pi_y + (0.810)5\sigma] + 35.81$$

A similar attempt has been made for a series of Mn carbonyls containing isocyanide groups.[863]

TABLE III-51. Carbonyl Orbital Occupancies[a] and Force Constants

Compound	Structure	5σ	$2\pi_x$	$2\pi_y$	k (mdyn/Å)[b]
$Cr(CO)_5Cl^-$	trans	1.407	0.355	0.355	14.07
$Cr(CO)_5Br^-$	trans	1.405	0.353	0.353	14.10
$Mn(CO)_4I_2^-$	trans	1.354	0.302	0.330	15.48
$Mn(CO)_4IBr^-$	trans	1.355	0.302	0.327	15.48
$Mn(CO)_4Br_2^-$	trans	1.357	0.302	0.325	15.50
$Cr(CO)_5Br^-$	cis	1.456	0.261	0.282	15.56
$Cr(CO)_5Cl^-$	cis	1.457	0.261	0.276	15.58
$Mn(CO)_5Cl$	trans	1.352	0.286	0.286	16.28
$Mn(CO)_5Br$	trans	1.350	0.286	0.286	16.32
$Mn(CO)_5I$	trans	1.349	0.286	0.286	16.37
$Mn(CO)_4I_2^-$	cis	1.402	0.251	0.251	16.75
$Mn(CO)_4IBr^-$	cis	1.404	0.241	0.252	16.77
$Mn(CO)_4Br_2^-$	cis	1.406	0.242	0.242	16.91
$Mn(CO)_5I$	cis	1.394	0.213	0.240	17.29
$Mn(CO)_5Br$	cis	1.394	0.212	0.228	17.39
$Fe(CO)_4I_2$	trans	1.293	0.252	0.285	17.43
$Mn(CO)_5Cl$	cis	1.395	0.211	0.218	17.46
$Fe(CO)_4Br_2$	trans	1.295	0.250	0.272	17.53
$Fe(CO)_5Br^+$	trans	1.287	0.233	0.233	17.93
$Fe(CO)_5Cl^+$	trans	1.289	0.233	0.233	17.95
$Fe(CO)_4I_2$	cis	1.337	0.221	0.221	17.95
$Fe(CO)_4Br_2$	cis	1.338	0.205	0.205	18.26
$Fe(CO)_5Cl^+$	cis	1.325	0.171	0.177	18.99
$Fe(CO)_5Br^+$	cis	1.325	0.171	0.193	19.00

[a] The cis and trans designations of the CO groups are made with respect to the position of the halogen or halogens.
[b] C–K force constants (see Ref. 862).

(5) Hydrocarbonyls

Hydrocarbonyls exhibit bands characteristic of both M—H and M—CO groups. Kaesz and Saillant[864] reviewed the vibrational spectra of metal carbonyls containing the hydrido group. Vibrational spectra of hydrido complexes containing other groups will be discussed in Sec. III-22. In general, the terminal M–H group exhibits a relatively sharp- and medium-intensity $\nu(MH)$ band in the 2200–1800 cm^{-1} region. The MH stretching band can be distinguished easily from the CO stretching band by the deuteration experiment.

Edgell and co-workers[865] assigned the infrared bands at 1934 and 704 cm^{-1} of $HCo(CO)_4$ to $\nu(CoH)$ and $\delta(CoH)$, respectively, and proposed structure I of Fig. III-59, in which the H atom is on the C_3 axis. Stammreich et al.[766] reported the Raman spectrum of $HFe(CO)_4^-$, which is expected to have a structure similar to that of $HCo(CO)_4$. According to X-ray analysis,[866] the $Mn(CO)_5$ skeleton of $HMn(CO)_5$ takes the $\mathbf{C}_{4v}$ structure shown in structure II of Fig. III-59. Kaesz and co-workers[867,868] assigned the infrared spectrum of $HMn(CO)_5$ in the $\nu(CO)$ region on the basis of this structure. The Raman spectra of $HMn(CO)_5$ and $HRe(CO)_5$ exhibit their $\nu(MH)$ at 1780 and 1824 cm^{-1}, respectively.[869]

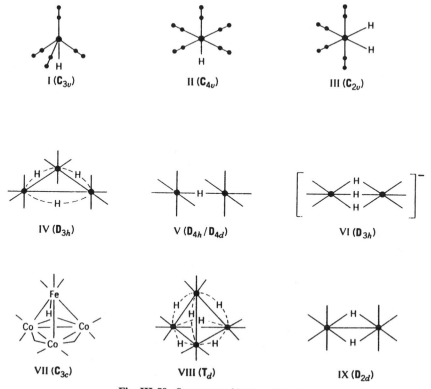

Fig. III-59. Structures of hydrocarbonyls.

A complete vibrational assignment of gaseous $HMn(CO)_5$ has been made by Edgell et al.[870] The infrared spectrum of $H_2Fe(CO)_4$ in hexane at $-78°C$ exhibits three or more $\nu(CO)$ above 2000 cm^{-1} and a weak, broad $\nu(FeH)$ at 1887 cm^{-1}. Thus, Farmery and Kilner[871] suggested structure III of Fig. III-59. Table III-52 lists the observed frequencies of other hydrocarbonyl compounds.

TABLE III-52. Vibrational Frequencies of Metal Hydrocarbonyl Compounds (cm^{-1})^a

Compound	$\nu(CO)$	$\nu(MH)$	$\delta(MH)$	Ref.
RhH(CO)(PPh$_3$)$_3$	1926	2004	784	872
IrH(CO)(PPh$_3$)$_3$	1930	2068	822	872
IrHCl$_2$(CO)(PEt$_2$Ph)$_2$	2101	2008	—	873
IrHBr$_2$(CO)(PEt$_2$Ph)$_2$	2035	2232	—	873
IrHCl$_2$(CO)(PPh$_3$)$_2$	2027	2240	—	874
OsHCl(CO)(PPh$_3$)$_2$	1912	2097	—	874
OsH$_2$(CO)$_2$(PPh$_3$)$_2$	2014	1928	—	875
	1990	1873		

^{a}For the configurations of these molecules, see the original references.

It is rather difficult to locate the bridging $\nu(MH)$ in polynuclear hydrocarbonyls. These vibrations appear in the region from 1600 to 800 cm^{-1}, and are rather broad at room temperature although they are sharpened at low temperatures. Higgins et al.[876] were the first to suggest the presence of bridging hydrogens in $Re_3H_3(CO)_{12}$ (structure IV of Fig. III-59) since no terminal $\nu(ReH)$ bands were observed. Smith et al.[877] observed a very weak and broad band at 1100 cm^{-1} in the Raman spectrum of $Re_3H_3(CO)_{12}$ and assigned it to the bridging $\nu(ReH)$ since it shifted to 787 cm^{-1} upon deuteration.

Although structure V was proposed for $[M_2H(CO)_{10}]^-$ (M = Cr, Mo and W),[878] the W–H–W angle of the tungsten complex was found to be 137°.[879] Shriver and co-workers[880] located the antisymmetric and symmetric stretching vibrations of the W–H–W bridge at 1683 and ~900 cm^{-1}, respectively. At ~80 K, the latter splits into four bands at 960, 869, 832, and 702 cm^{-1}. Although the origin of this splitting is not clear, the possibility of Fermi resonance with an overtone or a combination band involving the $\nu(WC)$ or $\delta(WCO)$ was ruled out based on $CO–C^{18}O$ isotopic shifts.[880]

Ginsberg and Hawkes[881] suggested structure VI for $[Re_2H_3(CO)_6]^-$ since they could not observe any terminal $\nu(ReH)$ vibrationals. The bridging $\nu(FeH)$ band of $FeHCo_3(CO)_{12}$ in the infrared was finally located at 1114 cm^{-1} by Mays and Simpson,[882] using a highly concentrated KBr pellet. This band shifts to 813 cm^{-1} upon deuteration. On the basis of mass spectroscopic and infrared evidence, they proposed structure VII, in which the H atom is located inside the metal atom cage. From the spectra in the $\nu(CO)$ region, together with X-ray evidence, Kaesz et al.[883] proposed the T_d skeleton (structure VIII) for $[Re_4H_6(CO)_{12}]^{2-}$. It showed no terminal $\nu(ReH)$, but a broad bridging $\nu(ReH)$ centered at 1165 cm^{-1} was observed in its Raman spectrum. This band shifts to 832 cm^{-1} with less broadening upon deuteration. Bennett et al.[884] found no terminal $\nu(ReH)$ in the infrared spectrum of $Re_2H_2(CO)_8$. However, its Raman spectrum exhibits bands at 1382 and 1272 cm^{-1}, which shift to 974 and 924 cm^{-1}, respectively, upon deuteration. The D_{2h} structure (IX) was proposed for this compound.

Figure III-60 shows the Raman spectra of $Ru_4H_4(CO)_{12}$ and $Ru_4D_4(CO)_{12}$ obtained by Knox et al.[885] Two $\nu(RuH)$ bands at 1585 and 1290 cm^{-1} of the former compound are shifted to 1153 and 909 cm^{-1}, respectively, upon deuteration. Its infrared spectrum exhibits five $\nu(CO)$ instead of the two expected for T_d symmetry. Thus a structure of D_{2d} symmetry, which lacks two H atoms from structure VIII, was proposed.[886] The $\nu(OsH)$ vibrations of the analogous Os complex have also been assigned.[887] In the $[Ru_6H(CO)_{18}]^-$ ion, the H atom is located at the center of an octahedron consisting of six Ru atoms.[888] Oxton et al.[889] located its $\nu(R–H)$ at 845 and 806 cm^{-1} (95 K) which are probably split by Fermi resonance.

As stated in Sec. III-7 (aquo complexes), the inelastic neutron scattering (INS) technique is very effective in locating hydrogen vibrations. White and Wright[890] found two hydrogen vibrations at 608 and 312 cm^{-1} in the INS spectrum of $Mn_3H_3(CO)_{12}$. However, the nature of these vibrations is not clear.

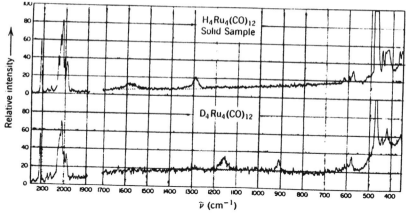

Fig. III-60. Raman spectra of $Ru_4H_4(CO)_{12}$ and its deuterated analog.[885]

(6) Metal Carbonyls in Inert Gas Matrices

A number of unstable and transient metal carbonyls have been synthesized and their structures determined by vibrational spectroscopy in inert gas matrices. In most cases, only $\nu(CO)$ vibrations have been measured to determine the structures of these compounds since it is rather difficult to observe low-frequency modes in inert gas matrices.

These transient carbonyls can be prepared by two methods. The first involves direct reaction of metal vapor with CO diluted in inert gas matrices: $M + xCO \rightarrow M(CO)_x$. As discussed in Sec. I-24 of Part A, DeKock[891] first succeeded in preparing the $Ni(CO)_x$ $(x = 1–3)$ series by this method. Although the method yields a mixture of carbonyls of various stoichiometries, bands characteristic of each species can be determined in several ways: warm-up experiments, concentration-dependence studies, isotope substitutions, and so on.[892] Table III-53 lists the number of infrared-active $\nu(CO)$ predicted for possible structures.

TABLE III-53. Number of Infrared-Active CO Stretching Vibrations for $M(CO)_x$

Molecule	Symmetry and Structure		IR-Active $\nu(CO)$
$M(CO)$	$C_{\infty v}$	Linear	Σ^+
$M(CO)_2$	$D_{\infty h}$	Linear	Σ_u^+
$M(CO)_2$	C_{2v}	Bent	$A_1 + B_2$
$M(CO)_3$	D_{3h}	Trigonal-planar	E'
$M(CO)_3$	C_{3v}	Trigonal-pyramidal	$A_1 + E$
$M(CO)_4$	T_d	Tetrahedral	F_2
$M(CO)_4$	D_{4h}	Square-planar	E_u
$M(CO)_5$	C_{4v}	Tetragonal-pyramidal	$2A_1 + E$
$M(CO)_5$	D_{3h}	Trigonal-bipyramidal	$A_2'' + E'$
$M(CO)_6$	O_h	Octahedral	F_{1u}

The structures of $M(CO)_2$, $M(CO)_3$, and $M(CO)_4$ (M = Ni,[891], Pd,[893,894] and Pt[895]) have been found to be linear, trigonal-planar, and tetrahedral, respectively, since all of these compounds exhibit only one $\nu(CO)$. In the case of the $M(CO)_{1-6}$ series (M = Ta,[891] U,[896] Pr, etc.[897]), it was more difficult to determine the structures of the transient species because the spectra were more complicated. More recently, matrix cocondensation reactions mentioned above have been extended to Li,[898,899] B,[900] Al,[901,902] and Ga.[903]

Moskovits and Ozin[892] determined the structures of a large number of metal carbonyls produced via matrix cocondensation reactions. They proposed an unusual isocarbonyl structure, $O\equiv C-Au-O\equiv C$ for $Au(CO)_2$,[904] Krishnan et al.[898] observd the $\nu(CO)$ of LiCO and LiOC at 1806 and 1614 cm^{-1}, respectively, in Kr matrices. The IR spectra of lanthanoid and actinoid carbonyls prepared in inert gas matrices have been reviewed by Sheline and Slater.[905]

The second method utilizes *in situ* photolysis of stable metal carbonyls in inert gas matrices. For example, Poliakoff and Turner[906] carried out UV photolysis of ^{13}CO-enriched $Fe(CO)_5$ in SF_6 and Ar matrices [$Fe(CO)_5 \xrightarrow{h\nu} Fe(CO)_4 + CO$], and concluded that the structure of $Fe(CO)_4$ is C_{2v} since it exhibits four $\nu(CO)$ ($2A_1 + B_1 + B_2$) in the infrared spectrum. Graham et al.[907] proposed the C_{4v} structure for $Cr(CO)_5$ produced by the photolysis of $Cr(CO)_6$ in inert gas matrices. On the other hand, Kündig and Ozin[908] proposed the D_{3h} structure for $Cr(CO)_5$ prepared by co-condensation of Cr atoms with CO in inert gas matrices. They derived a general rule that $M(CO)_5$ species take the D_{3h} structure when the number of valence-shell electrons is even [Cr (16), Fe (18)], and the C_{4v} structure when it is odd [V (15), Mn (17)]. However, the D_{3h} structure of $Cr(CO)_5$ has been questioned by Black and Braterman[909] and Perutz and Turner.[910]

The UV photolysis of $Fe_2(CO)_9$ in Ar matrices produces $Fe_2(CO)_8$ of C_{2v} symmetry having two bridging CO groups.[911] Photolysis of $Mn_2(CO)_{10}$ by plane-polarized light in Ar matrices produces $Mn_2(CO)_9$, which exhibits the $\nu(CO)$ at 1764 cm^{-1},[912,913]. Dunkin et al.[912] proposed the semibridging structure based on its polarization properties:

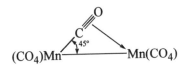

UV photolysis of $MnRe(CO)_{10}$ in Ar matrices yields $MnRe(CO)_9$ with the $\nu(CO)$ at 1759.8 cm^{-1},[914]. A semibridging structure similar to that proposed above may be expected.

Charged carbonyls such as $Cr(CO)_5^-$, $Fe(CO)_4^-$, and $Ni(CO)_3^-$ can also be prepared in inert gas matrices using several techniques.[915] For example, the $Fe(CO)_4^-$ ion has been obtained via photolysis of cocondensation products of $Fe(CO)_5$ with Na in Ar matrices. According to Timney,[916] it is not the $Fe(CO)_4^-$

ion of C_{3v} symmetry as originally proposed but rather the $Fe(CO)_3^-$ ion of C_{2v} or C_s symmetry. McNeish et al.[917] noted that the Nernst glower of an IR spectrophotometer accelerates the formation of $Fe(CO)_4(CH_4)$ via the reaction of $Fe(CO)_4$ with methane in a methane matrix. Turner[918] reviewed photochemical fragmentation of hexacarbonyls of Cr, Mo, and W. Finally, Weitz and coworkers studied the IR spectra of transient metal carbonyls obtained by laser photolysis in the gas phase.[919]

Carbonyl complexes of the type MX_2CO are formed by reacting metal halide vapor directly with CO in inert gas matrices.[920,921] In this case, $\nu(CO)$ shifts to higher frequencies by complexation, since the bonding is dominated by the donation of σ-electrons to the metal. On the other hand, $\nu(MX)$ shifts to lower frequencies because the oxidation state of the metal is lowered by accepting σ-electrons from CO. Figure III-61 shows infrared spectra of the PbF_2—L system (L = CO, NO, and N_2) in Ar matrices obtained by Tevault and Nakamoto.[921]

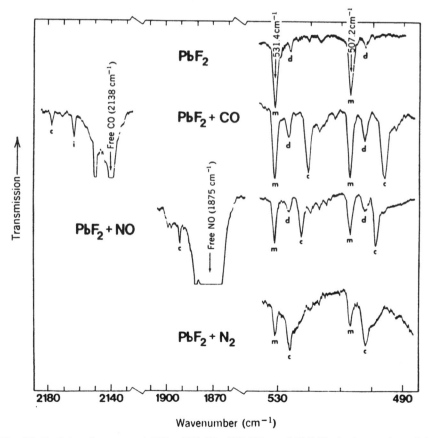

Fig. III-61. Infrared spectra of PbF_2, PbF_2CO, PbF_2NO, and PbF_2N_2 in Ar matrices: (m) monomeric PbF_2; (d) dimeric PbF_2; (c) complex; (i) impurity (HF–CO).[921]

In this series, the magnitudes of the shifts of the PbF_2 and L stretching bands (cm^{-1}) relative to the free state are as follows:

	PbF_2CO	PbF_2NO	PbF_2N_2
$\nu_s(PbF_2)$	−10.8	−8.8	−5.8
$\nu_a(PbF_2)$	−10.9	−8.5	−5.0
$\nu(L)$	+38.4	+16.4	—

This result definitely indicates that CO is the best, NO is the next best, and N_2 is the poorest σ-donor.

Other work involves the direct deposition of stable carbonyls in inert gas matrices, mainly to study the effect of matrix environments on the structure. Both $Fe(CO)_5$[922] and $M_3(CO)_{12}$ (M = Ru and Os)[923] were found to be distorted from D_{3h} symmetry in inert gas matrices. If a thick deposit is made on a cryogenic window while maintaining a relatively high sample/inert gas dilution ratio, it is possible to observe low-frequency modes such as $\nu(MC)$ and $\delta(MCO)$. It was found that these bands show splittings due to the mixing of metal isotopes. For example, the $F_{1u}\nu(CrC)$ of $Cr(CO)_6$ in a N_2 matrix exhibits four bands due to ^{50}Cr, ^{52}Cr, ^{53}Cr, and ^{54}Cr (see Fig. I-36 of Part A). The magnitude of these isotope splittings may be used to estimate the degree of the $\nu(MC)$–$\delta(MCO)$ mixing in the low-frequency vibrations.[924]

(7)   CO Adducts of Metalloporphyrins

Vibrational spectra of CO adducts of metalloporphyrins have been reviewed by Kitagawa and Ozaki[201] and Yu.[925] In general, these compounds exhibit the $\nu(CO)$ in the 2100–1900 cm^{-1} region. In the $M(TPP)(CO)_2$ series, the $\nu_a(CO)$ follows the order:

$Co(TPP)(CO)_2$[926] $Fe(TPP)(CO)_2$[927] $Ru(TPP)(CO)_2$[928]

2078   >   2042   >   2005   (cm^{-1})

This result indicates that the degree of the M $\rightarrow$ CO π-back donation increases in going from Co(II) to Fe(II) to Ru(II). The 1:1 adduct, Fe(TPP)(CO), exhibits the $\nu(CO)$ at 1973 cm^{-1}, which is lower than that of $Fe(TPP)(CO)_2$ because the net M $\rightarrow$ CO π-back donation decreases in the latter due to competition between the two CO ligands.[927]

Yu and co-workers carried out an extensive RR study on CO adducts of metalloporphyrins by using isotopic ligands, $^{13}C^{16}O$, $^{12}C^{18}O$, and $^{13}C^{18}O$. In the 700–100 cm^{-1} region, $Fe(T_{piv}PP)(CO)(1\text{-MeIm})$ ($T_{piv}PP$: "picket-fence" porphyrin shown in Fig. III-62a) shows only one isotope-sensitive band at 489 cm^{-1}. This band has been assigned to the $\nu(Fe—C)$ because it shifts to 485 to

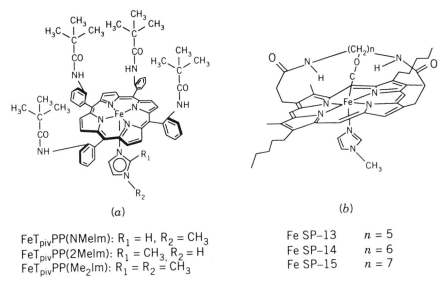

$$FeT_{piv}PP(NMeIm): R_1 = H, R_2 = CH_3$$
$$FeT_{piv}PP(2MeIm): R_1 = CH_3, R_2 = H$$
$$FeT_{piv}PP(Me_2Im): R_1 = R_2 = CH_3$$

Fe SP–13 $n = 5$
Fe SP–14 $n = 6$
Fe SP–15 $n = 7$

Fig. III-62. Structures of (a) "picket-fence" and (b) "strapped" porphyrins.

481 to 477 cm^{-1} in order shown above of the isotopic CO ligands. It is also sensitive to the nature of the *trans*-ligand (L); the weaker the M–L bond is, the stronger the Fe–CO bond. However, no bands assignable to the $\delta(FeCO)$ were resonance enhanced.[929]

In CO adducts of simple metalloporphyrins such as Fe(TPP)(CO), the Fe—C—O bond is linear and normal to the porphyrin plane. In the "strapped" porphyrins (Fig. III-62b), the Fe—C—O (linear) bond is tilted because of steric hindrance of the "strap." According to Yu,[925] this tilting increases the electron donation from a pyrrole ring (π) to the CO (π^*) orbital because of a better overlap between these orbitals. This would decrease the CO bond order and increase the Fe–C bond order. Thus, the following trends are observed as the "strap" is shortened:[930]

	SP-15		SP-14		SP-13
$\nu(CO)$ (cm^{-1})	1945	>	1939	>	1932
$\nu(Fe\!-\!C)$ (cm^{-1})	509	<	512	<	514
	—		504		506

In the latter two complexes, the $\nu(Fe–C)$ bands are split into two bands; the higher- and lower-frequency components were attributed to the "tilted" and "upright" conformers, respectively. Similar trends in frequency were observed for a hybrid of the "picket-fence" and the "basket-handle" porphyrins.[930a]

The RR spectra of these "strapped" porphyrins exhibit the $\delta(FeCO)$ which shows the "zigzag" isotope shift pattern. For example, SP-14 exhibits this band at 578 cm^{-1}, which is shifted to 563 ($^{13}C^{16}O$), 575 ($^{12}C^{18}O$), and 561 ($^{13}C^{18}O$) cm^{-1} by the isotopic substitutions indicated in the parentheses. In contrast, the $\nu(Fe-C)$ vibration near 510 cm^{-1} shows a normal (monotonous) isotopic shift pattern. This difference has often been used to distinguish these two modes. It should be noted, however, that the observation of a "zigzag" isotope shift pattern does not necessarily indicate the bending mode (Sec. V-2(3)). Yu et al.[930] also observed that the degree of resonance enhancement of the $\delta(FeCO)$ relative to the $\nu(Fe-C)$ mode increases as the distortion of the $Fe-C\equiv O$ linkage increases by shortening the "strap" length.

As described above, Yu and co-workers originally assigned the $\delta(FeCO)$ near 560 cm^{-1}, which are higher than the $\nu(Fe-C)$ near 510 cm^{-1}, and their assignments have been followed by many other workers. However, an alternative assignment has recently been proposed for the CO adducts of heme proteins (Sec. V-2(1)).

(8) Complexes of Carbon Dioxide (CO_2)

Although CO_2 is highly inert, a few complexes with metal atoms and metal ions in the low oxidation state are known. These complexes have been a subject of considerable interest because they have the potential to become catalysts in activating CO_2, which is the most abundant source of C_1 compounds.

Thus far, vibrational studies on metal complexes of CO_2 are limited to a small number of compounds, because stable complexes of CO_2 are rare. The CO_2 ligand may coordinate to a metal in any one of the following schemes:

end-on(C_s)   C-bonded(C_{2v})   side-on(C_s)

Different from the linear CO_2 molecule in the free state, the CO_2 ligand in metal complexes is always bent. Furthermore, the bond orders of the two CO bonds change markedly upon coordination. Thus, the three vibrations observed for free CO_2 [1337 (ν_1), 667 (ν_2), and 2349 cm^{-1} (ν_3), Table II-2b of Part A] show large downshifts in metal complexes. Table III-54 lists the observed frequencies of typical complexes and their modes of coordination. The CO_2 vibrations in metal complexes can easily be identified since they are sensitive to $^{13}C^{12}O$ and $^{12}C^{18}O$ substitutions. According to Jegat et al.,[932,933] it is possible to distinguish the three modes of coordination based on the $\nu_3 - \nu_1$ values and the magnitudes of isotopic shifts due to ^{13}C and ^{18}O substitutions.

TABLE III-54. Observed CO_2 Stretching Frequencies (cm^{-1}) and Modes of Coordination

Compound	ν_3	ν_1	Mode of Coordination	Ref.
$Cu(CO_2)^a$	1716	1215	End-on	931
$Al(CO_2)^a$	1780	1146	End-on	931
$Fe(CO_2)^a$	1565	1210	C-bonded	931
$Cp_2Ti(CO_2)(PMe_3)$	1671	1187	C-bonded	932
$Mo(CO_2)_2 (PMe_3)_4$	1668	1153	Side-on	933
		1102		
$Fe(CO_2) (PMe_3)_4$	1623	1106	Side-on	933
$Ni(CO_2) (PCy_3)_2{}^b$	1741	1150	Side-on	934
		1093		
$Ni(CO_2) (PEt_3)_2$	1660	1203	Side-on	935
	1635	1009		
$RhCl(CO_2) (PBu_3)_2$	1668	1165	Side-on	936
	1630	1120		

^{a}In inert gas matrix.
^{b}Cy: cyclohexyl.

III-18. NITROSYL COMPLEXES

Like CO, NO acts as a σ-donor and a π-acceptor. The NO contains one more electron than CO, and this electron is in the $2p\pi^*$ orbital. The loss of this electron gives the nitrosonium ion, $(NO)^+$ which is much more stable than NO. Thus, the $\nu(NO)$ of the nitrosonium ion (2273 cm^{-1}) is much higher than that of the latter (1880 cm^{-1}). On the other hand, the addition of one electron to this orbital produces the $(NO)^-$ ion, which is less stable and gives a lower frequency (~ 1366 cm^{-1}) than NO. Such a charge effect has already been discussed in Sec. II-1 of Part A.

In nitrosyl complexes, $\nu(NO)$ ranges from 1900 to 1500 cm^{-1}. Recent X-ray studies on nitrosyl complexes have revealed the presence of linear and bent M—NO groups:

$$M-N\equiv O: \qquad M-\ddot{N} \diagdown \overset{..}{\underset{.}{O}}$$

$$\text{I} \qquad\qquad \text{II}$$

In the valence-bond theory, the hybridizations of the N atom in (I) and (II) are sp and sp^2, respectively. If the pair of electrons forming the M–N bond is counted as the ligand electrons, the nitrosyl groups in (I) and (II) are regarded as NO^+ and NO^-, respectively. Thus, one is tempted to correlate $\nu(NO)$ with the charge on NO and the MNO angle. It was not possible, however, to find simple relationships between them since $\nu(NO)$ is governed by several other factors (electronic effects of other ligands, nature of the metal, structure, and charge of the whole complex etc.).[937] According to Haymore and Ibers,[938] the distinction of linear and bent geometry can be made by using properly corrected ν(NO) values; the MNO group is linear or bent, respectively, if this value is above or below 1620–1610 cm^{-1}. Several review articles are available for vibrational spectra of nitrosyl complexes.[937,939–943]

(1)   Inorganic Nitrosyl Complexes

Table III-55 lists the vibrational frequencies of typical nitrosyl complexes. Although the M—NO group is expected to show ν(NO), ν(MN), and δ(MNO), only ν(NO) have been observed in most cases. The latter two modes are often coupled since their frequencies are close to each other. Jones et al.[948] carried out a complete analysis of the vibrational spectra of Co(CO)$_3$(NO) and its ^{13}C, ^{18}O, and ^{15}N analogs. According to Quinby-Hunt and Feltham,[956] vibrational spectra of a wide variety of nitrosyl complexes can be accounted for on the basis of the simple three-body (M–N–O) model as long as the complex does not contain two or more NO groups attached to the metal.

Vibrational spectra of nitroprusside salts have been studied extensively.[957] Khanna et al.[958,959] assigned the IR and Raman spectra of the Na$_2$[Fe(CN)$_5$-(NO)]·2H$_2$O crystal and its deuterated analog. On the basis of a comparison of ν(CN), δ(FeCN), and ν(Fe–CN) between the Fe(II) and Fe(III) complexes of the [Fe(CN)$_5$X]$^{n-}$-type ions. Brown[960] suggested that the Fe–NO bonding of the [Fe(CN)$_5$(NO)]$^{2-}$ ion be formulated as Fe(III)–NO and not as Fe(II)–NO$^+$. Tosi and Danon[961] studied the IR spectra of [Fe(CN)$_5$X]$^{n-}$ ions (X = H$_2$O, NH$_3$, NO$_2^-$, NO$^-$, and SO$_3^{2-}$). The ν(CN) of the nitroprusside (2170, 2160, and 2148 cm^{-1}) are unusually high in this series because the Fe–CN π-back bonding in

TABLE III-55. Vibrational Frequencies of Inorganic Nitrosyl Complexes (cm^{-1})

Compound	ν(NO)	ν(MN)	δ(MNO)	Ref.
Cr(NO)$_4$	1721	650	496	944
Co(NO)$_3$	1860	—	—	945
	1795			
[Mo(NO)$_5$]$^{5+}$	1912, 1816,	665, 633,	470, 320,	946
	1675	560, 516,	186	
		380		
Cr(CO)$_3$ (NO)$_2$	1705	—	—	947
Co(CO)$_3$ (NO)	1822	609	566	948
Mn(CO)$_4$ (NO)	1781	524	657	949
Mn(PF$_3$) (NO)$_3$	1836, 1744	—	—	950
cis[MoCl$_4$(NO)$_2$]$^{2-}$	1720, 1600	—	—	951
NiCl$_2$(NO)$_2$	1872, 1842	—	—	952
[RuCl$_5$(NO)]$^{2-}$	1904	606	588	953
[RuBr$_5$(NO)]$^{2-}$	1870	572	300	954
[Tc(NO) (CNCMe$_3$)$_5$]$^{2+,a}$	1865	—	—	955

^{a}CNCMe$_3$: *tert*-butyl isocyanide. This complex is formulated as [Tc(I)(NO)$^+$(CNCMe$_3$)$_5$]$^{2+}$ based on its high ν(NO).

this ion is much less than in other compounds owing to extensive Fe–NO π-back bonding. Vibrational spectra of nitroprusside salts of various forms have been assigned.[962–964] Finally, the IR spectra of $K_3[Mn(CN)_5(NO)]$ and its ^{15}NO analog have been reported.[965,966]

According to X-ray analysis, $RuCl(NO)_2(PPh_3)_2PF_6$ contains one linear and one bent M—NO group which exhibit the $\nu(NO)$ at 1845 and 1687 cm^{-1}, respectively.[967] $CoCl_2(NO)L_2[L = P(CH_3)Ph_2]$ exists in two isomeric forms:

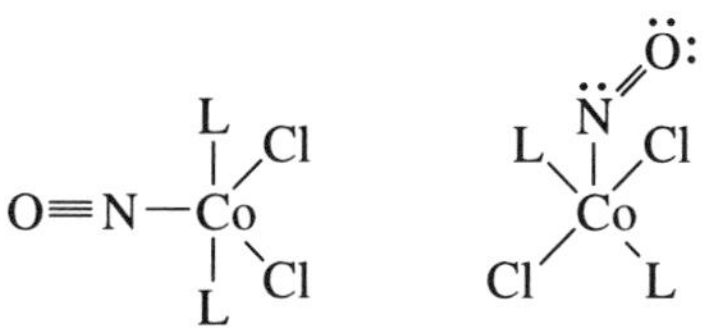

The $\nu(NO)$ of the former is at 1750 cm^{-1}, whereas that of the latter is at 1650 cm^{-1}.[968]

The $\nu(NO)$ of the bridging nitrosyl group is much lower than that of the terminal nitrosyl group. For example, $(C_5H_5)_2Cr_2(NO)_3(NXY)$ (X = OH and Y = t-Bu) shown below exhibits the terminal $\nu(NO)$ at 1683 and 1625 cm^{-1} and the bridging $\nu(NO)$ at 1499 cm^{-1}.[969]

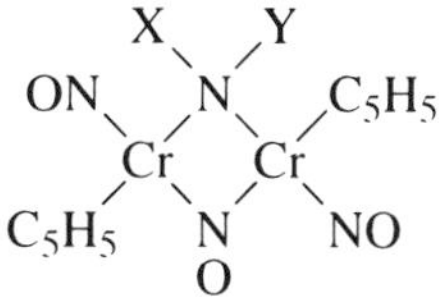

Similar frequencies are reported for an analogous compound [X = Et and Y = $B(Et)_2$].[970] The structure of $M_3(CO)_{10}(NO)_2$ (M = Ru and Os) resembles that of $Fe_3(CO)_{12}$ (structure VIII in Fig. III-53) with double nitrosyl bridges in place of the double carbonyl bridges in the latter. As expected, $\nu(NO)$ of these nitrosyl groups are very low: 1517 and 1500 cm^{-1} for the Ru compound, and 1503 and 1484 cm^{-1} for the Os compound.[971]

The $\nu(NO)$ is spin-state sensitive in Fe(NO)(salphen) [salphen: N,N'-o-phenylenebis(salicylideneimine)]: 1724 cm^{-1} for the high-spin (room temperature) and 1643 cm^{-1} for the low-spin (liquid-N_2 temperature) state.[972] Photolysis of $Cr(NO)_4$ in Ar matrices produces $Cr(NO)_3(NO^*)$ where NO* denotes a bent NO group with an unusually low $\nu(NO)(1450$ cm^{-1}).[973] Similar observations were made for the photolysis products of Mn(CO)-(NO)$_3$[974] and Ni(C$_5$H$_5$)(NO).[975]

(2) NO Adducts of Metalloporphyrins

Table III-56 lists the $\nu(NO)$ of metalloporphyrins in which the NO groups take linear geometry. In $Fe(TPP)(NO)_2$, however, the $\nu(NO)$ at 1870 and 1690 cm^{-1}

TABLE III-56. NO Stretching Frequencies of Metalloporphyrins (cm^{-1})

Compound	IR/Raman	$\nu(NO)$	Ref.
Cr(TPP) (NO)	IR	1700	976
Mn(TPP) (NO)	IR	1760	976
Fe(TPP) (NO)	IR	1700	977
Co(TPP) (NO)	IR (in matrix)	1693	926
Fe(TPP) (NO)$_2$	IR	1870	977
		1690	
Fe(PPDME) (NO)	IR	1660	978
Fe(PPDME) (NO)-(N-MeIm)	IR	1676	978
Fe(TPP) (NO)	RR (in THF)	1681	979
Fe(TPP) (NO$^-$)	RR (in THF)	1496	979

^{a}PPDME: Protoporphyrin IX dimethyl ester.

have been assigned to the linear Fe(II)—(NO)$^+$ and the bent Fe(II)—(NO)$^-$, respectively.[977]

The low-frequency modes, such as ν(M–NO) and δ(MNO), have been observed by RR studies (Soret excitation). For example, a "strapped" porphyrin, Mn(SP-15)(NO)(N-MeIm) (Fig. III-62) exhibits the ν(NO), ν(M–NO), and δ(MNO) at 1727, 631, and 578 cm^{-1}, respectively.[980] A simple porphyrin such as Mn(PPDME)(NO)(N-MeIm) (PPDME: protoporphyrin IX dimethylester) shows the ν(NO) and ν(M–NO) at 1733 and 628 cm^{-1}, respectively. Thus, introduction of steric hindrance lowers the ν(NO) and raises the ν(M–NO). Lipscomb et al.[981] observed the RR spectra of NO adducts of iron porphyrins by Soret excitation. The ν(Fe(II)–NO) and ν(Fe(III)–NO) are at ~527 and ~600 cm^{-1}, respectively. The NO adducts of heme proteins exhibit the ν(NO) at ~554 cm^{-1}, which is much higher than that of simple porphyrins (~527 cm^{-1}). Thus, the cage effect of proteins raises the ν(NO) in this case. It was also noted that the ν(Fe(II)–NO) is insensitive to the nature of the *trans* ligand. This is markedly different from the ν(Fe(II)–CO), ν(Fe(II)–O$_2$), and ν(Fe(II)–CN), which are sensitive to the *trans* ligand. The δ(MNO) vibrations were not observed for iron porphyrins.

(3) Metastable States of Nitroprussides

When a sample of Na$_2$[Fe(CN)$_5$(NO)]·2H$_2$O is irradiated by the 488.0-nm line of an Ar–ion laser at 20 K, two electronically excited metastable states (MS$_1$ and MS$_2$) are produced. Güida et al.[982,983] have measured the IR spectra of these metastable states using an orientated single crystal. The upper traces of Fig. III-63 show the polarized IR spectra of the ground state complex (GS) with the electric vectors parallel to the a and c axes, respectively. The ν(NO) near 1950 cm^{-1} is strong and broad in the former but rather weak in the latter, because the linear Fe–N–O axis[984] is on the ab plane of the orthorhombic crystal. The δ(FeNO) and ν(Fe–NO) bands are seen at 667 and 658 cm^{-1}, respectively.

Upon irradiation at 20 K, two sets of new bands appear as shown in the lower

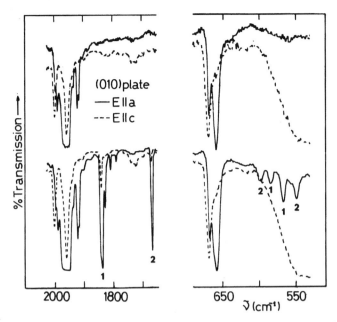

Fig. III-63. Low-temperature polarized IR spectra of ground-state (top traces) and metastable (bottom traces) anions in $Na_2[Fe(CN)_5NO]\cdot2H_2O$ single-crystal plate which was cut along the 010 plane. 1 and 2 denote the peaks resulting from the MS_1 and MS_2 states, respectively (reproduced with permission from Ref. 982).

traces of Fig. III-63. The bands at 1834, 583, and 565 cm^{-1} (marked by 1) are assigned to the $\nu(NO)$, $\delta(FeNO)$, and $\nu(Fe–NO)$ of MS_1, and those at 1663, 597, and 547 cm^{-1} (marked by 2) are assigned to the corresponding modes of MS_2. Thus, all three bands are shifted markedly in going from GS to MS_1 and MS_1 to MS_2. Similar red-shifts are observed for the $\nu(CN)$ near 2150 cm^{-1}, although the magnitudes of their shifts are much smaller than those observed for the FeNO group vibrations. This result indicates that the electronic transitions mainly involve the FeNO bond. The MS_1 is stable below 200 K, whereas MS_2 is stable only below 150 K. X-Ray analysis shows that the Fe–N–O bond is practically linear in MS_1.[985] It may also be linear in MS_2, as suggested from a comparison of angular dependence of the IR intensity of the $\nu(NO)$ between MS_1 and MS_2.[982] The RR spectra of GS and MS_1 have been reported by Krasser et al.[986]

In the case of the analogous osmium complex, $Na_2[Os(CN)_5(NO)]\cdot2H_2O$,[987] the GS, MS_1, and MS_2 states exhibit the $\nu(NO)$ at 1897, 1790, and 1546 cm^{-1}, respectively. The MS_1 and MS_2 states can selectively be populated by irradiating the sample with an Ar-ion laser (457.9 nm) and a mercury lamp (280–340 nm), respectively, at 80 K. Figure III-64 shows the effect of heating on the IR spectrum of MS_1 thus produced. It is seen that the onset decay temperature (T_2) of MS_2 (~220 K) is higher than that of MS_1 (T_1, ~190 K). This is opposite to

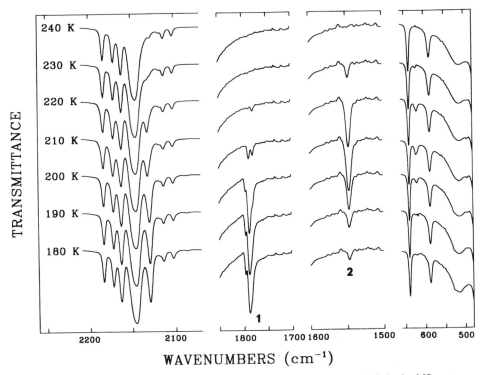

Fig. III-64. Effect of heating on the IR spectrum of $Na_2[Os(CN)_5NO]\cdot 2H_2O$ in the MS_1 state which was produced by irradiation with an Ar–ion laser (457.9 nm) at 80 K. 1 and 2 denote the $\nu(NO)$ bands of the MS_1 and MS_2 states, respectively (reproduced with permission from Ref. 987).

the case of the nitroprusside discussed above. Furthermore, the population of the MS_2 state begins to increase at the expense of the MS_1 state near the decay temperature, T_1.

III-19. COMPLEXES OF DIOXYGEN

Dioxygen (molecular oxygen) adducts of metal complexes have been studied extensively because of their importance as oxygen carriers in biological systems (Sec. V-2) and as catalytic intermediates in oxidation reactions of organic compounds. A number of review articles are available on the chemistry of dioxygen adducts.[937,988–997]

As discussed in Sec. II-1 of Part A, the bond order of the O–O linkage decreases as the number of electrons in the antibonding $2p\pi^*$ orbital increases in the following order:

	$[O_2^+]AsF_6$		O_2		$K[O_2^-]$		$Na_2[O_2^{2-}]$
Bond order	2.5	>	2.0	>	1.5	>	1.0
Bond distance (Å)	1.123	<	1.207	<	1.28	<	1.49
$\nu(O_2)$ (cm^{-1})	1858	>	1555	>	1108	>	~760

The decrease in the bond order causes an increase in the O–O distance and a decrease in the $\nu(O_2)$. In fact, there is a good linear relationship between the O–O bond order and the $\nu(O_2)$ of these simple dioxygen compounds.

Dioxygen adducts of more complex molecules are generally classified into two groups; complexes which exhibit $\nu(O_2)$ in the 1200–1100 cm^{-1} region are called "superoxo" because their frequencies are close to that of KO_2, and complexes whose $\nu(O_2)$ are in the 920–750 cm^{-1} region are called "peroxo" because their frequencies are close to that of Na_2O_2. As will be shown later, there are many compounds which exhibit $\nu(O_2)$ outside of these regions. Thus, this distinction of dioxygen adducts is not always clear-cut.

Structurally, the dioxygen adducts are classifed into three types:

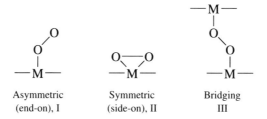

| Asymmetric (end-on), I | Symmetric (side-on), II | Bridging III |

In I, the two oxygen atoms are not equivalent, whereas they are equivalent in II and III. Thus, the $\nu(^{16}O^{18}O)$ splits into two bands in I but shows no splitting in II and III. The $^{16}O^{18}O$ gas is easily prepared by isotope scrambling of a mixture of $^{16}O_2$ and $^{18}O_2$. As will be shown later, vibrational studies with $^{16}O^{18}O$ have proved to be very useful in distinguishing these two types.

In the following, we classify dioxygen adducts according to the type of ligand involved, and review typical results obtained for each type.

(1) Dioxygen Adducts of Metal Atoms

As stated in Sec. I-24 of Part A, a number of stable and unstable complexes of the ML_n type have been synthesized via matrix cocondensation reactions of metal vapor(M) with gaseous ligands(L). Table III-57 lists typical results obtained for the $M(O_2)$-type compounds. It is seen that the $\nu(O_2)$ of these dioxygen adducts scatter over a wide range from 1120 to 920 cm^{-1}. Previously, we noted that the $\nu(O_2)$ decreases as the negative charge on the O_2 increases. Thus, these results seem to suggest that the negative charge on the O_2 can be varied continuously by changing the metal. In fact, Lever et al.[1006] noted that there is a linear relationship between the electron affinity of the M^{2+} ion and the $M–O_2$ CT transition energy in the $M(O_2)_2$ series and that the latter is linearly related to the $\nu(O_2)$.

The dioxygen ligand may coordinate to a metal in the end-on or side-on fashion. These two structures can be distinguished by using the isotope scrambling technique. Andrews[1007] first applied this method to the structure determination of the ion-pair complex $Li^+O_2^-$; a mixture of $^{16}O_2$, $^{16}O^{18}O$, and $^{18}O_2$ was prepared by Tesla coil discharge of a $^{16}O_2–^{18}O_2$ mixture, and reacted with Li vapor in an Ar matrix. Three $\nu(O_2)$ were observed in the Raman spectrum:

This result clearly indicates side-on coordination since four bands are expected for end-on coordination (see above). Using the same technique, Ozin and co-workers[1002,1004] showed that, in all cases they studied, O_2 coordinates to a metal in the side-on fashion and that, in $M(O_2)_2$ (M = Ni, Pd, and Pt), the complexes take the spiro $\mathbf{D}_{2d}$ structure. Some metal superoxides and peroxides are prepared by ordinary methods and their $\nu(O_2)$ are reported by Evans[1008] and Eysel and Thym.[1009]

As mentioned in Sec. II-2 of Part A, some dioxygen adducts produced by sputtering techniques in inert gas matrices take linear or bent $O{=}M{=}O$ structures which are close to the dioxo compounds (Sec. III-20).

TABLE III-57. Vibrational Frequencies of M(O$_2$)-Type Compounds (cm^{-1})

Compound	$\nu(O_2)$	$\nu_s(MO)$	$\nu_a(MO)$	Ref.
$^6LiO_2$	1097.4	743.8	507.3	998
$^7LiO_2$	1096.9	698.8	492.4	998
NaO_2	1094	390.7	332.8	998
KO_2	1108	307.5	—	998
RbO_2	1111.3	255.0	282.5	998
CsO_2	1115.6	236.5	268.6	998
AgO_2	1082/1077	—	—	999
RhO_2	900	—	422	1000
InO_2	1084	332	277.7	1001
GaO_2	1089	380	285.5	1001
AuO_2	1092	—	—	1002
TlO_2	1082	296	250	1003
PdO_2	1024.0	427	—	1004
NiO_2	966.2	504	—	1004
FeO_2	946	—	—	1005
PtO_2	926.6	—	—	1004

(2) Dioxygen Adducts of Transition Metal Complexes

A number of peroxo complexes of transition metal ions have been isolated. The dioxygens in these compounds take the symmetric side-on structures, and exhibit the $\nu(O_2)$ in the 900–800 cm^{-1} region and the $\nu_a(MO_2)$ and $\nu_s(MO_2)$ in the 650–430 cm^{-1} region. In general, the $\nu_a(MO_2)$ is higher than the $\nu_s(MO_2)$. Table III-58 lists the observed frequencies of typical peroxo complexes. Some of these assignments have been confirmed by $^{16}O_2/^{18}O_2$ substitution. Nakamura et al.[1021] have made normal coordinate calculations on peroxo complexes.

According to X-ray analysis,[1022] [Cu(HB(3,5-R pz)$_3$)]$_2$(O$_2$) (R = i-Pr and Ph; pz, pyrazole) contains a rare symmetrical side-on bridge:

The O–O distance (1.412 Å) is typical of peroxo complexes and all the Cu–O distances are essentially the same (1.90–1.93 Å). The RR spectrum exhibits the $\nu(O_2)$ at 741 cm^{-1}, which is much lower than those shown in Table III-58. This compound serves as a model for dioxygen binding in hemocyanin (Sec. V-5).

(3) Dioxygen Adducts of Cobalt Ammine and Schiff-Base Complexes

Extensive vibrational studies have been made on dioxygen adducts of cobalt ammine and Schiff-base complexes. Table III-59 lists the $\nu(O_2)$ and $\nu(CoO)$ of representative compounds.

The $\nu(O_2)$ of dinuclear cobalt complexes such as $\{[Co(NH_3)_5]_2O_2\}^{n+}$ ($n = 4$ or 5) are markedly different depending upon whether the O$_2$ group is of superoxo or peroxo type. The $\nu(O_2)$ of the $\{[Co(NH_3)_5]_2O_2\}^{5+}$ ion appears strongly

TABLE III-58. Observed Frequencies of Peroxo Complexes (cm^{-1})

Complex	$\nu(O_2)$	$\nu(MO_2)$	Ref.
(NH$_4$)$_3$[Ti(O$_2$)F$_5$]	905	600, 530	1010
K$_2$[Ti(O$_2$)(C$_2$O$_4$)]	895	611, 536	1011
Zr(O$_2$)(H$_2$EDTA)	840	650, 600	1012
(NH$_4$)$_2$ [ZrO(O$_2$)F$_2$]	850	640, 585	1013
(NH$_4$)$_3$ [Zr(O$_2$)F$_5$]	837	550, 471	1014
K$_3$[V(O$_2$)$_4$]	854	620, 567	1015
K$_3$[Ta(O$_2$)F$_4$]	866	592, 518	1016
A$_3$[PO$_4$\{WO(O$_2$)$_2$\}$_4$]^a	843	591, 526	1017
[Fe(EDTA)(O$_2$)]$^{3-}$ (aq.)	815	—, —	1018
Na$_2$[UO$_2$(O$_2$)(CO$_3$)]	980	615, 550	1019
Pt(O$_2$) (PPh$_3$)$_2$	821	460, 437	1020

^{a}A: [N(C$_6$H$_{13}$)$_4$]$^+$ ion.

TABLE III-59. Vibrational Frequencies and Structures of Dioxygen Adducts of Cobalt Ammine and Schiff-Base Complexes (cm^{-1})

Compound	Structure	$\nu(O_2)$	$\nu(Co—O)$	Ref.
Co(J-en)(py)O$_2$ [a]	Superoxo end-on	1146	—	1023
Co(salen)(py)O$_2$	Superoxo end-on	1144	527	1024
{[Co(NH$_3$)$_5$]$_2$O$_2$}Cl$_5$·3H$_2$O	Superoxo bridging	1122	620, 441	1025
{[Co(NH$_3$)$_5$]$_2$O$_2$}(NO$_3$)$_5$	Superoxo bridging	1122	—	1026
K$_5${[Co(CN)$_5$]$_2$O$_2$}H$_2$O	Superoxo bridging	1104	493	1027
[Co(salen)]$_2$O$_2$	Superoxo bridging	1011	533	1028 1029
[Co(salen)(pyO)]$_2$O$_2$ [b]	Peroxo bridging	910	535	1030
[Co(salen)(py)]$_2$O$_2$	Peroxo bridging	884	543	1029
[Co(J-en)(py)]$_2$O$_2$	Peroxo bridging	841	562	1023
[Co(DMG)(PPh$_3$)]$_2$O$_2$ [c]	Peroxo bridging	818	551	1031
K$_6${[Co(CN)$_5$]$_2$O$_2$}H$_2$O	Peroxo bridging	804	602	1027
{[Co(NH$_3$)$_5$]$_2$O$_2$}(NO$_3$)$_4$	Peroxo bridging	805	642, 547	1025
{[Co(NH$_3$)$_5$]$_2$O$_2$}(NCS)$_4$	Peroxo bridging	786	—	1026

[a] J-en: N,N'-ethylenebis(2,2'-diacetylethylideneaminato) anion.
[b] pyO: pyridine N-oxide.
[c] DMG: dimethylglyoximato anion.

in Raman spectra (1122 cm^{-1}) but is forbidden in IR spectra because the O–O bridge is centrosymmetric. However, the $\nu(O_2)$ of a dibridged complex ion, [(NH$_3$)$_4$Co(NH$_2$)(O$_2$)Co(NH$_3$)$_4$]$^{4+}$, is observed as 1068 cm^{-1} in IR spectra.[1025] N,N'-ethylenebis(salicylideniminato)cobalt, Co(salen), binds dioxygen

reversibility in the solid state.[1032] Figure III-65 shows the resonance Raman spectra of [Co(salen)]$_2$O$_2$ at ~100 K.[1028] The bands at 1011 and 533 cm^{-1} are

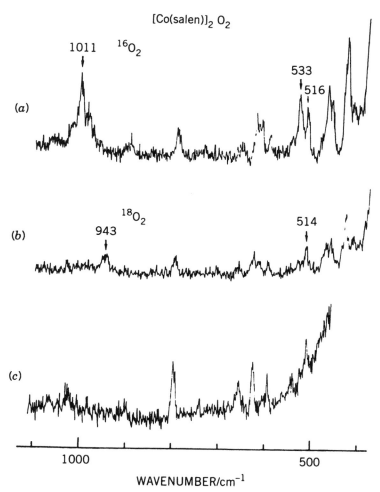

Fig. III-65. The RR spectra of (A) [Co(salen)]$_2$ ^{16}O$_2$, (B) [Co(salen)]$_2$ ^{18}O$_2$, and (C) Co(salen) (579 nm excitation, ~100 K).[1028]

shifted to 943 and 514 cm^{-1}, respectively, by ^{16}O$_2$–^{18}O$_2$ substitution, and thus assigned to the ν(O$_2$) and ν(CoO), respectively. The former frequency is unique in that it is between those of superoxo and peroxo complexes. However, this band is shifted to the normal peroxo range when the base ligands are coordinated *trans* to the dioxygen. Evidently, electron donation from the base to the bridging dioxygen is responsible for the shift of ν(O$_2$) to a lower frequency.

When a Co Schiff-base (SB) complex in a nonaqueous solvent absorbs oxygen in the presence of a base (B), the following equilibria are established:

$$[Co(SB)B] + O_2 \rightleftharpoons [Co(SB)B]O_2$$
$$[Co(SB)B]O_2 + [Co(SB)B] \rightleftharpoons [Co(SB)B]_2O_2$$

The $\nu(O_2)$ of the 1:1 (Co/O_2) adduct is near 1140 cm^{-1}, whereas that of the 1:2 adduct is between 920 and 820 cm^{-1} Using these bands as the markers, it is possible to examine the effects of oxygen pressure, temperature, and solvent polarity on the above equilibria. Figure III-66 shows the RR spectra of Co(J-en) in CH$_2$Cl$_2$ containing pyridine which were saturated with oxygen at various oxygen pressures and temperatures.[1023] It is seen that the concentration of the 1:1 adduct (1143 cm^{-1}) increases and that of the 1:2 adduct (836 cm^{-1}) decreases as the oxygen pressure increases (A → B) and as the temperature decreases (D → C → B). It was also noted that 1:1 adduct is favored in a polar solvent containing a relatively strong base.

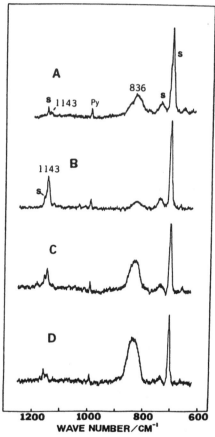

Fig. III-66. The RR spectra of Co(J-en) in CH$_2$Cl$_2$ containing 3% pyridine which was saturated with O$_2$ at various O$_2$ pressures and temperatures (580 nm excitation): (A) 1 atm, $-78°$C, (B) ~3 atm, $-80°$C; (C) ~3 atm, $-30°$C; (D) ~3 atm, $+20°$C. S and py denote the solvent and pyridine bands, respectively.[1023] For the structure of J-en, see the footnote in Table III-59.

(4) "Base-Free" Dioxygen Adducts of Metalloporphyrins and Related Compounds

Table III-60 lists the structures and observed frequencies of "base-free" dioxygen adducts of metalloporphyrins which were mostly prepared by matrix cocondensation reactions at low temperatures. The $\nu(O_2)$ varies continuously from the superoxo to the peroxo region. The $\nu(O_2)$ is the highest in $Co(TPP)O_2$ (superoxo) and the lowest in $[MoO(TPP)O_2]^-$ (peroxo) although some complexes exhibit the $\nu(O_2)$ in the intermediate region. The $\nu(O_2)$ of "base-free" $Co(TPP)O_2$ is 133 cm^{-1} higher than that of "base-bound" $Co(TPP)(1\text{-MeIm})O_2$ (1143 cm^{-1}) discussed in the following section. A shift of similar magnitude (127 cm^{-1}) is observed in going from $[Co(salen)]_2O_2$ to its pyridine adduct. Evidently, these shifts are caused by the base ligands, which increase the negative charge on the dioxygen ("base-ligand effect").

Table III-60 also indicates the "metal-ion effect"; the $\nu(O_2)$ is lowered and the mode of coordination is shifted from the end-on to the side-on as the metal-ion is changed in the following order:

$$\begin{array}{lcccc} & Co(TPP)O_2 & & Fe(TPP)O_2 & Mn(TPP)O_2 \\ \nu(O_2)\,(\text{cm}^{-1}) & 1278\ (\text{end-on}) & > & 1195\ (\text{end-on}) & \\ & & & 1106\ (\text{side-on}) & > & 983\ (\text{side-on}) \end{array}$$

TABLE III-60. Structures and Observed Frequencies (cm^{-1}) of "Base-Free" Dioxygen Adducts

Complex	Structure	$\nu(^{16}O_2)$	$\nu(^{18}O_2)$	Δ^a	$\nu(MO_2)$	Refs.
Co(TPP)O$_2$	End-on	1278	1209	69	345	1033
Co(OEP)O$_2$	End-on	1275	1202	73		1034
Co(J-en)O$_2$	End-on	1260	1192	68		1023
Co(salen)O$_2$	End-on	1235	1168	67		1035
[Co(salen)]$_2$O$_2$	Bridging	1011	943	68		1028
Fe(TPP)O$_2$	End-on	1195	1127	68	509	1036, 1037
	Side-on	1106	1043	63		
Fe(OEP)O$_2$	End-on	1190	1124	66		1036
	Side-on	1104	1042	62		
Fe(Pc)O$_2$	End-on	1207	1144	63	488	1036, 1038
Fe(salen)O$_2$	Side-on	1106	1043	63		1036
[Fe(salen)]$_2$O$_2$	Bridging	1001	943	58		1039
Ru(TPP)O$_2$	End-on	1167	1101	66		1040
[Ru(TPP)]$_2$O$_2$	Bridging	1114	1057	57		1040
[Os(TPP)]$_2$O$_2$	Bridging	1090	1030	70		1040
Mn(TPP)O$_2$	Side-on	983	933	50	433	1041, 1042
Mn(Pc)O$_2$	Side-on	992	935	57		1043
[MoO(TPP)O$_2$]$^-$	Side-on	876	—	—	521	1044
					490	

$^a\Delta = \nu(^{16}O_2) - \nu(^{18}O_2)$.

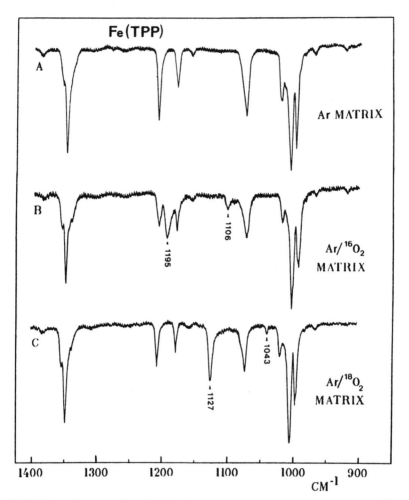

Fig. III-67. Infrared spectra of (A) Fe(TPP) in Ar matrix, (B) Fe(TPP) cocondensed with $^{16}O_2$/Ar (1/10), and (C) Fe(TPP) cocondensed with $^{18}O_2$/Ar (1/10) at ~15 K.[1036]

The IR spectra of Fe(TPP)O_2 shown in Fig. III-67 are of particular interest since it exhibits two $\nu(O_2)$ at 1195 and 1106 cm^{-1}, corresponding to the end-on and side-on isomers, respectively.[1036]

Table III-60 reveals another interesting trend; the $\nu(O_2)$ becomes lower as the in-plane ligand is changed in the order

	Fe(Pc)O_2		Fe(TPP)O_2		Fe(salen)O_2
$\nu(O_2)$ (cm^{-1})	1207	>	1195, 1106	≥	1106

Here, Pc denotes the phthalocyanato ion. This result indicates that the larger the π-conjugated system of the in-plane ligand is, the less the negative charge on the dioxygen, and the higher the $\nu(O_2)$ ("in-plane ligand effect").

(5) "Base-Bound" Dioxygen Adducts of Metalloporphyrins

"Base-bound" dioxygen adducts of metalloporphyrins are highly important as models of respiratory heme proteins (Sec. V-1, 2). Several reviews[1045,937,201] are available on vibrational spectra of "base-bound" dioxygen adducts of metalloporphyrins. Table III-61 lists the $\nu(^{16}O_2)$, $\nu(M-^{16}O_2)$, and $^{16}O_2/^{18}O_2$ isotope shifts observed for end-on, "base-bound" dioxygen adducts.

As mentioned previously, the $\nu(O_2)$ of these "base-bound" adducts are lower than those of the corresponding "base-free" adducts because of the "base-ligand effect." In a series of base-bound adducts such as $Co(TPP-d_8)(B)O_2$ (B: nitrogen-donor base), the $\nu(O_2)$ decreases linearly as the pK_a of the base increases.[1052] Thus, the $\nu(O_2)$ is the highest (1167 cm^{-1}) for the most acidic base (4-cyanopyridine, $pK_a = 1.90$) and the lowest (1151 cm^{-1}) for the most basic base (4-dimethylaminopyridine, $pK_a = 9.70$). If the pK_a value is regarded as a rough measure of σ-donation, this result suggests that the $\nu(O_2)$ is governed largely by the degree of σ-donation of these bases.

The $\nu(O_2)$ is shifted markedly to lower frequency when a thiolate ligand such as $SC_6H_5^-$ coordinates to a metal. For example, the $\nu(O_2)$ of the $[Co(TPP)(SC_6H_5)O_2]^-$ ion (1122 cm^{-1}) is 21 cm^{-1} lower than that of $Co(TPP)(py)O_2$ (1144 cm^{-1}) in the same solvent (CH_2Cl_2).[1053] As shown in Table III-61, a similar downshift is noted for Fe(II) porphyrins. Since the pK_a of $SC_6H_5^-$ (6.5) does not differ appreciably from that of py (5.25), the observed shift must be attributed largely to an increase in π-donation. This π-donation is promoted by two factors: (1) relative to py, the thiolate ligand possesses an extra electron on the $2p$ orbital which overlaps on the $d\pi$ orbital of the metal, and (2) the $SC_6H_5^-$ ligand tends to take an orientation which maximizes the $p_\pi - d\pi$ overlap and minimizes the steric repulsion from the mesophenyl groups. The biological significance of thiolate coordination will be discussed in conjunction with cytochrome P-450 (Sec. V-3(2)).

Table III-61 also shows the "metal-ion effect"—the $\nu(O_2)$ of the Fe(II)

TABLE III-61. Observed Frequencies of "Base-Bound" Dioxygen Adducts of Metalloporphyrins (cm^{-1}, in solution)

Compound[a]	$\nu(^{16}O_2)$	$\Delta\nu(O_2)^b$	$\nu(M-^{16}O_2)$	$\Delta\nu(MO_2)^c$	Ref.
$Co(TPP)(py)O_2$	1144	60	519	21	1046
$Co(TPP)(pip)O_2$	1142	64	509	20	1047
$Fe(TPP)(pip)O_2$	1157	64	575	24	1047
$Co(T_{piv}PP)(1,2-Me_2Im)O_2$	1153	65	—	—	1047
$Co(T_{piv}PP)(1-MeIm)O_2$	—	—	517	23	1047
$[Co(T_{piv}PP)(SC_6HF_4)O_2]^-$	1126	66	—	—	1048
$Fe(T_{piv}PP)(1,2-Me_2Im)O_2$	1159	66	—	—	1049
$Fe(T_{piv}PP)(1-MeIm)O_2$	—	—	568	23	1050
$[Fe(T_{piv}PP)(SC_6HF_4)O_2]^-$	1140	60	—	—	1051

[a]T_{piv}PP: picket-fence porphyrin (see Fig. III-62a).
[b]$\Delta\nu(O_2) = \nu(^{16}O_2) - \nu(^{18}O_2)$.
[c]$\Delta\nu(M-O_2) = \nu(M-^{16}O_2) - \nu(M-^{18}O_2)$.

adduct is higher than that of the corresponding Co(II) adduct. Although this is opposite to the case of "base-free" adducts discussed previously, the "metal-ion effect" on the $\nu(M-O_2)$ is the same in both cases; the $\nu(Fe-O_2)$ is always higher than the $\nu(Co-O_2)$. These results can be accounted for in terms of the following bonding schemes:[1039]

$$\overset{|}{\underset{|}{Co}} \xrightarrow{\sigma} O_2 \qquad B \xrightarrow{\sigma} \overset{|}{\underset{|}{Co}} \longrightarrow O_2$$

$$\overset{|}{\underset{|}{Fe}} \underset{\pi}{\overset{\sigma}{\rightleftharpoons}} O_2 \qquad B \xrightarrow{\sigma} \overset{|}{\underset{|}{Fe}} \underset{\pi}{\overset{\sigma}{\rightleftharpoons}} O_2$$

When the O_2 is bound to Co(II) (d^7), the Co–O_2 bond is formed mainly by σ-donation from Co(d_{z^2}) to the antibonding O_2 (π_g^*) orbital (i.e., Co(III)–O_2^-). In the case of Fe(II)(d^6), however, the Fe–O_2 bond is formed by σ-donation from O_2 (π_g^*) to Fe(d_{z^2}), which is counteracted by a stronger π-donation in the opposite direction. This would strengthen the Fe–O_2 bond and weaken the O–O bond relative to those of the corresponding Co(II) adduct. In fact, "base-free" Fe(II) adducts exhibit lower $\nu(O_2)$ and higher $\nu(M-O_2)$ than the corresponding Co(II) adducts. In "base-bound" Co(II) complexes, σ-donation from the base ligand causes a marked increase in the negative charge on O_2, thus causing a large downshift in $\nu(O_2)$ relative to "base-free" complexes. However, the base-ligand effect is much smaller in Fe(II) complexes because σ-donation from the base is opposed by σ-donation from the O_2. As a result, the O_2 in Fe(II) adducts is less negative than that of Co(II) complexes. Thus, the $\nu(O_2)$ of Fe(II) adducts become higher than those of Co(II) adducts. The $\nu(Fe-O_2)$ is always higher than the $\nu(Co-O_2)$ because of the multiple bond character of the former.

It is well established that the O_2 bound to myoglobin and hemoglobin is stabilized by forming the $N—H\cdots O_2$ hydrogen bond with the distal imidazole of the peptide chain (Sec. V-2). To mimic this "cavity effect," "protected" porphyrins such as "picket-fence" and "strapped" porphyrins (Fig. III-62) have been prepared. Figure III-68 compares the structures of the pickets in picket-fence porphyrin, Co(α_4-T_{piv}PP) and its derivative, Co(α_4-T_{neo}PP). It was found that the $\nu(O_2)$ decreases in the following order:

	Co(TPP-d_8)(B)O_2		Co(T_{piv}PP)(B)O_2		Co(T_{neo}PP)(B)O_2
$\nu(O_2)$ (cm^{-1})	1167	>	1161	>	1148

Here, B is 4-cyanopyridine and the spectra were measured in toluene.[1052,1054] Presumably, the $N—H\cdots O_2$ hydrogen bonding is weaker in Co(T_{piv}PP) than in Co(T_{neo}PP) because the repulsive force between the C(CH$_3$)$_3$ group and the

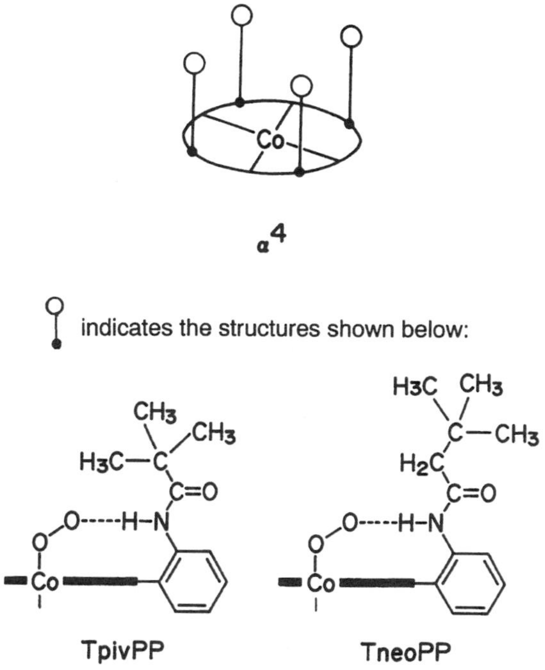

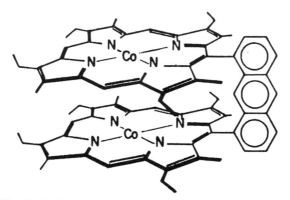

$_a{}^4$

⚬ indicates the structures shown below:

TpivPP   TneoPP

Fig. III-68. Structures of O_2 adducts of "picket-fence" Co(II) porphyrin and its derivative. Also see Fig. III-62(*a*).

bound dioxygen tends to push the pivalamide group outward. This repulsive force would be decreased when the pivalamide group is replaced by the neo-pentylcarboxamide group (T_{neo}). Odo et al.[1054] carried out an extensive RR study on $\nu(O_2)$ of Co(II) complexes of a variety of picket-fence porphyrins.

Pillard dicobalt cofacial diporphyrin, shown in Fig. III-69, is another type

Fig. III-69. Structure of "Pillard" dicobalt cofacial diporphyrin.

of modified porphyrin which is of great interest in structural chemistry. Since a large base ligand such as 4-(dimethylamino)pyridine cannot enter the inter-porphyrin cavity, it forms a bridging O_2 adduct which exhibits the $\nu(O_2)$ at 1098 cm^{-1} (superoxo-type) in RR spectra. This band is not observed if a small base such as γ-picoline is first added to the Co–Co complex solution and then the solution is oxygenated. This result indicates that a small base occupies the interporphyrin space so that formation of the Co–O_2–Co bridge is blocked.[1055] However, the 1098-cm^{-1} band is observed if the Co–Co complex solution is first saturated by O_2 and then a small base is added. Apparently, the Co–O_2–Co bond once formed is too stable to be cleaved by the addition of a base ligand.

The RR spectrum of Co(TPP-d_8)(py)O_2 in CH$_2$Cl$_2$ exhibits a single $\nu(O_2)$ band at 1143 cm^{-1}. However, this band becomes a doublet (1155 and 1139 cm^{-1}) when 1,2-dimethylimidazole is used as the base. An extensive study involving a variety of base ligands[1046] has shown that vibrational coupling between $\nu(O_2)$ and a nearby base ligand vibration of the same symmetry is responsible for the observed doublet structure and resonance enhancement. In

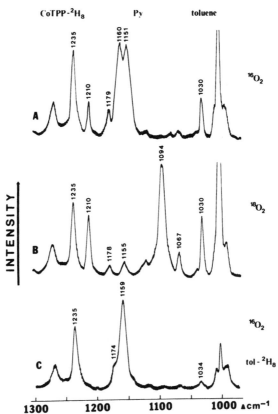

Fig. III-70. The RR spectra of Co(TPP-d_8) in toluene containing 3% pyridine at $-85°$C under $\sim$4 atm O_2 pressure. (A) $^{16}O_2$, (B) $^{18}O_2$, and (C) $^{16}O_2$ in toluene-d_8.[1052]

the case of Co(TPP-d_8)(py)$^{18}O_2$ in CH$_2$Cl$_2$, a weak py band appears at 1067 cm^{-1} in addition to the $\nu(^{18}O_2)$ at 1084 cm^{-1}. The 1067 cm^{-1} band of py is not observed in the case of the $^{16}O_2$ adduct. This is another example of vibrational coupling between the py mode and the $\nu(^{18}O_2)$.[1046]

The internal mode of a solvent molecule can also be resonance-enhanced via a similar mechanism.[1052] As shown in Fig. III-70, the RR spectrum of Co(TPP-d_8)(py)O_2 in toluene exhibits two strong bands at 1160 and 1151 cm^{-1} where the $\nu(^{16}O_2)$ band is expected (trace A). This doublet structure does not appear in toluene-d_8 (trace C) and is not observed in the $\nu(^{18}O_2)$ region (trace B). Toluene exhibits three bands at 1210 (T$_1$), 1178 (T$_2$), and 1155 cm^{-1} (T$_3$) with an intensity ratio of ca. 6 : 1 : 1 (trace B). Thus, it is reasonable to attribute the observed splitting to a strong vibrational coupling between $\nu(^{16}O_2)$ and T$_3$ which are very close in frequency. If $\nu(^{16}O_2)$ is shifted between T$_2$ and T$_3$ by using a weaker base (4-cyanopyridine), both internal modes of toluene are resonance-enhanced, as seen in Fig. III-71A. In this case, the magnitudes of frequency perturbation and resonance enhancement are less, relative to the previous case, since $\nu(^{16}O_2)$ is further from the solvent modes. As seen in Fig. III-71B, the multiple structure observed for Co(TPP-d_8) disappears completely when picket-fence porphyrin, Co(T$_{piv}$PP), is employed. This result indicates that the vibrational coupling observed for "unprotected porphyrin" cannot occur in picket-fence porphyrin because the four pivaloyl groups prevent the access of toluene

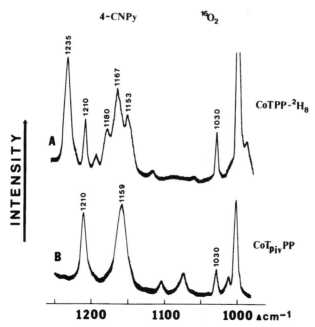

Fig. III-71. The RR spectra of (A) Co(TPP-d_8) and (B) Co(T$_{piv}$PP) in toluene containing 3% 4-cyanopyridine at $-85°$C under ~4 atom O$_2$ pressure.[1052]

to bound dioxygen. Thus, not only "frequency matching" but also "direct O_2-solvent association" is necessary to cause such vibrational coupling. Vibrational couplings between $\nu(O_2)$ of bound dioxygen and internal modes of base ligands and/or solvents have been found in many other systems.[1056,1057] Thus, RR spectra of dioxygen adducts of metalloporphyrins must be interpreted with caution. Proniewicz and Kincaid[1058] carried out quantitative treatments of these vibrational couplings using a Fermi resonance scheme.

III-20. METAL COMPLEXES CONTAINING OXO GROUPS

(1)   Metal Complexes Containing Monooxo Groups

There are many compounds containing monooxo groups ($M{=}O$) in which relatively heavy metal atoms are bonded to oxygen via double bonds. In most cases, their $\nu(M{=}O)$ vibrations can be assigned without difficulties since they are relatively free from vibrational couplings and appear strongly in the 1100–900 cm^{-1} region of IR spectra. Examples of $M{=}O$ group vibrations in inorganic compounds are found in the $\nu(ZX)$ vibrations of ZXY_3 (Table II-6g), ZXY_4 (Table II-7b), and ZXY_5 (Table II-8c)-type compounds discussed in Part A.

Recently, a new type of isomerism involving monooxo groups was found by Wieghardt et al.[1059] For example, the crystals of $[PF_6][W(O)LCl_2]$ (L: a tridentate ligand) can be obtained in the blue and green forms. X-Ray analyses show that the structures of these two forms are identical except for the $W{=}O$ and $W{-}N$ (*trans* to $W{=}O$) distances; the $W{=}O$ bond length in the blue form (1.72 Å) is shorter than that in the green form (1.89 Å). Correspondingly, the $\nu(W{=}O)$ of the former (980 cm^{-1}) is higher than that of the latter (960 cm^{-1}).

Similar isomerism has been reported for complexes containing the $Nb{=}O$ and $Nb{=}S$ groups.[1060] Thus, the yellow form of $Nb(O)Cl_3(PMe_3)_3$ exhibits the $\nu(Nb{=}O)$ at 882 cm^{-1} ($Nb{=}O$ distance, 1.78 Å), whereas its green isomer shows it at 871 cm^{-1} ($Nb{=}O$ distance, 1.93 Å). In $Nb(S)Cl_3(PMe_3)_3$, the orange form exhibits the $\nu(Nb{=}S)$ at 455 cm^{-1} ($Nb{=}S$ distance, 2.196 Å), while the green form shows it at 489 cm^{-1}, although its $Nb{=}S$ distance (2.296 Å) is longer than that of the orange form. The origin of this anomaly is not clear. Moreover, the origin of this new type of isomerism ("bond-stretch" isomerism) is not understood.

(2)   Metal Complexes Containing Dioxo Groups*

As stated in Sec. II-2 of Part A, the dioxo groups ($O{=}M{=}O$) such as $Mo(O)_2$, $Ru(O)_2$, $W(O)_2$, $Re(O)_2$, $Os(O)_2$, and $U(O)_2$ exhibit strong- to medium-intensity IR bands in the 1100–850 cm^{-1} region. Although the *trans* (lin-

---

*To avoid confusion with dioxygen adducts (MO_2), the dioxo groups are written as $M(O)_2$.

ear) dioxo group exhibits only the $\nu_a(O{=}M{=}O)$ vibration in IR spectra and only the $\nu_s(O{=}M{=}O)$ vibration in Raman spectra, the *cis* (bent) dioxo group is expected to show both vibrations in either spectra.[1061] Thus, *trans*-$[Os(O)_2(bipy)_2]$ (bipy: 2,2′-bipyridine) exhibits only one band at 872 cm^{-1}, whereas its *cis*-isomer shows two bands at 833 (ν_s) and 863 cm^{-1} (ν_a) in IR spectra.[1062] However, the *trans*-$[Re(O)_2(py)_4]^+$ ion exhibits both symmetric and antisymmetric $\nu(O{=}Re{=}O)$ at 907 and 822 cm^{-1}, respectively, in RR spectra (CH_3CN solution).[1063] The reason for this anomaly is not clear. The *cis*-$V(O)_2$ groups show the ν_s and $\nu_a(O{=}Ru{=}O)$ at 922–910 and 907–876 cm^{-1}, respectively.[1064] Similar results are reported for *cis*-$Mo(O)_2$[1065–1067] and *cis*-$W(O)_2$ groups.[1068,1069] In the case of $Ru(TPP)(O)_2$, the $\nu_s(O{=}Ru{=}O)$ vibration is observed at 808 cm^{-1} in RR spectra in solution.[1070] The corresponding vibration appears at 821 cm^{-1} in IR spectra.[1071]

(3) Oxoferryl Porphyrins and Related Compounds

Formation of oxoferryl porphyrin, FeO(porphyrin), from oxyironporphyrin is the most crucial step in the reaction cycle of cytochrome P-450 (Sec. V-3). Bajdor and Nakamoto[1072] first prepared FeO(TPP) via laser photolysis of $Fe(TPP)O_2$ in pure O_2 matrices at ~15 K. As shown in Fig. III-72, a new band appears at 852 cm^{-1} upon laser irradiation (406.7 nm, 1~2 mW) of $Fe(TPP)O_2$, and its intensity reaches the maximum after about 20 min. This band is shifted to 818 cm^{-1} by $^{16}O_2$–$^{18}O_2$ substitution. Similar experiments with scrambled dioxygen ($^{16}O_2/^{16}O^{18}O/^{18}O_2 \cong 1/2/1$) produce only two bands at 852 and 818 cm^{-1}. These results clearly indicate that the bands at 852 and 818 cm^{-1} are due to the $\nu(Fe{=}^{16}O)$ and $\nu(Fe{=}^{18}O)$, respectively, of FeO(TPP) which were formed by the cleavage of the bound dioxygen in $Fe(TPP)O_2$. ^{54}Fe–^{56}Fe substitution experiments further confirmed these assignments. A simple diatomic approximation gives a FeO stretching force constant of 5.32 mdyn/Å, which is much larger than that of the FeO bond in $[Fe(TPP)]_2O$ (3.8 mdyn/Å).[1073] A more detailed study by Proniewicz et al.[1074] shows that the Fe atom in oxoferryl porphyrin is Fe(IV) and low-spin, and that the FeO bond should be formulated as Fe(IV) $\rightleftharpoons$ O^{2-}. Here, the arrowed line indicates a σ-bond formed via the d_{z^2}-p_z overlap and the broken lines represent two π-bonds formed via the d_{xz}-p_x and d_{yz}-p_y overlaps. It is conventionally written as Fe$=$O. Similar experiments readily produced FeO(OEP) and FeO(salen) but not FeO(Pc). These results suggest that the O–O bond strength decreases in the order $Fe(Pc)O_2 > Fe(TPP)O_2 > Fe(salen)O_2$ as indicated in their $\nu(O_2)$ (Table III-60).

Oxoferrylporphyrin, FeO(TMP) (TMP: tetramesitylporphyrin), can also be produced by electrooxidation of Fe(TMP)(OH) in CH_2Cl_2 at $-40°C$. According to Czernuszewicz and Macor,[1075] it exhibits the $\nu(Fe{=}O)$ at 841 cm^{-1}. Cooling is necessary because it is unstable and readily reacts with CH_2Cl_2 to form Fe(TMP)Cl at higher temperature.

The $\nu(M{=}O)$ of other oxo porphyrins are $(V{=}O)TPP$ (1007 cm^{-1}), $(Cr{=}O)$ (TPP) (1025 cm^{-1}), and $(Mn{=}O)(TPP)$ (754 cm^{-1}). The sudden drop

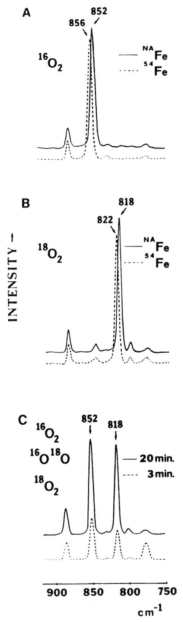

Fig. III-72. The RR spectra of Fe(TPP) cocondensed with O_2 at ~15 K (406.7 nm excitation): (A) ^{NA}Fe(TPP) with $^{16}O_2$; (B) ^{NA}Fe(TPP) with $^{18}O_2$, and (C) ^{NA}Fe(TPP) with isotopically scrambled O_2. The broken lines in (A) and (B) denote the spectra of ^{54}Fe(TPP) cocondensed with respective gases. All the spectra in (A), (B), and (C) (solid line) were obtained after 20-min laser irradiation. The dotted line in (C) indicates the spectrum obtained only after 3-min laser irradiation. ^{NA}Fe(Fe in natural abundance) contains 92% ^{56}Fe.[1072]

in $\nu(M{=}O)$ in going from the $V{=}O$ and $Cr{=}O$ to $Mn{=}O$ and $Fe{=}O$ may be accounted for by the d electron configurations discussed in Sec. III-21(3).

The $\nu(Fe{=}O)$ is lowered as much as 30–70 cm^{-1} if a base ligand is coordinated at the *trans* position to the $Fe{=}O$ group. For example, the $\nu(Fe{=}O)$ of FeO(TPP)(1-MeIm) is observed at 820 cm.$^{-1,}$[1076,1077] As will be discussed in Sec. V-3, the $\nu(Fe{=}O)$ of Horseradish Peroxidase Compound II(HRP-II) is at ~780 cm^{-1}, partly due to the presence of the proximal histidine at the *trans* position.

As mentioned in Sect. I-23 of Part A, the highest occupied molecular orbitals of metalloporphyrins are of a_{1u} or a_{2u} symmetry, and are nearly degenerate under $\mathbf{D}_{4h}$ symmetry. Thus, one-electron oxidation of a metalloporphyrin produces a π-cation radical of $^2A_{1u}$ or $^2A_{2u}$ symmetry. Figure III-73 shows the coefficients of the atomic p_z orbitals represented by the size of the circles.[1078] The shaded and nonshaded circles indicate their signs. It is seen that the $C_\beta{-}C_\beta$ bond is antibonding in the a_{1u} orbital, whereas it is bonding in the a_{2u} orbital. As shown previously (Table III-10), the ν_2 band of a metalloporphyrin is largely due to the $C_\beta{-}C_\beta$ stretching mode. Then, ν_2 is expected to downshift in an $^2A_{2u}$ radical and to upshift in an $^2A_{1u}$ radical. Using this and other criteria, Spiro and co-workers[1079,1080] concluded that Cu(TPP$^{\cdot+}$) and Ni(TPP$^{\cdot+}$) are $^2A_{2u}$ types while Ni(OEP$^{\cdot+}$) is an $^2A_{1u}$ type. Radical types of other metalloporphyrins have been studied by Babcock and co-workers.[1081,1082]

The π-cation radical of oxoferrylporphyrin is regarded as a model compound of HRP Compound I. Hashimoto et al.[1083] first assigned the $\nu(Fe{=}O)$

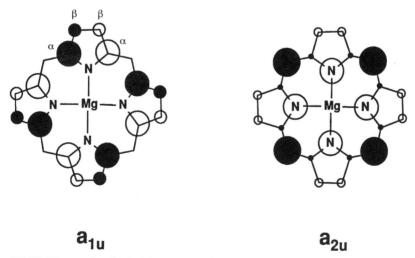

$$\mathbf{a_{1u}} \qquad\qquad \mathbf{a_{2u}}$$

Fig. III-73. The atomic orbital (AO) structure of Mg (Por) in the two highest occupied orbitals. The circle sizes are approximately proportional to the AO coefficients. The open circles represent negative signs of the upper lobe of the p_π AO's (reproduced with permission from Ref. 1078).

of FeO(TMP$^{\cdot+}$) at 828 cm^{-1} (in CH$_2$Cl$_2$/CH$_3$OH), whereas Kincaid et al.[1084] observed it at 801 cm^{-1} (in pure CH$_2$Cl$_2$). This discrepancy has been attributed to the difference in the solvents employed by these workers.[1085] In either case, the ν(Fe=O) is downshifted by 13–40 cm^{-1} upon formation of a π-cation radical. The ν(Fe=O) of FeO(TPP$^{\cdot+}$) was also observed at 815 cm^{-1} in O$_2$ matrices.[1086]

In the case of vanadyl porphyrins, the ν(V=O) is upshifted in (VO)(OEP$^{\cdot+}$)-($^2A_{1u}$), and downshifted in (VO)(TPP$^{\cdot+}$) and (VO)(TMP$^{\cdot+}$)($^2A_{2u}$).[1087,1088] A review article by Kitagawa and Mizutani[1089] provides more discussions on RR studies of highly oxidized metalloporphyrins including those discussed above.

In addition to RR studies mentioned above, IR studies are also useful in detecting the formation of π-cation radicals[1090,1091] and in distinguishing between the two radical types.[1092] The RR spectra of π-anion radicals of Zn(TPP) and its derivatives show that one-electron reduction lowers the symmetry of the porphyrin ring to $\mathbf{D}_{2d}$.[1093] The RR spectra of the highly reduced iron porphyrins [Fe(OEP)]$^-$ and [Fe(OEP)]$^{2-}$ in THF have been reported; the former contains high-spin Fe(I), whereas the latter is suggested to be the π-anion radical of the former.[1094]

Finally, the ν(Fe–OH) of the [Fe(OH)$_2$(TMPy-P2)]$^{3+}$ ion (TMPy-P2: tetrakis-(2N-methylpyridyl)porphyrinato anion, Sec. III-5(2)) in aqueous solution (pH = 12) has been observed at 447 and 498 cm^{-1}, respectively, for the high- and low-spin species.[1095] The ν(Fe–OC$_6$H$_5$) of a phenolate complex, Fe(OEP)(OC$_6$H$_5$) is at 607 cm^{-1}.[1096]

(4) Metal Complexes Containing Oxo Bridges

If the oxo bridge (M–O–M) is linear, the ν_a(MOM) is only IR-active and the ν_s(MOM) is only Raman-active. Although both become IR- and Raman-active

TABLE III-62. Structures and Vibrational Frequencies (cm^{-1}) of Oxo-Bridged Compounds

Compound	Structure	ν_a	ν_s	Ref.
[Fe(TPP)]$_2$O	Linear Fe—O—Fe	885	363	1073
[Fe(OEC)]$_2$O	Linear Fe—O—Fe	872	400	1097
[Cr(TPP)]$_2$O	Linear Cr—O—Cr	860	—	1098
(TPP)CrOFe(TPP)	Linear Cr—O—Fe	843	—	1099
[(WCl$_5$)]$_2$O	Linear W—O—W	—	223.3	1100
[(ReCl$_5$)]$_2$O	Linear Re—O—Re	—	229.8	1100
[HB(Pz)$_3$MoOCl]$_2$O^a	Linear Mo—O—Mo	786	380	1101
[HB(Pz)$_3$Fe(OAc)]$_2$O^a	Tribridged, bent	754	530	1102
[Fe$_2$O(phen)$_4$(H$_2$O)$_2$]$^{4+}$	Monobridged, bent (155°)	827	395	1103
[Fe$_2$O(tpa)$_2$(OC$_6$H$_5$)]$^{3+,b}$	Dibridged, bent (130°)	772	497	1103
[Fe$_2$O(tacn)$_2$(OAc)$_2$]$^{2+,c}$	Tribridged, bent (119°)	749	540	1103
[Mn$_2$O(tacn)$_2$(OAc)$_2$]$^{2+,c}$	Tribridged, bent	730	—	1104

^{a}HB(Pz)$_3$: tris(1-pyrazolyl)borato ion.

^{b}tpa: tris(2-pyridylmethyl)amine.

^{c}tacn: 1,4,7-triazacyclononane.

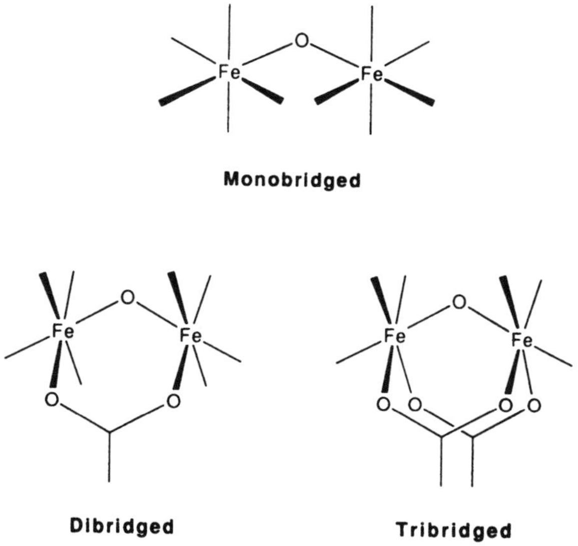

Monobridged

Dibridged   **Tribridged**

Fig. III-74. Structures of three types of oxo bridges (reproduced with permission from Ref. 1103).

in a bent geometry, the former is stronger in IR whereas the latter is stronger in Raman spectra. Table III-62 lists the structures and observed frequencies of typical oxo bridged complexes.

Three types of oxo bridges are illustrated in Fig. III-74. The tribridged complexes are of particular importance because they serve as model compounds of hemerythrin and other metalloproteins (Sec. V-4). Sanders–Loehr et al.[1103] carried out a systematic study on electronic and RR spectra of oxo-bridged dinuclear Fe(III) complexes in proteins and their model compounds. As shown in Appendix VII of Part A, the Fe—O—Fe angle can be calculated using the observed values of ν_a(FeOFe) and ν_s(FeOFe). These workers obtained excellent agreement between the observed (X-ray) and calculated Fe—O—Fe angles. They also noted that the molar Raman intensity of ν_s(FeOFe) is much larger in proteins than in model compounds, and suggested several possible reasons for this phenomenon.

III-21. COMPLEXES OF DINITROGEN AND RELATED LIGANDS

(1) Dinitrogen Complexes of Transition Metals

Since Allen and Senoff[1105] prepared the first stable dinitrogen (molecular nitrogen) compounds, $[Ru(N_2)(NH_3)_5]X_2$ ($X = Br^-$, I^-, BF_4^-, etc.), a large number of dinitrogen compounds have been synthesized. The chemistry and spectroscopy of these compounds have been reviewed extensively.[937,1106–1109] The structures of dinitrogen compounds are classified into three types:

$$M-N\equiv N \qquad M \overset{\overset{\displaystyle N}{\cdots}}{\underset{\displaystyle N}{\cdots}} \parallel \qquad M-N\equiv N-M$$

End-on	Side-on	Bridging
(linear)	(symmetrical)	(linear)

The terminal end-on coordination is most common. The $M-N_2$ bonding is interpreted in terms of the σ-donation and π-back-bonding, which were discussed in Secs. III-15 and III-17. Since N_2 is a weaker Lewis base than CO, π-back-bonding may be more important in nitrogen complexes than in CO complexes.[1110] Free N_2 exhibits $\nu(N\equiv N)$ at 2331 cm^{-1}, and this band shifts to 2220–1850 cm^{-1} upon coordination to the metal. Table III-63 lists the $\nu(N\equiv N)$ of typical complexes. Very little information is available for $\nu(M-N_2)$ and $\delta(M-N\equiv N)$ in the low-frequency region. Allen et al.[1110] assigned $\nu(Ru-N_2)$ of $[Ru(N_2)(NH_3)_5]^{2+}$-type compounds in the 508–474 cm^{-1} region, whereas other workers[1111,1112] attributed these bands to $\delta(Ru-N\equiv N)$. Figure III-75 shows the infrared spectrum of $[Ru(NH_3)_5N_2]Br_2$ obtained by Allen et al.

According to Srivastava and Bigorgne,[1121] $Co(N_2)H(PPh_3)_3$ exists in two forms in the solid state; one form exhibits $\nu(N\equiv N)$ at ca. 2087 cm^{-1}, and the other shows two bands of equal intensity at 2101 and 2085 cm^{-1}. However, their structural differences are unknown. Darensbourg[1122] obtained a linear relationship between $\nu(N\equiv N)$ and the absolute integrated intensity in a series of dinitrogen compounds.

Armor and Taube[1123] postulated the occurrence of the side-on structure as a possible transition state in linkage isomerization: $[(NH_3)_5Ru-^{14}N\equiv^{15}N]Br_2$ $\leftrightarrow [(NH_3)_5Ru-^{15}N\equiv^{14}N]Br_2$, Krüger and Tsay[1124] carried out X-ray analysis on $[\{(C_6H_5Li)_3Ni\}_2(N_2)\{(C_2H_5)_2O\}_2]_2$ and confirmed the presence of the side-on coordination in this compound; the $N-N$ distance was found to be extremely long (1.35 Å).

TABLE III-63. Observed $N\equiv N$ Stretching Frequencies (cm^{-1})

Complex	$\nu(N\equiv N)$	Ref.
$[Ru(N_2)(NH_3)_5]Br_2$	2105	1111
$[Ru(N_2)(NH_3)_5]I_2$	2124	1112
$[Os(N_2)(NH_3)_5]Cl_2$	2022, 2010	1113
$Co(N_2)(PPh_3)_3$	2093	1114
$Co(N_2)H(PPh_3)_3$	2105	1115
$Ir(N_2)Cl(PPh_3)_2$	2105	1116
$Ir(N_2)Cl(H)(PPh_3)_2(BF_4)$	2229	1117
trans-$Mo(N_2)_2(DPE)_2^b$	1970, (2020)	1118
cis-$W(N_2)_2(PMe_2Ph)_4$	1998, 1931	1119
$Co(N_2)(PR_3)(PR_2)_2^{2-a}$	1904 ~ 1864	1120

^{a}R = *n*-Bu or phenyl.
^{b}DPE = $Ph_2P-CH_2-CH_2-PPh_2$.

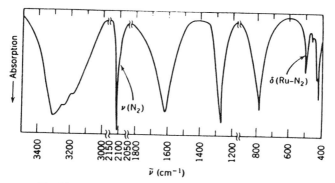

Fig. III-75. Infrared spectrum of $[Ru(NH_3)_5N_2]Br_2$.[1110]

The bridging $M-N\equiv N-M$ type complex should not show $\nu(N\equiv N)$ in the IR spectrum. However, it may show a strong $\nu(N\equiv N)$ in the Raman spectrum. Thus $[\{Ru(NH_3)_5\}_2(N_2)]^{4+}$ shows no infrared bands in the 2220–1920 cm^{-1} region, whereas a strong $\nu(N\equiv N)$ band appears at 2100 cm^{-1} in the Raman.[1125] If N_2 forms a bridge between two different metals, $\nu(N\equiv N)$ is observed in the infrared. For example, $\nu(N\equiv N)$ is at 1875 cm^{-1} in the infrared spectrum of $[(PMe_2Ph)_4ClRe-N_2-CrCl_3(THF)_2]$.[1126] According to X-ray analysis,[1127] an analogous compound, $[(PMe_2Ph)_4ClRe-N_2-MoCl_4(OMe)]$, has a $N\equiv N$ distance of 1.21 Å, and its $\nu(N\equiv N)$ is at 1660 cm^{-1}. As expected, the $\nu(N=N)$ of $[(CO)_5Cr-NH=NH-Cr(CO)_5]$ is very low (1415 cm^{-1}).[1128]

(2) Dinitrogen Adducts of Metal Atoms

Similar to $M(CO)_n$- and $M(O_2)_n$-type compounds discussed previously (Secs. III-17 and 19), it is possible to prepare simple $M(N_2)_n$-type adducts by reacting metal atoms with N_2 in inert gas matrices. Again the distinction of end-on and side-on geometry can be made by using the isotope scrambling techniques ($^{14}N_2$ + $^{14}N^{15}N$ + $^{15}N_2$). Figure III-76 shows the IR spectra of $Ni(N_2)$(end-on)[1129] and $Co(N_2)$(side-on).[1130] The observed frequencies (cm^{-1}) and assignments of the four bands of the former are as follows:

$Ni-^{14}N\equiv^{14}N$	$Ni-^{14}N\equiv^{15}N$	$Ni-^{15}N\equiv^{14}N$	$Ni-^{15}N\equiv^{15}N$
2089.9	2057.4	2053.6	2020.6

Table III-64 lists the $\nu(N_2)$ of $M(N_2)$-type complexes. All these adducts take the end-on structure except for $Co(N_2)$ and $Th(N_2)$.[1135] The structures of $M(N_2)_4$, $M(N_2)_3$, and $M(N_2)_2$ are tetrahedral, trigonal-planar, and linear, respectively, although slight distortions from these ideal symmetries occur due to the matrix effect. In all these compounds, the N_2 ligands take end-on geometry except $Pt(N_2)_2$, for which side-on coordination has been proposed.[1131]

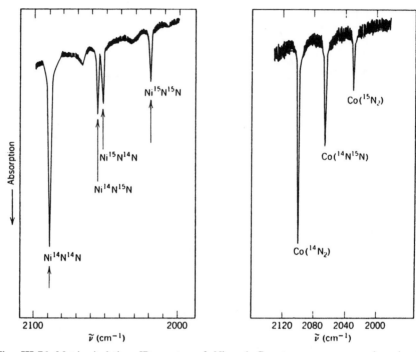

Fig. III-76. Matrix isolation IR spectra of Ni and Co atom vapors cocondensed with $^{14}N_2/^{14}N^{15}N/^{15}N_2/Ar$ at 10 K.[1129,1130]

Most of the studies described above have been made in the $\nu(N_2)$ region since low-frequency vibrations are generally weak and difficult to measure in inert gas matrics. However, the IR-active $\nu(M-N_2)$ vibrations of $Ni(N_2)_4$, $Rh(N_2)_4$, and $Pd(N_2)_2$ have been observed at 283, 345, and 340 cm^{-1}, respectively. The corresponding force constants are 0.81, 1.44, and 1.25 mdyn/Å, respectively.[1136] It should be noted that the Ni–CO stretching force constant of $Ni(CO)_4$ (1.80 mdyn/Å) is more than two times larger than that of $Ni(N_2)_4$ (0.81 mdyn/Å). Finally, UN_2 and PuN_2 prepared by the spattering techniques take the

TABLE III-64. Typical 1:1 (Metal/N_2) Adducts Prepared by Matrix Cocondensation Techniques

Adduct	$\nu(N_2)$	Ref.
$Ni(N_2)$	2088	1129
$Pd(N_2)$	2211	1129
$Pt(N_2)$	2173/2168	1131
$Co(N_2)$	2101	1130
$V(N_2)$	2216	1132
$Nb(N_2)$	1926/1931	1133
$Cr(N_2)$	2215	1134
$Th(N_2)$	1829	1135

linear N–M–N structures (Sec. II-2 of Part A). For more details on metal atom cocondensation reactions in inert gas matrices, see a review by Moskovits and Ozin.[1137]

(3) Nitrido complexes

If the N^{3-} ion coordinates to a metal, it is called a nitrido complex. Nitrido complexes of transition metals can be prepared by several methods, and their preparations, structures, and spectra have been reviewed by Griffith.[1138] The M≡N triple bonds are formed as a result of the strong π-donating property of the N^{3-} ion. Cleare and Griffith[1139] carried out an extensive study on vibrational spectra of nitrido complexes.

As shown in Table III-65, the $\nu(M\equiv N)$ of nonbridging nitrido complexes are generally found in the 1100–1000 cm^{-1} region. However, an exception was found for Fe(N)(OEP), which exhibits the $\nu(Fe\equiv N)$ at 876 cm^{-1}. This novel Fe(V) nitrido complex was prepared by laser photolysis of the corresponding azido complex at ~30 K[1146,1147]:

$$(OEP)Fe-N=N\equiv N \xrightarrow{h\nu} (OEP)Fe\equiv N + N_2$$

Figure III-77 shows the RR spectra of a thin film of $Fe(N_3)(OEP)$ which was irradiated by 488.0 nm line of Ar-ion laser. It is seen that, as the laser power increases, the bands characteristic of the azido group[233a] become weaker and a set of new bands at 876, 752, and 673 cm^{-1} become stronger. The 876 cm^{-1} band can be assigned to the $\nu(Fe\equiv N)$ based on the results of isotope substitution experiments involving ^{15}N N_2, $^{15}N_3$, and $^{54}Fe/^{56}Fe$. The remaining two bands are attributed to porphyrin core vibrations of the nitrido complex. Similar results have been obtained for the TPP analogs.

The large drop in the $\nu(M\equiv N)$ in going from nitrido porphyrins of Cr(V) and Mn(V) to Fe(V) may be accounted for in terms of the MO schemes shown

TABLE III-65. Vibrational Frequencies of Nonbridging Nitrido Complexes (cm^{-1})

Complex	$\nu(M\equiv N)$	Ref.
$[Nb(N)F_5]^{3-}$	1050	1140
$[Ta(N)Cl_5]^{3-}$	1040	1140
$Cr(N)(TTP)^a$	1017	1141
$Mo(N)(t\text{-BuO})_3$	1020	1142
$W(N)(t\text{-BuO})_3$	1010	1142
$Mn(N)(TPP)$	1052	1143
$[Tc(N)(py)_4]^{2+}$	1072	1144
$[Ru(N)Cl_4]^-$	1092	1145
$[Os(N)Cl_5]^{2-}$	1081	1139

^{a}TTP: tetra-p-tolylporphyrin.

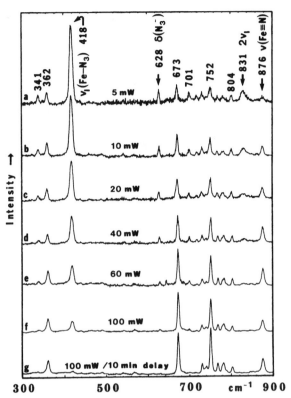

Fig. III-77. The RR spectra of a thin film of Fe(N$_3$)(OEP) at ~30 K, 488.0-nm excitation with different excitation power. (a) 5 mW, eight scans added; (b) 10 mW, four scans added; (c) 20 mW, two scans added; (d) 40 mW; (e) 60 mW; (f) 100 mW; (g) 100 mW, after 10-min preirradiation with 488.0 nm (100 mW).[114]

in Fig. III-78a.[1147] The electron configuration shown was proposed by Czernuszewicz et al.[1148] to explain a similar drop in ν(M=O) in going from O=V(TPP) and O=Cr(TPP) to O=Mn(TPP) and O=Fe(TPP) (Sec. III-20). According to their scheme, the bond order is low in the latter two compounds since electrons enter the antibonding d_{xz} and d_{yz} orbitals. Since the M(V)≡N^{3-} system is one-electron deficient relative to the M(IV)=O^{2-} system, the Fe(V)≡N^{3-} bond is isoelectronic with the Mn(IV)=O^{2-} bond. Thus, the electronic configuration of the Fe(V)≡N system may be $(d_{xy})^1(d_{xz})^1(d_{yz})^1$ (high spin) or $(d_{xy})^2(d_{xz})^1$ (low spin). The former is preferred because of the relatively small Fe≡N stretching force constant (5.07 mdyn/Å). Structure-sensitive bands of OEP and TPP porphyrins (Sec. III-5) indicate the Fe(V) state for these nitrido complexes.

Dinuclear complexes containing linear and symmetrical nitrido bridges (M–N–M) exhibit the ν_a(MNM) in IR and ν_s(MNM) in Raman spectra. These vibrations are observed at 985 and 228 cm^{-1}, respectively, for [Ta$_2$NBr$_{10}$]$^{2-}$,[1149] and at 904 and 203 cm^{-1}, respectively, for [Nb$_2$NBr$_{10}$]$^{3-}$.[1149] For [W$_2$NCl$_{10}$]$^-$, the ν_a(WNW) was observed at 945 cm^{-1}.[1150] In (Fe(TPP))$_2$N containing low-

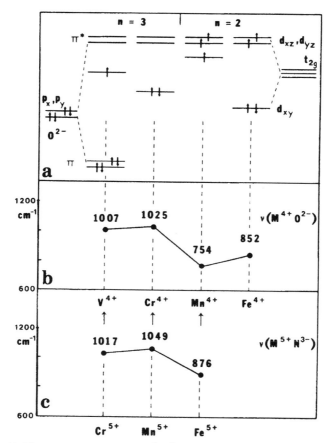

Fig. III-78. (a) Electron configuration of $M^{4+}O^{2-}$ porphyrin. (b) Variation of $\nu(M^{4+}O^{2-})$ in $M^{4+}O^{2-}$ porphyrins. (c) Variation of $\nu(M^{5+}N^{3-})$ in $M^{5+}N^{3-}$ porphyrins.[1147]

spin Fe(III) centers, the ν_s(FeNFe) has been observed at 424 cm^{-1} in RR spectra.[1151] The corresponding FeNFe stretching force constant (4.5 mdyn/Å) is slightly larger than the FeOFe stretching force constant (3.8 mdyn/Å) in (Fe(TPP))$_2$O.[1073] Thus, the FeNFe bridge may be expressed as Fe═N═Fe (bond order, 1.5).

Clear and Griffith[1139] list the vibrational frequencies of other nitrido bridges containing Ru, Os, and Ir. In a trinuclear complex ion, [Ir$_3$N(SO$_4$)$_6$(H$_2$O)$_3$]$^{4-}$, the nitrido atom is bonded to three Ir atoms trigonally and the ν_a(Ir$_3$N) is observed at 780 cm^{-1}.[1139]

III-22. COMPLEXES OF DIHYDROGEN AND RELATED LIGANDS

(1) Metal Complexes of Dihydrogen

The cocondensation reaction of Pd vapor with H$_2$ diluted with Kr produces two types of Pd(H$_2$)[1152]:

$$Pd—H \overset{H}{\diagup} \qquad Pd— \overset{H}{\underset{H}{|}}$$

End-on   Side-on

The former exhibits the $\nu(Pd–H_2)$ at 771 cm^{-1}, whereas the latter shows it as a doublet at 894.5/885.5 cm^{-1} in IR spectra.

In other complexes, dihydrogen is known to coordinate to a transition metal atom only in the side-on fashion. The metal-H_2 bonding is interpreted in terms of a delicate balance between σ-donation to the metal and back-donation to $\sigma*$, as illustrated below[1153]:

Photolysis of a mixture of $Fe(CO)_2(NO)_2$ with H_2 in liquid Xe ($-104°C$) produces $Fe(CO)(NO)_2(H_2)$, which exhibits the $\nu(H_2)$, $\nu_a(Fe–H_2)$, and $\nu_s(Fe–H_2)$ at 2973, 1374, and 870 cm^{-1}, respectively.[1154] These dihydrogen vibrations have been observed for $M(CO)_5(H_2)$ (M = Cr, Mo and W),[1155,1156] $M(CO)_3(Cp)(H_2)$ (M = V and Nb),[1157] and *cis*-$W(CO)_4(C_2H_4)(H_2)$,[1158] which were prepared by similar methods. In $V(CO)_3(Cp)(H_2)$, the $\nu(H_2)$ at 2642 cm^{-1} is shifted to 2377 and 1998 cm^{-1}, respectively, by HD and D_2 substitution. The fact that the IR spectrum of the HD compound exhibits the $\nu(HD)$ at an intermediate frequency between those of the H_2 and D_2 compounds and is not an overlap of the $\nu(V–H)$ and $\nu(V–D)$ bands (*vide infra*) provides definitive evidence that it is a dihydrogen compound and not a dihydride.

More stable dihydrogen complexes of the type, $M(CO)_3(PR_3)_2(H_2)$, where M is Mo and W and R is cyclohexyl (Cy) or isopropyl (*i*-Pr), were first prepared by Kubas et al.[1159,1160] Figure III-79 shows the IR spectra of $W(CO)_3(P(i\text{-}Pr)_3)_2(H_2)$ and its D_2 analog. The $\nu(H_2)$, $\nu_a(WH_2)$, $\nu_s(WH_2)$, and $\delta(WH_2)$ are observed at 2695, 1567 (~1140), 953 (704), and 465 (312) cm^{-1}, respectively. The corresponding frequencies of the D_2 analog are given in the parentheses. It should be noted that these complexes are still labile and must be kept in an H_2-enriched atmosphere.

(2) Hydrido Complexes

Vibrational spectra of hydrocarbonyls have been discussed in Sec. III-17(5). Metal complexes containing terminal hydrido groups (M–H) exhibit the $\nu(M–H)$ and $\delta(M–H)$ in the 2250–1700 and 800–600 cm^{-1} regions, respectively; Table III-66 lists M–H frequencies of typical complexes.

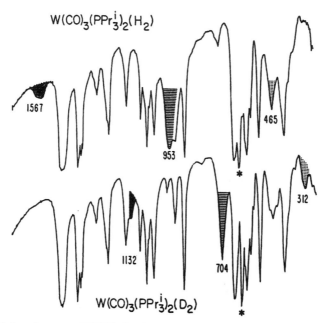

Fig. III-79. Infrared spectra of $W(CO)_3(P(i\text{-}Pr)_3)_2(H_2)$ and its D_2 analog (Nujol mull) (reproduced with permission from Refs. 1159 and 1160).

The $\nu(MH)$ is sensitive to other ligands, particularly those in the *trans*-position in the square-planar Pt(II) complexes. Thus Chatt et al.[1167] found that the order of $\nu(PtH)$ in *trans*-[Pt(H)X(PEt$_3$)$_2$] is as follows:

X =	NO$_3^-$	<	Cl$^-$	<	Br$^-$	<	I$^-$	<	NO$_2^-$	<	SCN$^-$	<	CN$^-$
$\nu(PtH)(cm^{-1})$	2242	>	2183	>	2178	>	2156	>	2150	>	2112	>	2041

This is the increasing order of *trans*-influence. Church and Mays[1168] found that the NMR Pt—H coupling constant and $\nu(PtH)$ decrease in the same order in

TABLE III-66. M—H Frequencies of Hydrido Complexes (cm^{-1})

Complex	$\nu(MH)$	$\delta(MH)$	Ref.
trans,trans-[Cr(H)(CO)$_2$(NO)(PEt$_3$)$_2$]	1661	—	1161
[Mo(H)(CN)$_7$]$^{4-}$	1805	—	1162
cis-[Fe(H)(CO)$_3$P(OC$_6$H$_5$)$_3$]$^-$	1895	—	1163
trans-[Fe(H)Cl{C$_2$H$_4$(PEt$_2$)$_2$}$_2$]	1849	656	1164
[Co(H)(CN)$_5$]$^{3-}$	1840	774	1165
[Rh(H)(CN)$_5$]$^{3-}$	1980	781	1162
[Ir(H)(CN)$_5$]$^{3-}$	2040	811	1162
Ir(H)(COD){As(C$_6$H$_5$)$_3$}$_2$ ^{a}	2030	—	1166

^{a}COD: 1,5-cyclooctadiene.

the *trans*-[Pt(H)L(PEt$_3$)$_2$]$^+$ series:

L =		py	<	CO	<	PPh$_3$	<	P(OPh)$_3$	<	P(OMe)$_3$	<	PEt$_3$
J(PtH)(Hz)		1106	>	967	>	890	>	872	>	846	>	790
ν(PtH)(cm^{-1})		2216	>	2167	>	2100	>	2090	>	2067	<	2090

In the series above, the σ-donor strength of L increases as the J(PtH) value decreases and ν(PtH) shifts to a lower frequency. Atkins et al.[1169] found linear relationships between the chemical shift of the hydride, the Pt–H coupling constant, ν(PtH), and the pK_a value of the parent carboxylic acid in a series of *trans*-[Pt(H)L(PEt$_3$)$_2$], where L is a carboxylate ligand.

X-Ray analyses have shown that both terminal and bridging hydrido groups exist in each of the three complexes shown in Fig. III-80. Compounds I and II exhibit the terminal and bridging ν(MH) in the 2100–1800 and 1200–950 cm^{-1} regions, respectively, while three ν(TaH) vibrations (1810, 1720, and 1650 cm^{-1}) are reported for Compound III.

Metal carbonyl ions such as [Co$_6$(CO)$_{15}$H]$^-$ contain rare interstitial hydrogens. The neutron diffraction study on its [N(P(C$_6$H$_5$)$_3$)$_2$]$^+$ salt indicates that the H atom is located at the center of the Co$_6$ octahedron.[1173] The same con-

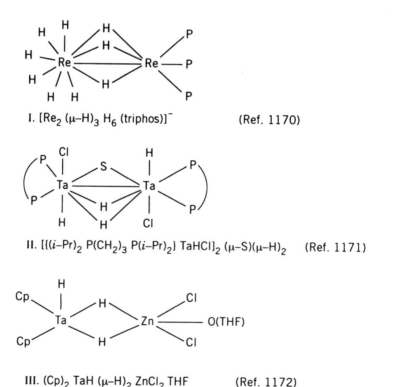

I. [Re$_2$ (μ–H)$_3$ H$_6$ (triphos)]$^-$ (Ref. 1170)

II. [{(*i*-Pr)$_2$ P(CH$_2$)$_3$ P(*i*-Pr)$_2$} TaHCl]$_2$ (μ–S)(μ–H)$_2$ (Ref. 1171)

III. (Cp)$_2$ TaH (μ–H)$_2$ ZnCl$_2$ THF   (Ref. 1172)

Fig. III-80. Structures of complexes containing both terminal and bridging hydrido groups.

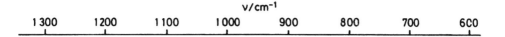

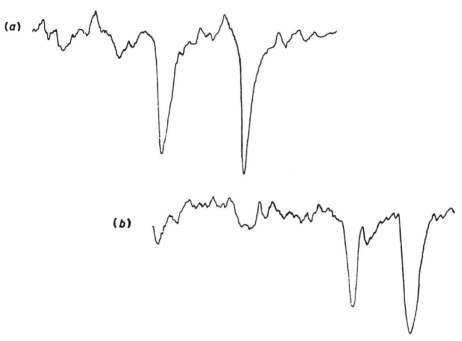

Fig. III-81. Infrared spectra of (*a*) K[Co$_6$(CO)$_{15}$H] and (*b*) K[Co$_6$(CO)$_{15}$D] in Nujol mull at ~110 K (reproduced with permission from Ref. 1175).

clusion has been reached by the inelastic neutron scattering (INS) study of the Cs$^+$ salt since it revealed the presence of a single ν(CoH) (triply degenerate) at 1056 cm^{-1},[1174]. Figure III-81 shows the low-temperature IR spectra of its K$^+$ salt obtained by Stanghellini and Longoni.[1175] It is seen that two bands at 1086 and 949 cm^{-1} are shifted to 772 and 677 cm^{-1}, respectively, by H/D substitution. Possible reasons for the observed splitting have been discussed by these workers. Recently, Corbett et al.[1176] observed two INS bands at 790 and 480 cm^{-1} for Li$_6$[Zr$_6$Cl$_{18}$H] at ~15 K, and assigned them to the E and A_1 vibrations, respectively, of the interstitial hydrogen at the trigonal (C$_{3v}$) site within the octahedral Zr$_6$ cluster.

III-23. HALOGENO COMPLEXES

Halogens (X) are the most common ligands in coordination chemistry. Several review articles[1177–1179] summarize the results of extensive infrared studies on halogeno complexes. Section II of Part A lists the vibrational frequencies of many halogeno complexes. Here the vibrational spectra of halogeno complexes containing other ligands are discussed. In most cases $\nu(MX)$ can readily be assigned by halogen or metal (isotope) substitution.

(1) Terminal Metal–Halogen Bond

Terminal MX stretching bands appear in the regions of 750–500 cm^{-1} for MF, 400–200 cm^{-1} for MCl, 300–200 cm^{-1} for MBr, and 200–100 cm^{-1} for MI. According to Clark and Williams,[124] the $\nu(MBr)/\nu(MCl)$ and $\nu(MI)/\nu(MCl)$ ratios are 0.77–0.74 and 0.65, respectively. Several factors govern $\nu(MX)$.[1180] If other conditions are equal, $\nu(MX)$ is higher as the oxidation state of the metal is higher. Examples have already been given for tetrahedral MX_4- and octahedral MX_6-type compounds, discussed in Section II of Part A. It is interesting to note, however, that in the $[M(dias)_2Cl_2]^{n+}$ series* $\nu(MCl)$ changes rather drastically in going from Ni(III) to Ni(IV) (Fig. III-89), while very little change is observed between Fe(III) and Fe(IV):

	d^4	d^5	d^6	d^7
	Fe(IV)	Fe(III)	Ni(IV)	Ni(III)
$\nu(MCl)$ (cm^{-1})	390	384	421	240

This was attributed to the presence of one electron in the antibonding e_g^* orbital in the Ni(III) complex.[1181]

If other conditions are equal, $\nu(MX)$ is higher as the coordination number of the metal is smaller. Table III-67 indicates the structure dependence of $\nu(NiX)$, obtained by Saito et al.[129] According to Wharf and Shriver,[1186] the SnX stretching force constants of halogenotin compounds are approximately proportional to the oxidation number of the metal divided by the coordination number of the complex.

It is interesting to note that the $\nu(SnCl)$ of free SnCl$_3^-$ ion [289 (A_1) and 252 (E) cm^{-1}] are shifted to higher frequencies upon coordination to a metal. Thus $\nu(SnCl)$ of $[Rh_2Cl_2(SnCl_3)_4]^{2-}$ are at 339 and 323 cm^{-1}. According to Shriver and Johnson,[1187] the L–X force constant of the LX_n-type ligand will increase upon coordination to a metal if X is significantly more electronegative than L. In the example above, chlorine is more electronegative than tin. In metal amine complexes (see Sec. III-1), $\nu(NH)$ shifts to lower frequencies because nitrogen is more electronegative than hydrogen. As expected, the $\nu(GeCl)$ of free GeCl$_3^-$ ion [303 (A_1) and 285 (E) cm^{-1}] are also shifted to higher frequencies in $[Pd(PhNC)(PPh_3)(GeCl_3)Cl]$ (384 and 360 cm^{-1}).[1188]

The MX vibrations are very useful in determining the stereochemistry of the

---

*dias: *o*-phenylenebis(dimethylarsine).

TABLE III-67. Structural Dependence of NiX Stretching Frequencies (cm^{-1})^a

Stretching Frequency	Linear Triatomic	*trans*-Planar	*cis*-Planar	Tetrahedral	*trans*-Octahedral
ν(NiCl)	NiCl$_2$^{b} 521	Ni(PEt$_3$)$_2$Cl$_2$^{c} 403	Ni(DPE)Cl$_2$^{d} 341, 328	Ni(PPh$_3$)$_2$Cl$_2$^{c} 341, 305	Ni(py)$_4$Cl$_2$ 207
ν(NiBr)	NiBr$_2$^{b} 414	Ni(PEt$_3$)$_2$Br$_2$^{c} 338	Ni(DPE)Br$_2$^{d} 290, 266	Ni(PPh$_3$)$_2$Br$_2$^{e} 265, 232	Ni(py)$_4$Br$_2$ 140
ν(NiI)			Ni(DPE)I$_2$^{d} 260, 212	Ni(PPh$_3$)$_2$I$_2$^{e} 215	Ni(py)$_4$I$_2$ 105
$\dfrac{\nu(\text{NiBr})}{\nu(\text{NiCl})}$	0.80	0.84	0.83^f	0.77^f	0.68
$\dfrac{\nu(\text{NiI})}{\nu(\text{NiCl})}$			0.70^f	0.67^f	0.51

^{a}DPE = 1,2-bis(diphenylphosphino)ethane.
^{b}Ref. 1182.
^{c}Ref. 1183.
^{d}Ref. 1184.
^{e}Ref. 1185.
^{f}This value was calculated by using average frequencies of two bands.

complex. Appendix V of Part A tabulates the number of infrared- and Raman-active vibrations of various MX$_n$Y$_m$-type compounds. Using these tables, it is possible to determine the stereochemistry of a halogeno complex simply by counting the number of ν(MX) fundamentals observed. Examples of this method will be given in the following sections.

(a) Square-Planar complexes. Vibrational spectra of planar M(NH$_3$)$_2$X$_2$ [M = Pt(II) and Pd(II)] were discussed in Sec. III-1. The *trans*-isomer ($\mathbf{D}_{2h}$) exhibits one ν(MX) (B_{3u}), whereas the *cis*-isomer ($\mathbf{C}_{2v}$) exhibits two ν(MX) (A_1 and B_2) bands in the infrared. The infrared spectra of *cis*- and *trans*-[Pd(NH$_3$)$_2$Cl$_2$] were shown in Fig. III-5. Similar results have been obtained for a pair of *cis*- and *trans*-[Pt(py)$_2$Cl$_2$][1189] and PtL$_2$X$_2$, where L is one of a variety of neutral ligands.[1190]

In planar Pt(II) and Pd(II) complexes, ν(MX) is sensitive to the ligand *trans* to the M–X bond. Thus the effect of "*trans*-influence"[1191] has been studied extensively by using infrared spectroscopy. In the [PtCl$_3$L]$^-$ series,[1192] ν(PtCl$_{trans}$) follows the order:

L =	CO	SMe$_2$	C$_2$H$_4$	SEt$_2$	AsEt$_3$	PPh$_3$	PMe$_3$	AsMe$_3$	PEt$_3$
ν(PtCl) (cm^{-1})	322 >	310 ~	309 ~	307 >	280 ~	279 ~	275 ~	272 ~	271

Their order represents an increasing degree of *trans*-influence, since ν(PtCl) becomes lower as a ligand of stronger *trans*-influence is introduced *trans* to the Pt–Cl bond. It was found that ν(PtCl$_{cis}$) is insensitive to the change in L. An order of *trans*-influence such as

$$Cl^- < Br^- < I^- \sim CO < CH_3 < PR_3 \sim AsR_3 < H$$

was noted from the order of $\nu(M-Cl_{trans})$ in a series of octahedral Rh(III) and Os(III) complexes.[1193]

Fujita et al.[1194] prepared two isomers of $PtCl(C_2H_4)$(L-ala), where L-ala is L-alanino anion:

(N-isomer)   (O-isomer)

I   II

Isomers I and II exhibit their ν(PtCl) at 360 and 340 cm^{-1}, respectively. Since the *trans*-influence of the N-donor is expected to be stronger than that of the O-donor, the structures of these two isomers have been assigned as shown above.

Complexes of the type $Ni(PPh_2R)_2Br_2$ (R = alkyl) exist in two isomeric forms: tetrahedral (green) and *trans*-planar (brown). Distinction between these two can be made easily since the numbers and frequencies of infrared-active ν(NiBr) and ν(NiP) are different for each isomer. Wang et al.[1195] studied the infrared spectra of a series of compounds of this type, and confirmed that ν(NiBr) and ν(NiP) are at ca. 330 and 260 cm^{-1}, respectively, for the planar form and at ca. 270–230 and 200–160 cm^{-1}, respectively, for the tetrahedral form. The presence or absence of the 330-cm^{-1} band is particularly useful in distinguishing these two isomers. According to X-ray analysis,[1196] the green form of $Ni(PPh_2Bz)_2Br_2$ (Bz = benzyl) is a mixture of the planar and tetrahedral molecules in a 1 : 2 ratio. Ferraro et al.[1197] studied the effect of high pressure on the infrared spectra of this compound, and found that all the bands characteristic of the tetrahedral form disappear as the pressure is increased to ca. 20,000 atm. This result indicates that the tetrahedral molecule can be converted to the planar form under high pressure if the energy difference between the two is relatively small. This conversion is completely reversible; the original green form is recovered as the pressure is reduced. High-pressure infrared spectroscopy has also been used to distinguish symmetric and antisymmetric MX stretching vibations. For example, Fig. III-82 shows the effect of pressure on ν_a(PtCl) and ν_s(PtCl) of Pt(NBD)Cl$_2$ (NBD: norbornadiene).[1198] It is seen that by increasing pressure, the intensity of ν_s(PtCl) is suppressed to a greater degree than that of ν_a(PtCl). For high-pressure vibrational spectroscopy, see a review by Ferraro.[1199,1200]

(b) Octahedral Complexes. cis-MX$_2$L$_4$ (**C**$_{2v}$) should exhibit two ν(MX), while trans-MX$_2$L$_4$ (**D**$_{4h}$) should give only one ν(MX) in the infrared. Thus cis-[IrCl$_2$(py)$_4$]Cl shows two ν(IrCl) at 333 and 327 cm^{-1}, while trans-

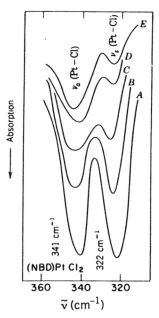

Fig. III-82. Effect of pressure on Pt–Cl stretching bands of Pt(NBD)Cl$_2$: (*A*) 1 atm; (*B*) 6000 atm; (*C*) 12,000 atm; (*D*) 18,000 atm, and (*E*) 24,000 atm.[1198]

[IrCl$_2$(py)$_4$]Cl exhibits only one ν(IrCl) at 335 cm^{-1}.[124] If MX$_3$L$_3$ is *fac* (**C**$_{3v}$), two ν(MX) are expected in the infrared. If it is *mer* (**C**$_{2v}$), three ν(MX) should be infrared-active. As is shown in Fig. III-13, *fac*-[RhCl$_3$(py)$_3$] gives two bands at 341 and 325 cm^{-1} and *mer*-[RhCl$_3$(py)$_3$] shows three bands at 355, 322, and 295 cm^{-1}.[1,124]

In MX$_4$L$_2$-type compounds, the number of IR-active ν(MX) is one for the *trans*-isomer (**D**$_{4h}$) and four for the *cis*-isomer (**C**$_{2v}$). For example, *trans*-[PtCl$_4$(NH$_3$)$_2$] exhibits one ν(PtCl) at 352 cm^{-1} (with a shoulder at 346 cm^{-1}), whereas *cis*-[PtCl$_4$(NH$_3$)$_2$] exhibits four ν(PtCl) at 353, 344, 330, and 206 cm^{-1}.[1,1201] Using Sn isotopes, Ohkaku and Nakamoto[1202] confirmed that *trans*-[SnCl$_4$L$_2$] (L = py, THF, etc.) exhibits one ν(SnCl) in the 342–370 cm^{-1} region, while *cis*-[SnCl$_4$(L—L)] (L—L = bipy, phen, etc.) shows four ν(SnCl) in the 340–280 cm^{-1} region. For MX$_5$L(**C**$_{4v}$), one expects three ν(MX) in the infrared. The ν(InCl) of [InCl$_5$(H$_2$O)]$^{2-}$ were observed at 280, 271, and 256 cm^{-1}.[1203]

(2) Bridging Metal–Halogen Bond

Halogens tend to form bridges between two metal atoms. In general, bridging MX stretching frequencies [ν_b(MX)] are lower than terminal MX stretching frequencies [ν_t(MX)]. Vibrational spectra of simple M$_2$X$_6$-type ions having bridging halogens were discussed in Sec. II-10 of Part A. Table III-68 lists the ν_t(MX) and ν_b(MX) of bridging halogeno complexes containing other ligands.

TABLE III-68. Terminal and Bridging Metal-Halogen Stretching Frequencies (cm^{-1})

Compound[a]	ν_t(MX)	ν_b(MX)	ν_b/ν_t [b]	Ref.
trans-Pd$_2$Cl$_4$L$_2$	360–339	308–294	0.86	1204
		283–241	0.75	
trans-Pt$_2$Cl$_4$L$_2$	368–347	331–317	0.91	1204
		301–257	0.78	
Pd$_2$Br$_4$L$_2$	285–265	220–185	0.74	1205
		200–165	0.66	
Pt$_2$Br$_4$L$_2$	260–235	230–210	0.89	1205
		190–175	0.74	
Pt$_2$I$_4$L$_2$	200–170	190–150	0.92	1205
		150–135	0.77	
Ni(py)$_2$Cl$_2$	—	193, 182	—	1206
Ni(py)$_2$Br$_2$	—	147	—	1206
Co(py)$_2$Cl$_2$				
Monomeric	347, 306	—		131
Polymeric	—	186, 174		131

[a]L = PMe$_3$, PEt$_3$, PPh$_3$, and so on.
[b]These values were calculated using average frequencies.

The *trans*-planar M$_2$X$_4$L$_2$-type compounds (C$_{2h}$) exhibit three infrared-active (B$_u$) ν(MX) modes: one ν(MX$_t$), and two ν(MX$_b$). For the latter two,

$$\text{L} \diagdown \underset{\text{X}_t}{\overset{\text{X}_b}{\text{M}}} \diagdown \underset{\text{X}_b}{\overset{\text{X}_t}{\text{M}}} \diagdown \text{X}_t$$

the higher-frequency band corresponds to ν(MX$_b$) *trans* to X, whereas the lower-frequency mode is assigned to ν(MX$_b$) *trans* to L since it is sensitive to the nature of L.[1204] Strong coupling is expected, however, among these modes since they belong to the same symmetry species.

The [Pt$_3$Br$_{12}$]$^{2-}$ ion takes a structure of nearly D$_{2h}$ symmetry in which two Pt(IV)Br$_6$ octahedra share edges with one planar Pt(II)Br$_4$ group:

$$\text{Br}_{t,eq} \diagup \underset{\text{Br}}{\overset{\text{Br}_{t,ax}}{\text{Pt}}} \diagdown \underset{\text{Br}_b}{\overset{\text{Br}_b}{\text{Pt}}} \diagdown \underset{\text{Br}_b}{\overset{\text{Br}_b}{\text{Pt}}} \diagdown \underset{\text{Br}}{\overset{\text{Br}_{t,ax}}{}} \text{Br}_{t,eq}$$

Here, ax and eq denote the axial and equatorial atoms, respectively. Figure III-83 shows the IR and Raman spectra of (TBA)$_2$[Pt$_3$Br$_{12}$] obtained by Hillebrecht et al.[1207] These spectra have been assigned completely via normal coordinate analysis. Four different Pt–Br stretching force constants were necessary to distinguish the Pt(IV)–Br$_{t,eq}$, Pt(IV)–Br$_{t,ax}$, Pt(II)–Br$_b$, and Pt(IV)–Br$_b$ bonds (1.75, 1.69, 1.10, and 1.05 mdyn/Å, respectively).

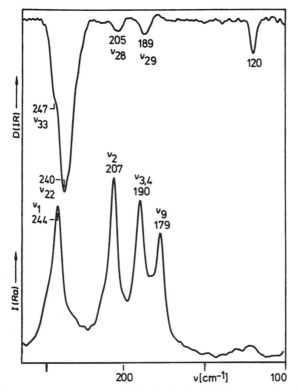

Fig. III-83. Infrared and Raman spectra of $(TBA)_2[Pt_3Br_{12}]$. TBA: tetra-*n*-butylammonium ion (reproduced with permission from Ref. 1207).

As discussed in Sec. III-3(1), $Co(py)_2Cl_2$ exists in two forms: the monomeric tetrahedral (blue) and the polymeric octahedral (lilac). The $\nu(Co-Cl_b)$ of the polymer is very low relative to that of the $\nu(Co-Cl_t)$ because of an increase in coordination number and the effect of bridging.[131] Polymeric $Ni(py)_2X_2$ also exhibits $\nu(Ni-X_b)$ below 200 cm^{-1} (Table III-68).[1206]

The mixed-valence $[Ru_2(NH_3)_6X_3]^{2+}$ ion (X = Cl and Br) contains a triple halogeno bridge:

$$-\overset{II}{Ru}\overset{\diagup\ X\ \diagdown}{\underset{\diagdown\ X\ \diagup}{-X-}}\overset{III}{Ru}-$$

The totally symmetric $\nu(Ru-X)$ and $\delta(RuX_3)$ vibrations (310 and 145 cm^{-1} for X = Cl, and 253 and 111 cm^{-1} for X = Br, respectively) are strongly enhanced in the RR spectrum by excitation in the visible region. Armstrong et al.[1208] were able to assign the electronic transition responsible for this resonance enhancement.

Vibrational spectra of metal cluster ions such as $[(M_6X_8)Y_6]^{2-}$ (M = Mo and W, X = a bridging halogen, and Y = a terminal halogen) and $[(M_6X_{12})Y_6]^{n-}$ (M = Nb and Ta) are discussed in Sec. II-12 of Part A. The low-frequency spectra of these compounds are difficult to assign empirically because of strong vibrational couplings among the $\nu(M–X)$, $\nu(M–Y)$, $\nu(M–M)$, and bending modes.

III-24. COMPLEXES CONTAINING METAL–METAL BONDS

A large number of complexes containing metal–metal (M–M) bonds are known, and their vibrational spectra have been reviewed extensively.[1209–1213] In Part A, we reviewed the vibrational spectra of the X_2Y_6-, X_2Y_8-, and X_2Y_{10}-type compounds containing M–M bonds (Secs. II-10 and 11) and metal clusters containing halogeno bridges (Sec. II-12). In this section, we discuss other complexes containing M–M bonds.

In general, $\nu(MM)$ appear in the low-frequency region (250–100 cm^{-1}) because the M–M bonds are relatively weak and the masses of metals are relatively large. However, the $\nu(MM)$ of some complexes are as high as 400 cm^{-1} due to the multiple bond character of their M–M bonds. If the dinuclear complex is centrosymmetric with respect to the M–M bond, the $\nu(MM)$ is forbidden in IR. However, the $\nu(M–M')$ of a heteronuclear complex is allowed in IR spectra. In contrast, Raman spectroscopy has distinct advantages in that both $\nu(MM)$ and $\nu(MM')$ appear strongly since large changes in polarizabilities are expected as a result of stretching covalent M–M(M') bonds. As shown in Sec. II-11 of Part A, a long series of overtones of $\nu(MM)$ can be observed under resonance conditions. Special caution must be taken, however, in measuring Raman spectra of metal–metal bonded compounds since they may undergo thermal and/or photochemical decomposition upon laser irradiation.

(1) Polynuclear Carbonyls

The $\nu(CO)$ of polynuclear carbonyls have been discussed in Sec. III-17(2). Here, we discuss the $\nu(MM)$ of polynuclear carbonyls in the low-frequency region. As an example, the Raman spectra of $Mn_2(CO)_{10}$, $MnRe(CO)_{10}$, and $Re_2(CO)_{10}$ are shown in Fig. III-84, where the $\nu(MM)$ are indicated for each compound.[1214] Risen and co-workers[1215–1217] carried out normal coordinate analyses on many dinuclear and trinuclear metal carbonyls. Table III-69 lists the observed $\nu(MM)$ and the corresponding force constants obtained by these and other workers. It is noted that the MM stretching force constants obtained by rigorous calculations are surprisingly close to those obtained by approximate calculations considering only metal atoms. There is a general trend that as the MM stretching force constant increases, the $\nu(MM)$ frequency decreases in going from lighter to heavier metals in the $M_2(CO)_{10}$ (M = Mn, Tc, and Re)[1214] and $[M_2(CO)_{10}]^{2-}$ (M = Cr, Mo, and W)[1219] series.

TABLE III-69. Metal–Metal Stretching Frequencies (cm^{-1}) and Force Constants

Compound	ν(MM)	Force Constant (mdyn/Å)		Ref.
		Rigorous Calculation	Approximate Calculation[a]	
$(CO)_5Mn\text{—}Mn(CO)_5$	160	0.59	0.41	1214
$(CO)_5Tc\text{—}Tc(CO)_5$	148	0.72	0.63	1214
$(CO)_5Re\text{—}Re(CO)_5$	122	0.82	0.82	1214
$(CO)_5Re\text{—}Mn(CO)_5$	157	0.81	0.62	1214
$(CO)_5Mn\text{—}W(CO)_5^-$	153	0.71	0.55	1215
$(CO)_5Mn\text{—}Mo(CO)_5^-$	150	0.60	0.47	1215
$(CO)_5Mn\text{—}Cr(CO)_5^-$	149	0.50	0.37	1215
$Cl_3Sn\text{—}Co(CO)_4$	204	1.23	0.97	1216
$Cl_3Ge\text{—}Co(CO)_4$	240	1.05	1.11	1216
$Cl_3Si\text{—}Co(CO)_4$	309	1.32	1.07	1216
$Br_3Ge\text{—}Co(CO)_4$	200	0.96	—	1217
$I_3Ge\text{—}Co(CO)_4$	161	0.52	—	1217
$Br_3Sn\text{—}Co(CO)_4$	182	1.05	—	1217
$I_3Sn\text{—}Co(CO)_4$	156	0.64	—	1217
$H_3Ge\text{—}Re(CO)_5$	209	—	1.34	1218
$H_3Ge\text{—}Mn(CO)_5$	219	—	0.88	1218
$H_3Ge\text{—}Co(CO)_4$	228	—	1.00	1218
$(CO)_4Co\text{—}Zn\text{—}Co(CO)_4$	170, 284[b]	1.30	—	814
$(CO)_4Co\text{—}Cd\text{—}Co(CO)_4$	163, 218[b]	1.28	—	814
$(CO)_4Co\text{—}Hg\text{—}Co(CO)_4$	163, 195[b]	1.26	—	814

[a]Calculations considering only metal atom skeletons.
[b]Under $\mathbf{D}_{3d}$ symmetry, these frequencies correspond to the A_{1g} (symmetric) and A_{2u} (antisymmetric) MCo stretching modes, respectively.

In Sec. III-17(3), we discussed the spectra of $M_2(CO)_4X_2$-type compounds (M = Mn, Tc, Re, Rh, etc.) in which the metals are bonded through halogen (X) bridges. Goggin and Goodfellow[1220] concluded, however, that the $[Pt_2(CO)_2X_4]^{2-}$ ion (X = Cl and Br) contains the direct Pt–Pt bond:

$$
\begin{array}{cc}
O & O \\
C & C \\
| & | \\
X\text{—}Pt\text{—}Pt\text{—}X \\
| & | \\
X & X
\end{array}
$$

They isolated two isomers of $[N(n - Pr)_4]_2[Pt_2(CO)_2Cl_4]$ which differ only in the angle of rotation about the Pt–Pt bond. Both isomers exhibit ν(PtPt) at ~170 cm^{-1}.

$Fe_2(CO)_9$ and $Fe_3(CO)_{12}$ exhibit very strong Raman bands at 225 and 219 cm^{-1}, respectively. San Filippo and Sniadoch[1221] assigned them to ν(FeFe). Later studies[1222] showed, however, that these bands are due to decomposition products resulting from strong laser irradiation. Thus, the appearance of strong

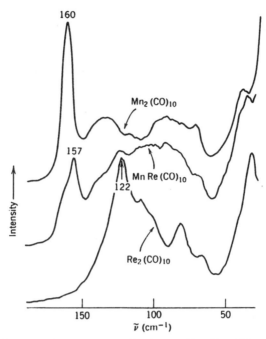

Fig. III-84. Low-frequency Raman spectra of polycrystalline $Mn_2(CO)_{10}$, $MnRe(CO)_{10}$ and $Re_2(CO)_{10}$ (632.8 nm excitation)[1214]

Raman bands in the low-frequency region does not necessarily mean that they are due to $\nu(MM)$. It is also noted that $Re_2(CO)_8X_2$ (X = Cl and Br), which does not contain Re–Re bonds, shows strong Raman bands at 125 cm^{-1} where $\nu(ReRe)$ of $Re_2(CO)_{10}$ appears.[1222] Cooper et al.[1223] were able to locate the $\nu(MM)$ of $Fe_2(CO)_9$ and $Fe_3(CO)_{12}$ at 260 and at 240 and 176 cm^{-1}, respectively. These assignments are based on the $^{54}Fe-^{56}Fe$ isotopic shifts observed in Raman spectra at ~10 K. Onaka and Shriver[1224] observed three $\nu(MM)$ bands at 235, 185, and 159 cm^{-1} in acetone solution of $Co_2(CO)_8$ which correspond to the three isomers discussed in Sec. III-17(2). They have shown that the $\nu(MM)$ is higher than 200 cm^{-1} for bridging carbonyls and between 190 and 140 cm^{-1} for single-bonded nonbridged complexes.

Trinuclear complexes such as $Ru_3(CO)_{12}$ and $Os_3(CO)_{12}$ contain a triangular M_3 skeleton for which two $\nu(MM)$ are expected under D_{3h} symmetry. Quicksall and Spiro[807] assigned the Raman bands at 185 and 149 cm^{-1} of the Ru complex to $\nu(RuRu)$ of the A_1' and E' species, respectively. The latter is coupled with other modes. The corresponding RuRu stretching force constant is 0.82 mdyn/Å. Kettle and co-workers[1225,1226] have assigned the $\nu(MM)$ of the $[Os_xRu_{3-x}(CO)_{12}]$- (x = 0, 1, 2, and 3) type complexes. In $Mn_3H_3(CO)_{12}$, Martin et al.[1227] assigned the $\nu(MnMn)$ at 163 (A_1') and 146 cm^{-1} (E'), and obtained the stretching force constant, $K(Mn–Mn)$ of 0.37 mdyn/Å with the interaction constant of $-0.08\ K$. On the other hand, Jayasooriya and Skinner[1228] assigned the RR bands at 198 (A_1') and 164 cm^{-1} (E') to the ν(MnMn), and obtained K(Mn–Mn) of 0.553 mdyn/Å with the interaction constant of −0.055 mdyn/Å. In $Re_3H_3(CO)_{12}$, the ν(ReRe) were observed at 126 (A_1') and 116/103 cm^{-1} (E') in the matrix-isolated IR spectrum.[1229]

Quicksall and Spiro[1230] assigned the Raman spectrum of $Ir_4(CO)_{12}$, which consists of a tetrahedral Ir skeleton; three ν(IrIr) bands were assigned at 207 (A_1), 161 (F_2), and 131 (E) cm^{-1}. The ratio of these three frequencies, $2:1.56:1.27$, is far from that predicted by a "simple cluster model" ($2:\sqrt{2}:1$),[1231] indicating the substantial coupling between the individual stretching modes. Their rigorous calculations gave K(Ir–Ir) of 1.69 mdyn/Å, together with interaction constants of −0.13 and +0.13 mdyn/Å for the adjacent and opposite Ir–Ir bonds, respectively. The ν(MM) of $Rh_4(CO)_{12}$[1232] and $Co_4(CO)_{12}$[1233] have been assigned, and the corresponding force constants calculated.[1233] As mentioned in Sec. III-17(5), the M_4 skeleton of $M_4H_4(CO)_{12}$ (M = Ru and Os) take $\mathbf{D}_{2d}$ symmetry. Then, the six ($3 \times 4 - 6$) normal vibrations are classified into $2A_1(R) + B_1(R) + B_2(IR, R) + E(IR, R)$. Kettle and Stanghellini[1234] have made complete assignments of these modes.

(2) Compounds Containing Metal–Metal Multiple Bonds

A number of compounds containing unusually short M–M bonds exhibit unusually high ν(MM). For example, the Mo–Mo distance of $Mo_2(OAc)_4$ is only 2.09 Å and its ν(MM) is at 406 cm^{-1}. According to Cotton,[1235] this Mo–Mo bond consists of one σ-bond, two π-bonds and a δ-bond (bond order 4). Such a quadruple bond is also expected for $[Re_2Cl_8]^{2-}$ which exhibits the ν(ReRe) at 272 cm^{-1} with the Re–Re distance of 2.22 Å.[1236,1237] Table III-70 lists ν(MM) of typical compounds. It is seen that the ν(MM) of dimolybdenum compounds of bond order 4 scatter over a wide range. In contrast, dirhenium compounds exhibit a nice ν(MM)-bond order relationship as demonstrated by Fig. III-85.

Table III-70 also shows that the ν(RhRh) is sensitive to the nature of the axial ligand and is downshifted by ~60 cm^{-1} when the acetato group is replaced by the thioacetato group. The ν(MoMo) of $Mo_2(O_2CCH_3)_4$ is upshifted by 9 cm^{-1} when Mo in natural abundance (mainly ^{96}Mo) is replaced by the ^{92}Mo isotope. Such a metal–isotope shift provides definitive assignment for the metal–metal vibration.[1248] Normal coordinate analyses are reported for $M_2(O_2CCH_3)_4$ and $M_2X_8^{n-}$ (M = Mo and Re; X = Cl and Br).[1249,1250] The ν(MM) of porphyrin dimers are listed in Sec. III-5(4).

(3) Metal Cluster Compounds

Vibrational spectra of metal clusters including $[(M_6X_8)Y_6]^{2-}$ (M = Mo and W) and $[(M_6X_{12})Y_6]^{n-}$ (M = Nb and Ta) where X and Y are bridging and terminal halogens, respectively, have been discussed in Sec. II-12 of Part A. The ν(MM) are also reported for metal clusters of other types. In the mixed-valence ion, $[Pt_4(NH_3)_8(C_5H_4NO)_4]^{5+}$, shown in Fig. III-86a, the intradimer (Pt_1-Pt_2) and

TABLE III-70. Bond Orders, Bond Distances, and Stretching Frequencies (cm^{-1}) of Metal–Metal Multiply Bonded Compounds

Compound	Bond Order	Bond Distance	ν(MM)	Ref.
$Mo_2(O_2CCH_3)_4$	4	2.09	406	1238, 1239
$Mo_2(O_2CCF_3)_4$	4	2.09	397	1240
$Mo_2(O_2CCF_3)_4(py)_2$	4	2.22	367	1240
$K_4[Mo_2Cl_8]\cdot2H_2O$	4	2.14	345	1240
$K_3[Mo_2(SO_4)_4]\cdot3.5H_2O$	3.5	2.16	386	1241
			373	
$Re_2(O_2CCH_3)_4Cl_2$	4	2.24	289	1238
$[Bu_4N]_2[Re_2Cl_8]$	4	2.22	272	1236
$Re_2Cl_5(DTH)_2$ ^{a}	3	2.29	267	1238
$Re_2OCl_5(O_2CCH_2CH_3)_2(PPh_3)_2$	2	2.52	216	1238
$Re_2(CO)_{10}$	1	3.02	122	1238
$W_2(O_2C-t-Bu)_4(PPh_3)_2$	4	2.22	287	1242
$Os_2(O_2CCH_3)_4Cl_2$	3	2.31	229	1243
$Rh_2(O_2CCH_3)_4(PPh_3)_2$	4	2.45	289	1244, 1245
$Rh_2(O_2CCH_3)_4(AsPh_3)_2$	4	2.43	298	1246
$Rh_2(O_2CCH_3)_4(SbPh_3)_2$	4	2.42	306	1246
$Rh_2(OSC-CH_3)_4(PPh_3)_2$	4	—	226	1247

^{a}DTH = 2,5-dithiahexane.

interdimer (Pt_2–Pt_2) distances are 2.774 and 2.877 Å, respectively. Correspondingly, the $\nu(Pt_1$–$Pt_2)$ and $\nu(Pt_2$–$Pt_2)$ are observed at 149 and 69 cm^{-1}, respectively, in RR spectra.[1251] In the SO_4 bridged complexes, $[Pt_2(SO_4)_4(H_2O)L]^{n-}$ (n = 2 or 3), the ν(PtPt) are shifted markedly by changing the axial ligand L: 297, 236, and 224 cm^{-1}, respectively for L = H_2O, OH$^-$, and Cl$^-$.[1252] Similar sensitivity of the ν(AuAu) to the nature of the *trans* ligand is seen in the series of $[Au(CH_2)_2PPh_2]_2X_2$ shown in Fig. III-86b; 162, 132, and 103 cm^{-1}, respectively, for X = Cl, Br, and I.[1253]

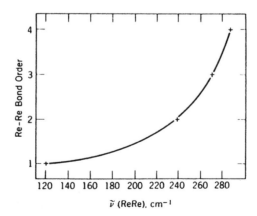

Fig. III-85. ν(Re–Re) vs. Re–Re bond order.

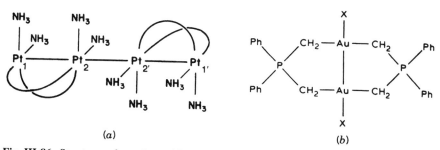

Fig. III-86. Structures of metal–metal bonded complex ions. (a) $[Pt_4(NH_3)_8(C_5H_4NO)_4]^{5+}$. The curves indicate α-pyridonate bridges. (b) $[Au(CH_2)_2PPh_2]_2X_2$.

(4) Metal–Metal Stretching Vibrations of Electronic Excited States

Using TR^3 spectroscopy (Sec. III-4(2)), it is possible to measure the $\nu(MM)$ of electronic excited states. As mentioned in Sec. II-11(2) of Part A, the $\nu(ReRe)$ of the $[Re_2Cl_8]^{2-}$ ion at 275 cm^{-1} in the ground state $((d\delta)^2, {}^1A_g)$ is shifted to 204 cm^{-1} in the excited state $((d\delta)(d\delta^*), {}^1A_{2u})$ since an electron is promoted from a bonding $(d\delta)$ to an antibonding $(d\delta^*)$ orbital.[1254] An opposite trend prevails when an electron is promoted from an antibonding to a bonding or less antibonding orbital. As an example, Figure III-87 shows the RR spectra of the $[Rh_2b_4]^{2+}$ ion (b: 1,3-diisocyanopropane) in the electronic ground $((d\sigma^*)^2, {}^1A_{1g})$ and excited $((d\sigma^*)(p\sigma), {}^3A_{2u})$ states obtained by Dallinger et al.[1255] It is seen that the $\nu(RhRh)$ at 79 cm^{-1} in the ground state is upshifted to 144 cm^{-1} in the excited state. Correspondingly, the Rh–Rh stretching force constant in the ground state (0.19 mdyn/Å) is increased ca. 3 times (0.63 mdyn/Å) by this electronic excitation. In the $[Pt_2(pop)_4]^{4-}$ ion (pop: $(P_2O_5H_2)^{2-}$ ion), the $\nu(PtPt)$ at 118 cm^{-1} in the ground state is upshifted to 156 cm^{-1} when an electron is promoted from the $d\sigma^*$ to the $p\sigma$ orbital.[1256] Similar trends are observed for $M_2(dppm)$ (M = Pd and Pt, dppm = bis(diphenylphosphino)methane).[1257] For more details, see a review by Morris and Woodruff.[1258]

III-25. COMPLEXES OF PHOSPHORUS AND ARSENIC LIGANDS

Ligands such as phosphines (PR_3) and arsines (AsR_3) (R = alkyl, aryl, halogen, etc.) form complexes with a variety of metals in various oxidation states. Vibrational spectroscopy has been used extensively to determine the structures of these compounds and to discuss the nature of the metal–ligand bonding.

(1) Complexes of Phosphorus Ligands

Vibrational frequencies of pyramidal XY_3-type ligands such as PH_3, PF_3, and their halogeno analogs are found in Sec. II-3 of Part A. The most simple phosphine ligand is PH_3. The vibrational spectra of $Ni(PH_3)_4$,[1259] $Ni(PH_3)(CO)_3$,[1260]

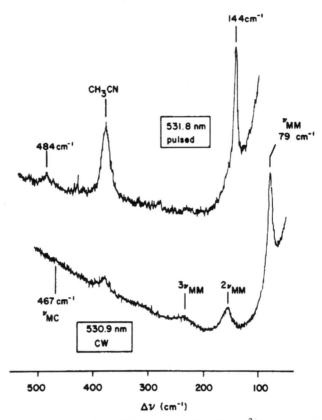

Fig. III-87. Lower trace: ground-state RR spectrum of the $[Rh_2b_4]^{2+}$ ion obtained by CW excitation at 530.9 nm. Upper trace: excited-state RR spectrum obtained by pulsed-laser excitation at 531.8 nm (reproduced with permission from Ref. 1255).

and Ni(PH$_3$)(PF$_3$)$_3$[1261] have been reported by Bigorgne and co-workers. All these compounds exhibit ν(PH), δ(PH$_3$), and ν(NiP) at 2370–2300, 1120–1000, and 340–295 cm^{-1}, respectively. A series of the Ni(PH$_3$)$_n$- (n = 1~4) type complexes have been prepared by matrix cocondensation reactions, and their ν(Ni–P) assigned at 390–395 cm^{-1}.[1262] Complete assignments based on normal coordinate calculations have been made on Ni(P(CH$_3$)$_3$)$_4$.[1263] The A_1 and F_2 ν(Ni–P) vibrations of this compound have been assigned at 296 and 343 cm^{-1}, respectively.

Trifluorophosphine (PF$_3$) forms a variety of complexes with transition metals. According to Kruck,[1264] the ν(PF) of free PF$_3$ (892, 860 cm^{-1}) are shifted slightly to higher frequencies (960–850 cm^{-1}) in M(PF$_3$)$_n$ (n = 4, 5, and 6) and HM(PF$_3$)$_4$ and to lower frequencies (850–750 cm^{-1}) in [M(PF$_3$)$_4$]$^-$ (M = Co, Rh, and Ir). These results have been explained by assuming that the P–F bond possesses a partial double-bond character which is governed by the oxidation state of the metal. For individual compounds, only references are given:

$M(PF_3)_4$ (M = Ni, Pd, and Pt),[1265] $M(PF_3)_5$,[1266] $V(PF_3)_6$,[1267] $Au(PF_3)Cl$,[1268] and cis-$MH_2(PF_3)_4$ (M = Fe, Ru, and Os).[1269] The $\delta(PF_3)$ and $\nu(MP)$ of some of these compounds are assigned at 590–280 and 250–180 cm^{-1}, respectively. Bénazeth et al.[1270] showed that the skeletal symmetry of $HCo(PF_3)_4$ is C_{3v}, while that of $[Co(PF_3)_4]^-$ is T_d. The $\nu(CoP)$ of these compounds are at 250–210 cm^{-1}. Woodward and co-workers[1271] carried out complete vibrational analyses of the $M(PF_3)_4$ (M = Ni, Pd, and Pt) series. Their results show the following trends:

		$Ni(PF_3)_4$	$Pd(PF_3)_4$	$Pt(PF_3)_4$
$\nu(MP)$ (cm^{-1})	A_1	195	204	213
	F_2	219	222	219
$K(M–P)$ (mdyn/Å)		2.71	3.17	3.82

For the $Ni(PX_3)_4$ (X: a halogen) series, Edwards et al. obtained the following:

		$Ni(PF_3)_4$	$Ni(PCl_3)_4$[1272]	$Ni(PBr_3)_4$[1273]	$Ni(PI_3)_4$[1274]
$\nu(MP)$ (cm^{-1})	A_1	195	135	78	(55)
	F_2	219	208	193	184
$K(M–P)$ (mdyn/Å)		2.71	2.35	2.05	—

The $\nu(NiP)$ of $Ni(PMe_3)_4$ are observed at 182 (A_1) and 197 cm^{-1} (F_2).[1275] In general, it is more difficult to assign the $\nu(MP)$ of alkyl and phenyl phosphine complexes than those of halogeno phosphine complexes because the former ligands exhibit many internal modes in the region where the $\nu(MP)$ are expected to appear. To overcome this difficulty, Shobatake and Nakamoto[1183] utilized the metal isotope technique (Sec. I-17 of Part A).

Figure III-88 shows the infrared spectra of trans-$[^{58,62}Ni(PEt_3)_2X_2]$ (X = Cl and Br), and Table III-71 lists the observed frequencies, metal isotope shifts, and band assignments. It is clear that the $\nu(NiP)$ of these complexes must be assigned near 270 cm^{-1}, in contrast to previous investigations, which placed these vibrations near 450–410 cm^{-1}.[1205,1276-1278]

Triphenylphosphine (PPh_3) is most common among phosphine ligands. It is not simple, however, to assign the $\nu(MP)$ of PPh_3 complexes since PPh_3 exhibits a number of ligand vibrations in the low-frequency region.[1279-1281] Using the metal isotope technique, Nakamoto et al.[1183] showed that tetrahedral $Ni(PPh_3)_2Cl_2$, for example, exhibits two $\nu(NiP)$ at 189.6 and 164.0 cm^{-1}, in agreement with the result of previous workers.[1282] In $Rh(PPh_3)_3Cl$, the $\nu(RhP)$ (550, 465, and 460 cm^{-1}) are higher than those of other $\nu(MP)$. This has been attributed to the effect of the $Rh(d\pi)–P(p\pi)$ bonding and the delocalization of the phenyl ring charge through the Rh and P atoms.[1283]

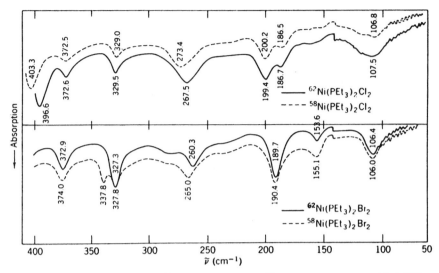

Fig. III-88. Far-IR spectra of $^{58}NiX_2(PEt_3)_2$ and its ^{62}Ni analogs (X = Cl and Br).[1183]

As stated in Sec. III-23, complexes of the type $Ni(PPh_2R)_2Br_2$ (R = alkyl) exist in two forms (tetrahedral and square-planar) which can be distinguished by the $\nu(NiBr)$ and $\nu(NiP)$.[1195] For R = Et, the $\nu(NiP)$ of the planar complex is at 243 cm^{-1}, whereas these vibrations are at 195 and 182 cm^{-1} in the tetrahedral complex. Udovich et al.[1184] studied the infrared spectra of $Ni(DPE)X_2$, where DPE is 1,2-bis(diphenylphosphino)ethane and X is Cl, Br, and I, by using the metal isotope technique. It was found that the $\nu(NiX)$ are always lower and $\nu(NiP)$ are always higher in the cis-$Ni(DPE)X_2$ than in the corresponding trans-

TABLE III-71. Infrared Frequencies, Isotopic Shifts, and Band Assignments of
$NiX_2(PEt_3)_2$ (X = Cl and Br) (cm^{-1})[1183]

PEt$_3$	$^{58}NiCl_2(PEt_3)_2$		$^{58}NiBr_2(PEt_3)_2$		
ν	ν	$\Delta\nu^a$	ν	$\Delta\nu^a$	Assignment^b
408	416.7	0.0	413.6	1.2	$\delta(CCP)$
—	403.3	6.7	337.8	10.5^c	$\nu(NiX)$
365	372.5	−0.1	374.0	1.1	$\delta(CCP)$
330	329.0	−0.5	327.8	0.5^c	$\delta(CCP)$
—	273.4	5.9	265.0	4.7	$\nu(NiP)$
245	(hidden)		(hidden)		$\delta(CCP)$
	200.2	0.8	190.4	0.7	$\delta(CPC)$
	186.2	−0.2	155.1	1.5	$\delta(NiX)$
	161.5	−0.5	(hidden)		$\delta(NiP)$

$^a\Delta\nu$ indicates metal-isotope shift, $\nu(^{58}Ni)-\nu(^{62}Ni)$.
^{b}Ligand vibrations were assigned according to Ref. 1278.
^{c}Since these two bands are overlapped (Fig. III-88), $\Delta\nu$ values are only approximate.

$Ni(PEt_3)_2X_2$. This difference has been attributed to the strong *trans*-influence of phosphine ligands.

A number of investigators have discussed the nature of the M–P bonding based on electronic, vibrational, and NMR spectra,[1284] and controversy has arisen about the degree of π-back bonding in the M–P bond. For example, Park and Hendra[1285] suggest the presence of a considerable degree of π-bonding in square-planar Pd(II) and Pt(II) complexes of PMe_3 and $AsMe_3$. On the other hand, Venanzi[1286] claims from NMR evidence that the Pt–P π-bonding is much less than originally predicted.[1287] It is rather difficult, however, to discuss the degree of π-bonding from vibrational spectra alone since the MP stretching frequency and force constant are determined by the net effect, which involves both σ- and π-bonding.

(2) Complexes of Arsenic Ligands

Complexes of the type $M(CO)_5L$, where L is arsine (AsH_3) and stibine (SbH_3) and M is Cr, Mo, and W, have been prepared by Fischer et al.[1288] $\nu(AsH)$ and $\delta(AsH_3)$ are near 2200 and 900 cm^{-1}, respectively. Complexes of trimethylarsine ($AsMe_3$) have been studied by several investigators. Goodfellow et al.[1289] and Park and Hendra[1285] measured the infrared spectra of $M(AsMe_3)_2X_2$- (M = Pt and Pd; X = Cl, Br, and I) type complexes and assigned $\nu(MAs)$ in the 300–260 cm^{-1} region. The latter workers assigned $\nu(MSb)$ of analogous alkylstibine complexes at ca. 200 cm^{-1}. Konya and Nakamoto[1181] assigned $\nu(MAs)$ and $\nu(MX)$ of $[M(dias)_2]^{2+}$- and $[M(dias)_2X_2]Y_n$-type complexes by using the metal isotope technique. Figure III-89 shows the infrared spectra of $[^{58}Ni(dias)_2X_2]X$ and $[^{58}Ni(dias)_2X_2](ClO_4)_2$ (X = Cl and Br) and their ^{62}Ni analogs. Their results show that the $\nu(MAs)$ are very weak and appear at 325–295 cm^{-1} for the Ni, Co, and Fe complexes and at 270–210 cm^{-1} for the Pd and Pt complexes. For the $\nu(MX)$ of these complexes, see Sec. III-23.

Tertiary phosphine oxides and arsine oxides coordinate to a metal through their O atoms. The $\nu(P=O)$ of triphenylphosphine oxide (TPPO) at 1193 cm^{-1} is shifted by ca. 35 cm^{-1} to a lower frequency when it coordinates to Zn(II).[1290] The shift is much larger in $MX_4(TPPO)_2$ (160–120 cm^{-1}), where MX_4 is a tetrahalide of Pa, Np, and Pu.[1291] A similar observation has been made for $\nu(As=O)$ of arsine oxide and its complexes. Exceptions to this rule are found in $MnX_2(Ph_3AsO)_2$ (X = Cl and Br); their $\nu(As=O)$ are higher by 30–20 cm^{-1} than the frequency of the free ligand (880 cm^{-1}).[1292] Rodley et al.[1293] have assigned the $\nu(MO)$ of tertiary arsine oxide complexes at 440–370 cm^{-1}.

III-26. COMPLEXES OF SULFUR AND SELENIUM LIGANDS

A large number of metal complexes of ligands containing sulfur and selenium are known. Here the vibrational spectra of typical compounds will be reviewed

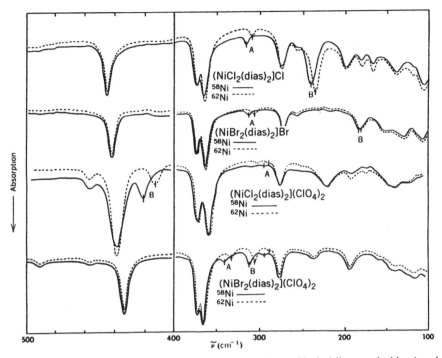

Fig. III-89. Far-IR spectra of octahedral nickel dias complexes. Vertical lines marked by A and B indicate Ni–As and Ni–X stretching bands, respectively.[1181]

briefly. For SO_3 and thiourea complexes that form metal–sulfur bonds, see Secs. III-12 and III-14, respectively.

(1) Complexes of S_n and Se_n ($n = 2\sim6$)

According to Müller et al.,[1294] most complexes containing the S_2^{2-} ligand take the following structures:

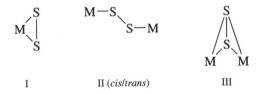

The $\nu(S_2)$ in the free state is 623 cm^{-1} (Table II-1b of Part A). Upon coordination, the $\nu(S_2)$ is shifted to 560–510 cm^{-1} in I and III, and to 510–480 cm^{-1} in II. It is rather difficult, however, to distinguish these structures by vibrational spectroscopy. An extensive compilation of structural and vibrational data of disulfur complexes is found in a review by Müller et al.[1294] Here, some recent references for various types of coordination and $\nu(S_2)$ are given: $MoO(S_2)_2(bipy)$ (I, 540 cm^{-1}),[1295] $Ti(TPP)(S_2)$ (I, 551 cm^{-1}),[1296] and $Co_2(CO)_6(S_2)$ (III, 615 cm^{-1}). The $\nu(S_2)$ of the last compound is unusually high because the S–S bond is unusually short (1.98 Å).[1297] The $Mo_3S_{13}^{2-}$ ion contains three type I and three type II S_2 ligands and one bridging S atom, and their $\nu(S_2)$ have been assigned using $^{92}Mo/^{100}Mo$ and $^{32}S/^{34}S$ isotope shift data.[1298] The $\nu(Se_2)$ of the free Se_2^{2-} ion is at 349 cm^{-1}, and this band is shifted to 310 cm^{-1} in $[Ir(Se_2)(DME)_2]Cl$ (DME: 1,2-bis(dimethylphosphino)ethane).[1299] In general, the $\nu(M\text{–}S)$ and $\nu(M\text{–}Se)$ are more difficult to assign empirically; the former is in the 380–300 cm^{-1} region and the latter is lower.

Although the S_n^{2-} and Se_n^{2-} ions ($n = 3\sim5$) take open-chain configurations in the free state (Sec. II-18 of Part A), they tend to form chelate rings in metal complexes. The $[M_2O_2(S_2)(S_4)]^{2-}$ (M = Mo and W) exhibits the $\nu(S_2)$ at 521–500 and 490–420 cm^{-1} for the S_2 and S_4 rings, respectively.[1300] The $\nu(Se_2)$ of the $[Zn(Se_4)_2]^{2-}$ ion are observed at 276 and 250 cm$^{-1,1301}$ while those of the $[Ni(Se_4)_2]^{2-}$ ion are reported to be 364 and 348 cm.$^{-1,1302}$ Figure III-90 shows the Raman spectra of $Cp_2M(Se_5)$ (M = Hf and Zr) obtained by Butler et al.[1303] The bands at 266(261), 247(245), and 200(214) cm^{-1} have been assigned to

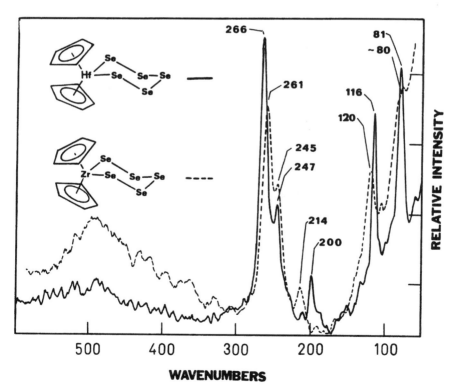

Fig. III-90. Raman spectra of solid Cp_2MSe_5 (M = Hf and Zr) in the 600–50 cm^{-1} region (514.5 nm excitation) (reproduced with permission from Ref. 1303).

$\nu_s(Se_2)$, $\nu_a(Se_2)$, and $\nu(MS)$, respectively. Raman frequencies are reported for the $[M(S_6)_2]^{2-}$ ion where M is Zn, Cd, and Hg.[1304]

(2) Complexes of S-Containing Diatomic Ligands

The $\nu(NS)$ of thionitrosyl (NS) complexes such as $CpCr(CO)_2(NS)$[1305] and $ReCl_2(NS)(PMe_2Ph)$[1306] are observed at 1180 cm^{-1}. This frequency is 40 cm^{-1} lower than that of free NS. Similarly, the $\nu(SO)$ of $RuCl(NO)(SO)(PPh_3)_2$ at 1061 cm^{-1} is 63 cm^{-1} lower than that of free SO (1124 cm^{-1}).[1307] Thiocarbonyl (CS) may form metal complexes of the following types:

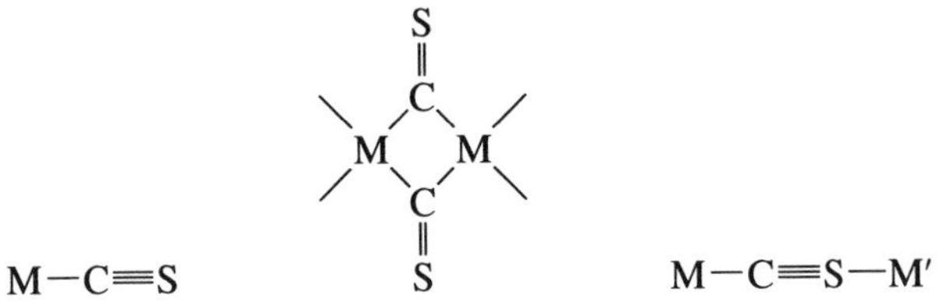

$$M-C\equiv S$$

I

$$
\begin{array}{c}
S \\
\| \\
\diagdown\;C\diagup \\
M\quad\; M \\
\diagup\;C\diagdown \\
\| \\
S
\end{array}
$$

II

$$M-C\equiv S-M'$$

III

The $\nu(CS)$ of free CS is at 1285 cm^{-1}. The $\nu(CS)$ of the C-bonded terminal CS group (Structure I) is higher than that of free CS (1360–1290 cm^{-1}).[1308-1310] The bridging C-bonded structure (II) was suggested for $Mn_2(\pi\text{-}Cp)_2(NO)_2(CS)_2$.[1311] The bridging C- and S-bonded structure (III) was suggested for $(DPE)_2(CO)W(CS)W(CO)_5$ since its $\nu(CS)$ is very low (1161 cm^{-1}).[1312] An extremely low $\nu(CS)$ (910 cm^{-1}) is reported for $Co_3Fe(Cp)(CO)_9(CS)$ in which the CS group acts as a six-electron donor to the three Co atoms.[1313] Normal coordinate analyses have been made on $M(CO)_5(CX)$ (M = Cr and W; X = S and Se)[1314] and $(C_6H_6)Cr(CO)_2(CX)$ (X = O, S, and Se).[1315]

(3) Complexes of SO_2, CS_2, and Related Ligands

Sulfur dioxide (SO_2) may take one of the following structures when it coordinates to a metal:

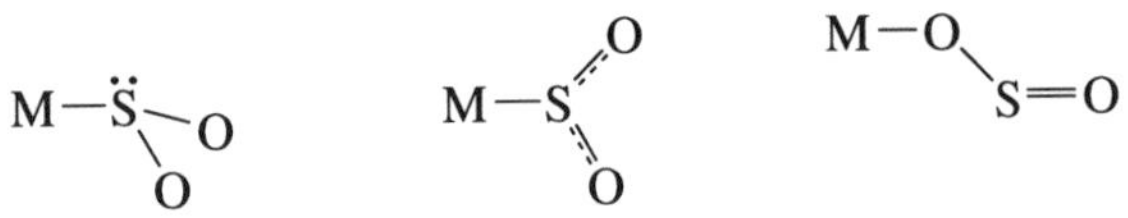

$$
M-\ddot{S}\diagdown_O^{\diagup\,O}
$$

I

$$
M-S\diagup^{O}_{\diagdown O}
$$

II

$$
\begin{array}{c}
M-O \\
\diagdown \\
S=O
\end{array}
$$

III

$$
\begin{array}{c}
O \\
\| \\
M-S-M \\
\| \\
O
\end{array}
$$

IV

$$
M-\!\!\diagup S\diagdown^{\diagup O}_{O}
$$

V

$$
\begin{array}{c}
M-\!\!-M \\
\diagdown\;/ \\
S \\
\diagup\;\diagdown \\
O\quad O
\end{array}
$$

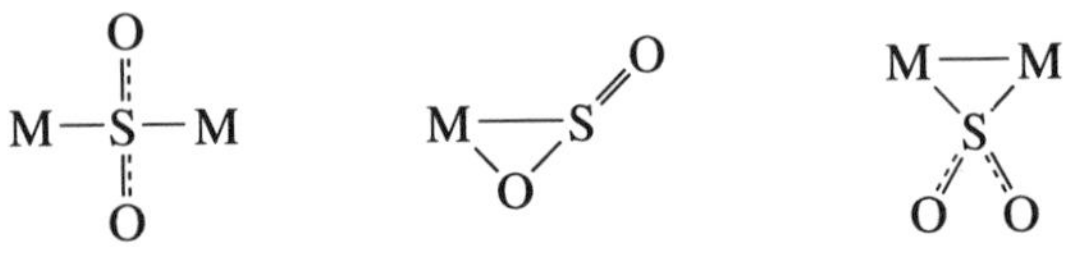

VI

Free SO_2 exhibits the $\nu_a(SO_2)$ and $\nu_s(SO_2)$ at 1351 and 1147 cm^{-1}, respectively (Sec. II-2 of Part A). Table III-72 lists the modes of coordination and stretching frequencies of SO_2 complexes. It is clearly not possible to determine the coordination geometry by vibrational data. In fact, most of the structures given in the table were determined by X-ray analysis. According to Kubas,[1330] the coordination geometry of the SO_2 ligand can be deduced by combining spectroscopic and chemical properties. The SO_2 stretching frequencies are useful, however, in distinguishing the O- and S-bonded complexes; a complex is O-bonded if $(\nu_a - \nu_s)$ is larger than 190 cm^{-1}, and S-bonded if it is smaller than 190 cm^{-1}.[1,1322]. Johnson and Dew[1331] observed a linkage isomerization (structure II → V) in [Ru(NH$_3$)$_4$(SO$_2$)Cl]Cl which was photochemically induced in the solid state. The ν_a and ν_s of structure II are at 1255 and 1110 cm^{-1}, respectively, whereas those of structure V are at 1165 and 940 cm^{-1}, respectively.

Vibrational spectra of transition-metal complexes of CS_2 and CS have been reviewed by Butler and Fenster.[1332] According to Wilkinson et al.,[1309,1333] CS_2 coordinates to the metal in four ways:

TABLE III-72. Structures and Observed Frequencies (cm^{-1}) of SO$_2$ Complexes

Complex	Structure	ν_a	ν_s	Ref.
IrCl(CO)(PPh$_3$)$_2$(SO$_2$)	I	1198–1185	1048	1316
PtBr(C$_6$H$_3$(CH$_2$NMe$_2$)$_2$-o,o')(SO$_2$)	I	1231	1074	1317
[Ru(NH$_3$)$_4$(SO$_2$)Cl]Cl	II	1301–1278	1100	1318
Fe(CO)$_2${P(OMe)$_3$)}$_2$(SO$_2$)	II	1225	1095	1319
RuCl(C$_5$Me$_5$){P(i-Pr)$_3$}(SO$_2$)	II	1249	1095	1320
Ni{P(C$_6$H$_5$)$_3$}$_3$(SO$_2$)	II	1195	1052	1321
Ni{P(C$_6$H$_5$)Me$_2$}$_3$(SO$_2$)	I	1170	1030	1321
SbF$_5$(SO$_2$)	III	1327	1102	1322
{(C$_5$H$_5$)Fe(CO)$_2$}$_2$(μ-SO$_2$)	IV	1135	993	1323
Fe$_2$(CO)$_8$(μ-SO$_2$)	IV	1203	1048	1324
RuCl(NO)(PPh$_3$)$_2$(SO$_2$)	V	—	895	1325
OsCl(NO)(PPh$_3$)$_2$(SO$_2$)	V	1133	846	1326
[Fe$_6$(CO)$_{15}$C(μ-SO$_2$)]$^{2-}$	VI	1180	1045	1327
[Pt$_3$Au(μ-CO)$_2$ (μ-SO$_2$){P(C$_6$H$_{11}$)$_3$}$_4$]$^+$	VI	1233	1075	1328
[Pd$_3$(μ-SO$_2$)$_2$ (μ-N$_3$)(PPh$_3$)$_3$]$^-$	VI	1202	1062, 1051	1329

The linear CS_2 molecule in the free state exhibits the $\nu_a(CS_2)(IR)$, $\nu_s(CS_2)$ (R), and $\delta(CS_2)(IR)$ at 1533, 658, and 397 cm^{-1}, respectively (Sec. II-2 of Part A). In metal complexes, the SCS bond is bent, except for structure II shown above. Most vibrational studies on CS_2 complexes report only the $\nu(CS_2)$ in the high-frequency region. The $\nu(CS_2)$ of structure I (π-bonded) and structure II (σ-bonded) are at 1100–1150 and 1510 cm^{-1}, respectively. The $\nu(CS_2)$ of the bridging CS_2 groups (structures III–IV) are in the 1155–1120 cm^{-1} region.[1333–1335]

From infrared and other evidence, [Ir(CS$_2$)(CO)(PPh$_3$)$_3$]BPh$_4$ was originally thought to be a six-coordinate complex with a π-bonded CS$_2$.[1336] However, X-ray analysis[1337] revealed an unexpected structure; a five-coordinate complex of the type [Ir(CO)(PPh$_3$)$_2$(S$_2$C—PPh$_3$)]BPh$_4$. Matrix cocondensation reactions of Ni atoms with CS$_2$/Ar produce a mixture of Ni(CS$_2$)$_n$ where n is 1, 2, and 3. It was not possible, however, to determine the mode of coordination from their IR spectra.[1338] A mixed carbon dichalcogenide such as SCSe can form a pair of geometrical isomers:

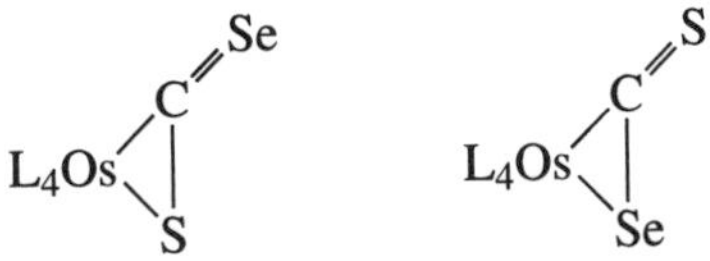

where L_4 represents (CO)(CNR)(PPh$_3$)$_2$ (R = p-tolyl). The $\nu(CSe)$ of the former and the $\nu(CS)$ of the latter have been observed at 1015 and 1066 cm^{-1}, respectively.[1339]

Allkins and Hendra[1340] carried out an extensive vibrational study on *cis*- and *trans*-[MX$_2$Y$_2$] and their halogen-bridged dimers, where M is Pd(II) and Pt(II), X is Cl, Br, and I, and Y is (CH$_3$)$_2$S, (CH$_3$)$_2$Se, and (CH$_3$)$_2$Te. The $\nu(MS)$, $\nu(MSe)$, and $\nu(MTe)$ were assigned in the ranges 350–300, 240–170, and 230–165 cm^{-1}, respectively. The vibrational spectra of PtX$_2$L$_2$,[1341] PdX$_2$L$_2$,[1342] AuX$_3$L and AuXL,[1343] where X is a halogen and L is a dialkylsulfide, have been assigned. Aires et al.[1344] reported the infrared spectra of MX$_3$L$_3$-type compounds, where M is Ru(III), Os(III), Rh(III), and Ir(III), X is Cl or Br, and L is Et$_2$S and Et$_2$Se. The $\nu(MS)$ and $\nu(MSe)$ of these compounds were assigned at 325–290 and 225–200 cm^{-1}, respectively, based on the *fac*-structure. On the other hand, Allen and Wilkinson[1345] proposed the *mer*-structure for these compounds, based on far-infrared and other evidence.

According to X-ray analysis, Pd$_2$Br$_4$(Me$_2$S)$_2$ is bridged via Br atoms, whereas Pt$_2$Br$_4$(Et$_2$S)$_2$ is bridged via S atoms:[1346]

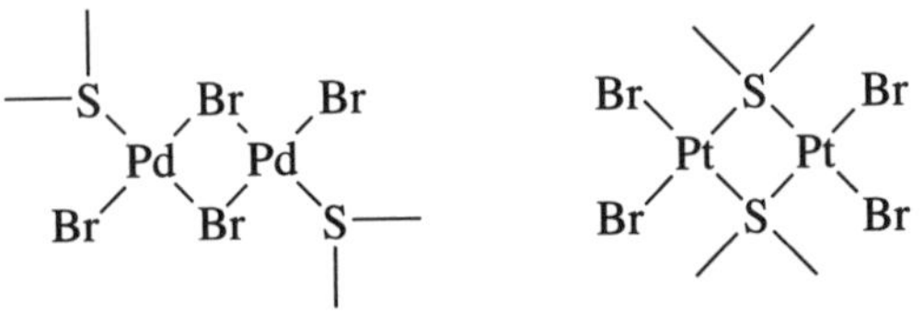

Adams and Chandler[1347] have shown that halogen-bridged $Pd_2Cl_4(SEt_2)_2$ exhibits $\nu(PdCl_t)$, $\nu(PdCl_b)$, and $\nu(PdS)$ at 366, 266, and 358 cm^{-1}, respectively, whereas sulfur-bridged $Pt_2Cl_4(SEt_2)_2$ exhibits $\nu(PtCl_t)$ and $\nu(PtS)$ at 365–325 and 422–401 cm^{-1}, respectively.

(4) Complexes of Thiometalates

Vibrational frequencies of thiometalate ions such as MS_4^{n-}, MS_3O^{n-}, and $MS_2O_2^{n-}$ are found in Sec. II-2 of Part A. These ions form trinuclear metal complexes such as shown below:

Müller and co-workers have carried out an extensive study on structures and spectra of transition metal complexes of thiometalates.[1348–1350] For example, they assigned the IR spectra of the $[Ni(MoS_4)_2]^{2-}$ ion based on normal coordinate analysis involving $^{58}Ni/^{62}Ni$ and $^{92}Mo/^{100}Mo$ isotopes.[1351] The IR-active metal–sulfur stretching frequencies follow the order:

	$\nu(MoS_t)$	$\nu(MoS_b)$	$\nu(NiS)$
$\tilde{\nu}(cm^{-1})$	494(B_{3u})	456(B_{2u})	332(B_{3u})*
		443(B_{3u})*	324(B_{2u})

Here, the asterisks indicate vibrational coupling between them. Figure III-91 shows the electronic and Raman/RR spectra of the $[Fe(WS_4)_2]^{3-}$ ion obtained by Müller and Hellmann.[1352] It is seen that the excitation at 647.1 nm gives a normal Raman spectrum, whereas those at 514.5 and 488.0 nm yield RR spectra. In the latter case, a series of overtones of ν_2 (totally symmetric $\nu(WS_t)$), as well as a combination with $\nu(FeS)$, are observed, because the absorption band near 500 nm originates in an electronic transition within the WS_4 ligand. Vibrational spectra are also reported for metal–sulfur bridging complexes formed by the ligands such as PS_4^{3-},[1353] and $PS_2R_2^{2-}$ (R = Me, Ph etc.).[1354]

The infrared spectra of thiocarbonato complexes of the type $[M(CS_3)_2]^{2-}$ [M = Ni(II), Pd(II), and Pt(II)] have been studied by Burke and Fackler[1355] and Cormier et al.[1356] The latter workers carried out normal coordinate analyses on the $[^{58}Ni(CS_3)_2]^{2-}$ ion and its ^{62}Ni analog. The $\nu(NiS)$ were assigned at 385 and 366 cm^{-1}, with a corresponding force constant of 1.41 mdyn/Å (UBF). Bis(dithioacetato)palladium, $Pd(CH_3CS_2)_2$, exists in three different crystalline forms. Piovesana et al.[1357] characterized each of these forms by IR spectroscopy.

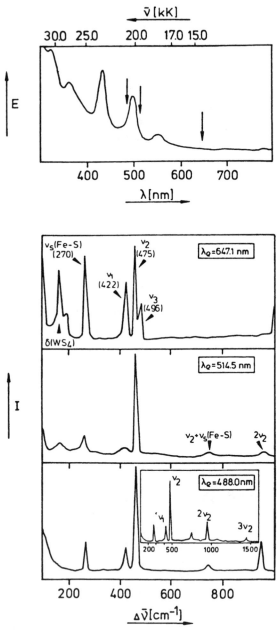

Fig. III-91. Electronic absorption (top trace) and Raman/RR spectra (bottom traces) of $[(Ph)_3PNP(Ph)_3]_2[N(Et)_4][Fe(WS_4)_2]2CH_3CN$. The terminal $\nu(WS_t)$ (ν_2 and ν_3) and bridging $\nu(WS_b)$ (ν_1) are strongly coupled in this case. The $\nu(FeS)$ is at 270 cm^{-1} (reproduced with permission from Ref. 1352).

(5) Complexes of Relatively Large Ligands

2,5-Dithiahexane(dth) forms metal complexes such as $[ReCl_3(dth)]_n$ and $Re_3Cl_9(dth)_{1.5}$. Cotton et al.[1358] showed from infrared spectra that dth of the former forms a chelate ring in the *gauche* conformation, whereas that of the latter forms a bridge between two metals by taking the *trans* conformation (see Sec. III-2(5)). Infrared spectra have been used to show that ethanedithiol forms a chelate ring of the *gauche* conformation in $Bi(S_2C_2H_4)X$, where X is Cl and Br.[1359] Schläpfer et al.[1360] assigned the ν(NiS) and ν(NiN) of dth, ete [2-(ethylthio)ethylamine], and mea [mercaptoethylamine] complexes with the metal isotope technique:

dth ete mea

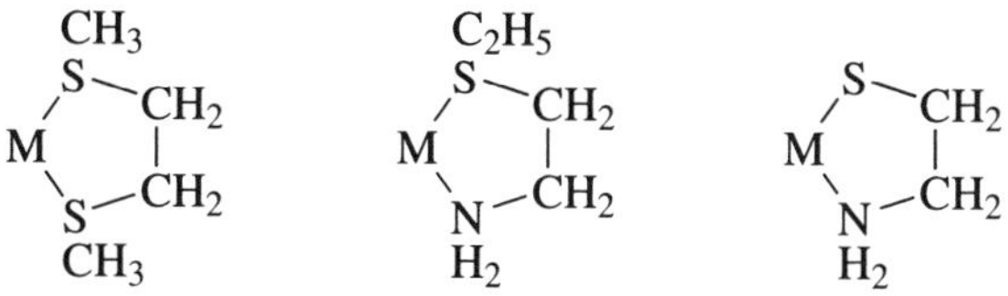

The infrared spectra of N,N-dialkyldithiocarbamato complexes have been studied extensively. All these compounds exhibit strong ν(C=N) bands in the 1600–1450 cm^{-1} region. These compounds are roughly classified into two types:

Bidentate coordination Unidentate coordination

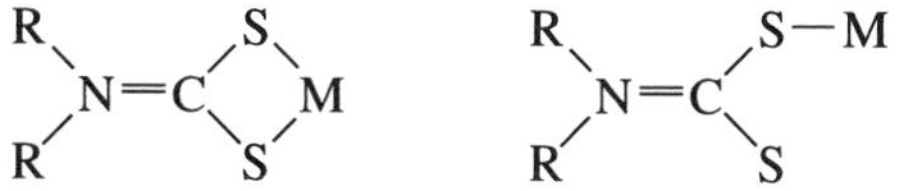

The former exhibits ν(CS) near 1000 cm^{-1} as a single band, whereas the latter shows a doublet in the same region.[1361] Also, the ν(C=N) of the former (above 1485 cm^{-1}) is higher than that of the latter (below 1485 cm^{-1}).[1362] The ν(MS) of the bidentate complexes are observed at 400–300 cm^{-1}.[1363] Dithiocarbamato complexes of Fe(III) undergo the high-spin ($^6A_1$) to low-spin ($^2T_2$) crossover. This change can be induced by applying high pressure[1364] or by lowering temperature.[1365] Sorai[1365] assigned the ν(FeS) of $Fe(S_2CN(Et)_2)_3$ of the high- and low-spin states to the IR bands at 355 and 552 cm^{-1}, respectively. Hutchinson et al.[1366] have shown by ^{54}Fe–^{57}Fe isotope shifts and variable-temperature studies that the ν(FeS) of Fe(III) dialkyldithiocarbamates appear at 250–205 and 350–305 cm^{-1}, respectively, for high- and low-spin states, and that intermediate-spin complexes show ν(FeS) in both regions. Nakamoto et al.[1367] performed normal coordinate analysis on the 1 : 1 (metal/ligand) model of dithiocarbamato Pt(II) complex $[Pt(S_2CNH_2)_2]$ and its ND_2 analog.

The infrared spectra of diselenocarbamato complexes have been reported,[1368] and assigned on the basis of normal coordinate analysis.[1369] In the Ni(II) complex, ν(NiSe) is assigned at 298 cm^{-1}, which is lower by 85 cm^{-1} than the ν(NiS) of the corresponding dithiocarbamato complex. The infrared spectra of xanthato complexes:

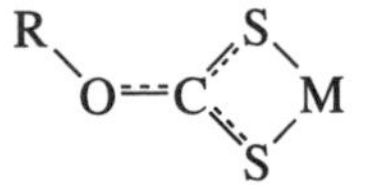

have been studied by Watt and McCormick.[1370] The ν(CO), ν(CS), and ν(MS) were assigned at 1325–1250, 760–540, and 380–340 cm^{-1}, respectively.

Savant et al.[1371] roughly classified monothiobenzoato complexes into three categories:

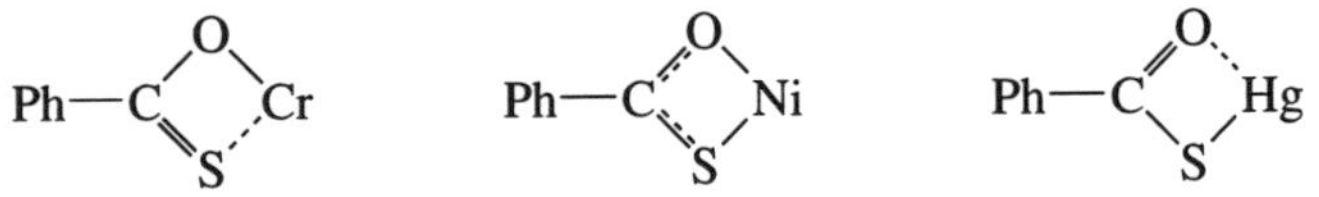

ν(CO)(cm^{-1}) 1465	1508	1630
ν(CS)(cm^{-1}) 982	958	912

In the Hg(II), Cu(I), and Ag(I) complexes, coordination occurs mainly via sulfur. In the Cr(III) complex, however, the Cr–O bond is stronger than the Cr–S bond. The Ni(II) complex is between these two cases and is close to symmetrical coordination. This is reflected in the frequency trends shown above.

In thiocarboxylato complexes of the type Ni(R—COS)$_2\cdot\frac{1}{2}$(EtOH) (R = CH$_3$, C$_2$H$_5$, and Ph), the ligand serves as a bridge between two metals as shown:

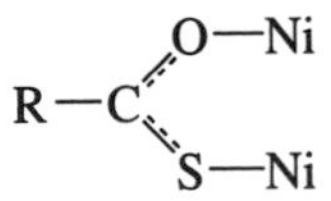

and their ν(CO) are reported to be at 1580–1520 cm^{-1}.[1372]

The infrared spectra of metal complexes of thiosemicarbazides

and thiosemicarbazones

have been reviewed by Campbell.[1373] For metal complexes of dithiocarbazic acid, infrared spectra support the N,S-chelated structure (shown below) rather than the S,S-chelated structure normally found for dithiocarbamato complexes.[1374]

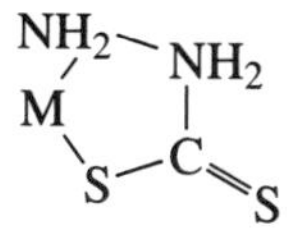

Dithiocarbazato
complex

In 1:1 complexes of N,N′-monosubstituted dithiooxamides, Desseyn et al.[1375] concluded from infrared spectra that metals such as Ni(II) and Cu(II) are primarily bonded to the N, whereas metals such as Hg(II), Pb(II), and Pd(II) are bonded to the S atom.

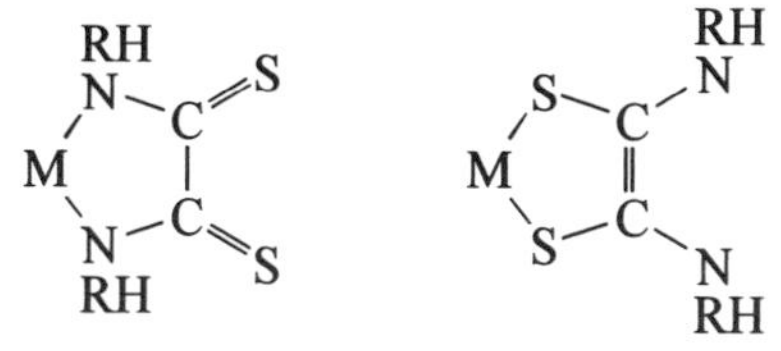

Infrared spectra are reported for tris-Co(III)(S-bonded) complexes of dithiooxamide and its analogs.[1376]

Fujita and Nakamoto[1377] assigned the IR spectra of the dithiooxalato (DTO) complexes $[M(DTO)_2]^{2-}$ [M = Ni(II), Pd(II), and Pt(II)] and $[Co(DTO)_3]^{3-}$ (all are S-bonded) based on normal coordinate analysis on the 1:1 (metal/ligand) model.

Coucouvanis et al.[1378] synthesized novel tin halide adducts of Ni(II) and Pd(II) dithiooxalto complexes:

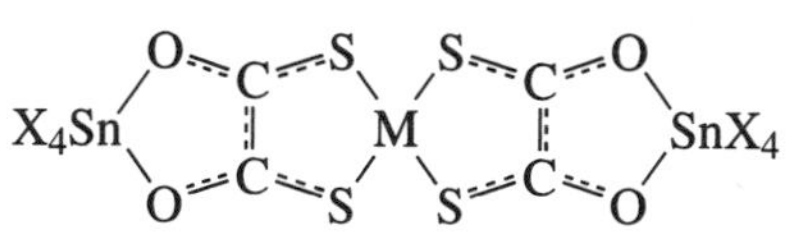

The ν(NiS) of the $[Ni(DTO)_2]^{2-}$ ion is at 349 cm^{-1}. In SnX$_4$ (X = Cl, Br, and I) adducts, this band shifts to 385–375 cm^{-1}, indicating a strengthening of the Ni–S bond by complexation. It was found that Cr(DTO)$_3$[Cu(PPh$_3$)$_2$]$_3$ exists in two isomeric forms;[1379] one in which the Cr atom is bonded to sulfur, and another in which it is bonded to oxygen of the DTO ion. As expected, ν(C=O), ν(CS), and ν[CrO(S)] are markedly different between these two isomers.

Metal complexes of 1,2-dithiolates (or dithienes) have been of great interest to inorganic chemists because of their redox properties.[1380] Schläpfer and Nakamoto[1381] prepared a series of complexes of the type $[Ni(S_2C_2R_2)_2]^n$, where R is H, Ph, CF_3, and CN and n is 0, -1, or -2, and carried out normal coordinate analysis to obtain rough estimates of the charge distribution based on the calculated force constants.

Infrared spectra of metal complexes with thio-β-diketones have been reviewed briefly.[1382,1383] Siimann and Fresco[1384] and Martin et al.[1385] carried out normal coordinate analysis on metal complexes of dithioacetylacetone, monothioacetylacetone, and related ligands. For dithioacetylacetonato complexes, two $\nu(MS)$ have been assigned at 390–340 and 300–260 cm^{-1}.

L-Cysteine has three potential coordination sites (S, N, and O), and infrared spectra have been used to determine the structures of its metal complexes. For example, the Zn(II) complex shows no $\nu(SH)$, and its carboxylate frequency indicates the presence of a free COO$^-$ group. Thus, the following structure was proposed:[1386]

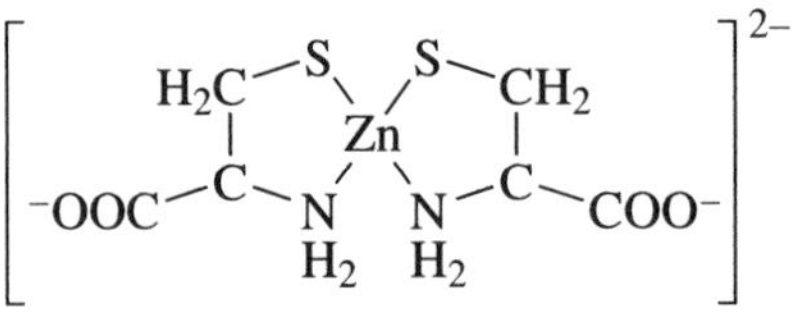

On the other hand, the (S, O) chelation has been suggested for the Pt(II) and Pd(II) complexes.[1387] IR spectra show that the Fe atom in Fe(cyst)(H$_2$O)$_{1.5}$ is bonded to the S, N, and O atoms whereas that in Na$_2$[Fe(cyst)$_2$]H$_2$O is bonded only to the S and N atoms.[1388] The $\nu(SS)$ of L-cystine complexes of Cu(II), Ni(II), and Zn(II), and so on are observed near 500 cm^{-1}.[1389]

McAuliffe et al.[1390] studied the IR spectra of metal complexes of methionine [CH$_3$—S—CH$_2$—CH$_2$—CH(NH$_2$)—COOH]. They found that most of the metals they studied [except Ag(I)] coordinate through the NH$_2$ and COO$^-$ groups, and that the CH$_3$S groups of these complexes are available for further coordination to other metals. McAuliffe[1391] suggested that complexes of the type M(methionine)Cl$_2$ [M = Pd(II) and Pt(II)] take the polymeric structure:

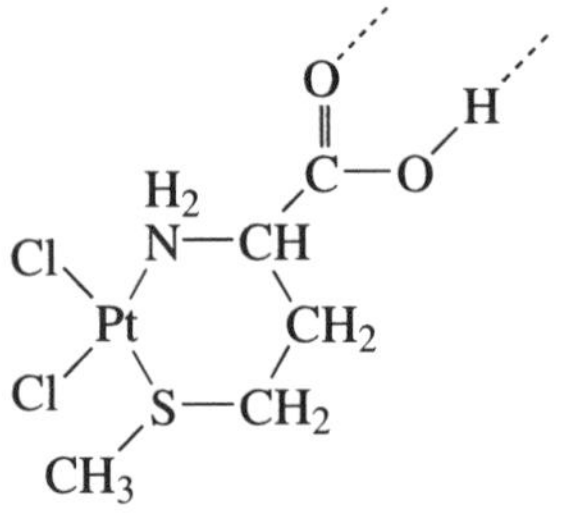

Infrared spectra of metal complexes of sulfur-containing ligands are reported for 2-methylthioaniline,[1392] 8-mercaptoquinoline,[1393] cyclic thioethers,[1394] N-alkylthiopicolinamides,[1395] and dithizone.[1396]

REFERENCES

1. K. H. Schmidt and A. Müller, *Coord. Chem. Rev.*, **19**, 41 (1976).

2. R. Plus, *J. Raman Spectrosc.*, **1**, 551 (1973); *Spectrochim. Acta*, **32A**, 263 (1976).

3. N. Tanaka, M. Kamada, J. Fujita, and E. Kyuno, *Bull. Chem. Soc. Jpn.*, **37**, 222 (1964).

4. T. V. Long, II and D. J. B. Penrose, *J. Am. Chem. Soc.*, **93**, 632 (1971).

5. K. H. Schmidt and A. Müller, *J. Mol. Struct.*, **22**, 343 (1974).

6. K. H. Schmidt and A. Müller, *Inorg. Chem.*, **14**, 2183 (1975).

7. A. Deak and J. L. Templeton, *Inorg. Chem.*, **19**, 1075 (1980).

8. W. P. Griffith, *J. Chem. Soc. A*, 899 (1966).

9. T. Shimanouchi and I. Nakagawa, *Inorg. Chem.*, **3**, 1805 (1964).

10. L. Sacconi, A. Sabatini, and P. Gans, *Inorg. Chem.*, **3**, 1772 (1964).

11. K. H. Schmidt, W. Hauswirth, and A. Müller, *J. Chem. Soc., Dalton Trans.*, 2199 (1975).

12. H. Siebert and H. H. Eysel, *J. Mol. Struct.*, **4**, 29 (1969).

13. T. V. Long, II, A. W. Herlinger, E. F. Epstein, and I. Bernal, *Inorg. Chem.*, **9**, 459 (1970).

14. J. M. Terrasse, H. Poulet, and J. P. Mathieu, *Spectrochim. Acta*, **20**, 305 (1964).

15. J. M. Janik, J. A. Janik, G. Pytasz, and J. Sokolowski, *J. Raman Spectrosc.*, **4**, 13 (1975).

16. M. J. Nolan and D. W. James, *J. Raman Spectrosc.*, **1**, 259, 271 (1973).

17. J. Hiraishi, I. Nakagawa, and T. Shimanouchi, *Spectrochim. Acta*, **24A**, 819 (1968).

18. J. Fujita, K. Nakamoto, and M. Kobayashi, *J. Am. Chem. Soc.*, **78**, 3295 (1956).

19. B. S. Ault, *J. Am. Chem. Soc.*, **100**, 5773 (1978).

20. K. Nakamoto, Y. Morimoto, and J. Fujita, *Proc. Int. Conf. Coord. Chem., 7th*, Stockholm, *1962*, p. 15 (1962).

21. I. Nakagawa and T. Shimanouchi, *Spectrochim. Acta*, **22**, 759 (1966).

22. R. Acevedo, G. Diaz, and C. D. Flint, *Spectrochim. Acta*, **41A**, 1397 (1985).

23. R. Acevedo, G. Diaz, M. M. Campos-Vallette, and B. Weiss, *Spectrochim. Acta*, **47A**, 355 (1991).

24. A. Müller, P. Christophliemk, and I. Tossidis, *J. Mol. Struct.*, **15**, 289 (1973).

25. K. Nakamoto, J. Takemoto, and T. L. Chow, *Appl. Spectrosc.*, **25**, 352 (1971).

26. D. N. Ishikawa, F. A. De Souza, and S. C. Tellez, *Spectrosc. Lett.*, **26**, 803 (1993).

27. M. Manfait, A. J. P. Alix, and J. Delaunay-Zeches, *Inorg. Chim. Acta*, **44**, L261 (1980).

28. L. A. Degen and A. J. Rowlands, *Spectrochim. Acta*, **47A**, 1263 (1991).

29. M. Manfait, A. J. P. Alix, L. Bernard, and T. Theophanides, *J. Raman Spectrosc.*, **7**, 143 (1978).

30. S. J. Cyvin, B. N. Cyvin, K. H. Schmidt, W. Wiegeler, A. Müller, and J. Brunvoll, *J. Mol. Struct.*, **30**, 315 (1976).

31. M. Manfait, A. J. P. Alix, and C. Kappenstein, *Inorg. Chim. Acta*, **50**, 147 (1981).

32. A. L. Geddes and G. L. Bottger, *Inorg. Chem.*, **8**, 802 (1969).

33. M. G. Miles, J. H. Patterson, C. W. Hobbs, M. J. Hopper, J. Overend, and R. S. Tobias, *Inorg. Chem.* **7**, 1721 (1968).

34. E. P. Bertin, I. Nakagawa, S. Mizushima, T. L. Lane, and J. V. Quagliano, *J. Am. Chem. Soc.*, **80**, 525 (1958).

35. S. C. Tellez, *Spectrosc. Lett.*, **21**, 871 (1988).

36. S. Süzer and L. Andrews, *J. Am. Chem. Soc.*, **109**, 300 (1987).

37. J. Szczepanski, M. Szczesniak, and M. Vala, *Chem. Phys. Lett.*, **162**, 123 (1989).

38. A. Loutellier, L. Manceron, and J. P. Perchard, *Chem. Phys.*, **146**, 179 (1990).

39. T. M. Loehr, J. Zinich, and T. V. Long, II, *Chem. Phys. Lett.*, **7**, 183 (1970).

40. M. M. Schmidke and M. Rosner, *Inorg. Chem.*, **28**, 2570 (1989).

41. A. F. Schreiner and J. A. McLean, *J. Inorg. Nucl. Chem.*, **27**, 253 (1965).

42. A. D. Allen and J. R. Stevens, *Can. J. Chem.*, **51**, 92 (1973).

43. M. W. Bee, S. F. A. Kettle, and D. B. Powell, *Spectrochim. Acta*, **30A**, 139 (1974).

44. C. Li, L. Jiang, and W. Tang, *Spectrochim. Acta*, **49A**, 339 (1993).

45. C. H. Perry, D. P. Athans, E. F. Young, J. R. Durig, and B. R. Mitchell, *Spectrochim. Acta*, **23A**, 1137 (1967).

46. P. J. Hendra, *Spectrochim. Acta*, **23A**, 1275 (1967).

47. K. Nakamoto, P. J. McCarthy, J. Fujita, R. A. Condrate, and G. T. Behnke, *Inorg. Chem.*, **4**, 36 (1965).

48. P. J. Hendra and N. Sadasivan, *Spectrochim. Acta*, **21**, 1271 (1965).

49. R. Layton, D. W. Sink, and J. R. Durig, *J. Inorg. Nucl. Chem.*, **28**, 1965 (1966).

50. J. R. Durig, R. Layton, D. W. Sink, and B. R. Mitchell, *Spectrochim. Acta*, **21**, 1367 (1965).

51. D. B. Powell, *J. Chem. Soc.*, 4495 (1956).

52. J. R. Durig and B. R. Mitchell, *Appl. Spectrosc.*, **21**, 221 (1967).

53. R. J. H. Clark and W. R. Trumble, *Inorg. Chem.*, **15**, 1030 (1976).

54. D. M. Adams and J. R. Hall, *J. Chem. Soc., Dalton Trans.*, 1450 (1973).

55. J. R. Hall and D. A. Hirons, *Inorg. Chim. Acta*, **34**, L277 (1979).

56. R. J. H. Clark, "Raman and Resonance Raman Spectroscopy of Linear Chain Complexes," in R. J. H. Clark and R. E. Hester, eds. *Advances in Infrared and Raman Spectroscopy*, Vol. 11, Wiley, New York, 1985, p. 95.

57. I. Nakagawa and T. Shimanouchi, *Spectrochim. Acta*, **22**, 1707 (1966).

58. T. Grzybek, J. M. Janik, A. Kulczycki, G. Pytasz, J. A. Janik, J. Ściesiński, and E. Ściesińska, *J. Raman Spectrosc.*, **1**, 185 (1973).

59. I. Nakagawa, *Bull. Chem. Soc. Jpn.*, **46**, 3690 (1973).

60. H. Poulet, P. Delorme, and J. P. Mathieu, *Spectrochim. Acta*, **20**, 1855 (1964).

61. J. A. Janik, W. Jacob, and J. M. Janik, *Acta Phys. Pol. A*, **38**, 467 (1970).

62. J. M. Janik, A. Magdal-Mikuli, and J. A. Janik, *Acta Phys. Pol. A*, **40**, 741 (1971).

63. S. Mizushima, I. Nakagawa, and D. M. Sweeny, *J. Chem. Phys.*, **25,** 1006 (1956); I. Nakagawa, R. B. Penland, S. Mizushima, T. J. Lane, and J. V. Quagliano, *Spectrochim. Acta*, **9,** 199 (1957).

64. K. Niwa, H. Takahashi, and K. Higashi, *Bull. Chem. Soc. Jpn.*, **44,** 3010 (1971).

65. K. Brodersen and H. J. Becher, *Chem. Ber.*, **89,** 1487 (1956).

66. A. Novak, J. Portier, and P. Bouvlier, *C.R. Hebd. Seances Acad. Sci.*, **261,** 455 (1965).

67. D. C. Bradley and M. H. Gitlitz, *J. Chem. Soc. A*, 980 (1969).

68. G. W. Watt, B. B. Hutchinson, and D. S. Klett, *J. Am. Chem. Soc.*, **89,** 2007 (1967).

69. Y. Y. Kharitonov, I. K. Dymina, and T. Leonova, *Izv. Akad. Nauk SSSR, Ser. Khim.*, 2057 (1966).

70. M. Goldstein and E. F. Mooney, *J. Inorg. Nucl. Chem.*, **27,** 1601 (1965).

71. J. Chatt, L. A. Duncanson, and L. M. Venanzi, *J. Chem. Soc.*, 4456, 4461 (1955); 2712 (1956); *J. Inorg. Nucl. Chem.*, **8,** 67 (1958).

72. L. A. Duncanson and L. M. Venanzi, *J. Chem. Soc.*, 3841 (1960).

73. F. D. Rochon, M. Doyon, and I. S. Butler, *Inorg. Chem.*, **32,** 2717 (1993).

73a. D. A. Thornton, *J. Coord. Chem.*, **24,** 261 (1991).

74. D. Nicholls and R. Swindells, *J. Inorg. Nucl. Chem.*, **30,** 2211 (1968).

75. K. Brodersen, *Z. Anorg. Allg. Chem.*, **290,** 24 (1957).

76. K. H. Linke, F. Dürholz, and P. Hädicke, *Z. Anorg. Allg. Chem.*, **356,** 113 (1968).

77. H. E. Oosthuizen, E. Singleton, J. S. Field, and G. C. van Niekerk, *J. Organomet. Chem.*, **279,** 433 (1985).

78. D. B. Musaev, M. N. Guseinov, N. G. Klyyuchnikov and R. Ya. Aliev, *Russ. J. Inorg. Chem. (Engl. Transl.)* **31,** 641 (1986).

79. L. Sacconi and A. Sabatini, *J. Inorg. Nucl. Chem.*, **25,** 1389 (1963).

80. Y. Y. Kharitonov, M. A. Sarukhanov, I. B. Baranovskii, and K. U. Ikramov, *Opt. Spektrosk.*, **19,** 460 (1965).

81. K. Nakatsu, M. Shiro, Y. Saito, and H. Kuroya, *Bull. Chem. Soc. Jpn.*, **30,** 158 (1957).

82. K. Nakatsu, *Bull. Chem. Soc. Jpn.*, **35,** 832 (1962).

83. R. E. Cramer and J. T. Huneke, *Inorg. Chem.*, **14,** 2565 (1975).

84. J. Gouteron, *J. Inorg. Nucl. Chem.*, **38,** 63 (1976).

85. G. D. Fleming and R. E. Shepherd, *Spectrochim. Acta*, **43A,** 1141 (1987).

86. G. Borch, P. H. Nielsen, and P. Klaeboe, *Acta Chem. Scand.*, **31A,** 109 (1977).

87. G. Borch, P. Klaeboe, and P. H. Nielsen, *Spectrochim. Acta*, **34A,** 87 (1978).

88. G. Borch, J. Gustavsen, P. Klaeboe, and P. H. Nielsen, *Spectrochim. Acta*, **34A,** 93 (1978).

89. B. E. Williamson, L. Dubicki, and S. E. Harnung, *Inorg. Chem.*, **27,** 3484 (1988).

90. C. D. Flint and A. P. Matthews, *Inorg. Chem.*, **14,** 1219 (1975).

91. Z. Gabelica, *Spectrochim. Acta*, **32A,** 327 (1976).

92. Y. Omura, I. Nakagawa, and T. Shimanouchi, *Spectrochim. Acta*, **27A,** 2227 (1971).

93. A. M. A. Bennett, G. A. Foulds, and D. A. Thornton, *Spectrochim. Acta,* **45A,** 219 (1989).

94. A. M. A. Bennett, G. A. Foulds, D. A. Thornton, and G. M. Watkins, *Spectrochim. Acta,* **46A,** 13 (1990).

95. P. Stein, V. Miskowski, W. H. Woodruff, J. P. Griffin, K. G. Werner, B. P. Gaber, and T. G. Spiro, *J. Chem. Phys.,* **64,** 2159 (1976).

96. M. Z. Zgierski, *J. Raman Spectrosc.,* **6,** 52 (1977).

97. A. B. P. Lever and E. Mantovani, *Can. J. Chem.,* **51,** 1567 (1973).

98. G. W. Rayner-Canham and A. B. P. Lever, *Can. J. Chem.,* **50,** 3866 (1972).

99. A. B. P. Lever and E. Mantovani, *Inorg. Chem.,* **10,** 817 (1971).

100. J. M. Rigg and E. Sherwin, *J. Inorg. Nucl. Chem.,* **27,** 653 (1965).

101. M. N. Hughes and W. R. McWhinnie, *J. Inorg. Nucl. Chem.,* **28,** 1659 (1966).

102. S. Kida, *Bull. Chem. Soc. Jpn.,* **39,** 2415 (1966).

103. E. B. Kipp and R. A. Haines, *Can. J. Chem.,* **47,** 1073 (1969).

104. D. B. Powell and N. Sheppard, *J. Chem. Soc.,* 3089 (1959).

105. N. Farrell, S. Gama de Almeida, and Y. Qu, *Inorg. Chim. Acta,* **178,** 209 (1990).

106. G. Newman and D. B. Powell, *J. Chem. Soc.,* 447 (1961); 3447 (1962).

107. K. Brodersen and T. Kahlert, *Z. Anorg. Allg. Chem.,* **348,** 273 (1966).

108. K. Brodersen, *Z. Anorg, Allg. Chem.,* **298,** 142 (1959).

109. T. Iwamoto and D. F. Shriver, *Inorg. Chem.,* **10,** 2428 (1971).

110. R. J. H. Clark and V. B. Croud, *Inorg. Chem.,* **25,** 1751 (1986).

111. G. C. Papavassiliou, T. Theophanides, and R. Rapsomanikis, *J. Raman Spectrosc.,* **8,** 227 (1979).

112. S. P. Love, S. C. Huckett, L. A. Worl, T. M. Frankcom, S. A. Ekberg, and B. I. Swanson, *Phys. Rev. B. Condens. Matter,* **47,** 11107 (1993).

113. Y. Omura, I. Nakagawa, and T. Shimanouchi, *Spectrochim. Acta,* **27A,** 1153 (1971).

114. R. W. Berg and K. Rasmussen, *Spectrochim. Acta,* **29A,** 37 (1973).

115. G. W. Watt and D. S. Klett, *Spectrochim. Acta,* **20,** 1053 (1964).

116. A. R. Gainsford and D. A. House, *Inorg. Chim. Acta,* **3,** 367 (1969).

117. H. H. Schmidtke and D. Garthoff, *Inorg. Chim. Acta,* **2,** 357 (1968).

118. K. W. Kuo and S. K. Madan, *Inorg. Chem.,* **8,** 1580 (1969).

119. J. H. Forsberg, T. M. Kubik, T. Moeller, and K. Gucwa, *Inorg. Chem.,* **10,** 2656 (1971).

120. D. A. Buckingham and D. Jones, *Inorg. Chem.,* **4,** 1387 (1965).

121. K. W. Bowker, E. R. Gardner, and J. Burgess, *Inorg. Chim. Acta,* **4,** 626 (1970).

122. S. Mizushima, I. Ichishima, I. Nakagawa, and J. V. Quagliano, *J. Phys. Chem.,* **59,** 293 (1955).

123. D. M. Sweeny, S. Mizushima, and J. V. Quagliano, *J. Am. Chem. Soc.,* **77,** 6521 (1955).

124. R. J. H. Clark and C. S. Williams, *Inorg. Chem.,* **4,** 350 (1965).

125. J. K. Wilmshurst and H. J. Bernstein, *Can. J. Chem.,* **35,** 1183 (1957).

126. J. H. S. Green, W. Kynaston, and H. M. Paisley, *Spectrochim. Acta*, **19,** 549 (1963).

127. C. Postmus, J. R. Ferraro, A. Quattrochi, K. Shobatake, and K. Nakamoto, *Inorg. Chem.*, **8,** 1851 (1969).

128. Y. Saito, M. Cordes, and K. Nakamoto, *Spectrochim. Acta*, **28A,** 1459 (1972).

129. Y. Saito, C. W. Schläpfer, M. Cordes, and K. Nakamoto, *Appl. Spectrosc.*, **27,** 213 (1973).

130. J. E. Ruede and D. A. Thornton, *J. Mol. Struct.*, **34,** 75 (1976).

131. D. A. Thornton, *Coord. Chem. Rev.*, **104,** 251 (1990).

132. M. Choca, J. R. Ferraro, and K. Nakamoto, *J. Chem. Soc., Dalton Trans.*, 2297 (1972).

133. R. H. Nuttall, A. F. Cameron, and D. W. Taylor, *J. Chem. Soc. A*, 3103 (1971).

134. M. Fleischmann, P. J. Hendra, and A. McQuillan, *Chem. Phys. Lett.*, **26,** 163 (1974); *J. Electroanal. Chem.*, **65,** 933 (1975).

135. D. J. Jeanmaire and R. P. Van Duyne, *J. Electroanal. Chem.*, **84,** 1 (1977).

136. R. P. Van Duyne, in C. B. Moore, ed., *Chemical and Biological Applications of Lasers*, Vol. 4, Academic Press, New York, 1979, p. 101.

137. M. G. Albrecht and J. A. Creighton, *J. Am. Chem. Soc.*, **99,** 5215 (1977).

138. J. A. Creighton, M. G. Albrecht, R. E. Hester, and J. A. D. Matthew, *Chem. Phys. Lett.*, **55,** 55 (1978).

139. H. Yamada and Y. Yamamoto, *J. Chem. Soc., Faraday Trans. 1*, **75,** 1215 (1979).

140. H. Yamada, *Appl. Spectrosc. Rev.*, **17,** 227 (1981).

141. S. A. Bilmes, *Chem. Phys. Lett.*, **171,** 141 (1990).

142. S. Farquharson, P. A. Lay, and M. J. Weaver, *Spectrochim. Acta*, **40A,** 907 (1984).

143. J. R. Allan, D. H. Brown, R. H. Nuttall, and D. W. A. Sharp, *J. Inorg. Nucl. Chem.*, **27,** 1305 (1965).

144. N. S. Gill and H. J. Kingdon, *Aust. J. Chem.*, **19,** 2197 (1966).

145. L. Cattalini, R. J. H. Clark, A. Orio, and C. K. Poon, *Inorg. Chim. Acta*, **2,** 62 (1968).

146. J. Burgess, *Spectrochim. Acta*, **24A,** 277 (1968).

147. M. Goldstein, E. F. Mooney, A. Anderson, and H. A. Gebbie, *Spectrochim. Acta*, **21,** 105 (1965).

148. A. B. P. Lever and B. S. Ramaswamy, *Can. J. Chem.*, **51,** 1582 (1973).

149. W. R. McWhinnie, *J. Inorg. Nucl. Chem.*, **27,** 2573 (1965).

150. D. E. Billing and A. E. Underhill, *J. Inorg. Nucl. Chem.*, **30,** 2147 (1968).

151. F. Farha and R. T. Iwamoto, *Inorg. Chem.*, **4,** 844 (1965).

152. D. G. Brewer and P. T. T. Wong, *Can. J. Chem.*, **44,** 1407 (1966).

153. J. Burgess, *Spectrochim. Acta*, **24A,** 1645 (1968).

154. L. El-Sayed and R. O. Ragsdale, *J. Inorg. Nucl. Chem.*, **30,** 651 (1968).

155. R. G. Garvey, J. H. Nelson, and R. O. Ragsdale, *Coord. Chem. Rev.*, **3,** 375 (1968).

156. C. P. Prabhakaran and C. C. Patel, *J. Inorg. Nucl. Chem.*, **34**, 3485 (1972).

157. I. S. Ahuja and P. Rastogi, *J. Chem. Soc. A*, 378 (1970).

158. F. A. Cotton and J. F. Gibson, *J. Chem. Soc. A*, 2105 (1970).

159. D. M. L. Goodgame, M. Goodgame, P. J. Hayward, and G. W. Rayner-Canham, *Inorg. Chem.*, **7**, 2447 (1968).

160. C. E. Taylor and A. E. Underhill, *J. Chem. Soc. A*, 368 (1969).

161. B. Cornilsen and K. Nakamoto, *J. Inorg. Nucl. Chem.*, **36**, 2467 (1974).

162. W. J. Eilbeck, F. Holmes, C. E. Taylor, and A. E. Underhill, *J. Chem. Soc. A*, 128 (1968).

163. D. M. L. Goodgame, M. Goodgame, and G. W. Rayner-Canham, *Inorg. Chim. Acta*, **3**, 399 (1969).

164. D. M. L. Goodgame, M. Goodgame, and G. W. Rayner-Canham, *Inorg. Chim. Acta*, **3**, 406 (1969).

165. W. J. Eilbeck, F. Holmes, C. E. Taylor, and A. E. Underhill, *J. Chem. Soc. A*, 1189 (1968).

166. G. A. Melson and R. H. Nuttall, *J. Mol. Struct.*, **1**, 405 (1968).

167. M. M. Cordes and J. L. Walter, *Spectrochim. Acta*, **24A**, 1421 (1968).

168. J. B. Hodgson, G. C. Percy, and D. A. Thornton, *J. Mol. Struct.*, **66**, 81 (1980).

169. S. Salama and T. G. Spiro, *J. Am. Chem. Soc.*, **100**, 1105 (1978).

170. D. S. Caswell and T. G. Spiro, *J. Am. Chem. Soc.*, **108**, 6470 (1986).

171. B. Hutchinson, J. Takemoto, and K. Nakamoto, *J. Am. Chem. Soc.*, **92**, 3335 (1970).

172. Y. Saito, J. Takemoto, B. Hutchinson, and K. Nakamoto, *Inorg. Chem.*, **11**, 2003 (1972); J. Takemoto, B. Hutchinson, and K. Nakamoto, *Chem. Commun.*, 1007 (1971).

173. E. König and E. Lindner, *Spectrochim. Acta*, **28A**, 1393 (1972).

174. R. Wilde, T. K. K. Srinivasan, and N. Ghosh, *J. Inorg. Nucl. Chem.*, **35**, 1017 (1973).

175. J. S. Strukl and J. L. Walter, *Spectrochim. Acta*, **27A**, 223 (1971).

176. R. J. H. Clark, P. C. Turtle, D. P. Strommen, B. Streusand, J. Kincaid, and K. Nakamoto, *Inorg. Chem.*, **16**, 84 (1977).

177. J. C. Rubim, *J. Electrochem. Soc.*, **140**, 1601 (1993).

178. S. M. Angel, M. K. Dearmond, R. J. Donohoe, and D. W. Wertz, *J. Phys. Chem.*, **89**, 282 (1985).

179. R. F. Dallinger and W. H. Woodruff, *J. Am. Chem. Soc.*, **101**, 1355 (1979).

180. P. G. Bradley, N. Kress, B. A. Hornberger, R. F. Dallinger, and W. H. Woodruff, *J. Am. Chem. Soc.*, **103**, 7441 (1981).

181. M. Forster and R. E. Hester, *Chem. Phys. Lett.*, **81**, 42 (1981).

182. S. McClanahan, T. Hayes, and J. Kincaid, *J. Am. Chem. Soc.*, **105**, 4486 (1983).

183. W. K. Smothers and M. S. Wrighton, *J. Am. Chem. Soc.*, **105**, 1067 (1983).

184. P. K. Mallick, D. P. Strommen, and J. R. Kincaid, *J. Am. Chem. Soc.*, **112**, 1686 (1990).

185. R. J. Donohoe, C. D. Tait, M. K. DeArmond, and D. W. Wertz, *Spectrochim. Acta*, **42A**, 233 (1986).

185a. P. K. Mallick, G. D. Danzer, D. P. Strommen, and J. R. Kincaid, *J. Phys. Chem.*, **92,** 5628 (1988).

185b. D. P. Strommen, P. K. Mallick, G. D. Danzer, R. S. Lumpkin, and J. R. Kincaid, *J. Phys. Chem.*, **94,** 1357 (1990).

185c. G. D. Danzer and J. R. Kincaid, *J. Phys. Chem.*, **94,** 3976 (1990).

186. J. Takemoto and B. Hutchinson, *Inorg. Nucl. Chem. Lett.*, **8,** 769 (1972).

187. E. König and K. J. Watson, *Chem. Phys. Lett.*, **6,** 457 (1970).

188. J. Takemoto and B. Hutchinson, *Inorg. Chem.*, **12,** 705 (1973).

189. J. R. Ferraro and J. Takemoto, *Appl. Spectrosc.*, **28,** 66 (1974).

190. P. F. B. Barnard, A. T. Chamberlain, G. C. Kulasingam, R. J. Dosser, and W. R. McWhinnie, *Chem. Commun.*, 520 (1970).

190a. R. H. Herber and L. M. Casson, *Inorg. Chem.*, **25,** 847 (1986).

190b. D. C. Figg and R. H. Herber, *Inorg. Chem.*, **29,** 2170 (1990).

190c. D. C. Figg, R. H. Herber, and J. A. Potenza, *Inorg. Chem.*, **31,** 2111 (1992).

191. J. J. McGarvey, S. E. J. Bell, and J. N. Bechara, *Inorg. Chem.*, **25,** 4325 (1986).

192. K. Nakamoto, *Advances in the Chemistry of the Coordination Compounds*, Macmillan, New York, 1961, p. 437.

193. H. van der Poel, G. van Koten, and K. Vrieze, *Inorg. Chem.*, **19,** 1145 (1980).

194. J. Takemoto, *Inorg. Chem.*, **12,** 949 (1973).

195. B. Hutchinson and A. Sunderland, *Inorg. Chem.*, **11,** 1948 (1972).

196. A. Bigotto, G. Costa, V. Galasso, and G. DeAlti, *Spectrochim. Acta*, **26A,** 1939 (1970).

197. A. Bigotto, V. Galasso, and G. DeAlti, *Spectrochim. Acta*, **27A,** 1659 (1971).

198. P. E. Rutherford and D. A. Thornton, *Spectrochim. Acta*, **35A,** 711 (1979).

199. N. Ohkaku and K. Nakamoto, *Inorg. Chem.*, **10,** 798 (1971).

200. B. Hutchinson, A. Sunderland, M. Neal, and S. Olbricht, *Spectrochim. Acta*, **29A,** 2001 (1973).

201. T. Kitagawa and Y. Ozaki, *Struct. Bonding (Berlin)*, **64,** 71 (1987).

202. H. Ogoshi, Y. Saito, and K. Nakamoto, *J. Chem. Phys.*, **57,** 4194 (1972).

203. H. Susi and J. S. Ard, *Spectrochim. Acta*, **33A,** 561 (1977).

204. S. Sunder and H. J. Bernstein, *J. Raman Spectrosc.*, **5,** 351 (1976).

205. L. L. Gladkov, A. T. Gradyushko, A. M. Shulga, K. N. Solovyov, and A. S. Starukhin, *J. Mol. Struct.*, **47,** 463 (1978); **45,** 267 (1978).

206. M. Abe, T. Kitagawa, and Y. Kyogoku, *J. Chem. Phys.*, **69,** 4526 (1978).

207. X.-Y. Li, R. S. Czernuszewicz, J. R. Kincaid, Y. O. Su, and T. G. Spiro, *J. Phys. Chem.*, **94,** 47 (1990).

208. X.-Y. Li, R. S. Czernuszewicz, J. R. Kincaid, P. Stein, and T. G. Spiro, *J. Phys. Chem.*, **94,** 31 (1990).

209. X.-Y. Li, R. S. Czernuszewicz, J. R. Kincaid, and T. G. Spiro, *J. Am. Chem. Soc.*, **111,** 7012 (1989).

210. J. R. Kincaid and K. Nakamoto, *J. Inorg. Nucl. Chem.*, **37,** 85 (1975).

211. V. L. DeVito and S. A. Asher, *J. Am. Chem. Soc.*, **111,** 9143 (1989).

212. V. L. DeVito, M. Z. Cai, H. Zhu, S. A. Asher, L. A. Kehres, and K. M. Smith, *J. Phys. Chem.*, **95,** 9320 (1991).

213. M. Atamian, R. J. Donohoe, J. S. Lindsey, and D. F. Bocian, *J. Phys. Chem.*, **93**, 2236 (1989).

214. A. D. Procyk, Y. Kim, E. Schmidt, H. N. Fonda, C. K. Chang, G. T. Babcock, and D. F. Bocian, *J. Am. Chem. Soc.*, **114**, 6539 (1992).

215. Y. Ozaki, K. Iriyama, H. Ogoshi, T. Ochiai, and T. Kitagawa, *J. Phys. Chem.*, **90**, 6105 (1986).

216. G. Chottard, P. Battioni, J.-P. Battioni, M. Lange, and D. Mansuy, *Inorg. Chem.*, **20**, 1718 (1981).

217. T. Kitagawa, H. Ogoshi, E. Watanabe, and Z. Yoshida, *J. Phys. Chem.*, **79**, 2629 (1979).

218. J. Teraoka and T. Kitagawa, *J. Phys. Chem.*, **84**, 1928 (1980).

219. T. Kitagawa, M. Abe, Y. Kyogoku, H. Ogoshi, E. Watanabe, and Z. Yoshida, *J. Phys. Chem.*, **80**, 1181 (1976).

220. L. M. Proniewicz, K. Bajdor, and K. Nakamoto, *J. Phys. Chem.*, **90**, 1760 (1986).

221. W.-D. Wagner and K. Nakamoto, *J. Am. Chem. Soc.*, **111**, 1590 (1989).

222. Y. Ozaki, K. Iriyama, H. Ogoshi, T. Ochiai, and T. Kitagawa, *J. Phys. Chem.*, **90**, 6113 (1986).

223. K. R. Rodgers, R. A. Reed, Y. O. Su, and T. G. Spiro, *Inorg. Chem.*, **31**, 2688 (1992).

224. K. Bütje and K. Nakamoto, *Inorg. Chim. Acta*, **167**, 97 (1990).

225. R. S. Czernuszewicz, X.-Y. Li, and T. G. Spiro, *J. Am. Chem. Soc.*, **111**, 7024 (1989).

226. U. Bobinger, R. Schweitzer-Stenner, and W. Dreybrodt, *J. Phys. Chem.*, **95**, 7625 (1991).

227. J. A. Shelnutt, C. J. Medforth, M. D. Berber, K. M. Barkigia, and K. M. Smith, *J. Am. Chem. Soc.*, **113**, 4077 (1991).

228. L. D. Sparks, C. J. Medforth, M. S. Park, J. R. Chamberlain, M. R. Ondrias, M. O. Senge, K. M. Smith, and J. A. Shelnutt, *J. Am. Chem. Soc.*, **115**, 581 (1993).

229. A. Stichternath, R. Schweitzer-Stenner, W. Dreybrodt, R. S. W. Mak, X. Y. Li, L. D. Sparks, J. A. Shelnutt, C. J. Medforth, and K. M. Smith, *J. Phys. Chem.*, **97**, 3701 (1993).

230. C. Piffat, D. Melamed, and T. G. Spiro, *J. Phys. Chem.*, **97**, 7441 (1993).

231. M. L. Mitchell, X.-Y. Li, J. R. Kincaid, and T. G. Spiro, *J. Phys. Chem.*, **91**, 4690 (1987).

232. H. Ogoshi, E. Watanabe, Z. Yoshida, J. R. Kincaid, and K. Nakamoto, *J. Am. Chem. Soc.*, **95**, 2845 (1973).

233. P. G. Wright, P. Stein, J. M. Burke, and T. G. Spiro, *J. Am. Chem. Soc.*, **101**, 3531 (1979).

233a. R. S. Czernuszewicz, W.-D. Wagner, G. B. Ray, and K. Nakamoto, *J. Mol. Struct.*, **242**, 99 (1991).

234. H. Oshio, T. Ama, T. Watanabe, and K. Nakamoto, *Inorg. Chim. Acta*, **96**, 61 (1985).

235. J. R. Kincaid and K. Nakamoto, *Spectrosc. Lett.*, **9**, 19 (1976).

236. T. Kitagawa, M. Abe, Y. Kyogoku, H. Ogoshi, E. Watanabe, and Z. Yoshida, *J. Phys. Chem.*, **80,** 1181 (1976).

237. T. G. Spiro, "The Resonance Raman Spectroscopy of Metalloporphyrins and Heme Proteins," in A. B. P. Lever and H. B. Gray, eds., *Iron Porphyrins*, Addison-Wesley, Reading, MA, 1983.

238. S. Asher and K. Sauer, *J. Chem. Phys.*, **64,** 4115 (1976).

239. H. Ogoshi, E. Watanabe, Z. Yoshida, J. Kincaid, and K. Nakamoto, *Inorg. Chem.*, **14,** 1344 (1975).

240. Y. Ozaki, T. Kitagawa, and H. Ogoshi, *Inorg. Chem.*, **18,** 1772 (1979).

241. C. D. Tait, J. M. Garner, J. P. Collman, A. P. Sattelberger, and W. H. Woodruff, *J. Am. Chem. Soc.*, **111,** 7806 (1989).

242. C. D. Tait, J. M. Garner, J. P. Collman, A. P. Sattelberger, and W. H. Woodruff, *J. Am. Chem. Soc.*, **111,** 9072 (1989).

243. J. K. Duchowski and D. F. Bocian, *Inorg. Chem.*, **29,** 4158 (1990).

244. J. K. Duchowski and D. F. Bocian, *J. Am. Chem. Soc.*, **112,** 8807 (1990).

245. J. A. Shelnutt, *J. Phys. Chem.*, **87,** 605 (1983).

246. R. G. Alden, M. R. Ondrias, and J. A. Shelnutt, *J. Am. Chem. Soc.*, **112,** 691 (1990).

247. C. A. Melendres and V. A. Maroni, *J. Raman Spectrosc.*, **13,** 319 (1984).

248. B. Hutchinson, B. Spencer, R. Thompson, and P. Neill, *Spectrochim. Acta*, **43A,** 631 (1987).

249. I. V. Aleksandrov, Y. S. Bobovich, V. G. Maslov, and A. N. Sidorov, *Opt. Spectrosc.*, **37,** 265 (1974).

250. R. Aroca, Z. O. Zeng, and J. Mink, *J. Phys. Chem. Solids*, **51,** 135 (1990).

251. A. J. Bovill, A. A. McConnell, J. A. Nimmo, and W. E. Smith, *J. Phys. Chem.*, **90,** 569 (1986).

252. D. Maheder and K. P. J. Williams, *J. Raman Spectrosc.*, **18,** 391 (1987).

253. B. Stymne, F. X. Sauvage, and G. Wettermark, *Spectrochim. Acta*, **35A,** 1195 (1979); **36A,** 397 (1980).

254. S. Sievertsen, H. Schlehahn, and H. Homborg, *Z. Anorg. Allg. Chem.*, **619,** 1064 (1993).

255. G. Terzian, B. Moubaraki, M. Mossoyan-Deneux, and D. Benlian, *Spectrochim. Acta*, **45A,** 675 (1989).

256. I. Nakagawa, T. Shimanouchi, and K. Yamasaki, *Inorg. Chem.*, **3,** 772 (1964); **7,** 1332 (1968).

257. M. Le Postolloe, J. P. Mathieu, and H. Poulet, *J. Chim. Phys.*, **60,** 1319 (1963).

258. M. J. Cleare and W. P. Griffith, *J. Chem. Soc. A*, 1144 (1967).

259. M. J. Nolan and D. W. James, *Aust. J. Chem.*, **23,** 1043 (1970).

260. J. T. Huneke, B. Meisner, L. Walford, and R. L. Bain, *Spectrosc. Lett.*, **7,** 91 (1974).

261. D. W. James and M. J. Nolan, *Aust. J. Chem.*, **26,** 1433 (1973).

262. K. Kanamori, A. Shimizu, H. Sekino, S. Hayakawa, Y. Seki, and K. Kawai, *J. Chem. Soc., Dalton Trans.*, 1827 (1988).

263. P. E. Merritt and S. E. Wiberley, *J. Phys. Chem.*, **59,** 55 (1955).

264. R. B. Hagel and L. F. Druding, *Inorg. Chem.*, **9,** 1496 (1970).

265. I. Nakagawa and T. Shimanouchi, *Spectrochim. Acta*, **23A,** 2099 (1967).

266. M. J. Nolan and D. W. James, *Aust. J. Chem.*, **26,** 1413 (1973).

267. K. Nakamoto, J. Fujita, and H. Murata, *J. Am. Chem. Soc.*, **80,** 4817 (1958).

268. D. M. L. Goodgame and M. A. Hitchman, *Inorg. Chem.*, **3,** 1389 (1964).

269. W. W. Fee, C. S. Garner, and J. N. M. Harrowfield, *Inorg. Chem.*, **6,** 87 (1967).

270. D. M. L. Goodgame, M. A. Hitchman, D. F. Marsham, and C. E. Souter, *J. Chem. Soc. A*, 2464 (1969).

271. D. M. L. Goodgame and M. A. Hitchman, *Inorg. Chem.*, **6,** 813 (1967).

272. L. El-Sayed and R. O. Ragsdale, *Inorg. Chem.*, **6,** 1640 (1967).

273. R. B. Penland, T. J. Lane, and J. V. Quagliano, *J. Am. Chem. Soc.*, **78,** 887 (1956).

274. I. R. Beattie and D. P. N. Satchell, *Trans. Faraday Soc.*, **52,** 1590 (1956).

275. A. M. Heyns and D. de Waal, *Spectrochim. Acta*, **45A,** 905 (1989).

276. J. L. Burmeister, *Coord. Chem. Rev.*, **3,** 225 (1968).

277. D. M. L. Goodgame and M. A. Hitchman, *Inorg. Chem.*, **4,** 721 (1965).

277a. K. L. Leighton, K. R. Grundy, and K. N. Robertson, *J. Organomet. Chem.*, **371,** 321 (1989).

277b. R. Birdy, D. M. L. Goodgame, J. C. McConway, and D. Rogers, *J. Chem. Soc., Dalton Trans.*, 1730 (1977).

278. D. M. L. Goodgame and M. A. Hitchman, *J. Chem. Soc. A*, 612 (1967).

279. D. M. L. Goodgame, M. A. Hitchman, and D. F. Marsham, *J. Chem. Soc. A*, 1933 (1970).

280. U. Thewalt and R. E. Marsh, *Inorg. Chem.*, **9,** 1604 (1970).

281. S. Kubo, T. Shibahara, and M. Mori, *Bull. Chem. Soc. Jpn.*, **52,** 101 (1979).

282. K. Wieghardt and H. Siebert, *Z. Anorg. Allg. Chem.*, **374,** 186 (1970).

283. D. M. L. Goodgame, M. A. Hitchman, D. F. Marsham, P. Phavanantha, and D. Rogers, *Chem. Commun.*, 1383 (1969).

284. D. M. L. Goodgame, M. A. Hitchman, and D. F. Marsham, *J. Chem. Soc. A*, 259 (1971).

285. M. Hass and G. B. B. M. Sutherland, *Proc. R. Soc. London, Ser. A*, **236,** 427 (1956).

286. V. P. Tayal, B. K. Srivastava, D. P. Khandelwal, and H. D. Bist, *Appl. Spectrosc. Rev.*, **16,** 43 (1980).

287. J. O. Lundgren and I. Olovsson, *Acta Crystallogr.*, **23,** 966 (1967).

288. A. C. Pavia and P. A. Giguère, *J. Chem. Phys.*, **52,** 3551 (1970).

289. A. Nakahara, Y. Saito, and H. Kuroya, *Bull. Chem. Soc. Jpn.*, **25,** 331 (1952).

290. J. M. Williams, *Inorg. Nucl. Chem. Lett.*, **3,** 297 (1967).

291. J. Roziere and J. Potier, *J. Inorg. Nucl. Chem.*, **35,** 1179 (1973).

292. I. Nakagawa and T. Shimanouchi, *Spectrochim. Acta*, **20,** 429 (1964).

293. V. Stefov, V. M. Petrusevski, and B. Soptrajanov, *J. Mol. Struct.*, **293,** 97 (1993).

294. C. R. Clark, C. E. F. Rickard, and M. J. Tayor, *Can. J. Chem.*, **64,** 1697 (1986).

295. D. M. Adams and P. J. Lock, *J. Chem. Soc. A*, 2801 (1971).

296. Y. Chen, D. H. Christenson, O. F. Nielsen, and E. Pedersen, *J. Mol. Struct.*, **294**, 215 (1993).

297. F. Agullo-Reuda, J. M. Calleja, M. Martini, G. Spinolo, and F. Cariati, *J. Raman Spectrosc.*, **18**, 485 (1987).

297a. S. P. Best, R. S. Armstrong, and J. K. Beattie, *J. Chem. Soc., Dalton Trans.*, 299 (1992).

298. C. W. Bauschlicher, S. R. Langhoff, H. Partridge, J. E. Rice, and A. Komornicki, *J. Chem. Phys.*, **95**, 5142 (1991).

299. H. L. Schlafer and H. P. Fritz, *Spectrochim. Acta*, **23A**, 1409 (1967).

300. T. G. Chang and D. E. Irish, *Can. J. Chem.*, **51**, 118 (1973).

301. R. E. Hester and R. A. Plane, *Inorg. Chem.*, **3**, 768 (1964).

302. H. Boutin, G. J. Safford, and H. R. Danner, *J. Chem. Phys.*, **40**, 2670 (1964); H. J. Prask and H. Boutin, *ibid.*, **45**, 699, 3284 (1966).

303. J. J. Rush, J. R. Ferraro, and A. Walker, *Inorg. Chem.*, **6**, 346 (1967).

304. B. S. Ault, *J. Am. Chem. Soc.*, **100**, 2426 (1978).

305. L. Manceron, A. Loutellier, and J. P. Perchard, *Chem. Phys.*, **92**, 75 (1985).

306. J. W. Kauffman, R. H. Hauge, and J. L. Margrave, *High Temp. Sci.*, **18**, 97 (1984).

307. T. Dupuis, C. Duval, and J. Lecomte, *C.R. Hebd. Seances Acad. Sci.*, **257**, 3080 (1963).

308. M. Maltese and W. J. Orville-Thomas, *J. Inorg. Nucl. Chem.*, **29**, 2533 (1967).

309. J. R. Ferraro and W. R. Walker, *Inorg. Chem.*, **4**, 1382 (1965).

310. W. R. McWhinnie, *J. Inorg. Nucl. Chem.*, **27**, 1063 (1965).

311. J. R. Ferraro, R. Driver, W. R. Walker, and W. Wozniak, *Inorg. Chem.*, **6**, 1586 (1967).

312. G. Blyholder and N. Ford, *J. Phys. Chem.*, **68**, 1496 (1964).

313. V. A. Maroni and T. G. Spiro, *J. Am. Chem. Soc.*, **89**, 45 (1967).

314. R. W. Adams, R. L. Martin, and G. Winter, *Aust. J. Chem.*, **20**, 773 (1967).

315. R. C. Mehrotra and J. M. Batwara, *Inorg. Chem.*, **9**, 2505 (1970).

316. L. M. Brown and K. S. Mazdiyasni, *Inorg. Chem.*, **9**, 2783 (1970).

317. G. Tatzel, M. Greune, J. Weidlein, and E. Jocob, *Z. Anorg. Allg. Chem.*, **533**, 83 (1986).

318. P. W. N. M. van Leeuwen, *Recl. Trav. Chim. Pays-Bas.*, **86**, 247 (1967).

319. D. Knetsch and W. L. Groeneveld, *Inorg. Chim. Acta*, **7**, 81 (1973).

320. S. Halut-Desportes and E. Musson, *Spectrochim. Acta*, **41A**, 661 (1985).

321. H. Wieser and P. J. Krueger, *Spectrochim. Acta*, **26A**, 1349 (1970).

322. G. W. A. Fowles, D. A. Rice, and R. A. Walton, *Spectrochim. Acta*, **26A**, 143 (1970).

323. S. Magull, K. Dehnicke, and D. Fenske, *Z. Anorg. Allg. Chem.*, **608**, 17 (1992).

324. A. Neuhaus, G. Frenzen, J. Pebler, and K. Dehnicke, *Z. Anorg. Allg. Chem.*, **618**, 93 (1992).

325. T. Lu, M. Tan, H. Su, and Y. Liu, *Polyhedron*, **12**, 1055 (1993).

326. H. Takeuchi, T. Arai, and I. Harada, *J. Mol. Struct.*, **146**, 197 (1986).

327. W. L. Driessen and W. L. Groeneveld, *Recl. Trav. Chim. Pays-Bas*, **88**, 977 (1969).

328. W. L. Driessen and W. L. Groeneveld, *Recl. Trav. Chim. Pays-Bas*, **90**, 258 (1971).

329. W. L. Driessen and W. L. Groeneveld, *Recl. Trav. Chim. Pays-Bas*, **90**, 87 (1971).

330. W. L. Driessen, W. L. Groeneveld, and F. W. Van der Wey, *Recl. Trav. Chim. Pays-Bas*, **89**, 353 (1970).

331. Z. Deng and D. E. Irish, *J. Chem. Soc., Faraday, Soc.*, **88**, 2891 (1992).

332. D. B. Powell and A. Woollins, *Spectrochim. Acta*, **41A**, 1023 (1985).

333. K. Itoh and H. J. Bernstein, *Can. J. Chem.*, **34**, 170 (1956).

334. G. B. Deacon and R. J. Phillips, *Coord. Chem. Rev.*, **33**, 227 (1980).

335. S. D. Robinson and M. F. Uttley, *J. Chem. Soc.*, 1912 (1973).

336. B. P. Straughan, W. Moore, and R. McLaughlin, *Spectrochim. Acta*, **42A**, 451 (1986).

337. N. W. Alcock, J. Culver, and S. M. Roe, *J. Chem. Soc., Dalton Trans.*, 1477 (1992).

338. G. Csontos, B. Heil, and C. Markó, *J. Organomet. Chem.*, **37**, 183 (1972).

339. G. Süss-Fink, G. Herrmann, P. Morys, J. Ellermann, and A. Veit, *J. Organomet. Chem.*, **284**, 263 (1985).

340. J. P. Bourke, R. D. Cannon, G. Grinter, and U. A. Jayasooriya, *Spectrochim. Acta*, **49A**, 685 (1993).

341. S. Pal, J. W. Gohdes, W. C. A. Wilisch, and W. H. Armstrong, *Inorg. Chem.*, **31**, 713 (1992).

342. T. A. Stephenson and G. Wilkinson, *J. Inorg. Nucl. Chem.*, **29**, 2122 (1967).

343. R. Kapoor, R. Sharma, and P. Kapoor, *Z. Naturforsch.*, **39B**, 1702 (1984).

344. H.-M. Gau, C.-T. Chen, T.-T. Jong, and M.-Y. Chien, *J. Organomet. Chem.*, **448**, 99 (1993).

345. D. Stoilova, G. Nikolov, and K. Balarev, *Izv. Akad. Nauk SSSR, Ser. Khim.*, **9**, 371 (1976).

346. G. Busca and V. Lorenzelli, *Miner. Chem.*, **7**, 89 (1982).

347. S. Baba and S. Kawaguchi, *Inorg. Nucl. Chem. Lett.*, **9**, 1287 (1973).

348. J. Vicente, M. T. Chicote, M. D. Bermudez, and M. Garcia-Garcia, *J. Organomet. Chem.*, **295**, 125 (1985).

349. K. Nakamoto, P. J. McCarthy, and B. Miniatus, *Spectrochim. Acta*, **21**, 379 (1965).

350. M. Tsuboi, T. Onishi, I. Nakagawa, T. Shimanouchi, and S. Mizuschima, *Spectrochim. Acta*, **12**, 253 (1958).

351. K. Fukushima, T. Onishi, T. Shimanouchi, and S. Mizushima, *Spectrochim. Acta*, **14**, 236 (1959).

352. A. J. Stosick, *J. Am. Chem. Soc.*, **67**, 365 (1945).

353. K. Nakamoto, Y. Morimoto, and A. E. Martell, *J. Am. Chem. Soc.*, **83**, 4528 (1961).

354. R. A. Condrate and K. Nakamoto, *J. Chem. Phys.*, **42**, 2590 (1965).

355. J. Kincaid and K. Nakamoto, *Spectrochim. Acta*, **32A**, 277 (1976).

356. M. L. Niven and D. A. Thornton, *Inorg. Chim. Acta*, **32**, 205 (1979).

357. J. B. Hodgson, G. C. Percy, and D. A. Thornton, *Spectrochim. Acta*, **35A**, 949 (1979).

358. G. C. Percy and H. S. Stenton, *J. Chem. Soc., Dalton Trans.*, 2429 (1976).

359. A. W. Herlinger, S. L. Wenhold, and T. V. Long, II, *J. Am. Chem. Soc.*, **92**, 6474 (1970).

360. A. W. Herlinger and T. V. Long, II, *J. Am. Chem. Soc.*, **92**, 6481 (1970).

361. J. A. Kieft and K. Nakamoto, to be published.

362. J. A. Kieft and K. Nakamoto, *J. Inorg. Nucl. Chem.*, **29**, 2561 (1967).

363. J. R. Kincaid, J. A. Larrabee, and T. G. Spiro, *J. Am. Chem. Soc.*, **100**, 334 (1978).

364. Y. Inomata, A. Shibata, Y. Yukawa, T. Takeuchi, and T. Moriwaki, *Spectrochim. Acta*, **44A**, 97 (1988).

365. D. H. Busch and J. C. Bailar, Jr., *J. Am. Chem. Soc.*, **75**, 4574 (1953); **78**, 716 (1956); M. L. Morris and D. H. Busch, *ibid.*, 5178; K. Swaminathan and D. H. Busch, *J. Inorg. Nucl. Chem.*, **20**, 159 (1961); R. E. Sievers and J. C. Bailar, Jr., *Inorg. Chem.*, **1**, 174 (1962).

366. D. Chapman, *J. Chem. Soc.*, 1766 (1955).

367. Y. Tomita and K. Ueno, *Bull. Chem. Soc. Jpn.*, **36**, 1069 (1963).

368. K. Krishnan and R. A. Plane, *J. Am. Chem. Soc.*, **90**, 3195 (1968).

369. A. A. McConnell and R. H. Nuttall, *Spectrochim. Acta*, **33A**, 459 (1977).

369a. W. S. Caughey, "Methods for Determining Metal Ion Environments," in D. W. Darnell and R. Wilkins, eds., *Proteins: Structure and Function of Metalloproteins*, Elsevier, New York, 1980, p. 95.

369b. E. G. Bartick and R. G. Messerschmidt, *Am. Lab.*, November, p. 56 (1984).

370. N. A. Marley, J. S. Gaffney, and M. M. Cunningham, *Spectroscopy*, **7**, 44 (1992).

371. S. Fronaeus and R. Larsson, *Acta Chem. Scand.*, **16**, 1433, 1447 (1962).

372. S. Fronaeus and R. Larsson, *Acta Chem. Scand.*, **14**, 1364 (1960).

373. R. Larsson, *Acta Chem. Scand.*, **19**, 783 (1965).

374. K. Nakamoto, Y. Morimoto, and A. E. Martell, *J. Am. Chem. Soc.*, **84**, 2081 (1962); **85**, 309 (1963).

375. Y. Tomita, T. Ando, and K. Ueno, *J. Phys. Chem.*, **69**, 404 (1965).

376. Y. Tomita and K. Ueno, *Bull. Chem. Soc. Jpn.*, **36**, 1069 (1963).

377. A. E. Martell and M. K. Kim, *J. Coord. Chem.*, **4**, 9 (1974).

378. M. K. Kim and A. E. Martell, *J. Am. Chem. Soc.*, **85**, 3080 (1963).

379. M. K. Kim and A. E. Martell, *Biochemistry*, **3**, 1169 (1964).

380. M. K. Kim and A. E. Martell, *J. Am. Chem. Soc.*, **88**, 914 (1966).

381. M. Tasumi, S. Takahashi, T. Nakata, and T. Miyazawa, *Bull. Chem. Soc. Jpn.*, **48**, 1595 (1975).

382. T. P. A. Kruck and B. Sarkar, *Can. J. Chem.*, **51**, 3563 (1973).

383. J. Fujita, A. E. Martell, and K. Nakamoto, *J. Chem. Phys.*, **36**, 324, 331 (1962).

384. R. D. Hancock and D. A. Thornton, *J. Mol. Struct.*, **6**, 441 (1970).

385. J. Gouteron, *J. Inorg. Nucl. Chem.*, **38**, 55 (1976).

386. H. G. M. Edwards and P. H. Hardman, *J. Mol. Struct.*, **273**, 73 (1992).

387. H. Homborg, W. Preetz, G. Barka, and G. Schätzel, *Z. Naturforsch.*, **35B**, 554 (1980).

388. K. L. Scott, K. Wieghardt, and A. G. Sykes, *Inorg. Chem.*, **12**, 655 (1973); K. Wieghardt, *Z. Anorg. Allg. Chem.*, **391**, 142 (1972).

389. D. Coucouvanis, K. D. Demadis, C. G. Kim, R. W. Dunham, and J. W. Kampf, *J. Am. Chem. Soc.*, **115**, 3344 (1993).

390. R. E. Hester and R. A. Plane, *Inorg. Chem.*, **3**, 513 (1964); E. C. Gruen and R. A. Plane, *ibid.*, **6**, 1123 (1967).

391. P. X. Armendarez and K. Nakamoto, *Inorg. Chem.*, **5**, 796 (1966).

392. P. Fischer, R. Graf, and J. Weidlein, *J. Organomet. Chem.*, **144**, 95 (1978).

393. F. Quaeyhagens, H. Hofmans, and H. O. Desseyn, *Spectrochim. Acta*, **43A**, 531 (1987).

394. B. B. Kedzia, P. X. Armendarez, and K. Nakamoto, *J. Inorg. Nucl. Chem.*, **30**, 849 (1968).

395. L. Cavalca, M. Nardelli, and G. Fava, *Acta Crystallogr.*, **13**, 594 (1960).

396. Y. Saito, K. Machida, and T. Uno, *Spectrochim. Acta*, **26A**, 2089 (1970).

397. T. J. Thamann and T. M. Loehr, *Spectrochim. Acta*, **36A**, 751 (1980).

398. K. Nakamoto, J. Fujuta, S. Tanaka, and M. Kobayashi, *J. Am. Chem. Soc.*, **79**, 4904 (1957).

399. C. G. Barraclough and M. L. Tobe, *J. Chem. Soc.*, 1993 (1961).

400. R. Eskenazi, J. Rasovan, and R. Levitus, *J. Inorg. Nucl. Chem.*, **28**, 521 (1966).

401. J. E. Finholt, R. W. Anderson, J. A. Fyfe, and K. G. Caulton, *Inorg. Chem.*, **4**, 43 (1965).

402. M. N. Bhattacharjee, M. K. Chaudhuri, and N. S. Islam, *Inorg. Chem.*, **28**, 2420 (1989).

403. W. R. McWhinnie, *J. Inorg. Nucl. Chem.*, **26**, 21 (1964).

404. I. S. Ahuja, *Inorg. Chim. Acta*, **3**, 110 (1969).

405. K. Wieghardt and J. Eckert, *Z. Anorg. Allg. Chem.*, **383**, 240 (1971).

406. R. Ugo, F. Conti, S. Cenini, R. Mason, and G. B. Robertson, *Chem. Commun.*, 1498 (1968).

407. R. W. Horn, E. Weissberger, and J. P. Collman, *Inorg. Chem.*, **9**, 2367 (1970).

408. M. K. Chaudhuri and N. S. Islam, *Inorg. Chem.*, **25**, 3749 (1986).

409. V. N. Krasil'nikov, *Russ. J. Inorg. Chem. (Engl. Transl.)*, **30**, 1499 (1985).

410. J. R. Ferraro and A. Walker, *J. Chem. Phys.*, **42**, 1278 (1965).

411. N. Tanaka, H. Sugi, and J. Fujita, *Bull. Chem. Soc. Jpn.*, **37**, 640 (1964).

412. J. A. Goldsmith, A. Hezel, and S. D. Ross, *Spectrochim. Acta*, **24A**, 1139 (1968).

413. M. R. Rosenthal, *J. Chem. Educ.*, **50**, 331 (1973).

414. B. J. Hathaway and A. E. Underhill, *J. Chem. Soc.*, 3091 (1961).

415. B. J. Hathaway, D. G. Holah, and M. Hudson, *J. Chem. Soc.*, 4586 (1963).

416. M. E. Farago, J. M. James, and V. C. G. Trew, *J. Chem. Soc. A*, 820 (1967).

417. A. E. Wickenden and R. A. Krause, *Inorg. Chem.*, **4**, 404 (1965).

418. L. E. Moore, R. B. Gayhart, and W. E. Bull, *J. Inorg. Nucl. Chem.*, **26,** 896 (1964).

419. M. Fourati, M. Chaabouni, J. L. Pascal, and J. Potier, *Can. J. Chem.*, **67,** 1693 (1989).

420. T. Chausse, A. Potier, and J. Potier, *J. Chem. Res. (S)*, 316 (1980).

421. M. Fourati, M. Chaabouni, H. F. Ayedi, J. L. Pascal, and J. Potier, *Can. J. Chem.*, **63,** 3499 (1985).

422. J. L. Pascal, J. Potier, and C. S. Zhang, *J. Chem. Soc., Dalton Trans.*, 297 (1985).

423. F. Favier and J. L. Pascal, *J. Chem. Soc., Dalton Trans.*, 1997 (1992).

424. L. S. Skogareva, V. P. Babaeva, and V. Ya. Rosolovskii, *Russ. J. Inorg. Chem. (Engl. Transl.)*, **31,** 500 (1986).

425. Z. K. Nikitina, N. V. Krivtsov, Yu. V. Chuprakov, and V. Ya. Rosolovskii, *Russ. J. Inorg. Chem. (Engl. Transl.)*, **33,** 1143 (1988).

426. S. F. Lincoln and D. R. Stranks, *Aust. J. Chem.*, **21,** 37 (1968).

427. T. A. Beech and S. F. Lincoln, *Aust. J. Chem.*, **24,** 1065 (1971).

428. R. Coomber and W. P. Griffith, *J. Chem. Soc.*, 1128 (1968).

429. S. D. Ross and N. A. Thomas, *Spectrochim. Acta*, **26A,** 971 (1970).

430. M. A. Soldatkina, L. B. Serezhkina, and V. N. Serezhkin, *Russ. J. Inorg. Chem. (Engl. Transl.)*, **30,** 1323 (1985).

431. R. I. Bickley, H. G. M. Edwards, R. E. Gustar, and J. K. F. Tait, *J. Mol. Struct.*, **273,** 61 (1992).

432. P. A. Tanner, S. T. Hung, T. C. W. Mak, and R. J. Wang, *Polyhedron*, **11,** 817 (1992).

433. D. S. Bohle and H. Vahrenkamp, *Inorg. Chem.*, **29,** 1097 (1990).

434. A. N. Freedman and B. P. Straughan, *Spectrochim. Acta*, **27A,** 1455 (1971).

435. C. F. Edwards, W. P. Griffith, and D. J. Williams, *J. Chem. Soc., Dalton Trans.*, 145 (1992).

436. P. A. Yeats, J. R. Sams, and F. Aubke, *Inorg. Chem.*, **12,** 328 (1973).

437. S. D. Brown and G. L. Gard, *Inorg. Chem.*, **17,** 1363 (1978).

438. P. C. Leung and F. Aubke, *Can. J. Chem.*, **62,** 2892 (1984).

439. B. M. Gatehouse, S. E. Livingstone, and R. S. Nyholm, *J. Chem. Soc.*, 3137 (1958).

440. J. Fujita, A. E. Martell, and K. Nakamoto, *J. Chem. Phys.*, **36,** 339 (1962).

441. R. E. Hester and W. E. L. Grossman, *Inorg. Chem.*, **5,** 1308 (1966).

442. J. A. Goldsmith and S. D. Ross, *Spectrochim. Acta*, **24A,** 993 (1968).

443. H. Elliott and B. J. Hathaway, *Spectrochim. Acta*, **21,** 1047 (1965).

444. M. R. Churchill, R. A. Lashewycz, K. Koshy, and T. P. Dasgupta, *Inorg. Chem.*, **20,** 376 (1981).

445. M. R. Churchill, G. Davies, M. A. El-Sayed, M. F. El-Shazly, J. P. Hutchinson, and M. W. Rupich, *Inorg. Chem.*, **19,** 201 (1980).

446. A. M. Greenaway, T. P. Dasgupta, K. C. Koshy, and G. G. Sadler, *Spectrochim. Acta*, **42A,** 954 (1986).

447. J. S. Ogden and S. J. Williams, *J. Chem. Soc., Dalton Trans.*, 456 (1981).

448. G. Busca and V. Lorenzelli, *Mater. Chem.*, **7**, 89 (1982).

449. C. C. Addison, N. Logan, S. C. Wallwork, and C. D. Barner, *Q. Rev., Chem. Soc.*, **25**, 289 (1971).

450. B. M. Gatehouse, S. E. Livinstone, and R. S. Nyholm, *J. Chem. Soc.*, 4222 (1957); *J. Inorg. Nucl. Chem.*, **8**, 75 (1958).

451. N. F. Curtis and Y. M. Curtis, *Inorg. Chem.*, **4**, 804 (1965).

452. C. C. Addison, R. Davis, and N. Logan, *J. Chem. Soc. A*, 3333 (1970).

453. B. Lippert, C. J. L. Lock, B. Rosenberg, and M. Zvagulis, *Inorg. Chem.*, **16**, 1525 (1977).

454. C. C. Addison and W. B. Simpson, *J. Chem. Soc.*, 598 (1965).

455. J. G. Allpress and A. N. Hambly, *Aust. J. Chem.*, **12**, 569 (1959).

456. R. J. Fereday, N. Logan, and D. Sutton, *J. Chem. Soc. A*, 2699 (1969).

457. D. W. Johnson and D. Sutton, *Can. J. Chem.*, **50**, 3326 (1972).

458. N. Logan and W. B. Simpson, *Spectrochim. Acta*, **21**, 857 (1965).

459. E. J. Duff, M. N. Hughes, and K. J. Rutt, *J. Chem. Soc. A*, 2126 (1969).

460. E. M. Briggs and A. E. Hill, *J. Chem. Soc. A*, 2008 (1970).

461. A. G. M. Al-Daher, K. W. Bagnall, E. Forsellini, F. Benetollo, and G. Bombieri, *J. Chem. Soc., Dalton Trans.*, 615 (1986).

462. J. C. G. Bünzli, J. P. Metabanzoulou, P. Froidevaux, and L. Jin, *Inorg. Chem.*, **29**, 3875 (1990).

463. A. B. P. Lever, E. Mantovani, and B. S. Ramaswamy, *Can. J. Chem.*, **49**, 1957 (1971).

464. J. R. Ferraro, A. Walker, and C. Cristallini, *Inorg. Nucl. Chem. Lett.*, **1**, 25 (1965).

465. C. C. Addison, D. W. Amos, D. Sutton, and W. H. H. Hoyle, *J. Chem. Soc. A*, 808 (1967).

466. R. H. Nuttall and D. W. Taylor, *Chem. Commun.*, 1417 (1968).

467. J. I. Bullock and F. W. Parrett, *Chem. Commun.*, 157 (1969).

468. J. R. Ferraro and A. Walker, *J. Chem. Phys.*, **42**, 1273 (1965); **43**, 2689 (1965); **45**, 550 (1966).

469. D. E. Irish and G. E. Walrafen, *J. Chem. Phys.*, **46**, 378 (1967).

470. P. M. Castro and P. W. Jagodzinski, *Spectrochim. Acta*, **47A**, 1707 (1991); *J. Phys. Chem.*, **96**, 5296 (1992).

471. R. E. Hester and K. Krishnan, *J. Chem. Phys.*, **46**, 3405 (1967); **47**, 1747 (1967).

472. F. A. Cotton and R. Francis, *J. Am. Chem. Soc.*, **82**, 2986 (1960).

473. G. Newman and D. B. Powell, *Spectrochim. Acta*, **19**, 213 (1963).

474. M. E. Baldwin, *J. Chem. Soc.*, 3123 (1961).

475. B. Nyberg and R. Larsson, *Acta Chem. Scand.*, **27**, 63 (1973).

476. J. P. Hall and W. P. Griffith, *Inorg. Chim. Acta*, **48**, 65 (1981).

477. A. D. Fowless and D. R. Stranks, *Inorg. Chem.*, **16**, 1271 (1977).

478. R. C. Elder and P. E. Ellis, Jr., *Inorg. Chem.*, **17**, 870 (1978).

479. E. Lindner and G. Vitzthum, *Chem. Ber.*, **102**, 4062 (1969).

480. G. Vitzthum and E. Lindner, *Angew. Chem., Int. Ed. Engl.*, **10**, 315 (1971).

481. K. Nakamoto and A. E. Martell, *J. Chem. Phys.*, **32**, 588 (1960).

482. M. Mikami, I. Nakagawa, and T. Shimanouchi, *Spectrochim. Acta*, **23A**, 1037 (1967).

483. H. Junge and H. Musso, *Spectrochim. Acta*, **24A**, 1219 (1968).

484. K. Nakamoto, C. Udovich, and J. Takemoto, *J. Am. Chem. Soc.*, **92**, 3973 (1970).

485. T. Schönherr, U. Rosellen, and H. H. Schmidtke, *Spectrochim. Acta*, **49A**, 357 (1993).

486. M. Handa, H. Miyamoto, T. Suzuki, K. Sawada, and Y. Yukawa, *Inorg. Chim. Acta*, **203**, 61 (1993).

487. R. C. Fay and R. N. Lowry, *Inorg. Nucl. Chem. Lett.*, **3**, 117 (1967).

488. W. D. Courrier, C. J. L. Lock, and G. Turner, *Can. J. Chem.*, **50**, 1797 (1972).

489. M. R. Caira, J. M. Haigh, and L. R. Nassimbeni, *J. Inorg. Nucl. Chem.*, **34**, 3171 (1972).

490. M. F. Richardson, W. F. Wagner, and D. E. Sands, *Inorg. Chem.*, **7**, 2495 (1968).

491. G. Schätzel and W. Preetz, *Z. Naturforsch.*, **31B**, 740 (1976).

492. J. C. Hammel, J. A. S. Smith, and E. J. Wilkins, *J. Chem. Soc. A*, 1461 (1969).

493. J. C. Hammel and J. A. S. Smith, *J. Chem. Soc. A*, 2883 (1969).

494. D. A. Thornton, *Coord. Chem. Rev.*, **104**, 173 (1990).

495. M. A. Bush, D. E. Fenton, R. S. Nyholm, and M. R. Truter, *Chem. Commun.*, 1335 (1970).

496. Y. Nakamura, N. Kanehisa, and S. Kawaguchi, *Bull. Chem. Soc. Jpn.*, **45**, 485 (1972).

497. F. A. Cotton and R. C. Elder, *J. Am. Chem. Soc.*, **86**, 2294 (1964); *Inorg. Chem.*, **4**, 1145 (1965).

498. P. W. N. M. van Leeuwen, *Recl. Trav. Chim. Pays-Bas*, **87**, 396 (1968).

499. Y. Nakamura and S. Kawaguchi, *Chem. Commun.*, 716 (1968).

500. S. Koda, S. Ooi, H. Kuroya, K. Isobe, Y. Nakamura, and S. Kawaguchi, *Chem. Commun.*, 1321 (1971).

501. Y. Nakamura, K. Isobe, H. Morita, S. Yamazaki, and S. Kawaguchi, *Inorg. Chem.*, **11**, 1573 (1972).

502. S. Koda, S. Ooi, H. Kuroya, Y. Nakamura, and S. Kawaguchi, *Chem. Commun.*, 280 (1971).

503. J. Lewis, R. F. Long, and C. Oldham, *J. Chem. Soc.*, 6740 (1965); D. Gibson, J. Lewis, and C. Oldham, *J. Chem. Soc. A*, 1453 (1966).

504. G. T. Behnke and K. Nakamoto, *Inorg. Chem.*, **6**, 433 (1967).

505. G. T. Behnke and K. Nakamoto, *Inorg. Chem.*, **6**, 440 (1967).

506. G. T. Behnke and K. Nakamoto, *Inorg. Chem.*, **7**, 330 (1968).

507. F. Bonati and G. Minghetti, *Angew. Chem., Int. Ed. Engl.*, **7**, 629 (1968).

508. D. Gibson, B. F. G. Johnson, and J. Lewis, *J. Chem. Soc. A*, 367 (1970).

509. S. Baba, T. Ogura, and S. Kawaguchi, *Inorg. Nucl. Chem. Lett.*, **7**, 1195 (1971).

510. G. Allen, J. Lewis, R. F. Long, and C. Oldham, *Nature (London)*, **202**, 589 (1964).

511. G. T. Behnke and K. Nakamoto, *Inorg. Chem.*, **7**, 2030 (1968).

512. J. Lewis and C. Oldham, *J. Chem. Soc. A*, 1456 (1966).

513. Y. Nakamura and K. Nakamoto, *Inorg. Chem.*, **14**, 63 (1975).

514. S. Okeya, T. Nakamura, S. Kawaguchi, and T. Hinomoto, *Inorg. Chem.*, **20**, 1576 (1981).

515. Z. Kanda, Y. Nakamura, and S. Kawaguchi, *Inorg. Chem.*, **17**, 910 (1978).

516. S. Kawaguchi, *Coord. Chem. Rev.*, **70**, 51 (1986).

517. L. G. Hulett and D. A. Thornton, *Spectrochim. Acta*, **27A**, 2089 (1971).

518. H. Junge, *Spectrochim. Acta*, **24A**, 1957 (1968).

519. B. Hutchinson, D. Eversdyk, and S. Olbricht, *Spectrochim. Acta*, **30A**, 1605 (1974).

520. F. Sagara, H. Kobayashi, and K. Ueno, *Bull. Chem. Soc. Jpn.*, **45**, 794 (1972).

521. R. B. Penland, S. Mizushima, C. Curran, and J. V. Quagliano, *J. Am. Chem. Soc.*, **79**, 1575 (1957).

522. A. Yamaguchi, T. Miyazawa, T. Shimanouchi, and S. Mizushima, *Spectrochim. Acta*, **10**, 170 (1957).

523. E. Giesbrecht and M. Kawashita, *J. Inorg. Nucl. Chem.*, **32**, 2461 (1970).

524. K. W. Bagnall, M. A. A. Ghany, G. Bombieri, and F. Benetollo, *Inorg. Chim. Acta*, **115**, 229 (1986).

525. A. Yamaguchi, R. B. Penland, S. Mizushima, T. J. Lane, C. Curran, and J. V. Quagliano, *J. Am. Chem. Soc.*, **80**, 527 (1958).

526. R. A. Bailey and T. R. Peterson, *Can. J. Chem.*, **45**, 1135 (1967).

527. K. Swaminathan and H. M. N. H. Irving, *J. Inorg. Nucl. Chem.*, **26**, 1291 (1964).

528. C. D. Flint and M. Goodgame, *J. Chem. Soc. A*, 744 (1966).

529. P. J. Hendra and Z. Jović, *J. Chem. Soc. A*, 735 (1967).

530. D. M. Adams and J. B. Cornell, *J. Chem. Soc. A*, 884 (1967).

531. R. Rivest, *Can. J. Chem.*, **40**, 2234 (1962).

532. T. J. Lane, A. Yamaguchi, J. V. Quagliano, J. A. Ryan, and S. Mizushima, *J. Am. Chem. Soc.*, **81**, 3824 (1959).

533. M. Schafer and C. Curran, *Inorg. Chem.*, **5**, 265 (1966).

534. R. K. Gosavi and C. N. R. Rao, *J. Inorg. Nucl. Chem.*, **29**, 1937 (1967).

535. G. B. Aitken, J. L. Duncan, and G. P. McQuillan, *J. Chem. Soc., Dalton Trans.*, 2103 (1972).

536. P. J. Hendra and Z. Jović, *Spectrochim. Acta*, **24A**, 1713 (1968).

537. R. J. Balahura and R. B. Jordan, *J. Am. Chem. Soc.*, **92**, 1533 (1970).

538. F. A. Cotton, R. Francis, and W. D. Horrocks, *J. Phys. Chem.*, **64**, 1534 (1960).

539. R. S. Drago and D. W. Meek, *J. Phys. Chem.*, **65**, 1446 (1961): D. W. Meek, D. K. Straub, and R. S. Drago, *J. Am. Chem. Soc.*, **82**, 6013 (1960).

540. B. B. Wayland and R. F. Schramm, *Inorg. Chem.*, **8**, 971 (1969); *Chem. Commun.*, 1465 (1968).

541. V. N. Krishnamarthy and S. Soundararajan, *J. Inorg. Nucl. Chem.*, **29**, 517 (1967); S. K. Ramalingam and S. Soundararajan, *Z. Anorg. Allg. Chem.*, **353**, 216 (1967).

542. C. G. Fuentes and S. J. Patel, *J. Inorg. Nucl. Chem.*, **32,** 1575 (1970).

543. F. Gaizer, *Polyhedron*, **4,** 1909 (1985).

544. C. V. Senoff, E. Maslowsky, Jr., and R. G. Goel, *Can. J. Chem.*, **49,** 3585 (1971).

545. D. A. Langs, C. R. Hare, and R. G. Little, *Chem. Commun.*, 1080 (1967).

546. W. Kitchings, C. J. Moore, and D. Doddrell, *Inorg. Chem.*, **9,** 541 (1970).

547. M. van Beusichem and N. Farrell, *Inorg. Chem.*, **31,** 634 (1992).

548. V. Yu. Kukushkin, V. K. Bel'skii, V. E. Konovalov, R. R. Shifrina, A. I. Moiseev, and R. A. Vlasova, *Inorg. Chim. Acta*, **183,** 57 (1991).

549. D. P. Riley and J. D. Oliver, *Inorg. Chem.*, **25,** 1814 (1986).

550. I. P. Evans, A. Spencer, and G. Wilkinson, *J. Chem. Soc., Dalton Trans.*, 204 (1973).

551. A. Mercer and J. Trotter, *J. Chem. Soc., Dalton Trans.*, 2480 (1975).

552. P. G. Antonov, Y. N. Kukushkin, V. I. Konnov, and B. I. Ionin, *Russ. J. Inorg. Chem. (Engl. Transl.)* **23,** 245 (1978).

552a. U. C. Sarma, K. P. Sarma, and R. K. Poddar, *Polyhedron*, **7,** 1727 (1988).

553. E. Alessio, G. Balducci, A. Lutman, G. Mestroni, M. Calligaris, and W. M. Attia, *Inorg. Chim. Acta*, **203,** 205 (1993).

554. A. Milicic-Tang and J. C. G. Bunzli, *Inorg. Chim. Acta*, **192,** 201 (1992).

555. R. Minkwitz and W. Molsbeck, *Z. Anorg. Allg. Chem.*, **612,** 35 (1992).

556. M. Tranquille and M. T. Forel, *Spectrochim. Acta*, **28A,** 1305 (1972).

557. C. V. Berney and J. H. Weber, *Inorg. Chem.*, **7,** 283 (1968).

558. G. Griffiths and D. A. Thornton, *J. Mol. Struct.*, **52,** 39 (1979).

559. B. R. James and R. H. Morris, *Spectrochim. Acta*, **34A,** 577 (1978).

560. P. W. N. M. van Leeuwen, *Recl. Trav. Chim. Pays-Bas*, **86,** 201 (1967); P. W. N. M. van Leeuwen and W. L. Groeneveld, *ibid.*, 721.

561. S. K. Madan, C. M. Hull, and L. J. Herman, *Inorg. Chem.*, **7,** 491 (1968).

562. K. A. Jensen and K. Krishnan, *Scand. Chim. Acta*, **21,** 1988 (1967).

563. A. G. Sharp, *The Chemistry of Cyano Complexes of the Transition Metals*, Academic Press, New York, 1976.

564. W. P. Griffith, *Coord. Chem. Rev.*, **17,** 177 (1975).

565. P. Rigo and A. Turco, *Coord. Chem. Rev.*, **13,** 133 (1974).

566. L. H. Jones and B. I. Swanson, *Acc. Chem. Res.*, **9,** 128 (1976).

567. H. Stammreich, B. M. Chadwick, and S. G. Frankiss, *J. Mol. Struct.*, **1,** 191 (1968).

568. B. M. Chadwick and S. G. Frankiss, *J. Mol. Struct.*, **31,** 1 (1976).

569. B. M. Chadwick and S. G. Frankiss, *J. Mol. Struct.*, **2,** 281 (1968).

570. G. J. Kubas and L. H. Jones, *Inorg. Chem.*, **13,** 2816 (1974).

571. W. P. Griffiths and J. R. Lane, *J. Chem. Soc., Dalton Trans.*, 158 (1972).

572. W. P. Griffith and G. T. Turner, *J. Chem. Soc. A*, 858 (1970).

573. P. W. Jensen, *J. Raman Spectrosc.*, **4,** 75 (1975).

574. H. Siebert and A. Siebert, *Angew. Chem., Int. Ed. Engl.*, **8,** 6009 (1969); *Z. Anorg. Allg. Chem.*, **378,** 160 (1970).

575. M. F. A. El-Sayed and R. K. Sheline, *J. Inorg. Nucl. Chem.*, **6,** 187 (1958).

576. R. Nast and D. Rehder, *Chem. Ber.*, **104,** 1709 (1971).

577. L. H. Jones and R. A. Penneman, *J. Chem. Phys.*, **22,** 965 (1954).

578. R. A. Penneman and L. H. Jones, *J. Chem. Phys.*, **24,** 293 (1956).

579. R. A. Penneman and L. H. Jones, *J. Inorg. Nucl. Chem.*, **20,** 19 (1961).

580. J. Fernandez-Beltran, J. Blanco-Pascual, and E. Reguera-Edilao, *Spectrochim. Acta*, **46A,** 685 (1990).

581. C. Kappenstein and R. P. Hugel, *Inorg. Chem.*, **16,** 250 (1977).

582. C. Kappenstein and R. P. Hugel, *Inorg. Chem.*, **17,** 1945 (1978).

583. B. M. Chadwick, D. A. Long, and S. U. Qureshi, *J. Mol. Struct.*, **63,** 167 (1980).

584. L. H. Jones and B. I. Swanson, *J. Chem. Phys.*, **63,** 5401 (1975).

585. G. R. Rossman, F.-D. Tsay, and H. B. Gray, *Inorg. Chem.*, **12,** 824 (1973).

586. P. M. Kiernan and W. P. Griffith, *Inorg. Nucl. Chem. Lett.*, **12,** 377 (1976).

587. W. P. Griffith, P. M. Kiernan, and J.-M. Brégeault, *J. Chem. Soc., Dalton Trans.*, 1411 (1978).

588. H. S. Trop, A. G. Jones, and A. Davison, *Inorg. Chem.*, **19,** 1993 (1980).

589. A. M. Soares, P. M. Kiernan, D. J. Cole-Hamilton, and W. P. Griffith, *Chem. Commun.*, 84 (1981).

590. J. L. Hoard, T. A. Hamor, and M. D. Glick, *J. Am. Chem. Soc.*, **90,** 3177 (1968).

591. H. Stammreich and O. Sala, *Z. Elektrochem.*, **64,** 741 (1960); **65,** 149 (1961).

592. K. O. Hartman and F. A. Miller, *Spectrochim. Acta*, **24A,** 669 (1968).

593. B. V. Parish, P. G. Simms, M. A. Wells, and L. A. Woodward, *J. Chem. Soc.*, 2882 (1968).

594. P. M. Kiernan and W. P. Griffith, *J. Chem. Soc., Dalton Trans.*, 2489 (1975).

595. T. V. Long, II and G. A. Vernon, *J. Am. Chem. Soc.*, **93,** 1919 (1971).

596. M. B. Hursthouse and A. M. Galas, *Chem. Commun.*, 1167 (1980).

597. K. N. Raymond, P. W. R. Corfield, and J. A. Ibers, *Inorg. Chem.*, **7,** 1362 (1968).

598. A. Terzis, K. N. Raymond, and T. G. Spiro, *Inorg. Chem.*, **9,** 2415 (1970).

599. L. J. Basile, J. R. Ferraro, M. Choca, and K. Nakamoto, *Inorg. Chem.*, **13,** 496 (1974).

600. E. Hellner, H. Ahsbahs, G. Dehnicke, and K. Dehnicke, *Ber. Bunsensenges. Phys. Chem.*, **77,** 277 (1973).

601. R. L. McCullough, L. H. Jones, and R. A. Penneman, *J. Inorg. Nucl. Chem.*, **13,** 286 (1960).

602. G. W. Chantry and R. A. Plane, *J. Chem. Phys.*, **33,** 736 (1960); **34,** 1268 (1961); **35,** 1027 (1961).

603. V. Caglioti, G. Sartori, and C. Furlani, *J. Inorg. Nucl. Chem.*, **13,** 22 (1960); **8,** 87 (1958).

604. L. H. Jones, *J. Mol. Spectrosc.*, **8,** 105 (1962); *J. Chem. Phys.*, **36,** 1209 (1962).

605. L. H. Jones, *J. Chem. Phys.*, **41,** 856 (1964).

606. D. Bloor, *J. Chem. Phys.*, **41,** 2573 (1964).

607. I. Nakagawa and T. Shimanouchi, *Spectrochim. Acta*, **18,** 101 (1962).

608. L. H. Jones, *Inorg. Chem.*, **2,** 777 (1963).

609. I. Nakagawa and T. Shimanouchi, *Spectrochim. Acta*, **26A,** 131 (1970).

610. L. H. Jones, B. I. Swanson, and G. J. Kubas, *J. Chem. Phys.*, **61,** 4650 (1974); B. I. Swanson and L. H. Jones, *Inorg. Chem.*, **13,** 313 (1974).

611. L. H. Jones, *J. Chem. Phys.*, **29,** 463 (1958).

612. H. Poulet and J. P. Mathieu, *Spectrochim. Acta*, **15,** 932 (1959).

613. L. H. Jones, *Spectrochim. Acta*, **17,** 188 (1961).

614. D. M. Sweeny, I. Nakagawa, S. Mizushima, and J. V. Quagliano, *J. Am. Chem. Soc.*, **78,** 889 (1956).

615. C. W. F. T. Pistorius, *Z. Phys. Chem.*, **23,** 197 (1960).

616. R. L. McCullough, L. H. Jones, and G. A. Crosby, *Spectrochim. Acta*, **16,** 929 (1960).

617. L. H. Jones and J. M. Smith, *J. Chem. Phys.*, **41,** 2507 (1964).

618. L. H. Jones, *J. Chem. Phys.*, **27,** 665 (1957).

619. L. H. Jones, *Spectrochim. Acta*, **19,** 1675 (1963).

620. L. H. Jones, *J. Chem. Phys.*, **26,** 1578 (1957); **25,** 379 (1956).

621. L. H. Jones, *J. Chem. Phys.*, **27,** 468 (1957); **21,** 1891 (1953); **22,** 1135 (1954).

622. V. Lorenzelli and P. Delorme, *Spectrochim. Acta*, **19,** 2033 (1963).

623. J. C. Coleman, H. Peterson, and R. A. Penneman, *Inorg. Chem.*, **4,** 135 (1965).

624. L. H. Jones, *Inorg. Chem.*, **3,** 1581 (1964); **4,** 1472 (1965); J. M. Smith, L. H. Jones, I. K. Kressin, and R. A. Penneman, *ibid.*, 369.

625. L. H. Jones and J. M. Smith, *Inorg. Chem.*, **4,** 1677 (1965).

626. M. N. Memering, L. H. Jones, and J. C. Bailar, Jr., *Inorg. Chem.*, **12,** 2793 (1973).

627. D. F. Shriver, *J. Am. Chem. Soc.*, **84,** 4610 (1962); **85,** 1405 (1963); D. F. Shriver and J. Posner, *ibid.*, **88,** 1672 (1966).

628. D. F. Shriver, S. A. Shriver, and S. E. Anderson, *Inorg. Chem.*, **4,** 725 (1965).

629. D. B. Brown, D. F. Shriver, and L. H. Schwartz, *Inorg. Chem.*, **7,** 77 (1968).

630. B. I. Swanson and J. J. Rafalko, *Inorg. Chem.*, **15,** 249 (1976).

631. B. I. Swanson, *Inorg. Chem.*, **15,** 253 (1976).

632. R. E. Hester and E. M. Nour, *J. Chem. Soc., Dalton Trans.*, 939 (1981).

633. H. G. Nadler, J. Pebler, and K. Dehnicke, *Z. Anorg. Allg. Chem.*, **404,** 230 (1974).

634. R. E. Wilde, S. N. Ghosh, and B. J. Marshall, *Inorg. Chem.*, **9,** 2513 (1970).

635. J. R. Ferraro, *Coord. Chem. Rev.*, **43,** 205 (1982).

636. J. R. Ferraro, L. J. Basile, J. M. Williams, J. I. McOmber, D. F. Shriver, and D. R. Greig, *J. Chem. Phys.*, **69,** 3871 (1978).

637. R. A. Walton, *Spectrochim. Acta*, **21,** 1795 (1965); *Can. J. Chem.*, **44,** 1480 (1966).

638. R. E. Clarke and P. C. Ford, *Inorg. Chem.*, **9,** 227 (1970).

639. J. C. Evans and G. Y.-S. Lo, *Spectrochim. Acta*, **21,** 1033 (1965).

640. J. Reedijk and W. L. Groeneveld, *Rec. Trav. Chim. Pays-Bas*, **86,** 1127 (1967).

641. R. Birk, H. Berke, H.-U. Hund, G. Huttner, L. Szolnai, L. Dahlenburg, U. Behrens, and T. Sielisch, *J. Organomet. Chem.*, **372,** 397 (1989).

642. T. C. Wright, G. Wilkinson, M. Motevalli, and M. B. Hursthouse, *J. Chem. Soc., Dalton Trans.*, 2017 (1986).

643. M. F. Farona and K. F. Kraus, *Inorg. Chem.*, **9**, 1700 (1970).

644. Y. Kinoshita, I. Matsubara, and Y. Saito, *Bull. Chem. Soc. Jpn.*, **32**, 741 (1959).

645. M. Kubota, D. L. Johnston, and I. Matsubara, *Inorg. Chem.*, **5**, 386 (1966).

646. Y. Kinoshita, I. Matsubara, and Y. Saito, *Bull. Chem. Soc. Jpn.*, **32**, 1216 (1959).

647. I. Matsubara, *Bull. Chem. Soc. Jpn.*, **34**, 1719 (1961); *J. Chem. Phys.*, **35**, 373 (1961).

648. J. K. Brown, N. Sheppard, and D. M. Simpson, *Philos. Trans. R. Soc., London*, **A247**, 35 (1954).

649. M. Kubota and D. L. Johnston, *J. Am. Chem. Soc.*, **88**, 2451 (1966).

650. I. Matsubara, *Bull. Chem. Soc. Jpn.*, **35**, 27 (1962).

651. F. A. Cotton and F. Zingales, *J. Am. Chem. Soc.*, **83**, 351 (1961).

652. A. Sacco and F. A. Cotton, *J. Am. Chem. Soc.*, **84**, 2043 (1962).

653. J. W. Dart, M. K. Lloyd, R. Mason, J. A. McCleverty, and J. Williams, *J. Chem. Soc., Dalton Trans.*, 1747 (1973).

654. P. M. Boorman, P. J. Craig, and T. W. Swaddle, *Can. J. Chem.*, **48**, 838 (1970).

655. J. L. Burmeister, *Coord. Chem. Rev.*, **3**, 225 (1968); **1**, 205 (1966).

656. R. A. Bailey, S. L. Kozak, T. W. Michelsen, and W. N. Mills, *Coord. Chem. Rev.*, **6**, 407 (1971).

657. A. H. Norbury, *Adv. Inorg. Chem. Radiochem.*, **17**, 231 (1975).

658. S. Ahrland, J. Chatt, and N. R. Davies, *Q. Rev. Chem. Soc.*, **12**, 265 (1958).

659. P. C. H. Mitchell and R. J. P. Williams, *J. Chem. Soc.*, 1912 (1960).

660. A. Turco and C. Pecile, *Nature (London)*, **191**, 66 (1961).

661. J. Lewis, R. S. Nyholm, and P. W. Smith, *J. Chem. Soc.*, 4590 (1961).

662. A. Sabatini and I. Bertini, *Inorg. Chem.*, **4**, 959 (1965).

663. C. Pecile, *Inorg. Chem.*, **5**, 210 (1966).

664. S. Fronaeus and R. Larsson, *Acta Chem. Scand.*, **16**, 1447 (1962).

665. R. A. Bailey, T. W. Michelsen, and W. N. Mills, *J. Inorg. Nucl. Chem.*, **33**, 3206 (1971).

666. R. Larsson and A. Miezis, *Acta Chem. Scand.*, **23**, 37 (1969).

667. R. J. H. Clark and C. S. Williams, *Spectrochim. Acta*, **22**, 1081 (1966).

668. D. Forster and D. M. L. Goodgame, *Inorg. Chem.*, **4**, 715 (1965).

669. M. A. Bennett, R. J. H. Clark, and A. D. J. Goodwin, *Inorg. Chem.*, **6**, 1625 (1967).

670. D. Forster and D. M. L. Goodgame, *Inorg. Chem.*, **4**, 823 (1965).

671. H. H. Schmidtke and D. Garthoff, *Helv. Chim. Acta*, **50**, 1631 (1967).

672. K. H. Schmidt, A. Müller, and M. Chakravorti, *Spectrochim. Acta*, **32A**, 907 (1976).

673. C. Engelter and D. A. Thornton, *J. Mol. Struct.*, **33**, 119 (1976).

674. H.-H. Fricke and W. Preetz, *Z. Naturforsch.*, **38B**, 917 (1983).

675. K. Bütje and W. Preetz, *Z. Naturforsch.*, **43B**, 371, 382 (1988).

676. W. Keim and W. Preetz, *Z. Anorg. Allg. Chem.*, **565**, 7 (1988).

677. T. Yamaguchi, K. Yamamoto, and H. Ohtaki, *Bull. Chem. Soc. Jpn.*, **58**, 3235 (1985).

678. I. Persson, A. Iverfeldt, and S. Ahrland, *Acta Chem. Scand.*, **A35**, 295 (1981).

679. M. M. Chamberlain and J. C. Bailar, Jr., *J. Am. Chem. Soc.*, **81**, 6412 (1959).

680. A. B. P. Lever, B. S. Ramaswamy, S. H. Simonsen, and L. K. Thompson, *Can. J. Chem.*, **48**, 3076 (1970).

681. J. J. MacDougall, J. H. Nelson, M. W. Babich, C. C. Fuller, and R. A. Jacobson, *Inorg. Chim. Acta*, **27**, 201 (1978).

682. G. Contreras and R. Schmidt, *J. Inorg. Nucl. Chem.*, **32**, 1295, 127 (1970).

683. F. Basolo, J. L. Burmeister, and A. J. Poe, *J. Am. Chem. Soc.*, **85**, 1700 (1963).

684. A. Sabatini and I. Bertini, *Inorg. Chem.*, **4**, 1665 (1965).

685. D. M. L. Goodgame and B. W. Malerbi, *Spectrochim. Acta*, **24A**, 1254 (1968).

686. J. L. Burmeister and F. Basolo, *Inorg. Chem.*, **3**, 1587 (1964).

687. T. E. Sloan and A. Wojcicki, *Inorg. Chem.*, **7**, 1268 (1968).

688. I. Stotz, W. K. Wilmarth, and A. Haim, *Inorg. Chem.*, **7**, 1250 (1968).

689. R. L. Hassel and J. L. Burmeister, *Chem. Commun.*, 568 (1971).

690. L. A. Epps and L. G. Marzilli, *Inorg. Chem.*, **12**, 1514 (1973).

691. I. Bertini and A. Sabatini, *Inorg. Chem.*, **5**, 1025 (1966).

692. G. R. Clark, G. J. Palenik, and D. W. Meek, *J. Am. Chem. Soc.*, **92**, 1077 (1970).

693. G. P. McQuillan and I. A. Oxton, *J. Chem. Soc., Dalton Trans.*, 1460 (1978).

694. K. K. Chow, W. Levason, and C. A. McAuliffe, *Inorg. Chim. Acta*, **15**, 79 (1975).

695. D. W. Meek, P. E. Nicpon, and V. I. Meek, *J. Am. Chem. Soc.*, **92**, 5351 (1970).

696. M. J. Coyer, M. Croft, J. Chen, and R. H. Herber, *Inorg. Chem.*, **31**, 1752 (1992).

697. S. M. Nelson and J. Rodgers, *Inorg. Chem.*, **6**, 1390 (1967).

698. J. L. Burmeister, R. L. Hassel, and R. J. Phelan, *Inorg. Chem.*, **10**, 2032 (1971).

699. J. Chatt and L. A. Duncanson, *Nature (London)*, **178**, 997 (1956).

700. J. Chatt, L. A. Duncanson, F. A. Hart, and P. G. Owston, *Nature (London)*, **181**, 43 (1958).

701. P. G. Owston and J. M. Rowe, *Acta Crystallogr.*, **13**, 253 (1960).

702. J. Chatt and F. A. Hart, *J. Chem. Soc.*, 1416 (1961).

703. B. R. Chamberlain and W. Moser, *J. Chem. Soc. A*, 354 (1969).

704. G. Liptay, K. Burger, E. Papp-Molnár, and Sz. Szebeni, *J. Inorg. Nucl. Chem.*, **31**, 2359 (1969).

705. J. M. Homan, J. M. Kawamoto, and G. L. Morgan, *Inorg. Chem.*, **9**, 2533 (1970).

706. R. A. Bailey and T. W. Michelsen, *J. Inorg. Nucl. Chem.*, **34**, 2671 (1972).

707. F. A. Cotton, A. Davison, W. H. Ilsley, and H. S. Trop, *Inorg. Chem.*, **18**, 2719 (1979).

708. F. A. Cotton, D. M. L. Goodgame, M. Goodgame, and T. E. Hass, *Inorg. Chem.*, **1**, 565 (1962).

709. J. Chatt and L. A. Duncanson, *Nature (London)*, **178**, 997 (1956).

710. M. E. Farago and J. M. James, *Inorg. Chem.*, **4**, 1706 (1965).

711. A. Turco, C. Pecile, and M. Nicolini, *J. Chem. Soc.*, 3008 (1962).

712. J. L. Burmeister and Y. Al-Janabi, *Inorg. Chem.*, **4,** 962 (1965).

713. D. Forster and D. M. L. Goodgame, *Inorg. Chem.*, **4,** 1712 (1965).

714. J. L. Burmeister and H. J. Gysling, *Inorg. Chim. Acta*, **1,** 100 (1967).

715. M. A. Jennings and A. Wojcicki, *Inorg. Chim. Acta*, **3,** 335 (1969).

716. J. L. Burmeister, H. J. Gysling, and J. C. Lim, *J. Am. Chem. Soc.*, **91,** 44 (1969).

717. K. Bütje and W. Preetz, *Z. Naturforsch.*, **43B,** 574 (1988).

718. U. Klopp and W. Preetz, *Z. Anorg. Allg. Chem.*, **619,** 1336 (1993).

719. V. Palaniappan and U. C. Agarwala, *Inorg. Chem.*, **25,** 4064 (1986).

720. F. A. Miller and G. L. Carlson, *Spectrochim. Acta*, **17,** 977 (1961).

721. D. Forster and W. D. Horrocks, *Inorg. Chem.*, **6,** 339 (1967).

722. D. Forster and D. M. L. Goodgame, *J. Chem. Soc.*, 262 (1965).

723. A. R. Chugtai and R. N. Keller, *J. Inorg. Nucl. Chem.*, **31,** 633 (1969).

724. D. Forster and D. M. L. Goodgame, *J. Chem. Soc.*, 1286 (1965).

725. E. J. Peterson, A. Galliart, and J. M. Brown, *Inorg. Nucl. Chem. Lett.*, **9,** 241 (1973).

726. R. A. Bailey and S. L. Kozak, *J. Inorg. Nucl. Chem.*, **31,** 689 (1969).

727. H.-D. Amberger, R. D. Fischer, and G. G. Rosenbauer, *Z. Naturforsch.*, **31B,** 1 (1976).

728. A. H. Norbury and A. I. P. Sinha, *J. Chem. Soc. A*, 1598 (1968).

729. S. J. Patel and D. G. Tuck, *J. Chem. Soc. A*, 1870 (1968).

730. H. G. M. Edwards, D. W. Farwell, I. R. Lewis, and N. Webb, *J. Mol. Struct.*, **271,** 27 (1992).

731. S. J. Anderson and A. H. Norbury, *J. Chem. Soc., Chem. Commun.*, 37 (1974).

732. K. Seppelt and H. Oberhammer, *Inorg. Chem.*, **24,** 1227 (1985).

733. J. Nelson and S. M. Nelson, *J. Chem. Soc. A*, 1597 (1969).

734. R. B. Saillant, *J. Organomet. Chem.*, **39,** C71 (1972).

735. T. Schönherr, *Inorg. Chem.*, **25,** 171 (1986).

736. W. Beck, *Chem. Ber.*, **95,** 341 (1962).

737. W. Beck, P. Swoboda, K. Feldl, and E. Schuierer, *Chem. Ber.*, **103,** 3591 (1970).

738. W. Beck and E. Schuierer, *Z. Anorg. Allg. Chem.*, **347,** 304 (1966).

739. W. Beck and E. Schuierer, *Chem. Ber.*, **98,** 298 (1965).

740. W. Beck, C. Oetker, and P. Swoboda, *Z. Naturforsch.*, **28B,** 229 (1973).

741. W. Weigand, U. Nagel, and W. Beck, *J. Organomet. Chem.*, **314,** C55 (1986).

742. W. Beck, *Organomet. Chem. Rev.*, **A7,** 159 (1971).

743. W. Beck and W. P. Fehlhammer, *J. Organomet. Chem.*, **279,** C22 (1985).

744. W. Beck, W. P. Fehlhammer, P. Pöllmann, E. Schuierer, and K. Feldl, *Chem. Ber.*, **100,** 2335 (1967).

745. W. Beck, W. P. Fehlhammer, P. Pöllmann, E. Schuierer, and K. Feldl, *Angew., Chem.*, **77,** 458 (1965).

746. D. Forster and W. D. Horrocks, *Inorg. Chem.*, **5,** 1510 (1966).

747. K. Steiner, W. Willing, U. Müller, and K. Dehnicke, *Z. Anorg. Allg. Chem.*, **555,** 7 (1987).

748. W.-M. Dyck, K. Dehnicke, F. Weller, and U. Müller, *Z. Anorg. Allg. Chem.*, **470**, 89 (1980).

749. D. Seybold and K. Dehnicke, *Z. Anorg. Allg. Chem.*, **361**, 277 (1968).

750. L. F. Druding, H. C. Wang, R. E. Lohen, and F. D. Sancilio, *J. Coord. Chem.*, **3**, 105 (1973).

751. I. Agrell, *Acta Chem. Scand.*, **25**, 2965 (1971).

752. W. Beck, W. P. Felhammer, P. Pöllman, and R. S. Tobias, *Inorg. Chim. Acta*, **2**, 467 (1968).

753. D. R. Herrington and L. J. Boucher, *Inorg. Nucl. Chem. Lett.*, **7**, 1091 (1971).

754. R. Vicente, A. Escuer, J. Ribas, M. S. El-Fallah, X. Solans, and M. Font-Bardia, *Inorg. Chem.*, **32**, 1920 (1993).

755. A. L. Balch, L. A. Fossett, R. R. Guimerans, M. M. Olmstead, P. E. Reedy, and F. E. Wood, *Inorg. Chem.*, **25**, 1248 (1986).

756. E. W. Abel and F. G. A. Stone, *Q. Rev., Chem. Soc.*, **23**, 325 (1969).

757. L. M. Haines and M. H. B. Stiddard, *Adv. Inorg. Chem. Radiochem.*, **12**, 53 (1969).

758. P. S. Braterman, *Metal Carbonyl Spectra*, Academic Press, New York, 1974.

759. M. Bigorgne, *J. Organomet. Chem.*, **94**, 161 (1975).

760. S. F. A. Kettle, *Top. Curr. Chem.*, **71**, 111 (1977).

761. C. P. Horwitz and D. F. Shriver, *Adv. Organomet. Chem.*, **23**, 219 (1984).

762. C. de la Cruz and N. Sheppard, *J. Mol. Struct.*, **224**, 141 (1990).

763. J. S. Kristoff and D. F. Shriver, *Inorg. Chem.*, **13**, 499 (1974).

764. G. Bouquet and M. Bigorgne, *Spectrochim. Acta*, **27A**, 139 (1971).

765. W. F. Edgell and J. Lyford, IV, *J. Chem. Phys.*, **52**, 4329 (1970).

766. H. Stammreich, K. Kawai, Y. Tavares, P. Krumholz, J. Behmoiras, and S. Bril, *J. Chem. Phys.*, **32**, 1482 (1960).

767. M. Bigorgne, *J. Organomet. Chem.*, **24**, 211 (1970).

768. L. H. Jones, R. S. McDowell, M. Goldblatt, and B. I. Swanson, *J. Chem. Phys.*, **57**, 2050 (1972).

769. L. H. Jones, R. S. McDowell, and M. Goldblatt, *Inorg. Chem.*, **8**, 2349 (1969).

770. E. W. Abel, R. A. N. McLean, S. P. Tyfield, P. S. Braterman, A. P. Walker, and P. J. Hendra, *J. Mol. Spectrosc.*, **30**, 29 (1969).

771. R. A. N. McLean, *Can. J. Chem.*, **52**, 213 (1974).

772. A. Terzis and T. G. Spiro, *Inorg. Chem.*, **10**, 643 (1971).

773. P. J. Hendra and M. M. Qurashi, *J. Chem. Soc. A*, 2963 (1968).

774. S. F. A. Kettle, I. Paul, and P. J. Stamper, *J. Chem. Soc., Chem. Commun.*, 1724 (1970).

775. S. F. A. Kettle, I. Paul, and P. J. Stamper, *Inorg. Chim. Acta*, **7**, 11 (1973).

776. S. F. A. Kettle and N. Luknar, *J. Chem. Phys.*, **68**, 2264 (1978).

777. M. R. Afiz, R. J. H. Clark, and N. R. D'Urso, *J. Chem. Soc., Dalton Trans.*, 250 (1977).

778. W. Scheuermann and K. Nakamoto, *J. Raman Spectrosc.*, **7**, 341 (1978).

779. D. Adelman and D. P. Gerrity, *J. Phys. Chem.*, **94**, 4055 (1990).

780. C. J. Jameson, D. Rehder, and M. Hoch, *Inorg. Chem.*, **27**, 3490 (1988).

781. J. E. Ellis, P. T. Barger, M. L. Winzenburg, and G. I. Warnock, *J. Organomet. Chem.*, **383**, 521 (1990).

782. J. E. Ellis, C. P. Parnell, and G. P. Hagen, *J. Am. Chem. Soc.*, **100**, 3605 (1978).

783. M. Bodenbinder, G. Balzer-Jollenbeck, H. Willner, R. J. Batchelor, F. W. B. Einstein, C. Wang, and F. Aubke, *Inorg. Chem.*, **35**, 82 (1996).

784. H. Willner, J. Schaebs, G. Hwang, F. Mistry, R. Jones, J. Trotter, and F. Aubke, *J. Am. Chem. Soc.*, **114**, 8972 (1992).

784a. G. Hwang, M. Bodenbinder, H. Willner, and F. Aubke, *Inorg. Chem.*, **32**, 4667 (1993).

785. W. F. Edgell, J. Lyford, R. Wright, W. M. Risen, Jr., and A. T. Watts, *J. Am. Chem. Soc.*, **92**, 2240 (1970); W. F. Edgell and J. Lyford, *ibid.*, **93**, 6407 (1971).

786. W. F. Edgell, S. Hegde, and A. Barbetta, *J. Am. Chem. Soc.*, **100**, 1406 (1978).

787. G. G. Summer, H. P. Klug, and L. E. Alexander, *Acta Crystallogr.*, **17**, 732 (1964).

788. F. A. Cotton and R. R. Monchamp, *J. Chem. Soc.*, 1882 (1960).

789. R. L. Sweany and T. L. Brown, *Inorg. Chem.*, **16**, 415 (1977).

790. R. K. Sheline and K. S. Pitzer, *J. Am. Chem. Soc.*, **72**, 1107 (1950).

791. H. M. Powell and R. V. G. Ewens, *J. Chem. Soc.*, 286 (1939).

792. L. F. Dahl and R. E. Rundle, *Acta Crystallogr.*, **16**, 419 (1963).

793. D. M. Adams, M. A. Hooper, and A. Squire, *J. Chem. Soc. A*, 71 (1971).

794. G. Bor, *Chem. Commun.*, 641 (1969).

795. R. A. Levenson, H. B. Gray, and G. P. Ceasar, *J. Am. Chem. Soc.*, **92**, 3653 (1970).

796. I. J. Hyams, D. Jones, and E. R. Lippincott, *J. Chem. Soc. A*, 1987 (1967).

797. N. Flitcroft, D. K. Huggins, and H. D. Kaesz, *Inorg. Chem.*, **3**, 1123 (1964).

798. G. D. Michels and H. J. Svec, *Inorg. Chem.*, **20**, 3445 (1981).

799. P. D. Harvey and I. S. Butler, *Can. J. Chem.*, **63**, 1510 (1985).

800. D. M. Adams, P. D. Hatton, and A. C. Shaw, *J. Phys., Condens. Matter*, **3**, 6145 (1991).

801. L. F. Dahl and J. F. Blount, *Inorg. Chem.*, **4**, 1373 (1965); C. H. Wei and L. F. Dahl, *J. Am. Chem. Soc.*, **91**, 1351 (1969).

802. N. E. Erickson and A. W. Fairhall, *Inorg. Chem.*, **4**, 1320 (1965).

803. F. A. Cotton and D. L. Hunter, *Inorg. Chim. Acta*, **11**, L9 (1974).

804. B. F. G. Johnson, *Chem. Commun.*, 703 (1976).

805. E. R. Corey and L. F. Dahl, *Inorg. Chem.*, **1**, 521 (1962).

806. D. K. Huggins, N. Flitcroft, and H. D. Kaesz, *Inorg. Chem.*, **4**, 166 (1965).

807. C. O. Quicksall and T. G. Spiro, *Inorg. Chem.*, **7**, 2365 (1968).

808. C. E. Anson and U. A. Jayasooriya, *Spectrochim. Acta*, **46A**, 967 (1990).

809. P. Corradini, *J. Chem. Phys.*, **31**, 1676 (1959).

810. G. Bor, G. Sbrignadello, and K. Noack, *Helv. Chim. Acta*, **58**, 815 (1975).

811. P. C. Steinhardt, W. L. Gladfelter, A. D. Harley, J. R. Fox, and G. L. Geoffroy, *Inorg. Chem.*, **19**, 332 (1980).

812. H. Stammreich, K. Kawai, O. Sala, and P. Krumholz, *J. Chem. Phys.*, **35**, 2175 (1961).

813. G. Bor, *Inorg. Chim. Acta*, **3**, 196 (1969).

814. R. J. Ziegler, J. M. Burlitch, S. E. Hayes, and W. M. Risen, Jr., *Inorg. Chem.*, **11**, 702 (1972).

815. F. A. Cotton, L. Kruczynski, and B. A. Frenz, *J. Organomet. Chem.*, **160**, 93 (1978).

816. W. D. Jones, M. A. White, and R. G. Bergman, *J. Am. Chem. Soc.*, **100**, 6770 (1978).

817. P. Braunstein, C. de Méric de Bellefon, and B. Oswald, *Inorg. Chem.*, **32**, 1649 (1993).

818. W. A. Herrmann, M. L. Ziegler, K. Weidenhammer, and H. Biersack, *Angew. Chem., Int. Ed., Engl.*, **18**, 960 (1979).

819. S. Martinengo, G. Ciani, and A. Sironi, *J. Chem. Soc., Chem. Commun.*, 1405 (1992).

820. J. D. Roth, G. J. Lewis, L. K. Safford, X. Jiang, L. F. Dahl, and M. J. Weaver, *J. Am. Chem. Soc.*, **114**, 6159 (1992).

821. J. D. Roth, G. J. Lewis, X. Jiang, L. F. Dahl, and M. J. Weaver, *J. Phys. Chem.*, **96**, 7219 (1992).

822. N. J. Nelson, N. E. Kime, and D. F. Shriver, *J. Am. Chem. Soc.*, **91**, 5173 (1969).

823. A. T. Bell and M. L. Hair, eds., *Vibrational Spectroscopies for Adsorbed Species*, ACS Symp. Ser., Vol. 137. Washington, DC: Am. Chem. Soc., 1980.

824. P. L. Stanghellini, M. J. Sailor, P. Kuznesof, K. H. Whitmire, J. A. Hriljac, J. W. Kolis, Y. Zheng, and D. F. Shriver, *Inorg. Chem.*, **26**, 2950 (1987).

825. C. E. Anson, D. B. Powell, A. G. Cowie, B. F. G. Johnson, J. Lewis, W. J. H. Nelson, J. M. Nicholls, and D. A. Welch, *J. Mol. Struct.*, **159**, 11 (1987).

826. G. Bor and P. L. Stanghellini, *Chem. Commun.*, 886 (1979).

827. P. F. Jackson, B. F. G. Johnson, J. Lewis, M. McPartlin, and W. J. H. Nelson, *Chem. Commun.*, 224 (1980).

828. I. A. Oxton, S. G. A. Kettle, P. F. Jackson, B. F. G. Johnson, and J. Lewis, *J. Mol. Struct.*, **71**, 117 (1981).

829. M. J. Sailor and D. F. Shriver, *J. Am. Chem. Soc.*, **109**, 5039 (1987).

830. P. Künding, M. Moskovits, and G. A. Ozin, *J. Mol. Struct.*, **14**, 137 (1972).

831. J. Mink and P. L. Goggin, *Can. J. Chem.*, **69**, 1857 (1991).

832. J. Browning, P. L. Goggin, R. J. Goodfellow, M. G. Norton, A. J. M. Rattray, B. Taylor, and J. Mink, *J. Chem. Soc., Dalton Trans.*, 2061 (1977).

833. C. Crocker, P. L. Goggin, and R. J. Goodfellow, *J. Chem. Soc., Chem. Commun.*, 1056 (1978).

834. M. Bruns and W. Preetz, *Z. Naturforsch.*, **41B**, 25 (1986).

835. I. J. Hyams and E. R. Lippincott, *Spectrochim. Acta*, **25A**, 1845 (1969).

836. D. K. Ottesen, H. B. Gray, L. H. Jones, and M. Goldblatt, *Inorg. Chem.*, **12**, 1051 (1973).

837. M. J. Cleare and W. P. Griffith, *J. Chem. Soc. A*, 372 (1969).

838. F. H. Johannsen, W. Preetz, and A. Scheffler, *J. Organomet. Chem.*, **102**, 527 (1975).

839. F. H. Johannsen and W. Preetz, *Z. Naturforsch.*, **32B**, 625 (1977).

840. M. A. El-Sayed and H. D. Kaesz, *Inorg. Chem.*, **2**, 158 (1963).

841. C. W. Garland and J. R. Wilt, *J. Chem. Phys.*, **36**, 1094 (1962).

842. L. F. Dahl, C. Martell, and D. L. Wampler, *J. Am. Chem. Soc.*, **83**, 1761 (1961).

843. B. F. G. Johnson, J. Lewis, P. W. Robinson, and J. R. Miller, *J. Chem. Soc. A*, 2693 (1969).

844. F. A. Cotton and B. F. G. Johnson, *Inorg. Chem.*, **6**, 2113 (1967).

845. D. F. Rieck, J. A. Gavney, R. L. Norman, R. K. Hayashi, and L. F. Dahl, *J. Am. Chem. Soc.*, **114**, 10369 (1992).

846. A. Loutellier and M. Bigorgne, *J. Chim. Phys.*, **67**, 78, 99, 107 (1970).

847. M. Bigorgne, *J. Organomet. Chem.*, **24**, 211 (1970).

848. J. Dalton, I. Paul, J. G. Smith, and F. G. A. Stone, *J. Chem. Soc. A*, 1195 (1968).

849. R. J. Angelici and M. D. Malone, *Inorg. Chem.*, **6**, 1731 (1967).

850. B. Hutchinson and K. Nakamoto, *Inorg. Chim. Acta*, **3**, 591 (1969).

851. H. Gäbelein and J. Ellermann, *J. Organomet. Chem.*, **156**, 389 (1978).

852. S. Schmitzer, U. Weis, H. Käb, W. Buchner, W. Malisch, T. Polzer, U. Posset, and W. Kiefer, *Inorg. Chem.*, **32**, 303 (1993).

853. A. M. English, K. R. Plowman, and I. S. Butler, *Inorg. Chem.*, **20**, 2553 (1981).

854. B. H. Weiller, *J. Am. Chem. Soc.*, **114**, 10910 (1992).

855. A. A. Chalmers, J. Lewis, and R. Whyman, *J. Chem. Soc. A*, 1817 (1967).

856. M. F. Farona, J. G. Grasselli, and B. L. Ross, *Spectrochim. Acta*, **23A**, 1875 (1967).

857. S. Singh, P. P. Singh, and R. Rivest, *Inorg. Chem.*, **7**, 1236 (1968).

858. F. A. Cotton and C. S. Kraihanzel, *J. Am. Chem. Soc.*, **84**, 4432 (1962); *Inorg. Chem.*, **2**, 533 (1963); **3**, 702 (1964).

859. F. A. Cotton, M. Musco, and G. Yagupsky, *Inorg. Chem.*, **6**, 1357 (1967).

860. L. H. Jones, *Inorg. Chem.*, **7**, 1681 (1968); **6**, 1269 (1967).

861. F. A. Cotton, *Inorg. Chem.*, **7**, 1683 (1968).

862. M. B. Hall and R. F. Fenske, *Inorg. Chem.*, **11**, 1619 (1972).

863. A. C. Sarapu and R. F. Fenske, *Inorg. Chem.*, **14**, 247 (1975).

864. H. D. Kaesz and R. B. Saillant, *Chem. Rev.*, **72**, 231 (1972).

865. W. F. Edgell, C. Magee, and G. Gallup, *J. Am. Chem. Soc.*, **78**, 4185, 4188 (1956); W. F. Edgell and R. Summitt, *ibid.*, **83**, 1772 (1961).

866. S. J. LaPlaca, W. C. Hamilton, and J. A. Ibers, *Inorg. Chem.*, **3**, 1491 (1964); *J. Am. Chem. Soc.*, **86**, 2288 (1964).

867. D. K. Huggins and H. D. Kaesz, *J. Am. Chem. Soc.*, **86**, 2734 (1964).

868. P. S. Braterman, R. W. Harrill, and H. D. Kaesz, *J. Am. Chem. Soc.*, **89**, 2851 (1967).

869. A. Davison and J. W. Faller, *Inorg. Chem.*, **6**, 845 (1967).

870. W. F. Edgell, J. W. Fisher, G. Asato, and W. M. Risen, Jr., *Inorg. Chem.*, **8**, 1103 (1969).

871. K. Farmery and M. Kilner, *J. Chem. Soc. A*, 634 (1970).

872. S. S. Bath and L. Vaska, *J. Am. Chem. Soc.*, **85**, 3500 (1963).

873. J. Chatt, N. P. Johnson, and B. L. Shaw, *J. Chem. Soc.*, 1625 (1964).

874. L. Vaska, *J. Am. Chem. Soc.*, **88**, 4100 (1966).

875. F. L'Eplattenier and F. Calderazzo, *Inorg. Chem.*, **7**, 1290 (1968).

876. D. K. Huggins, W. Fellman, J. M. Smith, and H. D. Kaesz, *J. Am. Chem. Soc.*, **86**, 4841 (1964).

877. J. M. Smith, W. Fellmann, and L. H. Jones, *Inorg. Chem.*, **4**, 1361 (1965).

878. R. G. Hayter, *J. Am. Chem. Soc.*, **88**, 4376 (1966).

879. R. Bau and T. F. Koetzle, *Pure Appl. Chem.*, **50**, 55 (1978).

880. C. B. Cooper, III, D. F. Shriver, D. J. Darensbourg, and J. A. Froelich, *Inorg. Chem.*, **18**, 1407 (1979); C. B. Cooper, III, D. F. Shriver, and S. Onaka, *Adv. Chem. Ser.*, **167**, 232 (1978).

881. A. P. Ginsberg and M. J. Hawkes, *J. Am. Chem. Soc.*, **90**, 5931 (1968).

882. M. J. Mays and R. N. F. Simpson, *J. Chem. Soc. A*, 1444 (1968); *Chem. Commun.*, 1024 (1967).

883. H. D. Kaesz, F. Fontal, R. Bau, S. W. Kirtley, and M. R. Churchill, *J. Am. Chem. Soc.*, **91**, 1021 (1969).

884. M. J. Bennett, W. A. G. Graham, J. K. Hoyano, and W. L. Hutcheon, *J. Am. Chem. Soc.*, **94**, 6232 (1972).

885. S. A. R. Knox, J. W. Koepke, M. A. Andrews, and H. D. Kaesz, *J. Am. Chem. Soc.*, **97**, 3942 (1975).

886. S. A. R. Knox and H. D. Kaesz, *J. Am. Chem. Soc.*, **93**, 4594 (1971).

887. C. E. Anson, U. A. Jayasooriya, S. F. A. Kettle, P. L. Stranghellini, and R. Rosetti, *Inorg. Chem.*, **30**, 2282 (1991).

888. C. R. Eady, B. F. G. Johnson, J. Lewis, M. C. Malatesta, P. Machin, and M. McPartlin, *Chem. Commun.*, 945 (1976).

889. I. A. Oxton, S. F. A. Kettle, P. F. Jackson, B. F. G. Johnson, and J. Lewis, *Chem. Commun.*, 687 (1979).

890. J. W. White and C. J. Wright, *Chem. Commun.*, 971 (1970).

891. R. L. DeKock, *Inorg. Chem.*, **10**, 1205 (1971).

892. M. Moskovits and G. A. Ozin, "Characterization of the Products of Metal Atom-Molecule Condensation Reactions by Matrix Infrared and Raman Spectroscopy," in J. R. Durig, ed., *Vibrational Spectra and Structure*, Vol. 4, Elsevier, Amsterdam, 1975. p. 187.

893. J. H. Darling and J. S. Ogden, *Inorg. Chem.*, **11**, 666 (1972); *J. Chem. Soc., Dalton Trans.*, 1079 (1973).

894. H. Huber, E. P. Kündig, M. Moskovits, and G. A. Ozin, *Nature (London), Phys. Sci.*, **235**, 98 (1972); E. P. Kündig, M. Moskovits, and G. A. Ozin, *Can. J. Chem.*, **50**, 3587 (1972).

895. E. P. Kündig, D. McIntosh, M. Moskovits, and G. A. Ozin, *J. Am. Chem. Soc.*, **95**, 7234 (1973).

896. J. L. Slater, R. K. Sheline, K. C. Lin, and W. Weltner, *J. Chem. Phys.*, **55**, 5129 (1971).

897. J. L. Slater, T. C. DeVore, and V. Calder, *Inorg. Chem.*, **12**, 1918 (1973); **13**, 1808 (1974).

898. C. N. Krishnan, R. H. Hauge, and J. L. Margrave, *J. Mol. Struct.*, **157**, 187 (1987).

899. C. Ayed, A. Loutellier, L. Manceron, and J. P. Perchard, *J. Am. Chem. Soc.*, **108**, 8138 (1986).

900. T. R. Burkholder and L. Andrews, *J. Phys. Chem.*, **96**, 10195 (1992).

901. C. Xu, L. Manceron, and J. P. Perchard, *J. Chem. Soc., Faraday Trans.*, **89**, 1291 (1993).

902. J. H. B. Chenier, C. A. Hampson, J. A. Howard, and B. Mile, *J. Chem. Soc., Chem. Commun.*, 730 (1986).

903. J. A. Howard, R. Sutcliffe, C. A. Hampson, and B. Mile, *J. Phys. Chem.*, **90**, 4268 (1986).

904. D. McIntosh and G. A. Ozin, *Inorg. Chem.*, **16**, 51 (1977).

905. R. K. Sheline and J. L. Slater, *Angew. Chem. Int. Ed. Engl.*, **14**, 309 (1975).

906. M. Poliakoff and J. J. Turner, *J. Chem. Soc., Dalton Trans.*, 1351 (1973); 2276 (1974).

907. M. A. Graham, M. Poliakoff, and J. J. Turner, *J. Chem. Soc. A*, 2939 (1971).

908. E. P. Kündig and G. A. Ozin, *J. Am. Chem. Soc.*, **96**, 3820 (1974).

909. J. D. Black and P. S. Braterman, *J. Am. Chem. Soc.*, **97**, 2908 (1975).

910. R. N. Perutz and J. J. Turner, *Inorg. Chem.*, **14**, 262 (1975).

911. S. C. Fletcher, M. Poliakoff, and J. J. Turner, *Inorg. Chem.*, **25**, 3597 (1986).

912. I. R. Dunkin, P. Härter, and C. J. Shields, *J. Am. Chem. Soc.*, **106**, 7248 (1984).

913. A. F. Hepp and M. J. Wrighton, *J. Am. Chem. Soc.*, **105**, 5934 (1983).

914. S. Firth, P. M. Hodges, M. Poliakoff, and J. J. Turner, *Inorg. Chem.*, **25**, 4608 (1986).

915. P. A. Breeze, J. K. Burdett, and J. J. Turner, *Inorg. Chem.*, **20**, 3369 (1981).

916. J. A. Timney, *J. Mol. Struct.*, **263**, 229 (1991).

917. A. McNeish, M. Poliakoff, K. P. Smith, and J. J. Turner, *J. Chem. Soc., Chem. Commun.*, 859 (1976).

918. J. J. Turner, *Angew. Chem., Int. Ed. Engl.*, **14**, 304 (1975).

919. For example, see P. L. Bogdan and E. Weitz, *J. Am. Chem. Soc.*, **111**, 3163 (1989).

920. D. A. Van Leirsburg and C. W. DeKock, *J. Phys. Chem.*, **78**, 134 (1974).

921. D. Tevault and K. Nakamoto, *Inorg. Chem.*, **15**, 1282 (1976).

922. B. I. Swanson, L. H. Jones, and R. R. Ryan, *J. Mol. Spectrosc.*, **45**, 324 (1973).

923. M. Poliakoff and J. J. Turner, *J. Chem. Soc. A*, 654 (1971).

924. D. Tevault and K. Nakamoto, *Inorg. Chem.*, **14**, 2371 (1975); A. Cormier, J. D. Brown, and K. Nakamoto, *ibid.*, **12**, 3011 (1973).

925. N.-T. Yu, *Methods in Enzymology*, Vol. 130, Academic Press, New York, 1986, p. 350.

926. M. Kozuka and K. Nakamoto, *J. Am. Chem. Soc.*, **103**, 2162 (1981).

927. B. B. Wayland, L. F. Mehne, and J. Swartz, *J. Am. Chem. Soc.*, **100**, 2379 (1978).

928. G. Eaton and S. Eaton, *J. Am. Chem. Soc.*, **97**, 235 (1975).

929. E. A. Kerr, H. C. Mackin, and N.-T. Yu, *Biochemistry*, **22**, 4373 (1983).

930. N.-T. Yu, E. A. Kerr, B. Ward, and C. K. Chang, *Biochemistry*, **22**, 4534 (1983).

930a. A. Desbois, M. Momenteau, and M. Lutz, *Inorg. Chem.*, **28**, 825 (1989).

931. J. Mascetti and M. Tranquille, *J. Phys. Chem.*, **92**, 2177 (1988).

932. C. Jegat, M. Fouassier, M. Tranquille, and J. Mascetti, *Inorg. Chem.*, **30**, 1529 (1991).

933. C. Jegat, M. Fouassier, and J. Mascetti, *Inorg. Chem.*, **30**, 1521 (1991).

934. C. Jegat, M. Fouassier, M. Tranquille, J. Mascetti, I. Tommasi, M. Aresta, F. Ingold, and A. Dedieu, *Inorg. Chem.*, **32**, 1279 (1993).

935. M. Aresta and C. F. Nobile, *J. Chem. Soc., Dalton Trans.*, 708 (1977).

936. M. Aresta and C. F. Nobile, *Inorg. Chim. Acta.*, **24**, L49 (1977).

937. J. A. McGinnety, *MTP Int. Rev. Sci. Inorg. Chem.*, **5**, 229 (1972).

938. B. L. Haymore and J. A. Ibers, *Inorg. Chem.*, **14**, 3060 (1975).

939. P. Gans, A. Sabatini, and L. Sacconi, *Coord. Chem. Rev.*, **1**, 187 (1966).

940. J. Masek, *Inorg. Chim. Acta, Rev.*, **3**, 99 (1969).

941. B. F. G. Johnson and J. A. McCleverty, *Progr. Inorg. Chem.*, **7**, 277 (1966).

942. W. P. Griffith, *Adv. Organomet. Chem.*, **7**, 211 (1968).

943. J. H. Enemark and R. D. Feltham, *Coord. Chem. Rev.*, **13**, 339 (1974).

944. M. Herberhold and A. Razavi, *Angew. Chem., Int. Ed. Engl.*, **11**, 1092 (1972).

945. I. H. Sabberwal and A. B. Burg, *Chem. Commun.*, 1001 (1970).

946. A. Keller, *Inorg. Chim. Acta*, **133**, 207 (1987).

947. J. A. Timney and C. A. Barnes, *Spectrochim. Acta*, **48A**, 953 (1992).

948. L. H. Jones, R. S. McDowell, and B. I. Swanson, *J. Chem. Phys.*, **58**, 3757 (1973).

949. G. Barna and I. S. Butler, *Can. J. Spectrosc.*, **17**, 2 (1972).

950. O. Crichton and A. J. Rest, *Inorg. Nucl. Chem. Lett.*, **9**, 391 (1973).

951. B. F. G. Johnson, *J. Chem. Soc. A*, 475 (1967).

952. Z. Iqbal and T. C. Waddington, *J. Chem. Soc. A*, 1092 (1969).

953. E. Miki, T. Ishimori, H. Yamatera, and H. Okuno, *J. Chem. Soc. Jpn.*, **87**, 703 (1966).

954. J. R. Durig, W. A. McAllister, J. N. Willis, Jr., and E. E. Mercer, *Spectrochim. Acta*, **22**, 1091 (1966).

955. K. E. Linder, J. C. Dewan, C. E. Costello, and S. Maleknia, *Inorg. Chem.*, **25**, 2085 (1986).

956. M. Quinby-Hunt and R. D. Feltham, *Inorg. Chem.*, **17**, 2515 (1978).

957. J. H. Swinebart, *Coord. Chem. Rev.*, **2**, 385 (1967).

958. R. K. Khanna, C. W. Brown, and L. H. Jones, *Inorg. Chem.*, **8**, 2195 (1969).

959. J. B. Bates and R. K. Khanna, *Inorg. Chem.*, **9**, 1376 (1970).

960. D. B. Brown, *Inorg. Chim. Acta*, **5**, 314 (1971).

961. L. Tosi and J. Danon, *Inorg. Chem.*, **3**, 150 (1964).

962. E. L. Varetti, M. M. Vergara, G. Rigotti, and A. Navaza, *J. Phys. Chem. Solids*, **51**, 381 (1990).

963. J. A. Güida, O. E. Piro, P. J. Aymonino, and O. Sala, *J. Raman Spectrosc.*, **23**, 131 (1992).

964. M. E. Chacón Villalba, E. L. Varetti, and P. J. Aymonino, *Vib. Spectrosc.*, **4**, 109 (1992).

965. A. Poletti, A. Santucci, and G. Paliani, *Spectrochim. Acta*, **27A**, 2061 (1971).

966. E. Miki, S. Kubo, K. Mizumachi, T. Ishimori, and H. Okuno, *Bull. Chem. Soc. Jpn.*, **44**, 1024 (1971).

967. C. G. Pierpont, D. G. Van Derveer, W. Durland, and R. Eisenberg, *J. Am. Chem. Soc.*, **92**, 4760 (1970).

968. C. P. Brock, J. P. Collman, G. Dolcetti, P. H. Farnham, J. A. Ibers, J. E. Lester, and C. A. Reed, *Inorg. Chem.*, **12**, 1304 (1973).

969. J. Müller and S. Schmitt, *J. Organomet. Chem.*, **160**, 109 (1978).

970. R. G. Ball, B. W. Hames, P. Legzdins, and J. Trotter, *Inorg. Chem.*, **19**, 3626 (1980).

971. J. R. Norton, J. P. Collman, G. Dolcetti, and W. T. Robinson, *Inorg. Chem.*, **11**, 382 (1972).

972. B. W. Fitzsimmons, L. F. Larkworthy, and K. A. Rogers, *Inorg. Chim. Acta*, **44**, L53 (1980).

973. S. K. Satija, B. I. Swanson, O. Crichton, and A. J. Rest, *Inorg. Chem.*, **17**, 1737 (1978).

974. O. Crichton and A. J. Rest, *J. Chem. Soc., Dalton Trans.*, 202, 208 (1978).

975. O. Crichton and A. J. Rest, *J. Chem. Soc., Chem. Commun.*, 403 (1973).

976. B. B. Wayland, L. W. Olson, and Z. U. Siddiqui, *J. Am. Chem. Soc.*, **98**, 94 (1976).

977. B. B. Wayland and L. W. Olson, *J. Am. Chem. Soc.*, **96**, 6037 (1974).

978. J. C. Maxwell and W. S. Caughey, *Biochemistry*, **15**, 388 (1976).

979. I. K. Choi, Y. Liu, D. W. Feng, K. J. Paeng, and M. D. Ryan, *Inorg. Chem.*, **30**, 1832 (1991).

980. N.-T. Yu, S. H. Lin, C. K. Chang, and K. Gersonde, *Biophys. J.*, **55**, 1137 (1989).

981. L. A. Lipscomb, B. S. Lee, and N.-T. Yu, *Inorg. Chem.*, **32**, 281 (1993).

982. J. A. Güida, P. J. Aymonino, O. E. Piro, and E. E. Castellano, *Spectrochim. Acta*, **49A**, 535 (1993).

983. J. A. Güida, O. E. Piro, and P. J. Aymonino, *Solid State Commun.*, **57**, 175 (1986).

984. F. Bottomly and P. S. White, *Acta Crystallogr.*, **B35**, 2193 (1979).

985. M. R. Pressprich, M. A. White, Y. Vekhter, and P. Coppens, *J. Am. Chem. Soc.*, **116**, 5233 (1994).

986. W. Krasser, Th. Wolke, S. Haussühl, J. Kuhl, and A. Breitschwerdt, *J. Raman Spectrosc.*, **17**, 83 (1986).

987. J. A. Güida, O. E. Piro, and P. J. Aymonino, *Inorg. Chem.*, **34**, 4113 (1995).

988. V. J. Choy and C. H. O'Connor, *Coord. Chem. Rev.*, **9**, 145 (1972/1973).

989. F. Basolo, B. M. Hoffman, and J. A. Ibers, *Acc. Chem. Res.*, **8**, 384 (1975).

990. L. Vaska, *Acc. Chem. Res.*, **9**, 175 (1976).

991. J. P. Collman, *Acc. Chem. Res.*, **10**, 265 (1977).

992. G. McLendon and A. E. Martell, *Coord. Chem. Rev.*, **19**, 1 (1976).

993. R. W. Erskine and B. O. Field, *Struct. Bonding (Berlin)*, **28**, 1 (1976).

994. R. D. Jones, D. A. Summerville, and F. Basolo, *Chem. Rev.*, **79**, 139 (1979).

995. T. D. Smith and J. R. Pilbrow, *Coord. Chem. Rev.*, **39**, 295 (1981).

996. M. H. Gubelmann and A. F. Williams, *Struct. Bonding (Berlin)*, **55**, 1 (1983).

997. E. C. Niederhoffer, J. H. Timmons, and A. E. Martell, *Chem. Rev.*, **84**, 137 (1984).

998. L. Andrews, "Infrared and Raman Spectroscopic Studies of Alkali-Metal-Atom Matrix-Reaction Products," in M. Moskovits and G. A. Ozin, eds., *Cryochemistry*, Wiley-Interscience, New York, 1976, p. 211.

999. D. McIntosh and G. A. Ozin, *Inorg. Chem.*, **16**, 59 (1977).

1000. A. J. L. Hanlan and G. A. Ozin, *Inorg. Chem.*, **16**, 2848 (1977).

1001. M. J. Zehe, D. A. Lynch, Jr., B. J. Kelsall, and K. D. Carlson, *J. Phys. Chem.*, **83**, 656 (1979).

1002. D. McIntosh and G. A. Ozin, *Inorg. Chem.*, **15**, 2869 (1976).

1003. B. J. Kelsall and K. D. Carlson, *J. Phys. Chem.*, **84**, 951 (1980).

1004. H. Huber, W. Klotzbücher, G. A. Ozin, and A. Vander Voet, *Can. J. Chem.*, **51**, 2722 (1973).

1005. S. Chang, G. Blyholder, and J. Fernandez, *Inorg. Chem.*, **20**, 2813 (1981).

1006. A. B. P. Lever, G. A. Ozin, and H. B. Gray, *Inorg. Chem.*, **19**, 1823 (1980).

1007. L. Andrews, *J. Chem. Phys.*, **50**, 4288 (1969).

1008. J. C. Evans, *J. Chem. Soc., Chem. Commun.*, 682 (1969).

1009. H. H. Eysel and S. Thym, *Z. Anorg. Allg. Chem.*, **411**, 97 (1975).

1010. M. K. Chaudhuri and B. Das, *Inorg. Chem.*, **25**, 168 (1986).

1011. E. M. Nour and S. Morsy, *Inorg. Chim. Acta*, **117**, 45 (1986).

1012. M. T. H. Trafder and A. A. M. A. Islam, *Polyhedron*, **8**, 109 (1989).

1013. C. R. Bhattacharjee, M. Bhattacharjee, M. K. Chaudhuri, and M. Choudhury, *Polyhedron*, **9**, 1653 (1990).

1014. R. Schmidt, G. Pausewang, and W. Massa, *Z. Anorg. Allg. Chem.*, **535**, 135 (1986).

1015. N. J. Campbell, A. C. Dengel, and W. P. Griffith, *Polyhedron*, **8**, 1379 (1989).

1016. M. C. Chakravorti, S. Ganguly, G. V. B. Subrahmanyam, and M. Bhattacharjee, *Polyhedron*, **12**, 683 (1993).

1017. A. C. Dengel, W. P. Griffith, and B. C. Parkin, *J. Chem. Soc., Dalton Trans.*, 2683 (1993).

1018. S. Ahmad, J. D. McCallum, A. K. Shiemke, E. H. Appelman, T. M. Loehr, and J. Sanders-Loehr, *Inorg. Chem.*, **27**, 2230 (1988).

1019. J. K. Basumatary, M. K. Chaudhuri, and R. N. Dutta Purkayastha, *J. Chem. Soc., Dalton Trans.*, 709 (1986).

1020. J. A. Crayston and G. Davidson, *Spectrochim. Acta*, **42A**, 1311 (1986).

1021. A. Nakamura, Y. Tatsuno, M. Yamamoto, and S. Otsuka, *J. Am. Chem. Soc.*, **93**, 6052 (1971).

1022. N. Kitajima, K. Fujisawa, C. Fujimoto, Y. Moro-oka, S. Hashimoto, T. Kitagawa, K. Toriumi, K. Tatsumi, and A. Nakamura, *J. Am. Chem. Soc.*, **114,** 1277 (1992).

1023. K. Nakamoto, Y. Nonaka, T. Ishiguro, M. W. Urban, M. Suzuki, M. Kozuka, Y. Nishida, and S. Kida, *J. Am. Chem. Soc.*, **104,** 3386 (1982).

1024. K. Bajdor, K. Nakamoto, H. Kanatomi, and I. Murase, *Inorg. Chim. Acta*, **82,** 207 (1984).

1025. T. Shibahara and M. Mori, *Bull. Chem. Soc. Jpn.*, **51,** 1374 (1978).

1026. C. G. Barraclough, G. A. Lawrence, and P. A. Lay, *Inorg. Chem.*, **17,** 3317 (1978).

1027. R. E. Hester and E. M. Nour, *J. Raman Spectrosc.*, **11,** 43 (1981).

1028. M. Suzuki, T. Ishiguro, M. Kozuka, and K. Nakamoto, *Inorg. Chem.*, **20,** 1993 (1981).

1029. E. M. Nour and R. E. Hester, *J. Mol. Struct.*, **62,** 77 (1980).

1030. K. Nakamoto, M. Suzuki, T. Ishiguro, M. Kozuka, Y. Nishida, and S. Kida, *Inorg. Chem.*, **19,** 2822 (1980).

1031. R. E. Hester and E. M. Nour, *J. Raman Spectrosc.*, **11,** 59 (1981).

1032. T. Tsumaki, *Bull. Chem. Soc. Jpn.*, **13,** 252 (1938).

1033. M. Kozuka and K. Nakamoto, *J. Am. Chem. Soc.*, **103,** 2162 (1981).

1034. M. W. Urban, K. Nakamoto, and J. Kincaid, *Inorg. Chem.*, **61,** 77 (1983).

1035. L. M. Proniewicz and K. Nakamoto, to be published.

1036. T. Watanabe, T. Ama, and K. Nakamoto, *J. Phys. Chem.*, **88,** 440 (1984).

1037. W.-D. Wagner, I. R. Paeng, and K. Nakamoto, *J. Am. Chem. Soc.*, **110,** 5565 (1988).

1038. K. Bajdor, H. Oshio, and K. Nakamoto, *J. Am. Chem. Soc.*, **106,** 7273 (1984).

1039. L. M. Proniewicz, T. Isobe, and K. Nakamoto, *Inorg. Chim. Acta*, **155,** 91 (1989).

1040. W. Lewandowski, L. M. Proniewicz, and K. Nakamoto, *Inorg. Chim. Acta*, **190,** 145 (1991).

1041. M. W. Urban, K. Nakamoto, and F. Basolo, *Inorg. Chem.*, **21,** 3406 (1982).

1042. A. Weselucha-Birczynska, L. M. Proniewicz, K. Bajdor, and K. Nakamoto, *J. Raman Spectrosc.*, **22,** 315 (1991).

1043. T. Watanabe, T. Ama, and K. Nakamoto, *Inorg. Chem.*, **22,** 2470 (1983).

1044. K. Hasegawa, T. Imamura, and M. Fujimoto, *Inorg. Chem.*, **25,** 2154 (1986).

1045. K. Nakamoto, *Coord. Chem. Rev.*, **100,** 363 (1990).

1046. K. Bajdor, J. R. Kincaid, and K. Nakamoto, *J. Am. Chem. Soc.*, **106,** 7741 (1984).

1047. K. Nakamoto, I. R. Paeng, T. Kuroi, T. Isobe, and H. Oshio, *J. Mol. Struct.*, **189,** 293 (1988).

1048. P. Doppelt and R. Weiss, *Nouv. J. Chim.*, **7,** 341 (1983).

1049. J. P. Collman, J. I. Brauman, T. R. Halbert, and K. S. Suslick, *Proc. Natl. Acad. Sci. U.S.A.*, **73,** 3333 (1976).

1050. J. M. Burke, J. R. Kincaid, S. Peters, R. R. Gagne, J. P. Collman, and T. G. Spiro, *J. Am. Chem. Soc.*, **100,** 6083 (1978).

1051. G. Chottard, M. Schappacher, L. Richard, and R. Weiss, *Inorg. Chem.*, **23,** 4557 (1984).

1052. J. R. Kincaid, L. M. Proniewicz, K. Bajdor, A. Bruha, and K. Nakamoto, *J. Am. Chem. Soc.*, **107,** 6775 (1985).

1053. K. Nakamoto and H. Oshio, *J. Am. Chem. Soc.*, **107,** 6518 (1985).

1054. J. Odo, H. Imai, E. Kyuno, and K. Nakamoto, *J. Am. Chem. Soc.*, **110,** 742 (1988).

1055. L. M. Proniewicz, J. Odo, J. Goral, C. K. Chang, and K. Nakamoto, *J. Am. Chem. Soc.*, **111,** 2105 (1989).

1056. L. M. Proniewicz, A. Bruha, K. Nakamoto, E. Kyuno, and J. R. Kincaid, *J. Am. Chem. Soc.*, **111,** 7050 (1989).

1057. L. M. Proniewicz, A. Bruha, K. Nakamoto, Y. Uemori, E. Kyuno, and J. R. Kincaid, *J. Am. Chem. Soc.*, **113,** 9100 (1991).

1058. L. M. Proniewicz and J. R. Kincaid, *J. Am. Chem. Soc.*, **112,** 675 (1990).

1059. K. Wieghardt, G. Backes-Dahmann, B. Nuber, and J. Weiss, *Angew. Chem., Int. Ed. Engl.*, **24,** 777 (1985).

1060. A. Bashall, V. C. Gibson, T. P. Kee, M. McPartlin, O. B. Robinson, and A. Shaw, *Angew. Chem., Int. Ed. Engl.*, **30,** 980 (1991).

1061. W. P. Griffith and J. D. Wickins, *J. Chem. Soc. A*, 400 (1968).

1062. J. C. Dobson, K. J. Takeuchi, D. W. Pipes, D. A. Geselowitz, and T. J. Meyer, *Inorg. Chem.*, **25,** 2357 (1986).

1063. C. S. Johnson, C. Mottley, J. T. Hupp, and G. D. Danzer, *Inorg. Chem.*, **31,** 5143 (1992).

1064. G. Pausewang and K. Dehnicke, *Z. Anorg. Allg. Chem.*, **369,** 265 (1969).

1065. F. W. Moore and R. E. Rice, *Inorg. Chem.*, **7,** 2510 (1968).

1066. A. Syamal and M. R. Maurya, *Transition Met. Chem.*, **11,** 255 (1986).

1067. K. Dreisch, C. Anderson, and C. Stalhandske, *Polyhedron*, **11,** 2143 (1992).

1068. B. Šoptrjanov, A. Nikolovski, and I. Petrov, *Spectrochim. Acta*, **24A,** 1617 (1968).

1069. W. Willing, F. Schmock, U. Müller, and K. Dehnicke, *Z. Anorg. Allg. Chem.*, **532,** 137 (1986).

1070. I. R. Paeng and K. Nakamoto, *J. Am. Chem. Soc.*, **112,** 3289 (1990).

1071. J. T. Groves and K. H. Ahn, *Inorg. Chem.*, **26,** 3831 (1987).

1072. K. Bajdor and K. Nakamoto, *J. Am. Chem. Soc.*, **106,** 3045 (1984).

1073. J. M. Burke, J. R. Kincaid, and T. G. Spiro, *J. Am. Chem. Soc.*, **100,** 6077 (1978).

1074. L. M. Proniewicz, K. Bajdor, and K. Nakamoto, *J. Phys. Chem.*, **90,** 1760 (1986).

1075. R. S. Czernuszewicz and K. A. Macor, *J. Raman Spectrosc.*, **19,** 553 (1988).

1076. R. T. Kean, W. A. Oertling, and G. T. Babcock, *J. Am. Chem. Soc.*, **109,** 2185 (1987).

1077. W. A. Oertling, R. T. Kean, R. Wever, and G. T. Babcock, *Inorg. Chem.*, **29,** 2633 (1990).

1078. D. Spangler, G. M. Maggiora, L. L. Shipman, and R. E. Christofferson, *J. Am. Chem. Soc.*, **99**, 7478 (1977).

1079. D. Kim, L. A. Miller, G. Rakhit, and T. G. Spiro, *J. Phys. Chem.*, **90**, 3320 (1986).

1080. R. S. Czernuszewicz, K. A. Macor, X.-Y. Li, J. R. Kincaid, and T. G. Spiro, *J. Am. Chem. Soc.*, **111**, 3860 (1989).

1081. W. A. Oertling, A. Salehi, Y. C. Chung, G. E. Leroi, C. K. Chang, and G. T. Babcock, *J. Phys. Chem.*, **91**, 5887 (1987).

1082. W. A. Oertling, A. Salehi, C. K. Chang, and G. T. Babcock, *J. Phys. Chem.*, **93**, 1311 (1989).

1083. S. Hashimoto, Y. Tatsuno, and T. Kitagawa, *J. Am. Chem. Soc.*, **109**, 8096 (1987).

1084. J. R. Kincaid, A. J. Schneider, and K.-J. Paeng, *J. Am. Chem. Soc.*, **111**, 735 (1989).

1085. S. Hashimoto, Y. Mizutani, Y. Tatsuno, and T. Kitagawa, *J. Am. Chem. Soc.*, **113**, 6542 (1991).

1086. L. M. Proniewicz, I. R. Paeng, and K. Nakamoto, *J. Am. Chem. Soc.*, **113**, 3294 (1991).

1087. Y. O. Su, R. S. Czernuszewicz, L. A. Miller, and T. G. Spiro, *J. Am. Chem. Soc.*, **110**, 4150 (1988).

1088. K. A. Macor, R. S. Czernuszewicz, and T. G. Spiro, *Inorg. Chem.*, **29**, 1996 (1990).

1089. T. Kitagawa and Y. Mizutani, *Coord. Chem. Rev.*, **135/136**, 685 (1994).

1090. E. T. Shimomura, M. A. Phillippi, and H. M. Goff, *J. Am. Chem. Soc.*, **103**, 6778 (1981).

1091. A. S. Hinman, B. J. Pavelich, and K. McGarty, *Can. J. Chem.*, **66**, 1589 (1988).

1092. K. Itoh, K. Nakahashi, and H. Toeda, *J. Phys. Chem.*, **92**, 1464 (1988).

1093. M. Atamian, R. J. Donohoe, J. S. Lindsay, and D. F. Bocian, *J. Phys. Chem.*, **93**, 2236 (1989).

1094. J. Teraoka, S. Hashimoto, H. Sugimoto, M. Mori, and T. Kitagawa, *J. Am. Chem. Soc.*, **109**, 180 (1987).

1095. K. R. Rodgers, R. A. Reed, Y. O. Su, and T. G. Spiro, *New J. Chem.*, **16**, 533 (1992).

1096. T. Uno, K. Hatano, Y. Nishimura, and Y. Arata, *Inorg. Chem.*, **29**, 2803 (1990).

1097. N. J. Boldt and D. F. Bocian, *J. Phys. Chem.*, **92**, 581 (1988).

1098. D. J. Liston and B. O. West, *Inorg. Chem.*, **24**, 1568 (1985).

1099. D. J. Liston, B. J. Kennedy, K. Murray, and B. O. West, *Inorg. Chem.*, **24**, 1561 (1985).

1100. K. S. K. Shin, R. J. H. Clark, and J. I. Zink, *J. Am. Chem. Soc.*, **112**, 3754 (1990).

1101. S. E. Lincoln and T. M. Loehr, *Inorg. Chem.*, **29**, 1907 (1990).

1102. R. S. Czernuszewicz, J. E. Sheats, and T. G. Spiro, *Inorg. Chem.*, **26**, 2063 (1987).

1103. J. Sanders-Loehr, W. D. Wheeler, A. K. Shiemke, B. A. Averill, and T. M. Loehr, *J. Am. Chem. Soc.*, **111**, 8084 (1989).

1104. K. Wieghardt, U. Bossek, D. Ventur, and J. Weiss, *J. Chem. Soc., Chem. Commun.*, 347 (1985).

1105. A. D. Allen and C. V. Senoff, *Chem. Commun.*, 621 (1965).

1106. A. D. Allen and F. Bottomley, *Acc. Chem. Res.*, **1**, 360 (1968).

1107. P. C. Ford, *Coord. Chem. Rev.*, **5**, 75 (1970).

1108. R. Murray and D. C. Smith, *Coord. Chem. Rev.*, **3**, 429 (1968).

1109. G. Henrici-Olive and S. Olive, *Angew. Chem., Int. Ed. Engl.*, **8**, 650 (1969).

1110. A. D. Allen, F. Bottomley, R. O. Harris, V. P. Reinsalu, and C. V. Senoff, *J. Am. Chem. Soc.*, **89**, 5595 (1967).

1111. S. Pell, R. H. Mann, H. Taube, and J. N. Armor, *Inorg. Chem.*, **13**, 479 (1974).

1112. M. W. Bee, S. F. A. Kettle, and D. B. Powell, *Spectrochim. Acta*, **30A**, 585 (1974).

1113. A. D. Allen and J. R. Stevens, *Chem. Commun.*, 1147 (1967).

1114. G. Speier and L. Markó, *Inorg. Chim. Acta*, **3**, 126 (1969).

1115. J. H. Enemark, B. R. Davis, J. A. McGinnety, and J. A. Ibers, *Chem. Commun.*, 96 (1968).

1116. J. P. Collman, M. Kubota, F. D. Vastine, J. Y. Sun, and J. W. Kang, *J. Am. Chem. Soc.*, **90**, 5430 (1968).

1117. H. Bauer and W. Beck, *J. Organomet. Chem.*, **308**, 73 (1986).

1118. M. Hidai, K. Tominari, Y. Uchida, and A. Misono, *Chem. Commun.*, 1392 (1969).

1119. B. Bell, J. Chatt, and G. J. Leigh, *Chem. Commun.*, 842 (1970).

1120. G. Speier and L. Markó, *J. Organomet. Chem.*, **21**, P46 (1970).

1121. S. C. Srivastava and M. Bigorgne, *J. Organomet. Chem.*, **19**, 241 (1969).

1122. D. J. Darensbourg, *Inorg. Chem.*, **11**, 1436 (1972).

1123. J. N. Armor and H. Taube, *J. Am. Chem. Soc.*, **92**, 2560 (1970).

1124. C. Krüger and Y.-H. Tsay, *Angew. Chem., Int. Ed. Engl.*, **12**, 998 (1973).

1125. J. Chatt, A. B. Nikolsky, R. L. Richards, and J. R. Sanders, *Chem. Commun.*, 154 (1969).

1126. J. Chatt, R. C. Fay, and R. L. Richards, *J. Chem. Soc. A*, 702 (1971).

1127. M. Mercer, R. H. Crabtree, and R. L. Richards, *Chem. Commun.*, 808 (1973).

1128. D. Sellman, A. Brandl, and R. Endell, *J. Organomet. Chem.*, **49**, C22 (1973).

1129. H. Huber, E. P. Kündig, M. Moskovits, and G. A. Ozin, *J. Am. Chem. Soc.*, **95**, 332 (1973).

1130. G. A. Ozin and A. Vander Voet, *Can. J. Chem.*, **51**, 637 (1973).

1131. E. P. Kündig, M. Moskovits, and G. A. Ozin, *Can. J. Chem.*, **51**, 2710 (1973).

1132. H. Huber, T. A. Ford, W. Klotzbücher, and G. A. Ozin, *J. Am. Chem. Soc.*, **98**, 3176 (1976).

1133. D. W. Green, R. V. Hodges, and D. M. Gruen, *Inorg. Chem.*, **15**, 970 (1976).

1134. T. C. DeVore, *Inorg. Chem.*, **15**, 1315 (1976).

1135. D. W. Green and G. T. Reedy, *J. Mol. Spectrosc.*, **74**, 423 (1979).

1136. G. A. Ozin and A. Vander Voet, *Can. J. Chem.*, **51**, 3332 (1973).

1137. M. Moskovits and G. A. Ozin, "Matrix Cryochemistry Using Transition Metal Atoms" in "Cryochemistry" (M. Moskovits and G. A. Ozin, eds.), John Wiley, New York, 1976, Chapter 8, p. 261.

1138. W. P. Griffith, *Coord. Chem. Rev.*, **8**, 369 (1972).

1139. M. J. Cleare and W. P. Griffith, *J. Chem. Soc. A*, 1117 (1970).

1140. S. M. Sinitsyn and N. A. Razorenova, *Russ. J. Inorg. Chem. (Engl. Transl.)*, **31**, 1618 (1986).

1141. J. T. Groves, T. Takahashi, and W. M. Butler, *Inorg. Chem.*, **22**, 884 (1983).

1142. D. M.-T. Chan, M. H. Chisholm, H. Folting, J. C. Huffman, and N. S. Marchant, *Inorg. Chem.*, **25**, 4170 (1986).

1143. C. Campochiara, J. A. Hofmann, and D. F. Bocian, *Inorg. Chem.*, **24**, 449 (1985).

1144. U. Abram, S. Abram, H. Spies, R. Kirmse, J. Stach, and K. Kohler, *Z. Anorg. Allg. Chem.*, **544**, 167 (1987).

1145. W. P. Griffith and D. Pawson, *J. Chem. Soc., Dalton Trans.*, 1315 (1973).

1146. W.-D. Wagner and K. Nakamoto, *J. Am. Chem. Soc.*, **111**, 1590 (1989).

1147. W.-D. Wagner and K. Nakamoto, *J. Am. Chem. Soc.*, **110**, 4044 (1988).

1148. R. S. Czernuszewicz, Y. O. Su, M. K. Stern, K. A. Macor, D. Kim, J. T. Groves, and T. G. Spiro, *J. Am. Chem. Soc.*, **110**, 4158 (1988).

1149. M. Horner, K. P. Frank, and J. Strahle, *Z. Naturforsch.*, **41B**, 423 (1986).

1150. T. Godemeyer, A. Berg, H.-D. Gross, U. Müller, and K. Dehnicke, *Z. Naturforsch.*, **40B**, 999 (1985).

1151. G. A. Schick and D. F. Bocian, *J. Am. Chem. Soc.*, **105**, 1830 (1983).

1152. G. A. Ozin and J. Garcia-Prieto, *J. Am. Chem. Soc.*, **108**, 3099 (1986).

1153. G. J. Kubas, *Comments Inorg. Chem.*, **7**, 17 (1988).

1154. G. E. Gadd, R. K. Upmacis, M. Poliakoff, and J. J. Turner, *J. Am. Chem. Soc.*, **108**, 2547 (1986).

1155. R. K. Upmacis, M. Poliakoff, and J. J. Turner, *J. Am. Chem. Soc.*, **108**, 3645 (1986).

1156. R. L. Sweany and A. Moroz, *J. Am. Chem. Soc.*, **111**, 3577 (1989).

1157. M. W. George, M. T. Haward, P. A. Hamley, C. Hughes, E. P. A. Johnson, V. K. Popov, and M. Poliakoff, *J. Am. Chem. Soc.*, **115**, 2286 (1993).

1158. S. A. Jackson, P. M. Hodges, M. Poliakoff, J. J. Turner, and F. W. Grevels, *J. Am. Chem. Soc.*, **112**, 1221 (1990).

1159. G. J. Kubas, R. R. Ryan, B. I. Swanson, P. J. Vergamini, and H. J. Wasserman, *J. Am. Chem. Soc.*, **106**, 451 (1984).

1160. G. J. Kubas, C. J. Unkefer, B. I. Swanson, and E. Fukushima, *J. Am. Chem. Soc.*, **108**, 7000 (1986).

1161. A. A. H. van der Zeijden, T. Burgi, and H. Berke, *Inorg. Chim. Acta*, **201**, 131 (1992).

1162. M. J. Mockford and W. P. Griffith, *J. Chem. Soc., Dalton Trans.*, 717 (1985).

1163. C. E. Ash, C. M. Kim, M. Y. Darensbourg, and A. L. Rheingold, *Inorg. Chem.*, **26,** 1357 (1987).

1164. J. Chatt and R. G. Hayter, *J. Chem. Soc.*, 5507 (1961).

1165. R. G. S. Banks and J. M. Pratt, *J. Chem. Soc. A*, 854 (1968).

1166. M. J. Fernandez, M. A. Esteruelas, M. Covarrubias, and L. A. Oro, *J. Organomet. Chem.*, **316,** 343 (1986).

1167. J. Chatt, L. A. Duncanson, and B. L. Shaw, *Chem. Ind. (London)*, 859 (1958).

1168. M. J. Church and M. J. Mays, *J. Chem. Soc.*, 3074 (1968); 1938 (1970).

1169. P. W. Atkins, J. C. Green, and M. L. H. Green, *J. Chem. Soc. A*, 2275 (1968).

1170. S. C. Abrahams, A. P. Ginsberg, T. F. Koetzle, S. P. Marsh, and C. R. Sprinkle, *Inorg. Chem.*, **25,** 2500 (1986).

1171. M. D. Fryzuk and D. H. McConville, *Inorg. Chem.*, **28,** 1613 (1989).

1172. T. M. Arkhireeva, B. M. Bulychev, T. A. Sokolova, G. L. Soloveichik, V. K. Belsky, and G. N. Boiko, *Inorg. Chim. Acta*, **141,** 221 (1988).

1173. D. W. Hart, R. G. Teller, C. Y. Wei, R. Bau, G. Longoni, S. Campanella, P. Chini, and T. F. Koetzle, *J. Am. Chem. Soc.*, **103,** 1458 (1981); *Angew. Chem., Int. Ed. Engl.*, **18,** 80 (1979).

1174. D. Graham, J. Howard, and T. C. Waddington, *J. Chem. Soc., Faraday Trans. 2*, **79,** 1713 (1983).

1175. P. L. Stanghellini and G. Longoni, *J. Chem. Soc., Dalton Trans.*, 685 (1987).

1176. J. D. Corbett, J. Eckert, U. A. Jayasooriya, G. J. Kearley, R. P. White, and J. Zhang, *J. Phys. Chem.*, **97,** 8386 (1993).

1177. R. J. H. Clark, in V. Gutmann, ed., *Halogen Chemistry*, Vol. 3, Academic Press, New York, 1967, p. 85.

1178. R. H. Nuttall, *Talanta*, **15,** 157 (1968).

1179. A. J. Carty, *Coord. Chem. Rev.*, **4,** 29 (1969).

1180. R. J. H. Clark, *Spectrochim. Acta*, **21,** 955 (1965).

1181. K. Konya and K. Nakamoto, *Spectrochim. Acta*, **29A,** 1965 (1973).

1182. K. Thompson and K. Carlson, *J. Chem. Phys.*, **49,** 4379 (1968).

1183. K. Shobatake and K. Nakamoto, *J. Am. Chem. Soc.*, **92,** 3332 (1970).

1184. C. Udovich, J. Takemoto, and K. Nakamoto, *J. Coord. Chem.*, **1,** 89 (1971).

1185. P. M. Boorman and A. J. Carty, *Inorg. Nucl. Chem. Lett.*, **4,** 101 (1968).

1186. I. Wharf and D. F. Shriver, *Inorg. Chem.*, **8,** 914 (1969).

1187. D. F. Shriver and M. P. Johnson, *Inorg. Chem.*, **6,** 1265 (1967).

1188. B. Crociani, T. Boschi, and M. Nicolini, *Inorg. Chim. Acta*, **4,** 577 (1970).

1189. F. H. Herbelin, J. D. Herbelin, J. P. Mathieu, and H. Poulet, *Spectrochim. Acta*, **22,** 1515 (1966).

1190. D. M. Adams, J. Chatt, J. Gerratt, and A. D. Westland, *J. Chem. Soc.*, 734 (1964).

1191. T. G. Appleton, H. C. Clark, and L. E. Manzer, *Coord. Chem. Rev.*, **10,** 335 (1973).

1192. R. J. Goodfellow, P. L. Goggin, and D. A. Duddell, *J. Chem. Soc. A*, 504 (1968).

1193. M. A. Bennett, R. J. H. Clark, and D. L. Milner, *Inorg. Chem.*, **6,** 1647 (1967).

1194. J. Fujita, K. Konya, and K. Nakamoto, *Inorg. Chem.*, **9,** 2794 (1970).

1195. J. T. Wang, C. Udovich, K. Nakamoto, A. Quattrochi, and J. R. Ferraro, *Inorg. Chem.*, **9,** 2675 (1970).

1196. B. T. Kilbourn and H. M. Powell, *J. Chem. Soc. A*, 1688 (1970).

1197. J. R. Ferraro, K. Nakamoto, J. T. Wang, and L. Lauer, *Chem. Commun.*, 266 (1973).

1198. C. Postmus, K. Nakamoto, and J. R. Ferraro, *Inorg. Chem.*, **6,** 2194 (1967).

1199. J. R. Ferraro, *Vibrational Spectroscopy at High External Pressures. The Diamond Anvil Cell*, Academic Press, New York, 1984.

1200. J. R. Ferraro, *Coord. Chem. Rev.*, **29,** 1 (1979).

1201. D. M. Adams and P. J. Chandler, *J. Chem. Soc. A*, 1009 (1967).

1202. N. Ohkaku and K. Nakamoto, *Inorg. Chem.*, **12,** 2440, 2446 (1973).

1203. D. M. Adams and D. C. Newton, *J. Chem. Soc., Dalton Trans.*, 681 (1972).

1204. R. J. Goodfellow, P. L. Goggin, and L. M. Venanzi, *J. Chem. Soc. A*, 1897 (1967).

1205. D. M. Adams and P. J. Chandler, *Chem. Commun.*, 69 (1966).

1206. M. Goldstein and W. D. Unsworth, *Inorg. Chim. Acta*, **4,** 342 (1970).

1207. H. Hillebrecht, G. Thiele, P. Hollmann, and W. Preetz, *Z. Naturforsch.*, **47B,** 1099 (1992).

1208. R. S. Armstrong, W. A. Horsfield, and K. W. Nugent, *Inorg. Chem.*, **29,** 4551 (1990).

1209. T. G. Spiro, *Prog. Inorg. Chem.*, **11,** 1 (1970).

1210. K. L. Watters and W. M. Risen, Jr., *Inorg. Chim. Acta, Rev.*, **3,** 129 (1969).

1211. E. Maslowsky, Jr., *Chem. Rev.*, **71,** 507 (1971).

1212. B. J. Bulkin and C. A. Rundell, *Coord. Chem. Rev.*, **2,** 371 (1967).

1213. D. F. Shriver and C. B. Cooper, III, *Adv. Infrared Raman Spectrosc.*, **6,** 127 (1980).

1214. C. O. Quicksall and T. G. Spiro, *Inorg. Chem.*, **8,** 2363 (1969).

1215. J. R. Johnson, R. J. Ziegler, and W. M. Risen, Jr., *Inorg. Chem.*, **12,** 2349 (1973).

1216. K. L. Watters, J. N. Britain, and W. M. Risen, Jr., *Inorg. Chem.*, **8,** 1347 (1969).

1217. K. L. Watters, W. M. Butler, and W. M. Risen, Jr., *Inorg. Chem.*, **10,** 1970 (1971).

1218. K. M. Mackay and S. R. Stobart, *J. Chem. Soc., Dalton Trans.*, 214 (1973).

1219. S. Onaka, C. B. Cooper, III, and D. F. Shriver, *Inorg. Chim. Acta*, **37,** L467 (1979).

1220. P. L. Goggin and R. J. Goodfellow, *J. Chem. Soc., Dalton Trans.*, 2355 (1973).

1221. J. San Filippo, Jr. and H. J. Sniadoch, *Inorg. Chem.*, **12,** 2326 (1973).

1222. B. I. Swanson, J. J. Rafalko, D. F. Shriver, J. San Filippo, Jr., and T. G. Spiro, *Inorg. Chem.*, **14,** 1737 (1975).

1223. C. B. Cooper, III, S. Onaka, D. F. Shriver, L. Daniels, R. L. Hance, B. Hutchinson, and R. Shipley, *Inorg. Chim. Acta*, **24,** L92 (1977).

1224. S. Onaka and D. F. Shriver, *Inorg. Chem.*, **15,** 915 (1976).

1225. S. F. A. Kettle and P. L. Stanghellini, *Inorg. Chem.*, **18,** 2749 (1979).

1226. G. A. Battiston, G. Bor, U. K. Dietler, S. F. A. Kettle, R. Rossetti, G. Sbrignadello, and P. L. Stanghellini, *Inorg. Chem.*, **19**, 1961 (1980).

1227. H. W. Martin, P. Skinner, R. K. Bhardwaj, V. A. Jayasooriya, D. B. Powell, and N. Sheppard, *Inorg. Chem.*, **25**, 2846 (1986).

1228. V. A. Jayasooriya and P. Skinner, *Inorg. Chem.*, **25**, 2850 (1986).

1229. V. A. Jayasooriya, S. J. Stotesbury, R. Grinter, D. B. Powell, and N. Sheppard, *Inorg. Chem.*, **25**, 2853 (1986).

1230. C. O. Quicksall and T. G. Spiro, *Inorg. Chem.*, **8**, 2011 (1969).

1231. C. O. Quicksall and T. G. Spiro, *Chem. Commun.*, 839 (1967).

1232. J. A. Creighton and B. T. Heaton, *J. Chem. Soc., Dalton Trans.*, 1498 (1981).

1233. C. Sourisseau, *J. Raman Spectrosc.*, **6**, 303 (1977).

1234. S. F. A. Kettle and P. L. Stanghellini, *Inorg. Chem.*, **26**, 1626 (1987).

1235. F. A. Cotton, *Chem. Soc. Rev.*, 27 (1975); *Acc. Chem. Res.*, **11**, 225 (1978).

1236. R. J. H. Clark and M. L. Franks, *J. Am. Chem. Soc.*, **98**, 2763 (1976).

1237. F. A. Cotton, B. A. Frenz, B. R. Stults, and T. R. Webb, *J. Am. Chem. Soc.*, **98**, 2768 (1976).

1238. J. San Filippo, Jr. and H. J. Sniadoch, *Inorg. Chem.*, **12**, 2326 (1973).

1239. A. J. Hempleman, R. J. H. Clark, and C. D. Flint, *Inorg. Chem.*, **25**, 2915 (1986).

1240. C. L. Angell, F. A. Cotton, B. A. Frenz, and T. R. Webb, *Chem. Commun.*, 399 (1973).

1241. A. Loewenschuss, J. Shamir, and M. Ardon, *Inorg. Chem.*, **15**, 238 (1976).

1242. D. J. Santure, J. C. Huffman, and A. P. Sattelberger, *Inorg. Chem.*, **24**, 371 (1985).

1243. R. J. H. Clark, A. J. Hempleman, and D. A. Tocher, *J. Am. Chem. Soc.*, **110**, 5968 (1988).

1244. R. J. H. Clark, A. J. Hempleman, and C. D. Flint, *J. Am. Chem. Soc.*, **108**, 518 (1986).

1245. R. J. H. Clark and A. J. Hempleman, *Inorg. Chem.*, **27**, 2225 (1988).

1246. R. J. H. Clark, A. J. Hempleman, H. M. Dawes, M. B. Hursthouse, and C. D. Flint, *J. Chem. Soc., Dalton Trans.*, 1775 (1985).

1247. R. J. H. Clark, D. J. West, and R. Withnall, *Inorg. Chem.*, **31**, 456 (1992).

1248. B. Hutchinson, J. Morgan, C. B. Cooper, III, Y. Mathey, and D. F. Shriver, *Inorg. Chem.*, **18**, 2048 (1979).

1249. W. K. Bratton, F. A. Cotton, M. Debeau, and R. A. Walton, *J. Coord. Chem.*, **1**, 121 (1971).

1250. A. P. Ketteringham, C. Oldham, and C. J. Peacock, *J. Chem. Soc., Dalton Trans.*, 1640 (1976).

1251. H. K. Mahtani and P. Stein, *J. Am. Chem. Soc.*, **111**, 1505 (1989).

1252. T. G. Appleton, J. R. Hall, and D. W. Neale, *Inorg. Chim. Acta*, **104**, 19 (1985).

1253. R. J. H. Clark, J. H. Tocher, J. P. Fackler, R. Neira, H. H. Murray, and H. Knackel, *J. Organomet. Chem.*, **303**, 437 (1986).

1254. R. F. Dallinger, *J. Am. Chem. Soc.*, **107**, 7202 (1985).

1255. R. F. Dallinger, V. M. Miskowski, H. B. Gray, and W. H. Woodruff, *J. Am. Chem. Soc.*, **103**, 1595 (1981).

1256. C.-M. Che, L. G. Butler, H. B. Gray, R. M. Crooks, and W. H. Woodruff, *J. Am. Chem. Soc.*, **105,** 5492 (1983).

1257. P. D. Harvey, R. F. Dallinger, W. H. Woodruff, and H. B. Gray, *Inorg. Chem.*, **28,** 3057 (1989).

1258. D. E. Morris and W. H. Woodruff, "Vibrational Spectra and the Structure of Electronically Excited Molecules in Solution," in R. J. H. Clark and R. E. Hester, eds., *Advances in Spectroscopy*, Vol. 14, Wiley, New York, 1987. p. 285.

1259. M. Trabelsi, A. Loutellier, and M. Bigorgne, *J. Organomet. Chem.*, **40,** C45 (1972).

1260. M. Bigorgne, A. Loutellier, and M. Pańkowski, *J. Organomet. Chem.*, **23,** 201 (1970).

1261. M. Trabelsi, A. Loutellier, and M. Bigorgne, *J. Organomet. Chem.*, **56,** 369 (1973).

1262. M. Trabelsi and A. Loutellier, *J. Mol. Struct.*, **43,** 151 (1978).

1263. A. Loutellier, M. Trabelsi, and M. Bigorgne, *J. Organomet. Chem.*, **133,** 201 (1977).

1264. Th. Kruck, *Angew. Chem., Int. Ed. Engl.*, **6,** 53 (1967).

1265. Th. Kruck and K. Bauer, *Z. Anorg. Allg. Chem.*, **364,** 192 (1969).

1266. Th. Kruck and A. Prasch, *Z. Anorg. Allg. Chem.*, **356,** 118 (1968).

1267. W. Collong and Th. Kruck, *Chem. Ber.*, **123,** 1655 (1990).

1268. W. Fuss and M. Ruhe, *Z. Naturforsch.*, **47B,** 591 (1992).

1269. Th. Kruck and A. Prasch, *Z. Anorg. Allg. Chem.*, **371,** 1 (1969).

1270. S. Bénazeth, A. Loutellier, and M. Bigorgne, *J. Organomet. Chem.*, **24,** 479 (1970).

1271. L. A. Woodward and J. R. Hall, *Spectrochim. Acta.*, **16,** 654 (1960); H. G. M. Edwards and L. A. Woodward, *ibid.*, **26A,** 897 (1970).

1272. H. G. M. Edwards and L. A. Woodward, *Spectrochim. Acta*, **26A,** 1077 (1970).

1273. H. G. M. Edwards, *Spectrochim. Acta*, **42A,** 431 (1986).

1274. H. G. M. Edwards, *J. Mol. Struct.*, **158,** 153 (1987).

1275. H. G. M. Edwards, *Spectrochim. Acta*, **42A,** 1401 (1986).

1276. P. L. Goggin and R. J. Goodfellow, *J. Chem. Soc. A*, 1462 (1966).

1277. G. D. Coates and C. Parkin, *J. Chem. Soc.*, 421 (1963).

1278. M. A. Bennett, R. J. H. Clark, and A. D. J. Goodwin, *Inorg. Chem.*, **6,** 1625 (1967).

1279. J. H. S. Green, *Spectrochim. Acta*, **24A,** 137 (1968).

1280. K. Shobatake, C. Postmus, J. R. Ferraro, and K. Nakamoto, *Appl. Spectrosc.*, **23,** 12 (1969).

1281. R. J. H. Clark, C. D. Flint, and A. J. Hempleman, *Spectrochim. Acta*, **43A,** 805 (1987).

1282. J. Bradbury, K. P. Forest, R. H. Nuttall, and D. W. A. Sharp, *Spectrochim. Acta*, **23A,** 2701 (1967).

1283. H. G. M. Edwards, A. F. Johnson, and I. R. Lewis, *Spectrochim. Acta*, **49A,** 707 (1993).

1284. J. G. Verkade, *Coord. Chem. Rev.*, **9,** 1 (1972).

1285. P. J. D. Park and P. J. Hendra, *Spectrochim. Acta*, **25A,** 227, 909 (1969).

1286. L. M. Venanzi, *Chem. Brit.*, **4,** 162 (1968).

1287. J. Chatt, G. A. Gamlen, and L. E. Orgel, *J. Chem. Soc.*, 486 (1958).

1288. E. O. Fischer, W. Bathelt, and J. Müller, *Chem. Ber.*, **103,** 1815 (1970).

1289. R. J. Goodfellow, J. G. Evans, P. L. Goggin, and D. A. Duddell, *J. Chem. Soc. A*, 1604 (1968).

1290. G. B. Deacon and J. H. S. Green, *Spectrochim. Acta*, **24A,** 845 (1968).

1291. D. Brown, J. Hill, and C. E. F. Richard, *J. Chem. Soc. A*, 497 (1970).

1292. D. M. L. Goodgame and F. A. Cotton, *J. Chem. Soc.*, 2298, 3735 (1961).

1293. G. A. Rodley, D. M. L. Goodgame, and F. A. Cotton, *J. Chem. Soc.*, 1499 (1965).

1294. A. Müller, W. Jaegermann, and J. H. Enemark, *Coord. Chem. Rev.*, **46,** 245 (1982).

1295. P. K. Chakrabarty, S. Bhattacharya, C. G. Pierpont, and R. Bhattacharyya, *Inorg. Chem.*, **31,** 3573 (1992).

1296. R. Guilard, C. Ratti, A. Tabard, P. Richard, D. Dubois, and K. M. Kadish, *Inorg. Chem.*, **29,** 2532 (1990).

1297. R. Minkwitz, H. Borrmann, and J. Nowicki, *Z. Naturforsch.*, **47B,** 915 (1992).

1298. V. P. Fedin, B. A. Kolesov, Yu. V. Mironov, and V. Ye. Fedorov, *Polyhedron*, **8,** 2419 (1989).

1299. A. P. Ginsberg and W. E. Lindsell, *J. Chem. Soc., Chem. Commun.*, 232 (1971).

1300. R. Bhattacharyya, P. K. Chakrabarty, P. N. Ghosh, A. K. Mukherjee, D. Podder, and M. Mukherjee, *Inorg. Chem.*, **30,** 3948 (1991).

1301. B. Neumüller, M. L. Ha-Eierdanz, U. Müller, S. Magull, G. Kräuter, and K. Dehnicke, *Z. Anorg. Allg. Chem.*, **609,** 12 (1992).

1302. A. Ahle, B. Neumüller, J. Pebler, M. Atanasov, and K. Dehnicke, *Z. Anorg. Allg. Chem.*, **615,** 131 (1992).

1303. I. S. Butler, P. D. Harvey, J. McCall, and A. Shaver, *J. Raman Spectrosc.*, **17,** 221 (1986).

1304. A. Müller, J. Schimanski, U. Schmanski, and H. Bögge, *Z. Naturforsch.*, **40B,** 1277 (1985).

1305. T. J. Greenhough, B. W. S. Kolthammer, P. Legzdins, and J. Trotter, *Inorg. Chem.*, **18,** 3548 (1979).

1306. M. W. Bishop, J. Chatt, and J. R. Dilworth, *Chem. Commun.*, 780 (1975).

1307. O. Heyke, G. Beuter, and I.-P. Lorenz, *J. Organomet. Chem.*, **440,** 197 (1992).

1308. W. Petz, *Inorg. Chim. Acta*, **201,** 203 (1992).

1309. M. C. Baird, G. Hartwell, and G. Wilkinson, *J. Chem. Soc. A*, 2037 (1967).

1310. J. D. Gilbert, M. C. Baird, and G. Wilkinson, *J. Chem. Soc. A*, 2198 (1968).

1311. A. Efraty, R. Arneri, and M. H. A. Huang, *J. Am. Chem. Soc.*, **98,** 639 (1976).

1312. B. D. Dombek and R. J. Angelici, *J. Am. Chem. Soc.*, **96,** 7568 (1974).

1313. L. Busetto, V. Zanotti, V. G. Albano, D. Braga, and M. Honari, *J. Chem. Soc., Dalton Trans.*, 1791 (1986).

1314. A. M. English, K. R. Plowman, and I. S. Butler, *Inorg. Chem.*, **20**, 2553 (1981).

1315. A. M. English, K. R. Plowman, and I. S. Butler, *Inorg. Chem.*, **21**, 338 (1982).

1316. L. Vaska and S. S. Bath, *J. Am. Chem. Soc.*, **88**, 1333 (1966).

1317. J. Terheijden, G. van Koten, W. P. Mul, D. J. Stufkens, F. Muller, and C. H. Stam, *Organometallics*, **5**, 519 (1986).

1318. L. H. Vogt, J. L. Katz, and S. E. Wiberley, *Inorg. Chem.*, **4**, 1157 (1965).

1319. F. Meier-Brocks, R. Albrecht, and E. Weiss, *J. Organomet. Chem.*, **439**, 65 (1992).

1320. W. A. Schenck and U. Karl, *Z. Naturforsch.*, **44B**, 988 (1989).

1321. U. Schimmelpfennig, R. Kalähne, K. D. Schleinitz, and E. Wenschuh, *Z. Anorg. Allg. Chem.*, **603**, 21 (1991).

1322. D. M. Byler and D. F. Shriver, *Inorg. Chem.*, **15**, 32 (1976).

1323. P. Reich-Rohrwig, A. C. Clark, R. L. Downs, and A. Wojcicki, *J. Organomet. Chem.*, **145**, 57 (1978).

1324. C. Sourisseau and J. Corset, *Inorg. Chim. Acta*, **39**, 153 (1980).

1325. R. D. Wilson and J. A. Ibers, *Inorg. Chem.*, **17**, 2134 (1978).

1326. M. Herberhold and A. F. Hill, *J. Organomet. Chem.*, **387**, 323 (1990).

1327. P. L. Bogdan, M. Sabat, S. A. Sunshine, C. Woodcock, and D. F. Shriver, *Inorg. Chem.*, **27**, 1904 (1988).

1328. D. M. P. Mingos and R. W. M. Wardle, *J. Chem. Soc., Dalton Trans.*, 73 (1986).

1329. A. D. Burrows, A. A. Gosden, C. M. Hill, and D. M. P. Mingos, *J. Organomet. Chem.*, **452**, 251 (1993).

1330. G. J. Kubas, *Inorg. Chem.*, **18**, 182 (1979).

1331. D. A. Johnson and V. C. Dew, *Inorg. Chem.*, **18**, 3273 (1979).

1332. I. S. Butler and A. E. Fenster, *J. Organomet. Chem.*, **66**, 161 (1974).

1333. M. C. Baird and G. Wilkinson, *J. Chem. Soc. A*, 865 (1967).

1334. D. S. Barratt and C. A. McAuliffe, *Inorg. Chim. Acta*, **97**, 37 (1985).

1335. Y. Yamamoto and H. Yamazaki, *J. Chem. Soc., Dalton Trans.*, 677 (1986).

1336. M. P. Yagupsky and G. Wilkinson, *J. Chem. Soc. A*, 2813 (1968).

1337. G. R. Clark, T. J. Collins, S. M. James, W. R. Roper, and K. G. Town, *Chem. Commun.*, 475 (1976).

1338. H. Huber, G. A. Ozin, and W. J. Power, *Inorg. Chem.*, **16**, 2234 (1977).

1339. P. J. Brothers, C. E. L. Headford, and W. R. Roper, *J. Organomet. Chem.*, **195**, C29 (1980).

1340. J. R. Allkins and P. J. Hendra, *J. Chem. Soc. A*, 1325 (1967); *Spectrochim. Acta*, **22**, 2075 (1966); **23A**, 1671 (1967); **24A**, 1305 (1968).

1341. E. A. Allen and W. Wilkinson, *Spectrochim. Acta*, **28A**, 725 (1972).

1342. R. J. H. Clark, G. Natile, U. Belluco, L. Cattalini, and C. Filippin, *J. Chem. Soc. A*, 659 (1970).

1343. E. A. Allen and W. Wilkinson, *Spectrochim. Acta*, **28A**, 2257 (1972).

1344. B. E. Aires, J. E. Fergusson, D. T. Howarth, and J. M. Miller, *J. Chem. Soc. A*, 1144 (1971).

1345. E. A. Allen and W. Wilkinson, *J. Chem. Soc., Dalton Trans.*, 613 (1972).

1346. P. L. Goggin, R. J. Goodfellow, D. L. Sales, J. Stokes, and P. Woodward, *Chem. Commun.*, 31 (1968).

1347. D. M. Adams and P. J. Chandler, *J. Chem. Soc. A*, 588 (1969).

1348. K. H. Schmidt and A. Müller, *Coord. Chem. Rev.*, **14**, 115 (1974).

1349. A. Müller, E. Diemann, R. Jostes, and H. Bögge, *Angew. Chem., Int. Ed. Engl.*, **20**, 934 (1981).

1350. A. Müller, "Thiometallato Complexes: Vibrational Spectra and Structural Chemistry," in J. R. Durig, ed., *Vibrational Spectra and Structure*, Vol. 15, Elsevier, New York, 1986. p. 251.

1351. E. Königer-Ahlborn, A. Müller, A. D. Cormier, J. D. Brown, and K. Nakamoto, *Inorg. Chem.*, **14**, 2009 (1975).

1352. A. Müller and W. Hellman, *Spectrochim. Acta*, **41A**, 359 (1985).

1353. U. Pätzmann, W. Brockner, S. N. Cyvin, and S. J. Cyvin, *J. Raman Spectrosc.*, **17**, 257 (1986).

1354. R. G. Cavell, W. Byers, E. D. Day, and P. M. Watkins, *Inorg. Chem.*, **11**, 1598 (1972).

1355. J. M. Burke and J. P. Fackler, *Inorg. Chem.*, **11**, 2744 (1972).

1356. A. Cormier, K. Nakamoto, P. Christophliemk, and A. Müller, *Spectrochim. Acta*, **30A**, 1059 (1974).

1357. O. Piovesana, C. Bellitto, A. Flamini, and P. F. Zanazzi, *Inorg. Chem.*, **18**, 2258 (1979).

1358. F. A. Cotton, C. Oldham, and R. A. Walton, *Inorg. Chem.*, **6**, 214 (1967).

1359. M. Ikram and D. B. Powell, *Spectrochim. Acta*, **28A**, 59 (1972).

1360. C. W. Schläpfer, Y. Saito, and K. Nakamoto, *Inorg. Chim. Acta*, **6**, 284 (1972); C. W. Schläpfer and K. Nakamoto, *ibid.*, 177.

1361. F. Bonati and R. Ugo, *J. Organomet. Chem.*, **10**, 257 (1967).

1362. C. O'Connor, J. D. Gilbert, and G. Wilkinson, *J. Chem. Soc. A*, 84 (1969).

1363. D. C. Bradley and M. H. Gitlitz, *J. Chem. Soc. A*, 1152 (1969).

1364. R. J. Butcher, J. R. Ferraro, and E. Sinn, *Inorg. Chem.*, **15**, 2077 (1976).

1365. M. Sorai, *J. Inorg. Nucl. Chem.*, **40**, 1031 (1978).

1366. B. Hutchinson, P. Neill, A. Finkelstein, and J. Takemoto, *Inorg. Chem.*, **20**, 2000 (1981).

1367. K. Nakamoto, J. Fujita, R. A. Condrate, and Y. Morimoto, *J. Chem. Phys.*, **39**, 423 (1963).

1368. K. A. Jensen and V. Krishnan, *Acta Chem. Scand.*, **24**, 1088 (1970).

1369. K. Jensen, B. M. Dahl, P. Nielsen, and G. Borch, *Acta Chem. Scand.*, **26**, 2241 (1972).

1370. G. W. Watt and B. J. McCormick, *Spectrochim. Acta*, **21**, 753 (1965).

1371. V. V. Savant, J. Gopalakrishnan, and C. C. Patel, *Inorg. Chem.*, **9**, 748 (1970).

1372. G. A. Melson, N. P. Crawford, and B. J. Geddes, *Inorg. Chem.*, **9**, 1123 (1970).

1373. M. J. M. Campbell, *Coord. Chem. Rev.*, **15**, 279 (1975).

1374. M. A. Ali, S. E. Linvingstone, and D. J. Phillips, *Inorg. Chim. Acta*, **5**, 119 (1971).

1375. H. O. Desseyn, W. A. Jacob, and M. A. Herman, *Spectrochim. Acta*, **25A**, 1685 (1969).

1376. S. P. Periepes, M. Bellaihou, and H. O. Desseyn, *Spectrosc. Lett.*, **26**, 751 (1993).

1377. J. Fujita and K. Nakamoto, *Bull. Chem. Soc. Jpn.*, **37**, 528 (1964).

1378. D. Coucouvanis, N. C. Baenziger, and S. M. Johnson, *J. Am. Chem. Soc.*, **95**, 3875 (1973).

1379. D. Coucouvanis and D. Piltingsrud, *J. Am. Chem. Soc.*, **95**, 5556 (1973).

1380. J. A. McCleverty, *Prog. Inorg. Chem.*, **10**, 49 (1968).

1381. C. W. Schläpfer and K. Nakamoto, *Inorg. Chem.*, **14**, 1338 (1975).

1382. M. Cox and J. Darken, *Coord. Chem. Rev.*, **7**, 29 (1971).

1383. S. E. Linvingstone, *Coord. Chem. Rev.*, **7**, 59 (1971).

1384. O. Siimann and J. Fresco, *Inorg. Chem.*, **8**, 1846 (1969); *J. Chem. Phys.*, **54**, 734, 740 (1971).

1385. C. G. Barraclough, R. L. Martin, and I. M. Stewart, *Aust. J. Chem.*, **22**, 891 (1969); G. A. Heath and R. L. Martin, *ibid.*, **23**, 1721 (1970).

1386. H. Shindo and T. L. Brown, *J. Am. Chem. Soc.*, **87**, 1904 (1965).

1387. M. Chandrasekharan, M. R. Udupa, and G. Aravamudan, *Inorg. Chim. Acta*, **7**, 88 (1973).

1388. R. Panossian, G. Terzian, and M. Guiliano, *Spectrosc. Lett.*, **12**, 715 (1979).

1389. R. J. Gale and C. A. Winkler, *Inorg. Chim. Acta*, **21**, 151 (1977).

1390. C. A. McAuliffe, J. V. Quagliano, and L. M. Vallarino, *Inorg. Chem.*, **5**, 1996 (1966).

1391. C. A. McAuliffe, *J. Chem. Soc. A*, 641 (1967).

1392. M. Ikram and D. B. Powell, *Spectrochim. Acta*, **27A**, 1845 (1971).

1393. Y. Mido and E. Sekido, *Bull. Chem. Soc. Jpn.*, **44**, 2130 (1971).

1394. J. A. W. Dalziel, M. J. Hitch, and S. D. Ross, *Spectrochim. Acta*, **25A**, 1055 (1969).

1395. W. W. Fee and J. D. Pulsford, *Inorg. Nucl. Chem. Lett.*, **4**, 227 (1968).

1396. D. Michalska and A. T. Kowal, *Spectrochim. Acta*, **41A**, 1119 (1985).

IV

APPLICATIONS IN ORGANOMETALLIC CHEMISTRY

Vibrational spectra of organometallic compounds have been reported extensively,[1] and comprehensive reviews are found in several monographs.[2-4] More limited reviews are available on specific metal elements or functional groups, and will be quoted in respective sections. Reference books on vibrational spectra of organic compounds[5-7] are useful in making band assignments since vibrational spectra of organometallic compounds in the high-frequency region are largely due to organic ligands or moieties. Spectral charts[8,9] and an index[10] of vibrational spectra of organic and organometallic compounds are also useful for this purpose. In the following, we review vibrational spectra of organometallic compounds with emphasis on metal–carbon streching vibrations in the low-frequency region since they provide structural and bonding information on the metal–carbon skeleton.

IV-1. METHYLENE, METHYL, AND ETHYL COMPOUNDS

The smallest organometallic compound may be a metal carbene in which the carbon atom of the methylene group is σ-bonded to a metal ($M-CH_2$). Such a compound can be prepared via cocondensation reaction of metal atom vapor with diazomethane in inert gas matrices:

$$M + CH_2N_2 \rightarrow M-CH_2 + N_2$$

Margrave and co-workers obtained the IR spectra of $CuCH_2$,[11] $FeCH_2$,[12] and $CrCH_2$[13] in Ar matrices. These molecules are planar and exhibit the six normal vibrations shown in Fig. III-29. In the case of $CrCH_2$, the $\nu_a(CH_2)$, $\nu_s(CH_2)$,

$\delta(CH_2)$, $\rho_r(CH_2)$ and $\nu(Cr-C)$ are observed at 2967, 2907, 688, 450, and 567 cm^{-1}, respectively. The $\nu(Cu-C)$ in $CuCH_2$ is at 614 cm^{-1}, and the $\nu(Fe-C)$ in $FeCH_2$ is at 624 cm^{-1}.

There are many compounds in which the carbon atom of the methyl group is σ-bonded to a metal ($M-CH_3$). Vibrational spectra of these methyl compounds can be interpreted in terms of the normal modes of a $1:1$ (metal/methyl) model shown in Fig. III-2. The $\nu_a(CH_3)$, $\nu_s(CH_3)$, $\delta_d(CH_3)$, $\delta_s(CH_3)$, and $\rho_r(CH_3)$, and $\nu(M-C)$ are at 3000–2800, 3000–2700, 1400–1350, 1300–1100, 950–700, and 700–400 cm^{-1}, respectively. The simplest is the $1:1$ (metal/methyl) complex such as $NaCH_3$ and KCH_3 which can be prepared in Ar matrices at 20 K.[14] Table IV-1 lists the Raman frequencies of typical $M(CH_3)_4$-type compounds. Figure IV-1 shows the Raman spectra of $Si(CH_3)_4$ and $Si(CD_3)_4$ obtained by Fischer et al.[17] In Fig. IV-2, the observed frequencies are plotted as a function of the atomic mass of the Group IVB elements.[18] It is seen that the $\delta_s(CH_3)$, $\rho_r(CH_3)$, $\nu(M-C)$, and $\delta(CMC)$ are shifted progressively to lower frequencies as the atomic mass increases. The $\rho_r(CH_3)$ and $\nu(M-C)$ are particularly sensitive to the nature of these metals. Under $\mathbf{T}_d$ symmetry, two $\nu(M-C)$ (A_1 and F_2) are expected for the $M(CH_3)_4$-type molecule. These vibrations are reported to be 508 and 530 cm^{-1}, respectively, for $Sn(CH_3)_4$,[19] and 598 and 696 cm^{-1}, respectively, for $Si(CH_3)_4$.[20]

Tables IV-2 and IV-3 list the MC stretching and CMC bending frequencies of various $M(CH_3)_n$-type molecules and ions observed in IR and/or Raman spectra. The number of IR- and Raman-active skeletal vibrations expected for each structure are found in Appendix V of Part A. As already seen in the $M(CH_3)_4$ series, both $\nu(M-C)$ bands are downshifted progressively as the mass of the central metal increases in the same family of the periodic table. Thus, the orders of $\nu(M-C)$ are:

$$Al(CH_3)_3 > Ga(CH_3)_3 > In(CH_3)_3$$

TABLE IV-1. Raman Frequencies^a of M(CH₃)₄-Type Molecules (cm⁻¹)[15,16]

Compound	$\nu_a(CH_3)$	$\nu_s(CH_3)$	$\delta_d(CH_3)$	$\delta_s(CH_3)$	$\rho_r(CH_3)$	$\nu(MC)$	$\delta(CMC)$
C(CH₃)₄	2959	2922	(1475)	—	926	733	418
	2963		1457		(926)	1260	332
Si(CH₃)₄	(2959)	2913	(1430)	1271	870	593	239
	2964		1421		(870)	698	190
	(2910)						
Ge(CH₃)₄	(2981)	2920	(1430)	1259	—	561	196
	2982		1420		(828)	599	188
Sn(CH₃)₄	(2984)	2920	(1447)	1211	—	509	137
	2988		—		(768)	527	133
Pb(CH₃)₄	2996	2924	1450	1170	767	478	145
	2924		1400	1154	700	459	130

^{a}( ) = IR frequency.

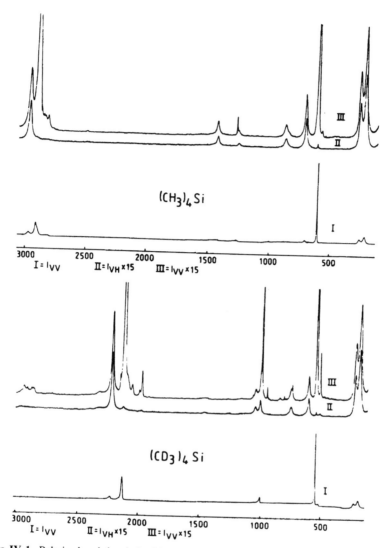

Figure IV-1. Polarized and depolarized Raman spectra of Si(CH₃)₄ and Si(CD₃)₄ (reproduced with permission from Ref. 17).

and

$$P(CH_3)_3 > As(CH_3)_3 > Sb(CH_3)_3 > Bi(CH_3)_3$$

An exception is found in the linear $M(CH_3)_2$ series:

	$Zn(CH_3)_2$		$Cd(CH_3)_2$		$Hg(CH_3)_2$
$\nu_s(M—C)$ (cm^{-1})	503	>	460	<	515
$\nu_a(M—C)$ (cm^{-1})	604	>	525	<	538

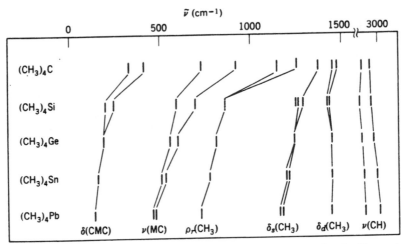

Figure IV-2. Vibrational frequencies of M(CH₃)₄-type compounds (reproduced with permission from Ref. 18).

TABLE IV-2. Metal–Carbon Skeletal Frequencies of M(CH₃)ₙ-Type Compounds (cm⁻¹)

Compound	Structure	ν_a(MC)	ν_s(MC)	δ(CMC)	Refs.
Be(CH₃)₂	Linear	1081	—	—	21
Zn(CH₃)₂	Linear	604	503	157	22–26
Cd(CH₃)₂	Linear	525	460	140	22, 24–27
Hg(CH₃)₂	Linear	538	515	160	22, 24–26
Se(CH₃)₂[a]	Bent	604	589	233	28, 29, 31
Te(CH₃)₂[a]	Bent	528	528	198	28–31
B(CH₃)₃	Planar	1177	680	341, 321	32–34
Al(CH₃)₃	Planar	760	530	170	35–37
Ga(CH₃)₃	Planar	577	521.5	162.5	36–40
In(CH₃)₃	Planar	500	467	132	39, 41, 42
P(CH₃)₃	Pyramidal	703	653	305, 263	43–45
As(CH₃)₃	Pyramidal	583	568	238, 223	43, 44, 46
Sb(CH₃)₃	Pyramidal	513	513	188	47
Bi(CH₃)₃	Pyramidal	460	460	171	47
Si(CH₃)₄	Tetrahedral	696	598	239, 202	48, 49
Ge(CH₃)₄	Tetrahedral	595	558	195, 175	16, 49
Sn(CH₃)₄	Tetrahedral	529	508	157	18, 49
Pb(CH₃)₄	Tetrahedral	476	459	120	49, 50
Ti(CH₃)₄	Tetrahedral	577	489	180	51
Sb(CH₃)₅	Trigonal-bipyramidal	514[b]	493[b]	213[b]	52
		456[c]	414[c]	199[b]	
				104[c]	
W(CH₃)₆	Octahedral	482	—	—	53

[a] New assignments have been proposed based on the **D₃d** model containing a linear C—M—C skeleton (Ref. 31).
[b] Equatorial.
[c] Axial.

TABLE IV-3. Metal–Carbon Skeletal Frequencies of $[M(CH_3)_n]^{m+}$-Type Compounds (cm^{-1})

Compound	Structure	$\nu_a(MC)$	$\nu_s(MC)$	$\delta(CMC)$	Refs.
$[Zn(CH_3)]^+$	—	—	557	—	54
$[In(CH_3)_2]^+$	Linear	566	502	—	55
$[Tl(CH_3)_2]^+$	Linear	559	498	114	56, 57
$[Sn(CH_3)_2]^{2+}$	Linear	582	529	180	58
$[Sn(CH_3)_3]^+$	Planar	557	521	152	59
$[Sb(CH_3)_3]^{2+}$	Planar	582	536	166	60
$[Se(CH_3)_3]^+$	Nonplanar	602	580	272	61
$[Te(CH_3)_3]^+$	Nonplanar	534	—	—	62
$[P(CH_3)_4]^+$	Tetrahedral	783	649	285	63
				170	64
$[As(CH_3)_4]^+$	Tetrahedral	652	590	217	47
$[Sb(CH_3)_4]^+$	Tetrahedral	574	535	178	47

Apparently, this is due to an irregular variation in the M—C bond order. Figure IV-3 shows the matrix-isolation IR spectra of the series above obtained by Bochmann et al.[26] The two strong bands below 800 cm^{-1} are due to the $\rho_r(CH_3)$ and $\nu_a(M—C)$ vibrations. In this case, the latter bands are observed at 613(Zn), 538(Cd), and 538(Hg) cm^{-1}.

Some metal alkyls are polymerized in condensed phases. $Li(CH_3)$ forms a tetramer containing $Li—CH_3—Li$ bridges in the solid state,[65] and its CH_3 fre-

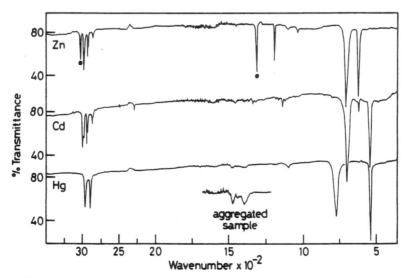

Figure IV-3. Infrared spectra of $M(CH_3)_2$ (M = Zn, Cd, and Hg) in Ar matrices. ● indicates bands due to methane in the $Zn(CH_3)_2$ spectrum (reproduced with permission from Ref. 26).

quencies are lower than those of nonbridging compounds [$\nu_a(CH_3)$ and $\nu_s(CH_3)$ are 2840 and 2780 cm^{-1}, respectively].[66]

Solid $Be(CH_3)_2$ and $Mg(CH_3)_2$ also form long-chain polymers through CH_3 bridges,[67] while $Al(CH_3)_3$ is dimeric in the solid state.[68,69] The infrared spectra of $Li[Al(CH_3)_4]$ and $Li_2[Zn(CH_3)_4]$ have been interpreted on the basis of linear polymeric chains in which the Al (or Zn) atom and the Li atom are bonded alternately through two CH_3 groups.[70] Normal coordinate analyses have been carried out on $M(CH_3)_2$- (M = Zn, Cd, and Hg),[22,25,71] dimeric $Al(CH_3)_3$-,[36,68,69] and linear $[M(CH_3)_2]^{n+}$-type cations.[55,58]

The ethyl group bonded to a metal ($M—CH_2—CH_3$) exhibits bands characteristic of both the CH_3 and CH_2 groups. It is difficult, however, to give complete assignments of the $M—C_2H_5$ group vibrations because of band overlapping and vibrational coupling. Table IV-4 lists the MC_n skeletal frequencies of typical $M(C_2H_5)_n$-type compounds. The MC stretching frequencies of the ethyl compounds are lower than those of the corresponding methyl compounds (Table IV-2) due to the larger mass of the ethyl, relative to the methyl group.

$Li(C_2H_5)$ is hexameric in hydrocarbon solvents[88] and is polymeric in the solid state.[89] The LiC stretching bands of these polymers are at 530–300 cm^{-1}.[90] The vibrational spectra of other polymeric ethyl compounds such as $Be(C_2H_5)_2$ (dimer),[91] $Mg(C_2H_5)_2$,[92] $Al(C_2H_5)_3$ (dimer),[93,94] and $Li[Al(C_2H_5)_4]$ (polymer)[95] have been reported. There are many other compounds containing higher alkyl groups. References for some typical compounds are as follows: $[Tl(n-C_3H_7)_2]Cl$ (57), $Al(n-C_3H_7)_3$ (93), $Ge(n-C_4H_9)_4$ (96), and $[Li(t-C_4H_9)]_4$ (97). Vibrational spectra have also been reported for cycloalkyl compounds

TABLE IV-4. Metal–Carbon Skeletal Frequencies of $M(C_2H_5)_n$-Type Compounds (cm^{-1})

Compound	Structure	$\nu_a(MC)$	$\nu_s(MC)$	$\delta(MCC)$	$\delta(CMC)$	Refs.
$Zn(C_2H_5)_2$	Linear	563	474	261	205	72, 73
$Cd(C_2H_4)_2$	Linear	498	445			37
$Hg(C_2H_5)_2$	Linear	515	488	267	140	74, 75
$^{10}B(C_2H_5)_3$	Planar	1135	—	—	287	32, 76
$Al(C_2H_5)_3$	Planar	662	489			36
$Ga(C_2H_5)_3$	Planar	496	—	—	—	77
$In(C_2H_5)_3$	Planar	457	447			44
$P(C_2H_5)_3$	Pyramidal	697, 669	619	410, 249	—	78, 79
$As(C_2H_5)_3$	Pyramidal	540	570, 563	—	—	79, 80
$Sb(C_2H_5)_3$	Pyramidal	505	505	—	—	78, 81
$Bi(C_2H_5)_3$	Pyramidal	450	450	253	124	82
$Si(C_2H_5)_4$	Tetrahedral	731	549	392, 233	170	63, 83
$Ge(C_2H_5)_4$	Tetrahedral	572	532	332	152	84, 85
$Sn(C_2H_5)_4$	Tetrahedral	508	490	272	132, 86	86, 87
$Pb(C_2H_5)_4$	Tetrahedral	461	443	243, 213	107	82, 86

such as $Zn(c\text{-}C_3H_5)_2$,[98] $M(c\text{-}C_3H_5)_4$ (M = Si, Ge, and Sn),[99] $Pb(c\text{-}C_3H_5)_4$,[100] and $Sb(c\text{-}C_3H_5)_5$.[101]

IV-2. VINYL, ALLYL, ACETYLENIC, AND PHENYL COMPOUNDS

Table IV-5 lists the CC and MC stretching frequencies of metal vinyl $(M-CH=CH_2)$, allyl $(M-CH_2-CH=CH_2)$, and acetylenic $(M-C\equiv CH)$ compounds in which the organic ligands are σ-bonded to the central metal. The $\nu(C=C)$ and $\nu(C\equiv C)$ are generally strong in the Raman. However, their infrared intensities depend on the metal involved. Vibrational spectra of halovinyl compounds have been reported for $Hg(CH=CHCl)_2$[119] and $M(CF=CF_2)_n$ (M = Hg, As, and Sn).[120] Complete vibrational assignments are available for metal allyl compounds such as $M(CH_2-CH=CH_2)_4$ (M = Si and Sn),[121] $(M(CH_2-CH=CH_2)_3$ (M = P and As),[122] and $Hg(CH_2-CH=CH_2)_2$.[123,124]

The phenyl group σ-bonded to a metal $(M-C_6H_5)$ exhibits bands characteristic of monosubstituted benzenes.[125] The $M-C_6H_5$ group exhibits 30 (3 × 12 − 6) fundamentals, only six of which, shown in Fig. IV-4, are sensitive to the change in metals. Table IV-6 lists the observed frequencies of these six modes

TABLE IV-5. Carbon–Carbon and Metal–Carbon Stretching Frequencies of Vinyl, Allyl, and Acetylenic Compounds (cm^{-1})

Compound	$\nu(C=C)$ or $\nu(C\equiv C)$	$\nu(M-C)$	Refs.
$Zn(CH=CH_2)_2$	1565	—	102
$Hg(CH=CH_2)_2$	1603	541, 513	103
$^{10}B(CH=CH_2)_3$	1604	1186, 651	104–107
$P(CH=CH_2)_3$	1590	715, 667	108
$Si(CH=CH_2)_4$	1591	725, 578, 541	109–111
$Ge(CH=CH_2)_4$	1595	600, 561	110
$Sn(CH=CH_2)_4$	1583	531, 490	110, 111
$Pb(CH=CH_2)_4$	1580	495, 479	110
$Si(CH_2-CH=CH_2)_4$	1631	707, 597, 526	112
$Sn(CH_2-CH=CH_2)_4$	1624	487, 475	112
$Hg(CH_2-CH=CH_2)_2$	1617	495, 475	113
$Si(C\equiv CH)_4$	2062, 2053	708, 534	114–116
$Ge(C\equiv CH)_4$	2062, 2057	523, 507	114–116
$Sn(C\equiv CH)_4$	2043	504, 447	114–116
$P(C\equiv CH)_3$	2061	646, 615	117
$As(C\equiv CH)_3$	2053	526, 517	117, 118
$Sb(C\equiv CH)_3$	2033	477, 450	117, 118
$(CH_3)_2Si(C\equiv CH)_2$ [a]	2041	548, 385, 377, 300	114–117
$(CH_3)_2Ge(C\equiv CH)_2$ [a]	2041	538, 521	114–117
$(CH_3)_2Sn(C\equiv CH)_2$ [a]	2016	454	114–117

[a]For these compounds, the $\nu(M-C)$ indicates that of the $M-C\equiv CH$ group.

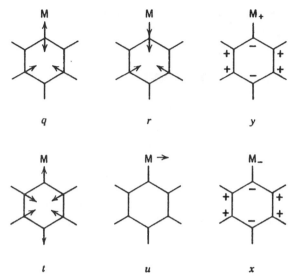

Figure IV-4. Metal-sensitive modes of the M—C_6H_5 moiety.

TABLE IV-6. Vibrational Frequencies of Metal-Sensitive Modes of Metal Phenyls (cm^{-1})

Compound	q	r	y	t	x	u	Refs.
$Hg(C_6H_5)_2$	1067	661	456	258	—	207	126–128
				252			
				248			
$^{10}B(C_6H_5)_3$	1285	893	600	650	245	408	129, 130
	1248						
$Al(C_6H_5)_3$	1085	670	460	420	207	332	130, 131
		643					
$Ga(C_6H_5)_3$	1085	665	453	315	180	245	130
			445			225	
$In(C_6H_5)_3$	1070	673	465	270	180	248	130, 132
						195	
$Si(C_6H_5)_4$	1108	709	519	435	185	261	133, 134
			511	239	171	223	
$Ge(C_6H_5)_4$	1091	—	481	332	187	232	134, 135
			465		168	214	
$Sn(C_6H_5)_4$	1075	616	459	268	193	225	134, 136, 137
			448	212	152		
$Pb(C_6H_5)_4$	1061	645	450	223	147	181	134
			440	201			
$P(C_6H_5)_3$	1089	—	501	428	248	209	136, 138
			509	398	268	190	
$As(C_6H_5)_3$	1082	667	474	313	237	192	136, 138
	1074					183	
$Sb(C_6H_5)_3$	1065	651	457	270	216	166	136, 138
				257			
$Bi(C_6H_5)_3$	1055	—	448	237	207	157	136, 138
				220			

for typical phenyl compounds. It is seen that most of these bands are shifted progressively to lower frequencies as the metal is changed in the order

$$Al > Ga > In$$

$$Si > Ge > Sn > Pb$$

$$P > As > Sb > Bi$$

Among the six modes, t and u are particularly metal-sensitive because they correspond to the $\nu(MC)$ and $\delta(CMC)$, respectively. The number of these bands reflects the local symmetry of the MC_n skeleton. For example, in the $M(C_6H_5)_3$ (M = P, As, Sb, and Bi) series, two $\nu(MC)$ bands (A_1 and E) are expected in IR as well as in Raman spectra since the symmetry of the MC_3 skeleton is C_{3v}. This is clearly demonstrated in Fig. IV-5 where the IR and Raman spectra of these compounds in benzene solution are shown.[138] The spectra obtained in the solid state[138] show further band splittings due to lowering of symmetry and intermolecular interaction.

IV-3. HALOGENO, PSEUDOHALOGENO, AND ACIDO COMPOUNDS

There are many organometallic compounds containing functional groups other than those discussed in the preceding sections. Vibrational spectra of these compounds can be interpreted approximately as the overlap of bands discussed previously and those described below.

(1) Halogeno Compounds

Table IV-7 lists metal–carbon and metal–halogen stretching frequencies of typical compounds. The $\nu(M-CH_3)$ and $\nu(M-C_6H_5)$ are observed in the 750–450 and 300–190 cm^{-1} regions, respectively, as previously found in methyl and phenyl compounds. The $\nu(M-X)$ (terminal) vibrations (Sec. III-23) appear in the regions following:

$\nu(MF)$	$\nu(MCl)$	$\nu(MBr)$	$\nu(MI)$	
800–400	550–200	450–140	260–100	(cm^{-1})

These ranges are rather wide because the $\nu(MX)$ varies markedly depending upon the structure. This is clearly demonstrated in Table IV-7. As shown previously for $\nu(MC)$, the $\nu(MX)$ also becomes lower as the mass of M increases

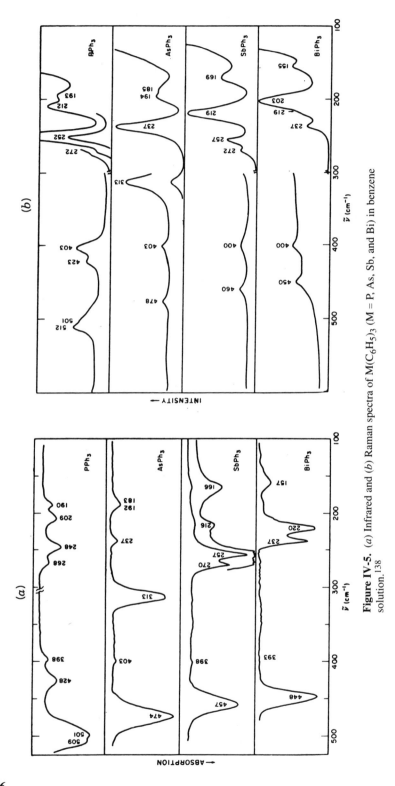

Figure IV-5. (a) Infrared and (b) Raman spectra of $M(C_6H_5)_3$ (M = P, As, Sb, and Bi) in benzene solution.[138]

TABLE IV-7. Metal–Carbon and Metal–Halogen Stretching Frequencies of Organometal Halogeno Compounds (cm⁻¹)

Compound	$\nu(MC)$	$\nu(MX)$	Refs.	Compound	$\nu(MC)$	$\nu(MX)$	Refs.
Cd(CH₃)Cl	476	247	139	Sb(CH₃)₃Cl₂[a]	577 538	282 272	146 147
Cd(CH₃)Br	475	206	139	Sb(CH₃)₃Br₂[a]	569 526	215 168	146 147
Cd(CH₃)I	482	167	139	Sb(CH₃)₃I₂[a]	559 508	144 122	146
Ti(CH₃)Cl₃	536	551 390	140	[Au(CH₃)₂Cl₂]⁻ (cis)	572	272	148
Al(CH₃)Cl₂	653	564 425	141	[Au(CH₃)₂Br₂]⁻ (cis)	563 558	268 197	148
Al(CH₃)₂Cl	603	691 453	141	Tl(C₆H₅)F₂	552 283	179 525	149
Si(CH₃)₃F	704 635	898	142	Tl(C₆H₅)Cl₂	273	500 382	150
Si(CH₃)₃Cl	704 635	472	143	Tl(C₆H₅)Br₂	206	332 270	150
Ge(CH₃)₃F	623 576	623	144	Bi(C₆H₅)₃F₂[a]	202	258 445	151
Ge(CH₃)₃Cl	612 569	378	145	Bi(C₆H₅)₃Cl₂[a]	213 195	253	151
Sn(CH₃)₃Cl	548 518	318	19	Bi(C₆H₅)₃Br₂[a]	215 197	153	151
Sb(CH₃)₃F₂[a]	591 546	484 465	146 147	Sn(C₆H₅)₃Cl	276 232	339	152

[a]The halogens occupy the axial positions of trigonal bipyramids.

in the same family of the periodic table:

	Si(CH$_3$)$_3$Cl		Ge(CH$_3$)$_3$Cl		Sn(CH$_3$)$_3$Cl
ν(M—Cl) (cm^{-1})	472	>	378	>	318

	Si(C$_6$H$_5$)Cl$_3$		Ge(C$_6$H$_5$)Cl$_3$		Sn(C$_6$H$_5$)Cl$_3$
ν(M—Cl) (cm^{-1})	596, 518	>	427, 407	>	383, 363

The same trends are observed for the ν(M—CH$_3$) and ν(M—C$_6$H$_5$) of these compounds.

Normal coordinate analyses have been carried out on Si(CH$_3$)$_3$Cl,[153] Si(CH$_3$)$_2$Cl$_2$,[154] Si(CH$_3$)Cl$_3$,[155] Sb(CH$_3$)$_3$X$_2$,[147] Ti(CH$_3$)Cl$_3$,[140] cis-[Au(CH$_3$)$_2$X$_2$]$^-$,[148] Si(C$_2$H$_5$)$_3$F,[156] and Tl(C$_6$H$_5$)X$_2$.[150] Detailed band assignments are reported for the Sn(CH$_3$)$_n$Cl$_{4-n}$[19] and M(C$_6$H$_5$)Cl$_3$ (M = Si, Ge, and Sn) series.[157]

In condensed phases, halogeno compounds tend to polymerize by forming halogeno bridges between two metal atoms. As discussed in Sec. III-23, the bridging frequencies are much lower than the terminal frequencies. For example, in [Al(CH$_3$)Cl$_2$]$_2$,

$$\begin{array}{ccc} Cl_t & Cl_b & CH_3 \\ \diagdown & \diagup \diagdown & \diagup \\ & Al \quad Al & \quad (C_{2h}) \\ \diagup & \diagdown \diagup & \diagdown \\ CH_3 & Cl_b & Cl_t \end{array}$$

the ν(Al—Cl$_t$) are at 495 (A_g) and 485 (B_u), whereas the ν(Al—Cl$_b$) are at 345 (A_g), 380 (A_u), and 322 cm^{-1} (B_u).[141] As shown in Table IV-7, the ν(Al—Cl) of monomeric Al(CH$_3$)Cl$_2$ are at 425 (ν_s) and 564 cm^{-1} (ν_a). Thus, it is possible to distinguish monomeric and polymeric structures by vibrational spectroscopy.

It was found that B(CH$_3$)$_2$X (X = F, Cl, and Br) is monomeric,[158] whereas Al(CH$_3$)$_2$F, Ga(CH$_3$)$_2$F, and In(CH$_3$)$_2$Cl are tetrameric,[159] trimeric,[160] and dimeric,[161] respectively, in benzene solution. Alkyl silicon and germanium halides tend to be monomeric, whereas alkyl tin and lead halides tend to be polymeric, in the liquid and solid phases. For example, Sn(CH$_3$)$_2$F$_2$ and Sn(CH$_3$)$_3$F are polymerized through the fluorine bridges.[162]

$$\begin{array}{cc} CH_3 & F \\ \diagdown F \mid F \diagup & H_3C \mid \\ Sn \quad Sn \quad Sn & Sn—CH_3 \\ \diagup F \mid F \diagdown & H_3C \mid \\ CH_3 & F \end{array}$$

The $\nu(\text{SnF})$ for terminal bonds are at 650–625 cm^{-1}, whereas those for bridging bonds are at 425–335 cm^{-1}. In the solid state, the coordination number of tin is five or six. Dialkyl compounds prefer six-coordinate structures, while trialkyl compounds tend to form five-coordinate structures. In both cases, the favored positions of the alkyl groups are those shown in the diagrams above. Normal coordinate calculations have been made on the *trans*-[Sn(CH$_3$)$_2$X$_4$]$^{2-}$ (X = F, Cl, and Br) series.[163]

Pb(CH$_3$)$_3$X (X = F, Cl, Br, and I) are monomeric in benzene but polymeric in the solid state; $\nu(\text{PbCl})$ of the monomer and polymer are observed at 281 and 191 cm^{-1}, respectively.[164] In the [Au(CH$_3$)$_2$X]$_2$ series, the Au atom takes a square-planar arrangement with two bridging halogens. The $\nu(\text{Au}-\text{X}_b)$ of these compounds are at 273 and 256 for X = Cl, 181 for X = Br, and 144 and 131 cm^{-1} for X = I.[165] Figure IV-6 shows the tetrameric structure of [Pt(CH$_3$)$_3$X]$_4$- type compounds. Vibrational spectra have been reported for [Pt(CH)$_3$X]$_4$ (X = Cl, Br, and I)[166,167] and [Pt(CH$_3$)$_2$X$_2$]$_n$ (X = Cl, Br, and I; n is probably 4).[168]

(2) Pseudohalogeno Compounds

Vibrational spectra of coordination compounds containing pseudohalogeno groups such as CN$^-$, SCN$^-$, CNO$^-$, and N$_3^-$ ions were discussed in Sec. III-16. The IR spectra of organometallic compounds involving these ligands have been reviewed by Thayer and West.[169] Although these compounds may have corresponding linkage isomers, only one of the isomers is generally stable for a given metal. The linear triatomic pseudohalogeno groups exhibit three vibrations in the regions 2300–2000 (ν_a), 1500–850 (ν_s), and 700–450 cm^{-1} (δ).

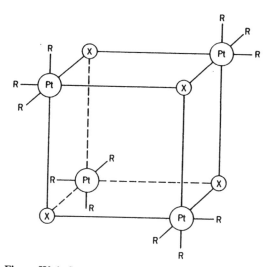

Figure IV-6. Structure of [Pt(CH$_3$)$_3$X]$_4$; R denotes CH$_3$.

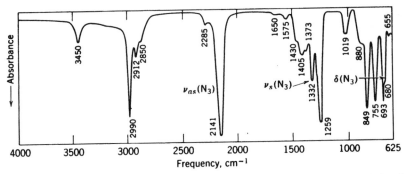

Figure IV-7. Infrared spectrum of $Si(CH_3)_3N_3$ (liquid film) (reproduced with permission from Ref. 170).

The ν_a is most useful as a diagnostic test of these groups because of its strong appearance in the IR region, which is free from interference by other bands. As an example, Fig. IV-7 shows the IR spectrum of $Si(CH_3)_3N_3$ obtained by Thayer and Strommen.[170]

Vibrational spectra of azido compounds are reported for $Zn(CH_3)N_3$,[171] $Hg(CH_3)N_3$,[172] $Al(CH_3)_2N_3$,[173] $Si(CH_3)_3N_3$,[174] and $Sn(CH_3)_3N_3$.[175] The $\nu(M—N)$ of these compounds are in the 600–400 cm^{-1} region. The NCO group is always bonded to the metal via the N atom (isocyanato compound). Vibrational spectra of monomeric $CH_3Hg—NCO$,[176] $(CH_3)_3Si—NCO$,[174,177] $(CH_3)_3 Ge—NCO$,[178] $(CH_3)_3Sb(NCO)_2$,[176] and $(C_6H_5)_3Bi(NCO)_2$[151] have been reported.

The NCS group may be bonded to a metal through the N or S atom or may form a bridge between two metals by using the N, the S, or both atoms. It is not easy to distinguish all these possible structures by vibrational spectra. Table IV-8 lists the modes of coordination and the vibrational frequencies of the M—NCS group. The NCSe group is N-bonded in $(CH_3)_3Si—NCSe$[184] and $(CH_3)_3Ge—NCSe$[185] but is Se-bonded in $(CH_3)_3Pb—SeCN$.[186] Only a very few compounds containing the M—CNO (fulminato) group are known. The spectrum of $(CH_3)_2Tl(CNO)$ is similar to that of $K[CNO]$, and the Tl—CNO bond may be ionic.[187] No isofulminato complexes are reported. The CN group is usually bonded to the metal through the C atom. In the case of $(CH_3)_3M(CN)$ (M = Si and Ge), however, the cyano and isocyano complexes are in equilibrium in the liquid phase, although the mole fraction of the latter is rather small. The CN stretching frequencies (cm^{-1}) of these isomers are as follows:

	$(CH_3)_3M—CN$	$(CH_3)_3M—NC$	
M = Si	2198	2095	(gas phase)[188]
M = Ge	2182	2090	(CHCl$_3$ solution)[189]

Complete band assignments for $(CH_3)_3GeCN$ and its deuterated analog have been made based on normal coordinate analysis.[190]

TABLE IV-8. Vibrational Frequencies of Thiocyanato and Isothiocyanato Compounds (cm^{-1})

Compound	Mode of Coordination	ν(CN)	ν(CS)	δ(NCS)	ν(MS) or ν(MN)	Ref.
$[CH_3Zn(SCN)]_\infty$	Zn, SCN—Zn, Zn	2190 ⎫ 2140 ⎬	685	455	—	179
$[CH_3Hg(SCN)_3]^{2-}$	Hg—SCN	2119	—	—	276	180
$[(CH_3)_3Al(SCN)]^-$	Al—SCN	2097	845	485	335	181
$[(CH_3)_2Al(SCN)]_3$	Al, SCN, Al	2075	627	501 ⎫ 438 ⎬	—	182
$(CH_3)_3Sn(NCS)$, solid (polymer)	Sn—NCS—Sn	2098 ⎫ 2079 ⎬ 2046 ⎭	779	474 ⎫ 467 ⎬	—	183
$(CH_3)_3Sn(NCS)$, CS$_2$ solution (monomer)	Sn—NCS	2050	781	485	478	183
$[(CH_3)_2Au(NCS)]_2$	Au, NCS, Au, SCN	2163	775	444 ⎫^{a} 430 ⎬	—	165

^{a}These bands may be assigned to ν(AuN).

(3) Acido Compounds

The free acetate ion (CH_3COO^-) exhibits the ν_a(COO) and ν_s(COO) at 1578 and 1414 cm^{-1}, respectively (Sec. III-8(4)). If it is covalently bonded to a metal as a unidentate ligand, the ν_a and ν_s are shifted to higher and lower frequencies, respectively.

$$CH_3-C\overset{O^{-1/2}}{\underset{O^{-1/2}}{\Big\langle}} \qquad M-O\overset{O}{\underset{\underset{CH_3}{|}}{\diagdown C \diagup}}$$

Figure IV-8 shows the IR spectra of the $Si(CH_3)_n(CH_3COO)_{4-n}$ ($n = 0 \sim 3$) obtained by Okawara et al.[191] The ν(C=O) are observed at 1765, 1748, 1732, and 1725 cm^{-1}, respectively, for $n = 0$, 1, 2, and 3. In contrast, analogous Sn compounds such as $Sn(CH_3)_3(CH_3COO)$ and $Sn(CH_3)_2Cl(CH_3COO)$ exhibit two ν(COO) near 1600 and 1420 cm^{-1}, indicating the ionic character of the tin–acetate interaction.[191,192] In $Sb(CH_3)_3(CH_3COO)_2$, however, the two acetate groups occupy the axial positions of the trigonal–bipyramidal structure, and are covalently bonded to the metal as unidentate ligands.[193]

As stated earlier, the trigonal–planar $Sn(CH_3)_3$ group in $Sn(CH_3)_3F$ is polymerized by forming the Sn—F—Sn bridges in the solid state. Similar polymerization occurs in $Sn(CH_3)_3L$ where L is $CH_3COO^{-,194}$, $NO_3^{-,195}$, $ClO_4^{-,196}$,

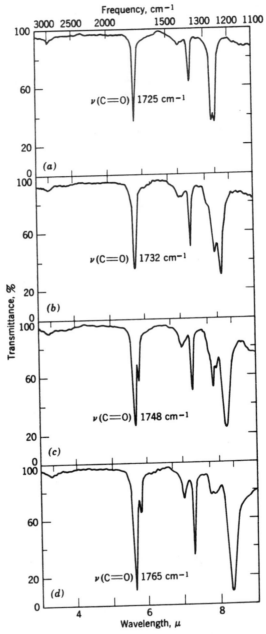

Figure IV-8. Infrared spectra of (*a*) Si(CH₃)₃(OCOCH₃), (*b*) Si(CH₃)₂(OCOCH₃)₂, (*c*) Si(CH₃)(OCOCH₃)₃, and (*d*) Si(OCOCH₃)₄ (reproduced with permission from Ref. 191).

and other acido groups:

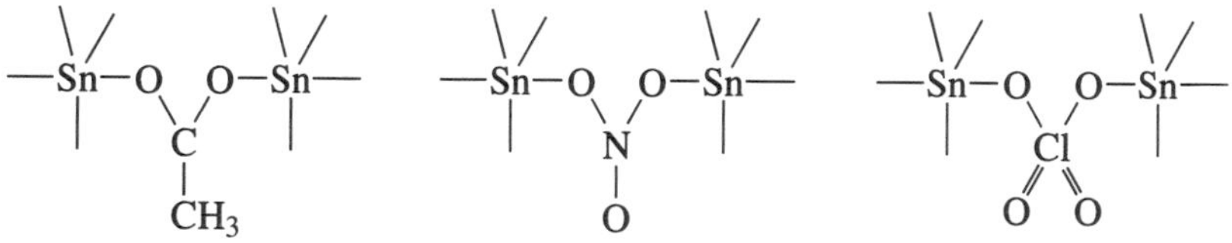

As expected for symmetric bidentate coordination, the separation of the ν_a (1600 cm^{-1}) and ν_s (1363 cm^{-1}) of the first compound (237 cm^{-1}) is much smaller than those observed for unidentate coordination ($\sim$ 460 cm^{-1}).[191] In agreement with the local symmetry of the NO$_3$ group in the structure shown above (C$_{2v}$ or lower), Sn(CH$_3$)$_3$(NO$_3$) exhibits three ν(NO) (1452, 1300, and 1021 cm^{-1}) in IR spectra.[195] Symmetric bidentate coordination of the ClO$_4$ group in Sn(CH$_3$)$_3$(ClO$_4$) is also confirmed by the observation of four ν(ClO) at 1212, 1112, 998, and 908 cm^{-1} in IR spectra.[196]

If Sn(CH$_3$)$_3$L is tetrahedral, two ν(Sn—C) vibrations should be IR-active. Thus, the tetrahedral Sn(CH$_3$)$_3$Cl molecule exhibits two ν(Sn—C) bands at 545 and 513 cm^{-1}.[1,191] On the other hand, only one ν(Sn—C) vibration is expected in IR spectra if it contains a trigonal–planar Sn(CH$_3$)$_3$ group. Examples of the latter are found in the polymeric Sn(CH$_3$)$_3$ L-type compounds discussed above. These compounds exhibit only one ν(Sn—C) band near 550 cm^{-1}.

Tetrahedral Sn(CH$_3$)$_2$X$_2$ (X = Cl, Br, and I) molecules exhibit two ν(Sn—C) vibrations at $\sim$560 and 515 cm^{-1} in IR spectra because their C—Sn—C groups are bent.[191] The C—Sn—C groups in polymeric (Sn(CH$_3$)$_2$)$_2$O(CO$_3$) and [Sn(CH$_3$)$_2$(NCS)]$_2$O may be bent since they show two ν(Sn—C) bands in IR spectra.[197] Only one ν(Sn—C) vibration is expected for a linear C—Sn—C group.

IV-4. COMPOUNDS CONTAINING OTHER FUNCTIONAL GROUPS

(1) Nitrogen Donors

The ν(MN) of ammine and related ligands have been discussed in Sec. III-1. The ν(MN) of [Hg(CH$_3$)(NH$_3$)]$^+$,[198], Ga(CH$_3$)$_3$(NH$_3$),[199] and [Sn(CH$_3$)$_3$(NH$_3$)]$^+$,[200] are observed at 585, 350, and 503 cm^{-1}, respectively. The IR spectrum of Al(CH$_3$)$_3$(NH$_3$) in Ar matrices shows significant nonplanarity of the AlC$_3$ skeleton.[201] In Sn(CH$_3$)$_2$(bipy)Cl$_2$, the Sn(CH$_3$)$_2$ group is linear because its IR spectrum shows only one ν(Sn—C) near 575 cm^{-1}.[202-204] The IR bands at 427 and 346 cm^{-1} were tentatively assigned to the ν(Sn—N).[202] The IR and Raman spectra of [Pt(CH$_3$)$_3$(NH$_3$)$_3$]$^+$ are interpreted on the basis of the *fac* structure (C$_{3v}$): the ν(PtN) are at 390 (A_1, Raman) and 410 and 377 cm^{-1} (E, IR).[205]

(2) Oxygen and Sulfur Donors

Compounds containing the hydroxo (OH) group exhibit $\nu(OH)$, $\delta(MOH)$, and $\nu(MO)$ at 3760–3000, 1200–700, and 900–300 cm^{-1}, respectively. As expected, these frequencies depend heavily on the metal and the strength of the hydrogen bond involved. The $\nu(SiO)$ of $Si(CH_3)_3(OH)$ is at 915 cm$^{-1,206}$, whereas the $\nu(SnO)$ of $Sn(CH_3)_3(OH)$ is at 370 cm$^{-1,207,208}$. Polymeric structures involving the trigonal–planar $Sn(CH_3)_3$ group and $Sn-OH-Sn$ bridges were proposed for the latter compound becuase it exhibits only one $\nu(Sn-C)$ at 540 cm^{-1}. Figure IV-9 shows the IR spectra of $Sn(CH_3)_3(OH)$ obtained by Okawara and Yasuda.[207] References on other hydroxo compounds are as follows: $Si(C_2H_5)_2(OH)_2$,[209] $Sb(CH_3)_4(OH)$,[210] $M(CH_3)_2(OH)_2$ (M = Ge, Sn, and Pb),[211,212] $[Ga(CH_3)_2(OH)]_4$,[213] and $[Pt(CH_3)_3(OH)]_4$.[167,214] $Si(CH_3)_3(SH)$ exhibits the $\nu(SH)$ and $\nu(SiS)$ at 2580 and 454 cm^{-1}, respectively, in the liquid state.[215]

The IR spectra of $Sn(CH_3)_2(OR)_2$ (R = Me, Et, etc.) exhibit the $\nu(C-O)$ and $\nu(Sn-O)$ in the 1070–1000 and 650–550 cm^{-1}, respectively.[216,217] Other pertinent references are $Al(CH_3)_2(OCH_3)$[218] and $Si(CH_3)_2(OCH_3)_2$.[219] The $\nu(M-S)$ of $Si(CH_3)_3(SC_2H_5)$,[220] $Si(CH_3)_3(SC_6H_5)$,[221] $Ge(CH_3)_3(SCH_3)$,[222] and $Sn(CH_3)_2(SCH_3)_2$[223] are assigned at 486, 459, 394, and 347 cm^{-1}, respectively.

The vibrational spectra of aquo complexes are characterized by the bands discussed in Sec. III-7(2). References are cited for $[Sn(CH_3)_3(OH_2)_2]^{+}$,[224] $[Hg(CH_3)(OH_2)]^{+}$,[225] and $[Pt(CH_3)_3(OH_2)_3]^{+}$.[226]

Vibrational spectra of O-bonded (chelated) acetylacetonato (acac) complexes were discussed in Sec. III-13. Some references for acac complexes of metal alkyls are $Ga(CH_3)_2(acac)$,[213] $Sn(CH_3)_2(acac)_2$,[227] $Pb(CH_3)_2(acac)_2$,[228,229] $Sb(CH_3)_2(acac)Cl_2$,[230] and $Au(CH_3)_2(acac)$.[231] The structure of $Sn(CH_3)_2(acac)_2$ was originally suggested to be *trans* in solution and in the solid state on the basis of NMR and vibrational spectra.[232] Later, the *cis* structure was proposed because of the large dipole moment (2.95 D) in benzene solution.[273] X-Ray analysis showed, however, that the compound is *trans* in the solid state.[234] Ramos and Tobias[227] suggest that the structure remains *trans* in solution and that the large dipole moment may originate in the nonplanarity of the SnO_4 plane with the remainder of the acac ring.

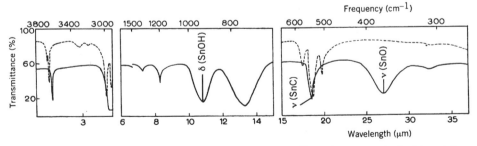

Figure IV-9. IR spectra of $Sn(CH_3)_3(OH)$. —— and ------ indicate the spectra obtained in nujol (or hexachlorobutadiene) mull and CCl$_4$ solution, respectively (reproduced with permission from Ref. 207).

TABLE IV-9. Metal–Hydrogen and Metal–Carbon Stretching Frequencies of Typical Hydrido Compounds (cm^{-1})

Compound	ν(MH)	ν(MC)	Ref.
CH_3SiH_3	2166, 2169	701	235
$(CH_3)_2SiH_2$	2145, 2142	728, 659	236
$(CH_3)_3SiH$	2123	711, 624	236
CH_3GeH_3	2086, 2085	604	237
$(CH_3)_2GeH_2$	2080, 2062	604, 590	238
$(CH_3)_3GeH$	2049	601, 573	238
CH_3SnH_3	1875	527	239
$(CH_3)_2SnH_2$	1869	536, 514	240
$(CH_3)_3SnH$	1837	521, 516	240
$(CH_3)_3PbH$	1709	—	241
$(CH_3)_2PH$	2288	703, 660	242
$(CH_3)_2AsH$	2080	580, 565	243

(3) Hydrido Compounds

Metal hydrido (H) complexes exhibit the ν(MH) and δ(MH) in the 2250–1700 and 800–600 cm^{-1} regions, respectively (Sec. III-22(2)). The ν(MH) and ν(MC) of typical compounds are listed in Table IV-9. It is seen that the ν(MH) decreases as the mass of the metal increases in the same family of the periodic table and as more methyl groups are bonded to the metal in the $M(CH_3)_nH_{4-n}$ ($n = 1–3$) series. Vibrational spectra of B_2H_6-type molecules were discussed in Sec. II-10 of Part A. Methyl-diboranes, $(CH_3)_nB_2H_{6-n}$ ($n = 1–4$) exhibit terminal and bridging ν(BH) at 2600–2500 and 2150–1525 cm^{-1}, respectively.[244] The dimeric $[(CH_3)_2AlH]_2$ species is predominant in the gaseous dimethylaluminum hydride: the ν_a(AlH) and ν_s(AlH) of the Al—H—Al bridges are observed at ~1353 and 1215 cm^{-1}, respectively.[245]

(4) Metal–Metal Bonded Compounds

As discussed in Sec. III-24, the ν(MM′) of metal–metal bonded compounds are generally strong in the Raman and weak in the infrared. Table IV-10 lists

TABLE IV-10. Metal–Metal Stretching Frequencies of Metal Alkyl Compounds (cm^{-1})

Compound	ν(MM′)	Ref.
$(CH_3)_3Si—Mn(CO)_5$	297 (R)	246
$(CH_3)_3Ge—Cr(CO)_3(\pi\text{-Cp})$	119 (R)	247
$(CH_3)_3Ge—Mn(CO)_5$	191 (R)	248
$(CH_3)_3Ge—Co(CO)_4$	192 (R)	249
$(CH_3)_3Sn—Mo(CO)_3(\pi\text{-Cp})$	172 (IR)	250
$(CH_3)_3Sn—Mn(CO)_5$	182 (IR, R)	250
$(CH_3)_3Sn—Re(CO)_5$	147 (R)	248
$(CH_3)_3Sn—Co(CO)_4$	176 (IR, R)	249

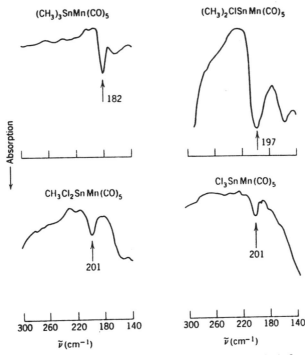

Figure IV-10. Far-IR spectra of $(CH_3)_{3-n}Cl_nSnMn(CO)_5$ where $n = 0, 1, 2,$ or 3. The arrow indicates the $\nu(Sn–Mn)$ band.[250]

the $\nu(MM')$ of some metal–metal bonded compounds. Figure IV-10 shows the $\nu(SnMn)$ bands observed in far-IR spectra of $(CH_3)_{3-n}Cl_nSn—Mn(CO)_5$-type compounds.[250] The vibrational spectra of $(CH_3)_3M—M(CH_3)_3$ (M = Si, Ge, Sn, and Pb)[251,252] and $M[Si(CH_3)_3]_4$ (M = Si, Ge, and Sn)[253] are reported.

IV-5. π-COMPLEXES OF OLEFINS, ACETYLENES, AND RELATED LIGANDS

Vibrational spectra of π-bonded complexes of ethylene, acetylene, and related ligands have been reviewed by Davidson.[254] In contrast to σ-bonded complexes (Sec. IV-2), the $\nu(C=C)$ and $\nu(C\equiv C)$ bands of π-bonded complexes show marked shifts to lower frequencies relative to those of free ligands.

(1) Complexes of Monoolefins

Ethylene and other olefins form π-complexes with transition metals. Free ethylene exhibits 12 ($3 \times 6 - 6$) normal modes which are classified into $3A_g(R) + A_u$ (i.a.) $+ 2B_{1g}(R) + B_{1u}(IR) + B_{2g}(R) + 2B_{2u}(IR) + 2B_{3u}(IR)$ under $\mathbf{D}_{2h}(\mathbf{V}_h)$

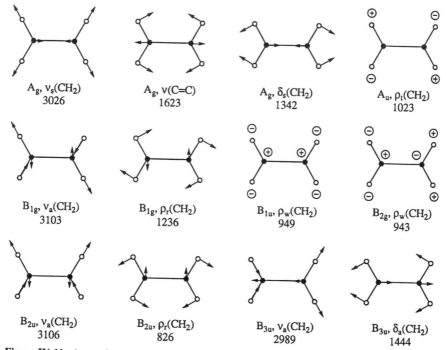

Figure IV-11. Approximate normal modes of vibration of ethylene. Symmetry, vibrational assignments, and observed frequencies (cm^{-1}) are given for each vibration. The $\nu(C=C)$ and $\delta_s(CH_2)$ are vibrationally coupled in the A_g species.

symmetry. Figure IV-11 shows the approximate normal modes and observed frequencies of these vibrations.[7]

The simplest and best-studied complex is Zeise's salt, $K[Pt(C_2H_4)Cl_3]H_2O$, in which the ethylene molecule replaces one of the Cl atoms of the square-planar $PtCl_4^{2-}$ ion with its $C=C$ axis perpendicular to the $PtCl_4$ plane. According to Chatt et al.,[255] the Pt(II)–ethylene interaction is described in terms of two bonding schemes: (A) a σ-type bond is formed by electron donation from the filled $2p\pi$ bonding orbital of the olefin to the vacant dsp^2 bonding orbital of the metal, and (B) a π-type bond is formed by back-donation of electron from a filled dp hybrid orbital of the metal to the $2p\pi^*$ antibonding orbital of the olefin. This is illustrated below:

The real bonding is somewhere between A and B, and the latter becomes more predominant as the oxidation state of the metal becomes lower. The results of X-ray analysis[256] as well as MO calculations[257] seem to indicate that bonding scheme A is predominant in the case of Zeise's salt.

The vibrational spectra of Zeise's salt have been studied by several investigators.[258-261] The effects of coordination on the vibrational frequencies of free ethylene are: (1) The $\nu(C{=}C)$ coupled with the $\delta_s(CH_2)$ is shifted markedly from 1623 to 1526[258] or 1243 cm^{-1},[260] (2) The $\rho_r(CH_2)$ are shifted to lower frequencies (~1030 to 840/720 cm^{-1}). (3) The $\rho_w(CH_2)$ and $\rho_t(CH_2)$ are shifted to higher frequencies (~945 to 1010/975 and 1023 to 1180 cm^{-1}, respectively).[260] The directions of these shifts are anticipated since (1) and (2) are in-plane, whereas (3) are out-of-plane, vibrations.

Approximate band assignments of IR spectra of Zeise's salt and its deuterated analog were first made by assuming that Zeise's anion is a composite of a perturbed C_2H_4 and the $PtCl_3$ moiety.[258] The band at 407 cm^{-1} was assigned to the $\nu(Pt{-}C_2H_4)$ (scheme A) since it did not belong to either components. More elaborate treatments employ a triangular model involving two Pt—C bonds (scheme B). Hiraishi[260] carried out normal coordinate analysis on such a model and assigned the Raman bands at 493 (*dp*) and 405 cm^{-1} (*p*) to the $\nu_a(Pt{-}C_2H_4)$ and $\nu_s(Pt{-}C_2H_4)$, respectively. The corresponding force constants for these two modes were calculated to be 1.92 and 1.45 mdyn/Å, respectively. The $\nu_a(Pt{-}C_2H_4)$ may be called the "tilt" mode since it involves a tilting motion of the ethylene against the rest of the anion. Crayston and Davidson[261] carried out similar calculations on several ethylene complexes. These workers assigned the bands at 455 and 380 cm^{-1} to the $\nu_a(Pt{-}C_2H_4)$ and $\nu_s(Pt{-}C_2H_4)$ of $Pt(0)(C_2H_4)(PPh_3)_2$. Both bands are downshifted on going from the Pt(II) to Pt(0) complexes because a "metallo-cyclopropane" form (scheme B) becomes more predominant in the latter.

Mink and co-workers[262,263] have assigned the IR and Raman spectra of $M(C_2H_4)_3$ (M = Pt, Pd, and Ni) based on normal coordinate analysis. These molecules take a $\mathbf{D}_{3h}$ structure, as shown in Fig. IV-12, and their 51 (3 × 19 − 6) normal vibrations are grouped into:

$$5A_1'(\text{R}) + 4A_2'(\text{i.a.}) + 10E'(\text{IR, R}) + 4A_1''(\text{i.a.}) + 4A_2''(\text{IR}) + 7E''(\text{R})$$

Figure IV-12 illustrates their skeletal modes, and Table IV-11 compares their band assignments and force constants for $Pt(C_2H_4)_3$ and Zeise's anion.

Figure IV-13 shows the IR and Raman spectra of $Pt(C_2H_4)_3$ obtained by Csaszar et al.[263] As shown in Table IV-11, the calculated force constants for the tilt and symmetric stretching modes are relatively close in Zeise's anion, whereas the former is larger in $Pt(C_2H_4)_3$. This result may suggest that the π-bonding (scheme B) is more predominant in the latter compound.[262]

Vibrational spectra are reported for many other complexes of ethylene. Some references are: *trans*-$M(C_2H_4)_2(CO)_4$ (M = Mo and W),[264]

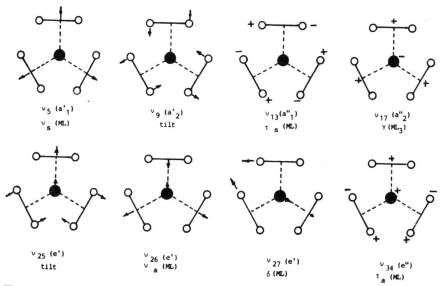

Figure IV-12. Skeletal modes of tris(ethylene)metal complexes. Symmetry and vibrational assignments are indicated for each mode. L denotes ethylene (reproduced with permission from Ref. 263).

$Fe(C_2H_4)(CO)_4$,[265] $[Rh(C_2H_4)_2Cl]_2$,[266] $Ir(C_2H_4)_4Cl$,[267] and $[Pt(C_2H_4)Cl_2]_2$.[258] Tetracyanoethylene (TCNE) also forms π-complexes with transition metals, and their vibrational spectra are reported for $M(TCNE)(CO)_5$ (M = Cr and W),[268] $Fe(TCNE)(CO)_4$,[269] and $Pt(TCNE)(PPh_3)_2$.[270]

(2) Allyl Complexes

According to X-ray analysis,[271] the two Pd atoms in $[Pd(\pi\text{-allyl})Cl]_2$ are bridged by two Cl atoms to form a square-planar $(PdCl)_2$ group and the allyl (C_3H_5)

TABLE IV-11. Observed Frequencies, Band Assignments, and Force Constants of Zeise's Anion and $Pt(C_2H_4)_3$[262,263]

$[Pt(C_2H_4)Cl_3]^-$ (C_{2v})	$Pt(C_2H_4)_3$ (D_{3h})	Band Assignment
Observed Frequencies (cm^{-1})		
1517 (A_1)	1617 (A_1')	$\nu(C{=}C) + \delta(CH_2)$
	1501 (E')	
501 (B_1)	448 (E')	Tilt
408 (A_1)	398 (A_1')	$\nu(Pt{-}C_2H_4)$
	332 (E')	
Force Constants (mdyn/Å)		
2.84	2.04	Tilt
2.54	1.66	Pt$-$C$_4$H$_4$ stretch

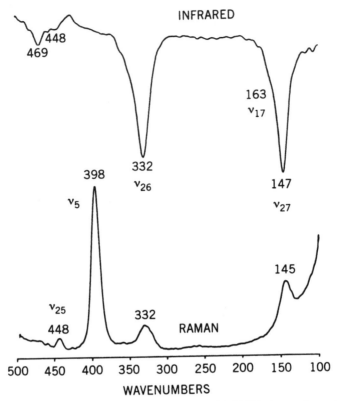

Figure IV-13. Far-IR and low-frequency Raman spectra of Pt(C$_2$H$_4$)$_3$ in petroleum ether (reproduced with permission from Ref. 263).

groups are bonded to the Pd atoms with their planes tilted by 112° with respect to the (PdCl)$_2$ plane so that the overall symmetry is C_{2h}:

Vibrational spectra of [Pd(π-allyl)Cl]$_2$ and related compounds in the low-frequency region were assigned based on isotope shifts due to ^{104}Pd/^{110}Pd substitution. The bands at 402 and 379 cm^{-1} were assigned to the ν_a(Pd-allyl) and ν_s(Pd-allyl), respectively, since both bands are downshifted by 3 cm^{-1} by such substitution.[272] References on other allyl complexes are: M(π-C$_3$H$_5$)$_2$ (M = Ni, Pd), M(π-C$_3$H$_5$)$_3$ (M = Rh, Ir),[273] [Pd(π-C$_3$H$_5$)X]$_2$ (X = Cl and Br),[274] Fe(π-C$_3$H$_5$)(CO)$_3$X (X = Br, NO$_3$),[275] and Mn(π-C$_3$H$_5$)(CO)$_4$.[276] Chenskaya et al.[277]

assigned the metal–olefin and metal–halogen vibrations of π-allyl complexes of transition metals.

(3) Complexes of Diolefins and Oligoolefins

Nonconjugated diolefins such as norbornadiene (NBD, C_7H_8) and 1,5-hexadiene (C_6H_{10}) form metal complexes via their $C{=}C$ double bonds (Figs. IV-14a, IV-14b). Complete vibrational assignments have been made for M(NBD)(CO)$_4$ (M

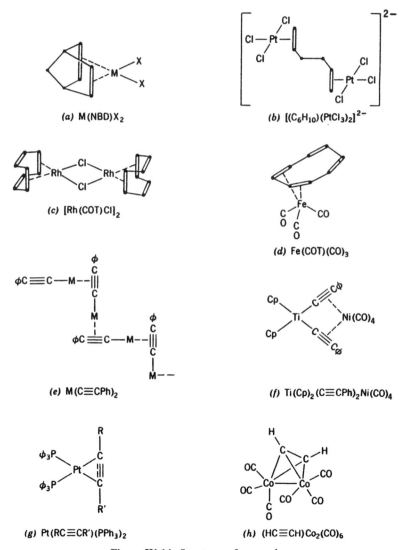

(a) M(NBD)X$_2$

(b) [(C$_6$H$_{10}$)(PtCl$_3$)$_2$]$^{2-}$

(c) [Rh(COT)Cl]$_2$

(d) Fe(COT)(CO)$_3$

(e) M(C≡CPh)$_2$

(f) Ti(Cp)$_2$(C≡CPh)$_2$Ni(CO)$_4$

(g) Pt(RC≡CR')(PPh$_3$)$_2$

(h) (HC≡CH)Co$_2$(CO)$_6$

Figure IV-14. Structures of π-complexes.

= Cr, Mo, and W),[278] Cr(NBD)(CO)$_4$,[279] and Pd(NBD)X$_2$ (X = Cl and Br).[279] The metal–olefin vibrations are assigned in the region from 305 to 200 cm^{-1}. The spectrum of K$_2$[(PtCl$_3$)$_2$(C$_6$H$_{10}$)] is similar to that of the free ligand in the *trans* conformation.[280] Thus, its structure may be shown as in Fig. IV-14*b*. However, the spectrum of Pt(C$_6$H$_{10}$)Cl$_2$ is more complicated than that of the free ligand and suggests a chelate structure such as that shown in Fig. IV-14*a*.

Free butadiene (C$_4$H$_6$) is *trans*-planar. However, it takes a *cis*-planar structure in Fe(C$_4$H$_6$)(CO)$_3$[281] and Fe(C$_4$H$_6$)$_2$CO.[282] For K$_2$[C$_4$H$_6$(PtCl$_3$)$_2$], the infrared spectrum indicates the *trans*-planar structure of the olefin.[283] In [Rh(COT)Cl]$_2$, cyclooctatetraene (COT) takes a tub conformation and coordinates to a metal via the 1,5 C=C double bonds, the 3,7 C=C double bonds being free (Fig. IV-14*c*). The C=C stretching bands of free COT are at 1630 and 1605 cm^{-1}, whereas those of the complex are at 1630 (free) and 1410 (bonded) cm^{-1}.[284] According to X-ray analysis,[285] only two of the four C=C double bonds of COT are bonded to the metal in Fe(COT)(CO)$_3$ (Fig. IV-14*d*). In this case, the C=C stretching band for free C=C double bonds is at 1562 cm^{-1}, whereas that for coordinated C=C double bonds is at 1460 cm^{-1}.[286] In [Rh(COD)Cl]$_2$ (COD, C$_8$H$_{12}$, 1,5-cyclooctadiene), the Rh atom is bonded to COD via the 1,5 C=C bonds in a manner similar to its COT analog (Fig. IV-14*c*). The Rh–olefin stretching vibrations were assigned in the range from 490 to 385 cm^{-1}.[287]

Figure IV-15 shows the far-IR spectra of ^{104}Pd(COT)Cl$_2$, ^{104}Pd(COD)Cl$_2$, and their ^{110}Pd analogs.[248] The COT complex exhibits four isotope-sensitive bands at 344, 319, 287, and 219 cm^{-1}. The first two are assigned to the ν_a(Pd—Cl) and ν_s(Pd—Cl), respectively, whereas the last two are attributed to the ν_a(Pd–olefin) and ν_s(Pd–olefin), respectively. Similar assignments can be made for the COD complex. These ν(Pd–olefin) frequencies are much lower than the ν(Pd—C$_2$H$_4$) (427 cm^{-1}) because of weaker metal–olefin bonds and larger olefin masses.

(4) Complexes of Acetylene and Related Ligands

Free HC≡C(C$_6$H$_5$) exhibits the C≡C stretching band at 2111 cm^{-1}. In the case of σ-bonded complexes (Sec. IV-2), this band shifts slightly to a lower frequency (2062–2016 cm^{-1}).[289] In M[—C≡C(C$_6$H$_5$)]$_2$ [M = Cu(I) and Ag(I)], it shifts to 1926 cm^{-1}. This relatively large shift was attributed to the formation of both σ- and π-type bonding, shown in Fig. IV-14*e*.[290,291] Ti[C≡C(C$_6$H$_5$)]$_2$(π-Cp)$_2$ reacts with Ni(CO)$_4$ to form the complex shown in Fig. IV-14*f*. The C≡C stretching band of the parent compound at 2070 cm^{-1} is shifted to 1850 cm^{-1} by complex formation.[292] According to Chatt and co-workers,[293] the C≡C stretching bands of disubstituted alkynes (2260–2190 cm^{-1}) are lowered to ca. 2000 cm^{-1} in Na[Pt(RC≡CR′)Cl$_3$] and [Pt(RC≡CR′)Cl$_2$]$_2$, and to ca. 1700 cm^{-1} in Pt(RC≡CR′)(PPh$_3$)$_2$. Here R and R′ denote various alkyl groups. The former represents a relatively weak π-bonding similar to that found for Zeise's salt, whereas the latter indicates strong π-bonding in which the C≡C triple bond is almost reduced to the double bond (Fig. IV-14*g*). Similar results were

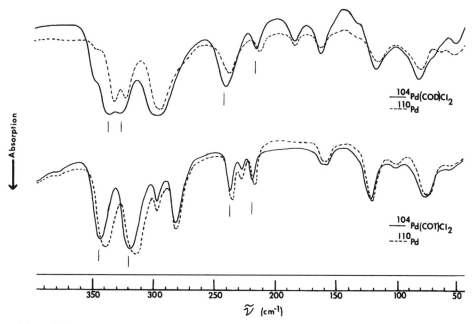

Figure IV-15. Far-IR spectra of Pd(COD)Cl$_2$ and Pd(COT)Cl$_2$. The solid and dashed lines indicate the spectra of the complexes containing ^{104}Pd and ^{110}Pd isotopes, respectively. Vertical lines show metal–isotope-sensitive bands.

found for (RC≡CR′)Co$_2$(CO)$_6$, which exhibits the C≡C stretching bands near 1600 cm^{-1}.[294] In the case of (HC≡CH)Co$_2$(CO)$_6$, the C≡C stretching band was observed at 1402 cm^{-1}, which is ca. 570 cm^{-1} lower than the value for free acetylene (1974 cm^{-1}). The spectrum of the coordinated acetylene in this complex is similar to that of free acetylene in its first excited state, at which the molecule takes a *trans*-bent structure. Considering possible steric repulsion between the hydrogens and the Co(CO)$_3$ moiety, a structure such as that shown in Fig. IV-14*h* was proposed.[295]

There are many other π-bonded acetylenic complexes for which vibrational data are available. For example, the chloro-bridged dimer, [WCl$_4$(IC≡Cl)]$_2$, exhibits the ν(C≡C) at 1619 and 1590 and the ν(WC) at 928 865 and 848 cm^{-1}. The former are ca. 510 cm^{-1} lower than that of free IC≡CI (2118 cm^{-1}).[296] In the case of (*t*-Bu—C≡C—*t*-Bu)$_2$Fe$_2$(CO)$_4$, the C≡C bonds are π-bonded to the Fe—Fe bridge as shown below:

$$
\begin{array}{ccc}
\text{R} & & \text{R} \\
\diagdown & & \diagup \\
\text{OC}\diagdown \quad \text{C}\!\!\equiv\!\!\text{C} \quad \diagup\text{CO} \\
\text{Fe}\!-\!\text{Fe} \\
\text{OC}\diagup \quad \text{C}\!\!\equiv\!\!\text{C} \quad \diagdown\text{CO} \\
\diagup & & \diagdown \\
\text{R} & & \text{R}
\end{array}
$$

R: *t*-Bu

Its IR spectrum shows a very weak $\nu(C\equiv C)$ at 1670 cm^{-1}. The weak band at 531 cm^{-1} and a strong band at 284 cm^{-1} were assigned to the ν(Fe-acetylene) coupled with δ(FeCO) and ν(Fe–Fe), respectively. According to normal coordinate analysis,[297] the Fe–Fe stretching-force constant (3.0 mdyn/Å) is about twice that of the Fe–Fe single bond (1.3 mdyn/Å) of Fe$_2$S$_2$(CO)$_6$.[298] Thus, the Fe–Fe bond of the above compound must be close to a double bond.

(5) Complexes of Nitriles

The $C\equiv N$ stretching frequency of $CF_3-C\equiv N$ is 2271 cm^{-1}. This band is shifted to 1734 cm^{-1} in Pt(CF$_3$CN)(PPh$_3$)$_2$ because of the formation of a Pt-nitrile π-bond.[299] A similar π-bonding has been proposed for Mn(CO)$_3$IL, where L is o-cyanophenyldiphenylphosphine:

In the latter case, the $C\equiv N$ stretching band of the free ligand at 2225 cm^{-1} is shifted to 1973 cm^{-1} in the complex.[300]

(6) Metal–Olefin Complexes in Inert Gas Matrices

A number of olefin complexes of the type M(olefin)$_n$ have been prepared via cocondensation reactions of metal vapors with olefins in inert gas matrices:

$$M + n(\text{olefin}) \longrightarrow M(\text{olefin})_n$$

Vibrational studies show that the olefins in these cocondensation products are all π-bonded to metal atoms. Moskovitz and Ozin[301] report the IR spectra of M(C$_2$H$_4$)$_n$ where M is Co, Ni, Cu, Pd, Ag, and Au. Andrews and coworkers measured and assigned the IR spectra of Li(C$_2$H$_4$)$_n$ (n = 1, 2, and 3)[302] and In(C$_2$H$_4$).[303] In the latter, the $\nu(C=C)$ coupled with the δ(CH$_2$) and ν(In—C) were observed at 1201 and 238 cm^{-1}, respectively. These frequencies are also reported for Ni(C$_2$H$_4$)$_n$ (n = 1, 2, and 3).[304] The first RR spectrum of such a cocondensation product was obtained for Cu(C$_2$H$_4$)$_3$, which exhibits the $\nu(C=C)$ and ν(Cu—C) at 1530 and 302 cm^{-1}, respectively.[305]

Similar studies have been extended to Li(C$_2$H$_2$),[306] Ni(C$_2$H$_2$),[307] Ni(C$_4$H$_6$),[308] (HgCl$_2$)(olefin),[309] and Fe(TPP)(C$_2$H$_4$).[310]

IV-6. CYCLOPENTADIENYL COMPOUNDS

The infrared spectra of cyclopentadienyl (C_5H_5 or Cp) complexes have been reviewed extensively by Fritz.[311] According to Fritz, they are roughly classified into four groups, each of which exhibits its own characteristic spectrum.

(1) Ionic Complexes

These are complexes such as MCp (M = K^+, Rb^+, and Cs^+) and MCp_2 (M = Sr^{2+}, Ba^{2+}, Mn^{2+}, and Eu^{2+}),[312–315] in which M^{n+} and Cp^- are ionically bonded. The spectra of these compounds are essentially the same as that of the $C_5H_5^-$ ion which takes a planar pentagonal structure of D_{5h} symmetry. The 24 (3 × 10 − 6) normal vibrations of this ion are classifed into:

$$2A_1'\ (R) + A_2'\ (i.a.) + A_2''\ (IR) + 3E_1'\ (IR) + E_1''\ (R) + 4E_2'\ (R) + 2E_2''\ (i.a.)$$

Figure IV-16 illustrates the approximate normal modes and observed frequencies (K^+Cp^-) of these vibrations. Four IR-active and seven Raman-active vibrations are expected for the Cp^- ion. In fact, the ionic complexes mentioned above exhibit four IR bands: ν(CH), 3100–3000 cm^{-1}, ν(CC), 1500–1400 cm^{-1}, δ(CH), 1010–1000 cm^{-1}, and π(CH), 750–650 cm^{-1}.

(2) Ion-Paired and Centrally σ-Bonded Complexes

These are complexes such as MCp (M = Li and Na), in which the metal ion is bonded to the center of the ring by forming a tight ion pair, or MCp_2 (M = Be, Mg, and Ca),[312,315–317] in which the metal is covalently bonded to the center of the ring. In this case, the local symmetry of the Cp ring is regarded as C_{5v}, and its 24 normal vibrations are classified into:

$$3A_1\ (IR, R) + A_2\ (i.a.) + 4E_1(IR, R) + 6E_2\ (R)$$

Thus, seven bands are expected to appear in IR spectra. These vibrations are observed in the following regions: ν(CH), 3100–3000 cm^{-1}; ν(CH), 2950–2900 cm^{-1}; ν(CC), 1450–1400 cm^{-1}; ν(CC), 1150–1100 cm^{-1}; δ(CH), 1010–990 cm^{-1}; two π(CH), 890–700 cm^{-1}. In addition, these complexes are expected to show one ν(M—Cp) (A_1) and one ring-tilt (E_1) vibration in the low-frequency region. The former is observed at 426 cm^{-1} for Li^+Cp^- and at 196 cm^{-1} for Na^+Cp^- in IR spectra,[318] and the latter in the 150–130 cm^{-1} region in Raman spectra.[319]

According to electron diffraction studies,[320] the two rings of $SnCp_2$ and $PbCp_2$ form angles of 45° and 55°, respectively, in the vapor state. On the assumption of angular structure in the solid state, two bands at 240 and 170 cm^{-1} of $SnCp_2$ have been assigned to the antisymmetric and symmetric M—Cp

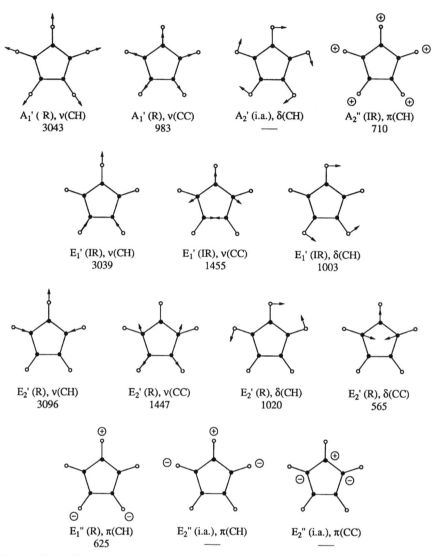

A_1' (R), $v(CH)$
3043

A_1' (R), $v(CC)$
983

A_2' (i.a.), $\delta(CH)$
—

A_2'' (IR), $\pi(CH)$
710

E_1' (IR), $v(CH)$
3039

E_1' (IR), $v(CC)$
1455

E_1' (IR), $\delta(CH)$
1003

E_2' (R), $v(CH)$
3096

E_2' (R), $v(CC)$
1447

E_2' (R), $\delta(CH)$
1020

E_2' (R), $\delta(CC)$
565

E_1'' (R), $\pi(CH)$
625

E_2'' (i.a.), $\pi(CH)$
—

E_2'' (i.a.), $\pi(CC)$
—

Figure IV-16. Normal modes of vibration of the cyclopentadienyl group. These figures are approximate, and only the displacements of the H or C atoms are shown. Symmetry, band assignments, and observed frequencies (cm^{-1}) are given for each mode.

stretching modes, respectively.[321] The IR spectrum of BeCp$_2$ in solution[322] exhibits the bands characteristic of the centrally σ-bonded ring (similar to CpBeCl) and those of the diene-type (σ-bonded) ring (HgCp$_2$) discussed in the later subsection. Thus, a structure such as (A) shown below, has been proposed. X-Ray analysis[323] as well as IR studies[324] show that BeCp$_2$ in the gaseous and solid phases take a "slip-sandwich" structure as shown in B. A highly symmetrical ferrocene-like structure ($\mathbf{D}_{5d}$) and a centrally σ-bonded structure ($\mathbf{C}_{5v}$) can be ruled out because seven IR bands are observed in the $\nu(CH)$ region.

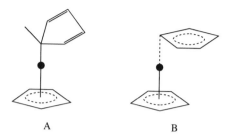

A   B

(3)   Centrally π-Bonded Complexes*

These are complexes such as $FeCp_2$ and $RuCp_2$, in which the transition metals are bonded to the center of the ring via the d-π-bond.

It is interesting to note that two rings in solid ferrocene take the staggered configuration ($\mathbf{D}_{5d}$), while those in ruthenocene take the eclipsed configuration ($\mathbf{D}_{5h}$):

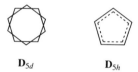

$\mathbf{D}_{5d}$ $\mathbf{D}_{5h}$

Since the number of infrared- or Raman-active fundamentals is the same for both conformations, they cannot be distinguished on the basis of the number of fundamentals observed.

Under $\mathbf{D}_{5d}$ symmetry, the 57 ($3 \times 21 - 6$) normal vibrations of ferrocene are classified into:

$$4A_{1g} \text{ (R)} + 2A_{1u} \text{ (i.a.)} + A_{2g} \text{ (i.a.)} + 4A_{2u} \text{ (IR)} + 5E_{1g} \text{ (R)}$$
$$+ 6E_{1u} \text{ (IR)} + 6E_{2g} \text{ (R)} + 6E_{2u} \text{ (i.a.)}$$

Thus, 10 vibrations are expected to be IR-active. These include the seven Cp bands discussed previously and three skeletal modes (ν_3, ν_5, and ν_6) illustrated in Fig. IV-17. Table IV-12 lists the observed IR frequencies of the centrally π-bonded MCp_2-type complexes, and Fig. IV-18 shows the IR spectra of $NiCp_2$ and $FeCp_2$. As shown in Table IV-12, the IR bands at 492 and 478 cm^{-1} of $FeCp_2$ have been assigned to the ring-tilt (ν_5) and $\nu(M-Cp)$ (ν_3), respectively. Both bands show marked isotope shifts (7–8 cm^{-1}) by $^{54}Fe/^{57}Fe$ substitution.[333]

---

*Or pentahapto (h^5) complexes.

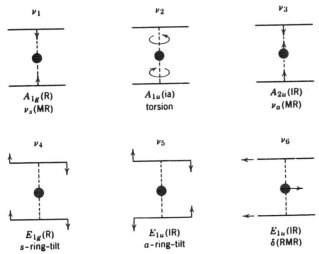

Figure IV-17. Skeletal modes of dicyclopentadienyl metal complexes (D_{5h} symmetry). R denotes the C_5H_5 ring in band assignments.

Lippincott and Nelson[329] carried out normal coordinate analysis on the C_5H_5 ring of ferrocene assuming D_{5h} symmetry. Fritz[311] calculated the approximate M–Cp stretching force constants using the equation available for the antisymmetric stretching vibration of a linear YXY-type molecule:

$$(5.89 \times 10^{-2})\tilde{\nu}_3^2 = \left(1 + \frac{2m_y}{m_x}\right) \frac{k}{m_y}$$

Here, $\tilde{\nu}_3$ is the observed $\nu_a(\text{M—Cp})$ in cm^{-1}, m_x and m_y are the point masses (atomic weight unit) of the metal and the Cp ring, respectively, and k is the

TABLE IV-12. Observed Infrared Frequencies and Band Assignments of Centrally π-Bonded MCp₂-Type Compounds (cm^{-1})

Compound	ν(CH)		ν(CC)		δ(CH)	π(CH)		Ring Tilt	ν(MR)[a]	δ(RMR)[a]	Refs.
FeCp₂	—	3077	1110	1410	1005	820	855	492	478	179	325–327
RuCp₂	—	3076	1095	1410	1005	808	834	450	380	170	327
OsCp₂	3061	3061	1098	1400	998	823	831	428	353	160	311, 328
CoCp₂	3041	3041	1101	1412	995	778	828	464	355	—	311, 328
NiCp₂	3075	3075	1110	1430	1000	773	839	355	355	—	325, 329
FeCp₂⁺	3108	3108	1116	1421	1017	805	860	501	423	—	330
	3100	3100	1110	1412	1001	779	841	490	405		
CoCp₂⁺	3094	3094	1113	1419	1010	860	895	495	455	172	331
IrCp₂⁺	3077	3077	1106	1409	1009	818	862	—	—	—	331

[a]R denotes the Cp ring. For the Raman spectra of MCp₂ (M = Mn, Cr, V, Ru, and Os), see Ref. 332.

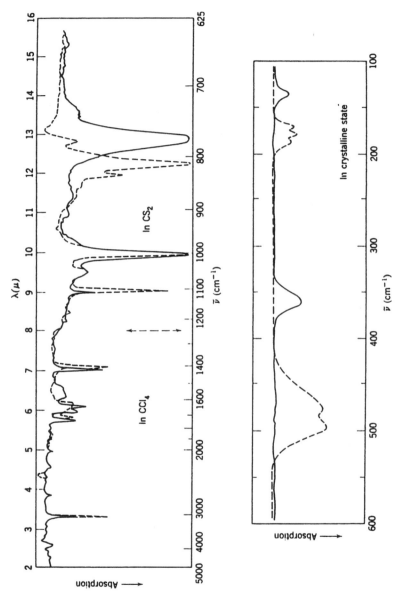

Figure IV-18. Infrared spectra of Ni(C$_5$H$_5$)$_2$ (solid line) and Fe(C$_5$H$_5$)$_2$ (dashed line).[325]

M—Cp stretching force constant in mdyn/Å. The results are:

	Os		Fe		Ru		Cr	Co	V	Ni	Zn
k(M—Cp)(mdyn/Å)	2.8	>	2.7	>	2.4	$\gg$	1.6	~1.5	~1.5	~1.5	~1.5
ν_a(M—Cp) (cm^{-1})	353		478		379		408	355	379	355	345

This may indicate the order of the M—Cp bond strength. Table IV-12 shows that the M—Cp stretching band of FeCp$_2$ at 478 cm^{-1} is shifted to a lower frequency when it is ionized to FeCp$_2^+$. Apparently, the deviation from the inert gas electronic configuration due to the ionization weakens the M—Cp bond. More accurate calculations on M—Cp stretching-force constants of ferrocene and its derivatives have been made by Phillips et al.,[334] who employed two observed frequencies (478 (ν_3) and 306 (ν_1) for ferrocene). The Fe—Cp stretching force constant was 3.11 mdyn/Å with the stretch–stretch interaction constant of 0.48 mdyn/Å. In another approach, Hyams[335] considered five Fe—C bonds between the Fe atom and the Cp ring, and obtained a "pseudoring"–Fe stretching force constant of 1.4 mdyn/Å. Schafer et al.[336,337] carried out the most complete normal coordinate analysis by assuming 10 Fe—C bonds between the Fe atom and the two Cp rings (**D**$_{5h}$ symmetry).

According to Yokoyama et al.,[338] the observed skeletal frequencies of FeCp$_2$ and NiCp$_2$ are as follows (cm^{-1}):

	ν_1 (R)	ν_3 (IR)	ν_4 (R)	ν_5 (IR)
FeCp$_2$	306	476	390	400
NiCp$_2$	245	355	198	355

These workers have explained the marked difference in ν_4 on the basis of their electronic structures. More references are available on vibrational spectra of FeCp$_2$[334,340] and RuCp.[341]

(4) Diene-Type (σ-Bonded) Complexes*

These are complexes such as HgCp$_2$ and CH$_3$HgCp,[342,343] in which the metal is σ-bonded to one of the C atoms of the Cp ring:

The spectra of these compounds are similar to that of C$_5$H$_6$ (cyclopentadiene), and are markedly different from those of the other groups discussed previously.

*Or monohapto (h^1) complexes.

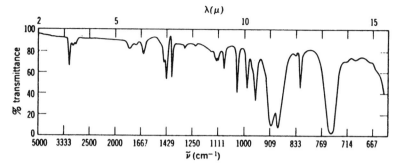

Figure IV-19. Infrared spectrum of $Hg(C_5H_5)_2$ in CS_2 (2–6 μ and 7.1–15.5 μ). $CHCl_3$ (6–6.6 μ), and CCl_4 (6.6–7.1 μ).[344]

Figure IV-19 shows the infrared spectrum of $HgCp_2$.[344] Band assignments of these compounds can be made based on those obtained for C_5H_6.[345] Infrared and NMR evidence suggests the presence of diene-type bonding for $(Cp)M(CH_3)_3$ (M = Si, Ge, and Sn).[346]

There are many other complexes in which the π-bonded (type 3) and the σ-bonded (type 4) cyclopentadienyl groups are mixed. As expected, these compounds exhibit bands characteristic of both groups. Typical examples are as follows: VCp_3 (two π and one σ),[347] $NbCp_4$ (two π and two σ),[348] $ZrCp_4$ (three π and one σ),[349–351] and $MoCp_4$ (three π and one σ).[352] Infrared,[348] X-ray,[350] and NMR[353] evidence indicates the presence of two π- and two σ-bonded Cp rings in $TiCp_4$.

(5) Complexes of Other Types

In addition to the complexes discussed above, recent X-ray studies have revealed the existence of other types. For example, an allylic (or a trihapto, h^3) bonding was found in $[Ni(h^3\text{-}Cp)(C_3H_4)]_2$, whose structure is shown in Fig. IV-20a.[354] In $TiCp_3$, two rings are π-bonded while the third is bonded to the metal through only two adjacent C atoms.[355] It is rather difficult, however, to distinguish these structures from other types by vibrational spectroscopy. In the case of $NbCp_2$, the very unusual structure shown in Fig. IV-20b was suggested from NMR and other evidence.[356] Here the Cp ring serves as a bridge between two metal atoms by forming one π- and one σ-bond.

IV-7. CYCLOPENTADIENYL COMPOUNDS CONTAINING OTHER GROUPS

(1) Carbonyl Compounds

The vibrational spectra of carbonyl compounds were discussed in Sec. III-17. Here we discuss only those containing cyclopentadienyl rings. It has been

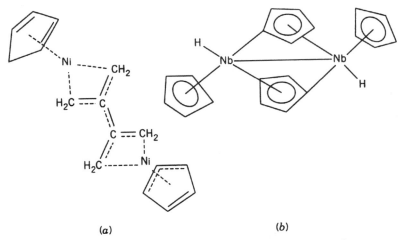

Figure IV-20. Structures of some cyclopentadienyl compounds.

well established that the number of CO stretching bands observed in the infrared depends on the local symmetry of the $M(CO)_n$ group in $M(Cp)_m(CO)_n$-type compounds.[311,357] For example, only two CO stretching bands have been observed for the following compounds, in accordance with the prediction from local symmetry:

C_{2v}	C_{3v}	C_{4v}
$1890(B_2)$	$1938(E)$	$1916(E)$
$1969(A_1)$	$2025(A_1)$	$2016(A_1)$

In the case of $MCp(CO)_3$ (M = Mn and Re), the breakdown of the C_{5v} selection rule for the Cp vibrations was noted in solution IR spectra.[358] The FT Raman spectra of $MnCp(CO)_3$ as well as $Cr(arene)(CO)_3$ are reported.[359] Other references for carbonyl compounds are $[M(Cp)(CO)_3]^-$ (M = Cr, Mo, and W),[360] $Mn(Cp)(CO)_3$,[361] $Re(Cp)(CO)_3$,[362] and $V(Cp)(CO)_4$.[363] In $M(Cp)(CO)_3$-type compounds,[364] the CO stretching frequencies increase in the order $V^{-1} < Cr^0 < Mn^{+1} < Fe^{2+}$. This indicates that the higher the oxidation state of the metal, the less the M—C π-back bonding and the higher the CO stretching frequency.

Originally, Fe(Cp)$_2$(CO)$_2$ was thought to contain two π-bonded Cp rings.[365] However, an infrared and NMR study[366] showed that one ring is π-bonded and the other σ-bonded to the metal. Later, X-ray analysis confirmed this structure.[367] The structure of Fe$_2$(Cp)$_2$(CO)$_4$ has been studied extensively. In the solid state, it takes a *trans*-bridged structure (Fig. IV-21a),[368] or a *cis*-bridged structure (Fig. IV-21b) if crystallized in polar solvents at lower temperatures.[369] The *cis*-isomer exhibits two terminal (1975 and 1933 cm^{-1}) and two bridging (1801 and 1766 cm^{-1}) bands. Although the *trans*-isomer also exhibits two terminal (1956 and 1935 cm^{-1}) and two bridging (1769 and 1755 cm^{-1}) bands, these splittings are probably due to the crystal-field effect.

The structure of Fe$_2$(Cp)$_2$(CO)$_4$ in solution has been controversial. Early infrared studies[370,371] suggested the presence of the *cis*-bridged structure (Fig. IV-21b) mixed with a trace of noncentrosymmetric, nonbridging isomer (Fig.

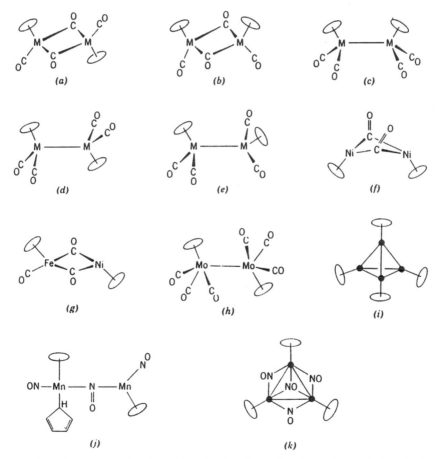

Figure IV-21. Structures of cyclopentadienyl carbonyl and nitrosyl complexes. The bridging CO groups are not shown in (*i*).

IV-21c). Manning[372] proposed, however, an equilibrium involving the three isomers, *a*, *b*, and *c* of Fig. IV-21. This was confirmed by Bullitt et al.,[373] who
gave the following assignments for the spectrum in a $CS_2-C_6D_5CD_3$ solution:
trans-isomer (*a*), 1954 and 1781 cm^{-1}; *cis*-isomer (*b*), 1998, 1954, 1810, and
1777 cm^{-1}. The frequencies of nonbridged species could not be determined
because of their very low concentration. In the case of $Ru_2(Cp)_2(CO)_4$, Bullitt
et al.[373] proposed an equilibrium containing four isomers: *a*, *b*, *d*, and *e* of Fig.
IV-21.

It is interesting to note that the bridging CO groups of $Fe_2(Cp)_2(CO)_4$ form
an adduct with trialkylaluminum[374] (see Sec. III-17):

$$
\begin{array}{c}
AlR_3 \\
| \\
O \\
\diagdown \; \underset{C}{\diagup} \; \diagup \\
Fe \qquad Fe \\
\diagup \; \underset{C}{\diagdown} \; \diagdown \\
O \\
| \\
AlR_3
\end{array}
$$

This indicates that the basicity of the bridging CO group is greater than
that of the terminal CO group. The CO stretching bands of the parent compound (R: isobutyl) are at 2005 and 1962 (terminal) and 1794 (bridging)
cm^{-1} in heptane solution. These bands are shifted to 2041 and 2003 (terminal) and 1680 (bridging) cm^{-1} by adduct formation. X-Ray analysis has been
carried out on $[Fe_2(Cp)_2(CO)_4][Al(C_2H_5)_3]_2$.[375] Formation of adducts such as
$[Fe_2(Cp)_2(CO)_4]BX_3$ (X = Cl and Br) and $[Fe(Cp)(CO)]_4 \cdot BX_3$ (X = F, Cl, and
Br) has also been confirmed.[376] These compounds exhibit bands at 1470–1290
cm^{-1} for bridging CO groups, which are bonded to a Lewis acid via the O
atom.

$Ni_2(Cp)_2(CO)_2$ exhibits two bridging CO stretching bands at 1854 and 1896
cm^{-1} in heptane solution. The structure shown in Fig. IV-21*f* with a puckered $Ni(CO)_2Ni$ bridge was proposed for this compound.[377] In heptane solution
$FeNi(Cp)_2(CO)_3$ shows a strong terminal CO stretching at 2004 cm^{-1} and two
bridging CO stretching bands at 1855 and 1825 cm^{-1}. Since the 1855 cm^{-1}
band (symmetric type) is very weak, the $Ni(CO)_2Fe$ bridge in this compound
was thought to be virtually planar, as shown in Fig. IV-21*g*.[377]

According to X-ray analysis,[378] the structure of $Mo_2(Cp)_2(CO)_6$ is *trans*-
centrosymmetric, as shown in Fig. IV-21*h*. The infrared spectrum in the CO
stretching region is consistent with this structure, both in the solid state and in
solution.[379] In solvents of high dielectric constants, however, the *trans*-rotamer
is rearranged into the *gauche*-rotamer.[380] For the infrared spectra of analogous
tungsten compounds, see Refs. 360 and 381. According to X-ray analysis,[382]
$Fe_4(Cp)_4(CO)_4$ takes a regular tetrahedral structure such as that shown in Fig.

IV-21*i*. It exhibits a bridging CO stretching band at 1649 cm^{-1},[374] in the infrared and a FeFe stretching band at 214 cm^{-1},[383] in the Raman.

(2) Halogeno Compounds

Cyclopentadienyl complexes containing metal–halogen bonds exhibit metal–halogen vibrations (Sec. III-23), together with those of the cyclopentadienyl rings. The low-frequency spectra of these compounds are complicated[384,385] because metal-ring skeletal modes couple with metal–halogen modes. The infrared spectra of $M(Cp)_2X_2$-type compounds (M = Ti, Zr, and Hf; X = Cl, Br, and I) have been studied by several investigators.[384–387] Also, infrared spectra have been reported for $Mo(Cp)(CO)_3X$[358] and $Mo(Cp)$—$(CO)_2X_3$ (X = Cl, Br, and I).[388]

(3) Nitrosyl Compounds

Vibrational spectra of nitrosyl compounds were discussed in Sec. III-18. The vibrational spectra of $Ni(Cp)(NO)$ and its deuterated and ^{15}N species have been assigned completely:[389] the NO stretching, NiN stretching, NiCp stretching, and NiCp tilt vibrations are at 1809, 649, 322, and 290 cm^{-1}, respectively. If this compound in an Ar matrix is irradiated by UV light, the bands near 1830 cm^{-1} disappear and a new band emerges at 1390 cm^{-1}.[390] This has been interpreted as indicating the following photoionization:

$$(Cp)NiNO \xrightarrow{h\nu} (Cp)Ni^+NO^-.$$

$Mn_2(Cp)_3(NO)_3$ exhibits two NO stretching bands at 1732 and 1510 cm^{-1}.[391] With the former attributed to the terminal and the latter to the bridging NO, the structure which is shown in Fig. IV-21*j* was proposed. The infrared spectrum of $Mo(Cp)(CO)_2(NO)$ has been reported.[392,358] Figure IV-21*k* shows the structure of $Mn_3(Cp)_3(NO)_4$, containing doubly and triply bridging NO groups. The bands at 1530 and 1480 cm^{-1} were assigned to the doubly bridged NO groups, whereas the 1320-cm^{-1} band was attributed to the triply bridged NO group.[393]

(4) Hydrido Complexes

Vibrational spectra of hydrido complexes were reviewed in Sec. III-22. The metal–hydrogen stretching bands for $Mo(Cp)_2H_2$,[394] $Re(Cp)_2H_2$, and $W(Cp)_2H_2$[395,396] have been observed in the 2100–1800 cm^{-1} region. X-Ray analysis on $Mo(Cp)_2H_2$[397] suggests that the coordination around the Mo atom is approximately tetrahedral. In polymeric $[Zr(Cp)_2H_2]_n$,[398] the bridging ZnH stretching vibration is observed as a strong, broad band at 1540 cm^{-1}.[399] A similar bridging TiH vibration is found at 1450 cm^{-1} for [Ti(Cp)$_2$H]$_2$.[400] In [{Rh(Cp')}$_2$HCl$_3$], where Cp' denotes the pentamethyl-Cp group, the bridging RhH vibration was assigned at 1151 cm^{-1}.[401]

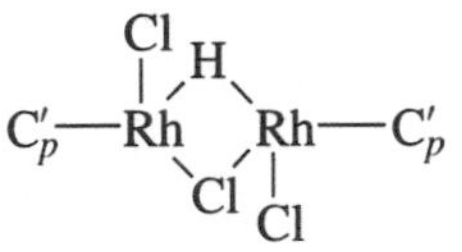

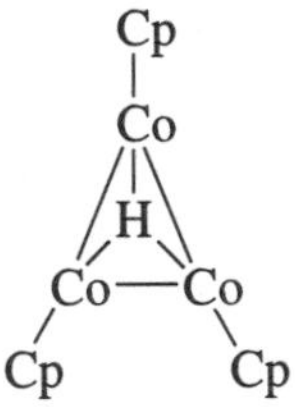

An extremely low CoH stretching frequency (950 cm^{-1}), together with an unusually high-field proton chemical shift observed for [Co(Cp)H]$_4$, was attributed to the triply bridged structure shown above (only one face of the tetrahedron is shown).[402]

(5) Complexes Containing Other Groups

As discussed in Sec. III-16, the mode of coordination of the pseudohalide ion can be determined by vibrational spectroscopy. Burmeister et al.[403] found that all NCS and NCSe groups are N-bonded in M(Cp)$_2$X$_2$-type compounds, where M is Ti, Zr, Hf, or V, and X is NCS or NCSe. In the case of analogous NCO complexes, Ti, Zr, and Hf form O-bonded complexes, whereas V forms an N-bonded complex. Later, Jensen et al.[404] suggested the N-bonded structure for the titanium complex.

The ν(CH) (3311 cm^{-1}) and ν(CN) (2097 cm^{-1}) of free HCN are shifted to 3188 and 2155 cm^{-1}, respectively, when it coordinates to the Ti atom in the [Ti(Cp)$_2$(HCN)$_2$]$^{2+}$ ion. This result has been attributed to the Ti ← HCN σ-donation.[405] The cis- and trans-isomers of the isonitrile complex, MoCp(CO)$_2$(t-BuNC)I, exhibit the ν(NC) at 2153 and 2138 cm^{-1}, respectively.

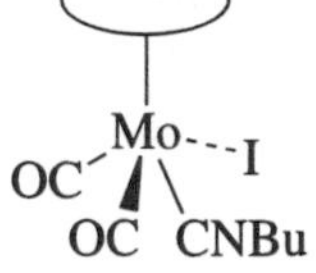

(cis) 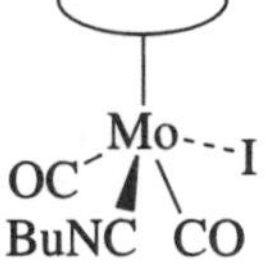   (trans)

The ν(NC) of the trans isomer is lower because its I → CN π-donation may be larger relative to the cis isomer.[406] The ν(NC) of MoCp(CO)$_2$(EtNC)I at 2168 cm^{-1} is shifted to 1869 cm^{-1} in Na[MoCp(CO)$_2$(EtNC)]. The marked downshift of ν(NC) may reflect a lower oxidation state (O) of the Mo atom in the latter since the Mo → CN-Et π-back donation increases as the oxidation state becomes lower.[407]

A strong $N \equiv N$ stretching band is observed at 1910 cm^{-1} in the Raman spectrum of $L_2(Cp)Mo-N \equiv N-Mo(Cp)L_2$ (L: PPh$_3$).[408] Thiocarbonyl complexes of the $(Cp)Mn(CO)_{3-n}(CS)_n$- ($n=1, 2, 3$) type exhibit the $C \equiv S$ stretching bands at 1340–1235 cm^{-1}.[409] In $(Cp)Nb(S_2)X$-type compounds (X = Cl, Br, I, and SCN), the S_2 is probably coordinated to the metal in a side-on fashion, and its SS stretching band may be assigned at 540 cm^{-1}.[410]

IV-8. COMPLEXES OF OTHER CYCLIC UNSATURATED LIGANDS

In addition to those discussed in the preceding sections, there are many other cyclic unsaturated ligands which form π-complexes with transition metals. Some of these complexes are discussed below.

(1)   Complexes of Cyclobutadiene (C$_4$H$_4$)

The local symmetry of the C_4H_4 ring in $Fe(C_4H_4)(CO)_3$ is regarded as C_{4v}, and its 18 ($3 \times 8 - 6$) vibrations are classified into:

$$3A_1 \text{ (IR, R)} + A_2 \text{ (i.a.)} + 4B_1 \text{ (i.a.)} + 2B_2 \text{ (i.a.)} + 4E \text{ (IR, R)}$$

Thus, seven vibrations are IR- as well as Raman-active. Two skeletal modes, ring tilt (E) and $\nu_s(Fe-C_4H_4)$ (A_1) should be added if the Fe atom is included. These vibrations were observed near 475 and 406 cm^{-1}, respectively.[411,412] Normal coordinate analysis on $Fe(C_4H_4)(CO)_3$ was carried out by Andrews and Davidson.[413] The IR spectra of $Ni(C_4(CH_3)_4)Cl_2$ and $M(C_4(C_6H_5)_4)X_2$ (M = Ni and Pd; X = Cl, Br, and I) are reported.[414]

(2)   Cyclopentadiene (C$_5$H$_6$) and Cyclohexadienyl (C$_6$H$_7$) Complexes

Cyclopentadiene forms π-complexes such as $MCp(C_5H_6)$ (M = Co and Rh) in which the two H atoms of the CH$_2$ group exhibit two separate bands; $\nu(CH_{endo})$ and $\nu(CH_{exo})$ at 2750 and 2945 cm^{-1}, respectively.[415]

A later study[416] shows, however, that the lower-frequency band at 2750 cm^{-1} must be assigned to $\nu(CH_{exo})$, since replacement of the exo hydrogen by the phenyl or perfluorophenyl group results in the disappearance of this band. In a cyclohexadienyl complex such as $Mn(C_6H_7)(CO)_3$, the bands at 2970 and 2830 cm^{-1} were assigned to $\nu(CH_{endo})$ and $\nu(CH_{exo})$, respectively.[417]

(3)   Complexes of Benzene (C_6H_6)

Under D_{6h} symmetry, the 30 ($3 \times 12 - 6$) normal vibrations of benzene are classified into:

$$2A_{1g} \text{ (R)} + A_{2g} \text{ (i.a.)} + A_{2u} \text{ (IR)} + 2B_{1u} \text{ (i.a.)} + 2B_{2g} \text{ (i.a.)} + 2B_{2u} \text{ (i.a.)}$$

$$+ E_{1g} \text{ (R)} + 3E_{1u} \text{ (IR)} + 4E_{2g} \text{ (R)} + 2E_{2u} \text{ (i.a.)}$$

Thus only four vibrations are IR-active and only seven vibrations are Raman-active. Figure IV-22 shows the normal modes and observed frequencies of these vibrations.

Dibenzene chromium, $Cr(C_6H_6)_2$, takes a ferrocene-like sandwich structure of D_{6h} symmetry (eclipsed form). Then, its 69 ($3 \times 25 - 6$) normal vibrations are grouped into:

$$4A_{1g} + 2A_{1u} + A_{2g} + 4A_{2u} + 2B_{1g} + 4B_{1u} + 4B_{2g} + 2B_{2u} + 5E_{1g} + 6E_{1u} + 6E_{2g} + 6E_{2u}$$

of which only A_{2u} and E_{1u} vibrations are IR-active and A_{1g}, E_{1g}, and E_{2g} vibrations are Raman-active. Thus 10- ($4 A_{2u}$ and 6 E_{1u}) bands should be observed in IR spectra. Table IV-13 lists these 10 frequencies including the ring tilt, $\nu_s(Cr{-}C_6H_6)$, and $\delta(C_6H_6{-}Cr{-}C_6H_6)$. Figure IV-23 shows the IR spectrum of $Cr(C_6H_6)_2$. Using the same approximation as used previously for ferrocene (Sec. IV-6(3)), the $Cr{-}C_6H_6$ stretching force constant of $Cr(C_6H_6)_2$ is calculated to be 2.43 mdyn/Å, which is smaller than that of ferrocene (2.7 mdyn/Å). Cyvin et al.[420] carried out normal coordinate analysis on $Cr(C_6H_6)_2$.

Infrared spectra have been reported for $V(C_6H_6)_2$[421] and $Fe(C_6H_6)_2$[422] which were prepared by matrix cocondensation techniques (Sec. IV-5(6)). Resonance Raman spectra of $V(C_6H_6)_2$ thus prepared exhibit the progression of the $\nu(V{-}C_6H_6)$ (257 cm^{-1}) up to the ninth overtone.[423] The CH out-of-plane vibrations of aromatic rings tend to shift to higher frequencies by forming sandwich complexes. Saito et al.[424] noted, however, that some vibrations are upshifted, while others are downshifted, when benzene forms the sandwich complex $[Cr(C_6H_6)_2]I$.

The IR and Raman spectra of $Cr(C_6H_6)(C_6F_6)$ (C_{6v} symmetry) have been assigned empirically.[425] Normal coordinate analyses of $Cr(C_6H_6)(CO)_3$ (C_{3v} symmetry) have been made by two groups of investigators.[426,427] The UV photolysis of $Cr(C_6H_6)(CO)_3$ in the gas phase produces $Cr(C_6H_6)(CO)_{1,2}$, which is characterized by $\nu(CO)$; $Cr(C_6H_6)(CO)_2$ is predominant upon 355 nm photolysis.[428]

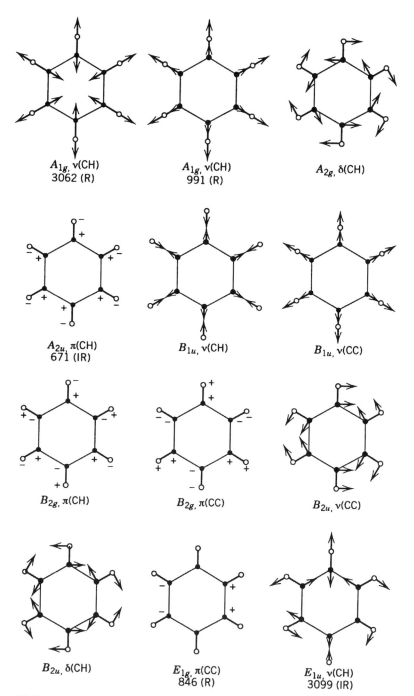

Figure IV-22. Approximate normal modes of vibration of benzene. Symmetry, band, assignments, and observed frequencies (cm^{-1}) of representative modes are given.

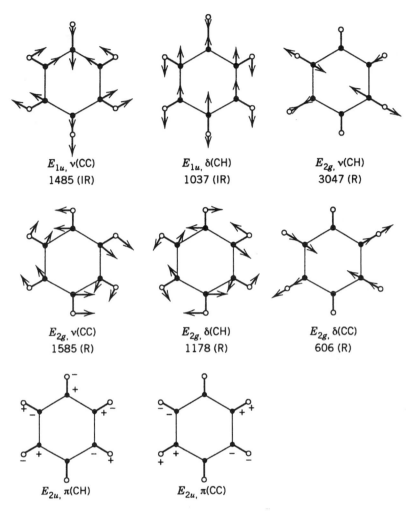

E_{1u}, ν(CC)
1485 (IR)

E_{1u}, δ(CH)
1037 (IR)

E_{2g}, ν(CH)
3047 (R)

E_{2g}, ν(CC)
1585 (R)

E_{2g}, δ(CH)
1178 (R)

E_{2g}, δ(CC)
606 (R)

E_{2u}, π(CH)

E_{2u}, π(CC)

Figure IV-22. (*Continued*)

TABLE IV-13. Infrared Frequencies of Dibenzene–Metal Complexes (cm^{-1})[311,418]

Complex	ν(CH)		ν(CC)	δ(CH)	ν(CC)	π(CH)		Ring Tilt	ν(MR)^a	δ(RMR)^a
Cr(C$_6$H$_6$)$_2$	3037	—	1426	999	971	833	794	490	459	(140)
Cr(C$_6$H$_6$)$_2^+$	3040	—	1430	1000	972	857	795	466	415	(144)
Mo(C$_6$H$_6$)$_2$	3030	2916	1425	995	966	811	773	424	362	—
W(C$_6$H$_6$)$_2$	3012	2898	1412	985	963	882	798	386	331	—
V(C$_6$H$_6$)$_2$	3058	—	1416	985	959	818	739	470	424	—

^{a}R denotes the C$_6$H$_6$ ring.

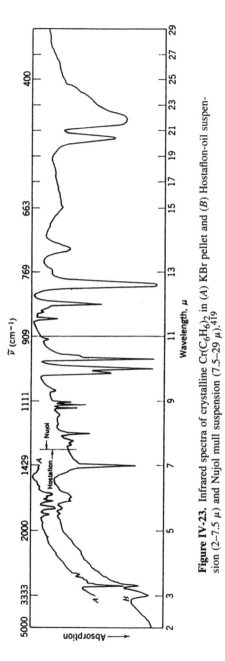

Figure IV-23. Infrared spectra of crystalline $Cr(C_6H_6)_2$ in (*A*) KBr pellet and (*B*) Hostaflon-oil suspension (2–7.5 μ) and Nujol mull suspension (7.5–29 μ).[419]

(4)   Tropylium Cation ($C_7H_7^+$) and π-C_7H_7 Metal Complexes

Under $\mathbf{D}_{7h}$ symmetry, the 36 ($3 \times 14 - 6$) normal vibrations of the planar tropylium cation are classified into:

$$2A_1' + A_2' + A_2'' + 3E_1' + E_1'' + 4E_2' + 2E_2'' + 4E_3' + 2E_3''$$

of which A_2'' and E_1' vibrations are IR-active, whereas A_1', E_1'', and E_2' vibrations are Raman-active. Thus, four IR and seven Raman bands are expected for the $C_7H_7^+$ cation. The four IR bands of $(C_7H_7)Br$ are observed at 3020 ($\nu(CH)$), 1477 ($\nu(CC)$), 992 ($\delta(CH)$), and 633 cm^{-1} ($\pi(CH)$).[429]

The IR frequencies of several metal π-complexes such as $M(C_7H_7)(CO)_3$ (M = Cr and Mo) are summarized by Fritz.[311] In these complexes, the symmetry of the $M(C_7H_7)$ moiety is regarded as $\mathbf{C}_{7v}$, and its 39 ($3 \times 15 - 6$) vibrations are grouped into:

$$4A_1 \text{ (IR, R)} + A_2 \text{ (i.a.)} + 5E_1 \text{ (IR, R)} + 6E_2 \text{ (R)} + 6E_3 \text{ (i.a.)}$$

Thus, nine vibrations are IR-active while 15 vibrations are Raman-active. These vibrations include the ring tilt (E_1) and $\nu_s(M{-}C_7H_7)$ (A_1) which are IR- as well as Raman-active. The Raman spectrum of $[Mo(C_7H_7(CO)_3]BF_4$ in the solid state exhibits the ring tilt at 331 and 324 cm^{-1}, and the $\nu_s(Mo{-}C_7H_7)$ at 309 cm^{-1}, with two shoulder bands at 302 and 295 cm^{-1}. The observed splitting of the former is due to lowering of symmetry in the crystalline state (site symmetry, $\mathbf{C}_1$).[430]

(5)   Complexes of Cyclooctadienyl Anion ($C_8H_8^{2-}$)

The $C_8H_8^{2-}$ ion takes an octagonal planar structure of $\mathbf{D}_{8h}$ symmetry, and its 42 ($3 \times 16 - 6$) vibrations are grouped into:

$$2A_{1g} + A_{2g} + A_{2u} + 2B_{1u} + 2B_{2g} + 2B_{2u} + E_{1g} + 3E_{1u} + 4E_{2g} + 2E_{2u} + 4E_{3g} + 2E_{3u}$$

of which A_{2u} and E_{1u} vibrations are IR-active, whereas A_{1g}, E_{1g}, and E_{2g} vibrations are Raman-active. Thus, four vibrations are IR-active and seven vibrations are Raman-active. The former bands of $K_2(C_8H_8)$ are observed at 2994 ($\nu(CH)$), 1431 ($\nu(CC)$), 880 ($\delta(CH)$), and 684 cm^{-1} ($\pi(CH)$).[311] Metal complexes such as $M(C_8H_8)_2$ (M = Th and U) take sandwich structures similar to that of ferrocene, but the two rings are eclipsed so that the overall symmetry becomes $\mathbf{D}_{8h}$. In this case, the 93 ($3 \times 33 - 6$) normal vibrations are classified as:

$$4A_{1g} + 2A_{1u} + A_{2g} + 4A_{2u} + 2B_{1g} + 4B_{1u} + 4B_{2g} + 2B_{2u} + 5E_{1g} + 6E_{1u}$$
$$+ 6E_{2g} + 6E_{2u} + 6E_{3g} + 6E_{3u}$$

Then, 10 vibrations ($4 A_{2u} + 6 E_{1u}$) are IR-active and 15 vibrations ($4 A_{1g} + 5 E_{1g} + 6 E_{2g}$) are Raman-active. The IR spectra of biscyclooctadienyl complexes mentioned above have been assigned by Hocks et al.[431] The ring tilt (E_{1u}) and ν_a(M–C_8H_8) (A_{2u}) of these complexes are observed at 695(698) and 250(240) cm^{-1}, respectively (the numbers in parentheses are for the uranium complex).

In Ti(C_8H_8)$_2$, however, one ring is symmetrically bonded (local symmetry, C_{8h}), while the other is asymmetrically bonded to the metal (local symmetry, C_s). Under C_{8h} symmetry, the 45 ($3 \times 17 - 6$) vibrations of the M(C_8H_8) moiety are classified into:

$$4A_1 + A_2 + 2B_1 + 4B_2 + 5E_1 + 6E_2 + 6E_3$$

Then, nine ($4A_1 + 5E_1$) vibrations are IR-active, whereas 15 ($4A_1 + 5E_1 + 6E_2$) vibrations are Raman-active. The IR spectra of M(C_8H_8)$_2$ (M = Ti and V) have been assigned partly on this basis.[431] Similar assignments can be made for the Ti(C_8H_8) moiety of Ti(C_8H_8)(C_5H_5).[432] The IR spectra of K[Ln(C_8H_8)$_2$] (Ln = Ce, Pr, Nd, and Sm) can be assigned on the basis of the sandwich structure (D_{8h}).[433]

Cyclooctatetrane (COT) takes a tub conformation in the free state. As discussed in Sec. IV-5(3), it takes a tub conformation in [Rh(COT)Cl]$_2$ and a chair conformation in Fe(COT)(CO)$_3$.

(6) Indenyl Complexes

The indenyl group may coordinate to the metal through a σ- or a π-bond:

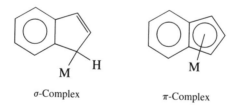

σ-Complex π-Complex

An example of the former is seen in Hg(C_9H_7)Cl, which exhibits an aromatic CH stretching at 3060–3050 and an aliphatic CH stretching band at 2920–2850 cm^{-1}. The latter band should be absent in the π-bonded complex.[434]

The IR spectra of π-bonded sandwich complexes such as Ru(C_9H_7)$_2$ (fully eclipsed) and Fe(C_9H_7)$_2$ (staggered) have been assigned. No appreciable differences were noted between these two complexes in the low-frequency region.[434]

(7) Complexes of Larger Ligands

Infrared spectra are reported for a mixed-valence-state complex, biferrocene (Fe^{2+}, Fe^{3+}) picrate[435] and bis(pentalenyl)Ni,[436] whose structures are shown in

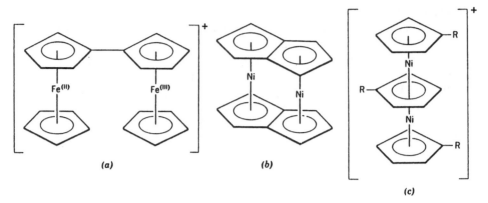

Figure IV-24. Structures of some metal sandwich compounds.

Fig. IV-24*a* and IV-24*b*, respectively. The two ferrocene moieties in the former are not independent since only one set of the skeletal modes is observed. The spectrum of a triple-decker compound, $[Ni_2(Cp')_3]BF_4(Cp': CH_3\text{-}Cp)$ (Fig. IV-24*c*), is similar to that of $Ni(Cp')_2$.[437] In the case of the $[Ni_2Cp_3]^+$ ion, the skeletal frequencies of the Ni–Cp (terminal) moiety was found to be 25–10 cm^{-1} higher than those of the Ni–Cp (bridging) moiety.[438]

IV-9. MISCELLANEOUS COMPOUNDS

There are many other organometallic compounds which have not been covered in the preceding sections. For these, the reader should consult general references cited previously. Other review articles on specific groups of compounds are listed below:

Alkyldiboranes: W. J. Lehmann and I. Shapiro, *Spectrochim. Acta,* **17,** 396 (1961).

Organoaluminum Compounds: E. G. Hoffman, *Z. Elektrochem;* **64,** 616 (1960).

Organosilicon Compounds: A. L. Smith, *Spectrochim. Acta,* **16,** 87 (1960).

Organogermanes: R. J. Cross and R. Glockling, *J. Organomet. Chem.,* **3,** 146 (1965).

Organotin Compounds: R. Okawara and W. Wada, *Adv. Organomet. Chem.,* **5,** 137 (1967).

Organophosphorus Compounds: D. E. C. Corbridge, *The Structural Chemistry of Phosphorus,* Elsevier, Amsterdam, 1974; L. C. Thomas, *Interpretation of the Infrared Spectra of Organophosphorus Compounds,* Heyden, London, 1974.

Organometallic Compounds of P, As, Sb, and Bi: E. Maslowsky, Jr., *J. Organomet. Chem,* **70,** 153 (1974).

REFERENCES

1. *Spectroscopic Properties of Inorganic and Organometallic Compounds*, Vol. 1 to present. The Chemical Society, London.
2. K. Nakamoto, "Characterization of Organometallic Compounds by Infrared Spectroscopy," in M. Tsutsui, ed., *Characterization of Organometallic Compounds*, Part I, Wiley, New York, 1969.
3. E. Maslowsky, Jr., *Vibrational Spectra of Organometallic Compounds*, Wiley, New York, 1977.
4. D. K. Higgins and H. D. Kaesz, "Use of Infrared and Raman Spectroscopy in the Study of Organometallic Compounds," in H. Reiss, ed., *Progress in Solid State Chemistry*, Vol. I, Macmillan, New York, 1964.
5. L. J. Bellamy, *The Infrared Spectra of Complex Molecules*, 3rd ed., Chapman and Hall, London, Vol. 1 (1975); Vol. 2 (1980).
6. D. Lin-Vien, N. B. Colthup, W. G. Fateley, and J. G. Grasselli, *The Handbook of Infrared and Raman Characteristic Frequencies of Organic Molecules*, Academic Press, San Diego, CA, 1991.
7. T. Shimanouchi, Tables of Molecular Vibrational Frequencies, Consolidated Volume, *Natl. Stand. Ref. Data Ser., U. S. Natl. Bur. Stand.*, **39**, June (1972).
8. B. Schrader, *Raman/IR Atlas of Organic Compounds*, VCH, New York, 1989.
9. *The Sadtler Standard Spectra*, Sadtler Research Laboratories, Division of Bio-Rad Laboratories, Inc. Philadelphia, PA.
10. N. N. Greenwood, E. J. F. Ross, and B. P. Straughan, *Index of Vibrational Spectra of Inorganic and Organometallic Compounds*, Vols. 1-3, Butterworth, London, 1972-1977.
11. S. C. Chang, Z. H. Kafafi, R. H. Hauge, W. E. Billups, and J. L. Margrave, *J. Am. Chem. Soc.*, **109**, 4508 (1987).
12. S. C. Chang, R. H. Hauge, Z. H. Kafafi, J. L. Margrave, and W. E. Billups, *J. Am. Chem. Soc.*, **110**, 7975 (1988).
13. W. E. Billups, S. C. Chang, R. H. Hauge, and J. L. Margrave, *Inorg. Chem.*, **32**, 1529 (1993).
14. K. Burczyk and A. J. Downs, *J. Chem. Soc., Dalton Trans.*, 2351 (1990).
15. E. Maslowsky, Jr., *Chem. Soc. Rev.*, **9**, 25 (1980).
16. A. M. Pyndyk, M. R. Aliev, and V. T. Aleksanyan, *Opt. Spectrosc. (Engl. Transl.)*, **36**, 393 (1974).
17. D. Fischer, K. Klostermann, and K.-L. Oehme, *J. Raman Spectrosc.*, **22**, 19 (1990).
18. W. F. Edgell and C. H. Ward, *J. Am. Chem. Soc.*, **77**, 6486 (1955).
19. W. F. Edgell and C. H. Ward, *J. Mol. Spectrosc.*, **8**, 343 (1962).
20. K. Shimizu and H. Murata, *J. Mol. Spectrosc.*, **5**, 44 (1960).
21. R. A. Kovar and G. L. Morgan, *Inorg. Chem.*, **8**, 1099 (1969).
22. I. S. Butler and M. L. Newbury, *Spectrochim. Acta*, **33A**, 669 (1977).
23. B. Nagel and W. Brüser, *Z. Anorg. Allg. Chem.*, **468**, 148 (1980).
24. J. R. Durig and S. C. Brown, *J. Mol. Spectrosc.*, **45**, 338 (1973).
25. C. Kippels, W. Thiel, D. C. McKean, and A. M. Coats, *Spectrochim. Acta*, **48A**, 1067 (1992).

26. M. A. Bochmann, M. A. Chesters, A. R. Coleman, R. Grinter, and D. R. Linder, *Spectrochim. Acta*, **48A,** 1173 (1992).

27. M. J. Almond, C. E. Jenkins, D. A. Rice, and C. A. Yates, *J. Mol. Struct.*, **222,** 219 (1990).

28. J. R. Allkins and P. L. Hendra, *Spectrochim. Acta*, **22,** 2075 (1966).

29. J. R. Durig, C. M. Plater, Jr., J. Bragin, and Y. S. Li, *J. Chem. Phys.*, **55,** 2895 (1971).

30. K. Aida, Y. Yamaguchi, and Y. Imai, *Spectrochim. Acta*, **43A,** 837 (1987).

31. K. Hamada and H. Morishita, *J. Mol. Struct.*, **44,** 119 (1978).

32. W. J. Lehmann, C. O. Wilson, and I. Shapiro, *J. Chem. Phys.*, **31,** 1071 (1959).

33. L. A. Woodward, J. R. Hall, R. N. Nixon and N. Sheppard, *Spectrochim. Acta*, **15,** 249 (1959).

34. H. J. Becher and F. Bramsiepe, *Spectrochim. Acta*, **35A,** 53 (1979).

35. R. J. O'Brien and G. A. Ozin, *J. Chem. Soc. A*, 1136 (1971).

36. S. Kvisle and E. Rytter, *Spectrochim. Acta*, **40A,** 939 (1984).

37. G. A. Atiyah, A. S. Grady, D. K. Russell, and T. A. Claxton, *Spectrochim. Acta*, **47A,** 467 (1991).

38. H. G. E. Coates and A. J. Downs, *J. Chem. Soc.*, 3353 (1964).

39. J. R. Hall, L. A. Woodward, and E. A. V. Ebsworth, *Spectrochim. Acta*, **20,** 1249 (1964).

40. J. R. Durig and K. K. Chatterjee, *J. Raman Spectrosc.*, **11,** 168 (1981).

41. Z. S. Huang, C. Park, and T. J. Anderson, *J. Organomet. Chem.*, **449,** 77 (1993).

42. A. J. Blake and S. Cradock, *J. Chem. Soc., Dalton Trans.*, 2393 (1990).

43. G. Bouquet and M. Bigorgne, *Spectrochim. Acta*, **23A,** 1231 (1967).

44. P. J. D. Park and P. J. Hendra, *Spectrochim. Acta*, **24A,** 2081 (1968).

45. H. Rojhantalab, J. W. Nibler, and C. J. Wilkins, *Spectrochim. Acta*, **32A,** 519 (1976).

46. H. Rojhantalab and J. W. Nibler, *Spectrochim. Acta*, **32A,** 947 (1976).

47. H. Siebert, *Z. Anorg. Allg. Chem.*, **273,** 161 (1953).

48. S. Sportouch, C. Lacoste, and R. Gaufrès, *J. Mol. Struct.*, **9,** 119 (1971).

49. F. Watari, *Spectrochim. Acta*, **34A,** 1239 (1978).

50. G. A. Crowder, G. Gorin, F. H. Kruse, and D. W. Scott, *J. Mol. Spectrosc.*, **16,** 115 (1965).

51. H. H. Eysel, H. Siebert, G. Groh, and H. J. Berthold, *Spectrochim. Acta*, **26A,** 1595 (1970).

52. A. J. Downs, R. Schmutzler, and I. A. Steer, *Chem. Commun.*, 221 (1966).

53. A. J. Shortland and G. Wilkinson, *J. Chem. Soc., Dalton Trans.*, 872 (1973).

54. J. W. Nibler and T. H. Cook, *J. Chem. Phys.*, **58,** 1596 (1973).

55. C. W. Hobbs and R. S. Tobias, *Inorg. Chem.*, **9,** 1998 (1970).

56. P. L. Goggin and L. A. Woodward, *Trans. Faraday Soc.*, **56,** 1591 (1960).

57. G. B. Deacon and J. H. S. Green, *Spectrochim. Acta*, **24A,** 885 (1968).

58. M. G. Miles, J. H. Patterson, C. W. Hobbs, M. J. Hopper, J. Overend, and R. S. Tobias, *Inorg. Chem.*, **7,** 1721 (1968).

59. H. Kriegsmann and S. Pischtschan, *Z. Anorg. Allg. Chem.*, **308,** 212 (1961).

60. A. J. Downs and I. A. Steer, *J. Organomet. Chem.*, **8,** P21 (1967).

61. K. J. Wynne and J. W. George, *J. Am. Chem. Soc.*, **91,** 1649 (1969).

62. M. T. Chen and J. W. George, *J. Am. Chem. Soc.*, **90,** 4580 (1968).

63. J. A. Creighton, G. B. Deacon, and J. H. S. Green, *Aust. J. Chem.*, **20,** 583 (1967).

64. R. Baumgärtner, W. Sawodny, and J. Goubeau, *Z. Anorg. Allg. Chem.*, **333,** 171 (1964).

65. E. Weiss and E. A. Lucken, *J. Organomet. Chem.*, **2,** 197 (1964).

66. R. West and W. Glaze, *J. Am. Chem. Soc.*, **83,** 3580 (1961).

67. P. Krohmer and J. Goubeau, *Z. Anorg. Allg. Chem.*, **369,** 238 (1969).

68. T. Ogawa, K. Hirota, and T. Miyazawa, *Bull. Chem. Soc. Jpn.*, **38,** 1105 (1965).

69. T. Ogawa, *Spectrochim. Acta*, **23A,** 15 (1968).

70. J. Yamamoto and C. A. Wilkie, *Inorg. Chem.*, **10,** 1129 (1971).

71. J. Mink and B. Gellai, *J. Organomet. Chem.*, **66,** 1 (1974).

72. S. Inoue and T. Yamada, *J. Organomet. Chem.*, **25,** 1 (1970).

73. B. Nagel and W. Brüser, *Z. Anorg. Allg. Chem.*, **468,** 148 (1980).

74. J. Mink and Y. A. Pentin *J. Organomet. Chem.*, **23,** 293 (1970).

75. J. L. Bribes and R. Gaufrès, *Spectrochim. Acta*, **27A,** 2133 (1971).

76. W. J. Lehmann, C. O. Wilson, and I. Shapiro, *J. Chem. Phys.*, **28,** 781 (1958).

77. J. Chouteau, G. Davidovics, F. D'Amato, and L. Savidan, *C. R. Hebd. Seances Acad. Sci.*, **260,** 2759 (1965).

78. J. H. S. Green, *Spectrochim. Acta*, **24A,** 137 (1968).

79. C. Crocker and P. L. Goggin, *J. Chem. Soc., Dalton Trans.*, 388 (1978).

80. A. E. Borisov, N. V. Novikova, N. A. Chumaevski, and E. B. Shkirtil, *Dokl. Akad. Nauk SSSR*, **173,** 855 (1967).

81. R. L. McKenney and H. H. Sisler, *Inorg. Chem.*, **6,** 1178 (1967).

82. J. A. Jackson and R. J. Nielson, *J. Mol. Spectrosc.*, **14,** 320 (1964).

83. M. I. Batuev, A. D. Petrov, V. A. Ponomarenko, and A. D. Mateeva, *Izv. Akad. Nauk SSSR, Otd. Kim. Nauk*, 1070 (1956).

84. L. A. Leites, Y. P. Egorov, J. Y. Zueva, and V. A. Ponomarenko, *Izv. Akad. Nauk SSSR, Otd. Kim. Nauk*, 2132 (1961).

85. W. R. Cullen, G. B. Deacon, and J. H. S. Green, *Can. J. Chem.*, **43,** 3193 (1965).

86. P. Taimasalu and J. L. Wood, *Trans. Faraday Soc.*, **59,** 1754 (1963).

87. C. R. Dillard and J. R. Lawson, *J. Opt. Soc. Am.*, **50,** 1271 (1960).

88. T. L. Brown, R. L. Gerteis, D. A. Bafus, and J. A. Ladd, *J. Am. Chem. Soc.*, **86,** 2134 (1964); T. L. Brown, J. A. Ladd, and C. N. Newmann, *J. Organomet. Chem.*, **3,** 1 (1965).

89. H. Dietrich, *Acta Crystallogr.*, **16,** 681 (1963).

90. T. L. Brown, in F. G. A. Stone and R. West, eds., *Advances in Organometallic Chemistry*, Vol. 3, Academic Press, New York, 1965, p. 374.

91. C. N. Atam, H. Müller, and K. Dehnicke, *J. Organomet. Chem.*, **37,** 15 (1972).

92. J. Kress and A. Novak, *J. Organomet. Chem.*, **121,** 7 (1976).

93. E. G. Hoffmann, *Z. Elektrochem.*, **64,** 616 (1960).

94. O. Yamamoto, *Bull. Chem. Soc. Jpn.*, **35,** 619 (1962).

95. C. A. Wilkie, *J. Organomet. Chem.*, **32,** 161 (1971).

96. R. J. Cross and F. Glocking, *J. Organomet. Chem.*, **3,** 146 (1965).

97. W. M. Scovell, B. Y. Kimura, and T. G. Spiro, *J. Coord. Chem.*, **1,** 107 (1971).

98. K. H. Thiele, S. Wilcke, and M. Ehrhardt, *J. Organomet. Chem.*, **14,** 13 (1968).

99. B. Busch and K. Dehnicke, *J. Organomet. Chem.*, **67,** 237 (1974).

100. L. Czuchajowski, J. Habdas, S. A. Kucharski, and K. Rogosz, *J. Organomet. Chem.*, **155,** 185 (1978).

101. A. H. Cowley, J. L. Mills, T. M. Leohr, and T. V. Long, *J. Am. Chem. Soc.*, **93,** 2150 (1971).

102. P. W. Jolly, M. I. Bruce, and P. G. A. Stone, *J. Chem. Soc.*, 5830 (1965).

103. J. Mink and Y. A. Pentin, *Acta Chim. Acad. Sci. Hung.*, **66,** 277 (1970).

104. A. K. Holliday. W. Reade, K. R. Seddon, and I. A. Steer, *J. Organomet. Chem.*, **67,** 1 (1974).

105. J. D. Odom, L. W. Hall, S. Riethmiller, and J. R. Durig, *Inorg. Chem.*, **13,** 170 (1974).

106. D. C. Andrews and G. Davidson, *J. Organomet. Chem.*, **36,** 349 (1972).

107. D. C. Andrews and G. Davidson, *J. Organomet. Chem.*, **76,** 373 (1974).

108. G. Davidson and S. Phillips, *Spectrochim. Acta*, **34A,** 949 (1978).

109. G. Davidson, *Spectrochim. Acta*, **27A,** 1161 (1971).

110. G. Masetti and G. Zerbi, *Spectrochim. Acta*, **26A,** 1891 (1970).

111. U. Kunze, E. Lindner, and J. Koola, *J. Organomet. Chem.*, **57,** 319 (1973).

112. R. S. P. Coutts and P. C. Wailes, *Aust. J. Chem.*, **22,** 1547 (1969).

113. C. L. Sloan and W. A. Barber, *J. Am. Chem. Soc.*, **81,** 1364 (1959).

114. D. I. Maclean and R. E. Sacher, *J. Organomet. Chem.*, **74,** 197 (1974).

115. J. G. Contreras and D. G. Tuck, *Inorg. Chem.*, **12,** 2596 (1973).

116. J. G. Contreras and D. G. Tuck, *J. Organomet. Chem.*, **66,** 405 (1974).

117. P. Krommes and J. Lorberth, *J. Organomet. Chem.*, **88,** 329 (1975).

118. W. M. A. Smit and G. Dijkstra, *J. Mol. Struct.*, **8,** 263 (1971).

119. A. N. Nesmeyanov, A. E. Borisov, N. V. Novikova, and E. I. Fedin, *J. Organomet. Chem.*, **15,** 279 (1968).

120. S. L. Stafford and F. G. A. Stone, *Spectrochim. Acta*, **17,** 412 (1961).

121. G. Davidson, P. G. Harrison, and E. M. Riley, *Spectrochim. Acta*, **29A,** 1265 (1973).

122. G. Davidson and S. Phillips, *Spectrochim. Acta*, **35A,** 83 (1979).

123. J. Mink and Y. A. Pentin, *J. Organomet. Chem.*, **23,** 293 (1970).

124. C. Souresseau and B. Pasquier, *J. Organomet. Chem.*, **39,** 51 (1972).

125. D. H. Whiffen, *J. Chem. Soc.*, 1350 (1956).

126. J. H. S. Green, *Spectrochim. Acta*, **24A,** 863 (1968).

127. J. Mink, G. Végh, and Y. A. Pentin, *J. Organomet. Chem.*, **35,** 225 (1972).

128. C. G. Barraclough, G. E. Berkovic, and G. B. Deacon, *Aust. J. Chem.*, **30,** 1905 (1977).

129. G. Costa, A. Camus, N. Marsich, and L. Gatti, *J. Organomet. Chem.*, **8,** 339 (1967).

130. A. N. Rodionov, N. I. Rucheva, I. M. Viktorova, D. N. Shigorin, N. I. Sheverdina, and K. A. Kocheshkov, *Izv. Akad. Nauk SSSR, Ser. Khim.*, 1047 (1969).

131. H. F. Shurvell, *Spectrochim. Acta*, **23A,** 2925 (1967).

132. N. Kumar, B. L. Kalsotra, and R. K. Multani, *J. Inorg. Nucl. Chem.*, **35,** 3019 (1973).

133. A. L. Smith, *Spectrochim. Acta*, **23A,** 1075 (1967).

134. A. L. Smith, *Spectrochim. Acta*, **24A,** 695 (1968).

135. J. R. Durig, C. W. Sink, and J. B. Turner, *Spectrochim. Acta*, **26A,** 557 (1970).

136. D. H. Brown, A. Mohammed, and D. W. A. Sharp, *Spectrochim. Acta*, **21,** 663 (1965).

137. N. S. Dance, W. R. McWhinnie, and R. C. Poller, *J. Chem. Soc., Dalton Trans.*, 2349 (1976).

138. K. Shobatake, C. Postmus, J. R. Ferraro, and K. Nakamoto, *Appl. Spectrosc.*, **23,** 12 (1969).

139. K. Cavanaugh and D. F. Evans, *J. Chem. Soc. A*, 2890 (1969).

140. D. C. McKean, G. P. McQuillan, I. Torto, N. C. Bednell, A. J. Downs, and J. M. Dickinson, *J. Mol. Struct.*, **247,** 73 (1991).

141. E. Rytter and S. Kvisle, *Inorg. Chem.*, **25,** 3796 (1986).

142. H. Kriegsmann, *Z. Anorg. Allg. Chem.*, **294,** 113 (1958).

143. H. Bürger, *Spectrochim. Acta*, **24A,** 2015 (1968).

144. K. Licht and P. Koehler, *Z. Anorg. Allg. Chem.*, **383,** 174 (1971).

145. J. R. Durig, K. K. Lau, J. B. Turner, and J. Bragin, *J. Mol. Spectrosc.*, **31,** 419 (1971).

146. B. A. Nevett and A. Perry, *J. Organomet. Chem.*, **71,** 399 (1974).

147. R. G. Goel, E. Maslowsky, Jr., and C. V. Senoff, *Inorg. Chem.*, **10,** 2572 (1971).

148. W. M. Scovell and R. S. Tobias, *Inorg. Chem.*, **9,** 945 (1970).

149. B. P. Asthana and C. M. Pathak, *Spectrochim. Acta*, **41A,** 595 (1985).

150. B. P. Asthana and C. M. Pathak, *Spectrochim. Acta*, **41A,** 1235 (1985).

151. R. G. Goel and H. S. Prasad, *Spectrochim. Acta*, **32A,** 569 (1976).

152. D. Tudela and J. M. Calleja, *Spectrochim. Acta*, **49A,** 1023 (1993).

153. K. Shimizu and H. Murata, *J. Mol. Spectrosc.*, **4,** 201 (1960).

154. K. Shimizu and H. Murata, *J. Mol. Spectrosc.*, **4,** 214 (1960).

155. K. Shimizu and H. Murata, *Bull. Chem. Soc. Jpn.*, **32,** 46 (1959).

156. T. Tanaka and S. Murakami, *Bull. Chem. Soc. Jpn.*, **38,** 1465 (1965).

157. J. R. Durig, C. W. Sink, and S. F. Bush, *J. Chem. Phys.*, **45,** 66 (1966).

158. H. J. Becher, *Z. Anorg. Allg. Chem.*, **294,** 183 (1958).

159. J. Weidlein and V. Krieg. *J. Organomet. Chem.*, **11,** 9 (1968).

160. H. Schmidbaur, J. Weidlein, H. F. Klein, and K. Eiglmeier, *Chem. Ber.*, **101,** 2268 (1968).

161. H. C. Clark and A. L. Pichard, *J. Organomet. Chem.*, **8,** 427 (1967).

162. L. E. Levchuk, J. R. Sams, and F. Aubke, *Inorg. Chem.*, **11,** 43 (1972).

163. C. W. Hobbs and R. S. Tobias, *Inorg. Chem.*, **9**, 1037 (1970).

164. E. Amberger and R. Honigschmid-Grossich, *Chem. Ber.*, **98**, 3795 (1965).

165. W. M. Scovell, G. C. Stocco, and R. S. Tobias, *Inorg. Chem.*, **9**, 2682 (1970).

166. D. E. Clegg and J. R. Hall, *J. Organomet. Chem.*, **22**, 491 (1970).

167. P. A. Bulliner, V. A. Maroni, and T. G. Spiro, *Inorg. Chem.*, **9**, 1887 (1970).

168. J. R. Hall and G. A. Swile, *J. Organomet. Chem.*, **56**, 419 (1973).

169. J. S. Thayer and R. West, *Adv. Organomet. Chem.*, **5**, 169 (1967).

170. J. S. Thayer and D. P. Strommen, *J. Organomet. Chem.*, **5**, 383 (1966).

171. H. Müller and K. Dehnicke, *J. Organomet. Chem.*, **10**, P1 (1967).

172. K. Dehnicke and D. Seybold, *J. Organomet. Chem.*, **11**, 227 (1968).

173. J. Müller and K. Dehnicke, *J. Organomet. Chem.*, **12**, 37 (1968).

174. J. R. Durig, J. F. Sullivan, A. W. Cox, Jr., and B. J. Streusand, *Spectrochim. Acta*, **34A**, 719 (1978).

175. J. Müller, *Z. Naturforsch.*, **34B**, 536 (1979).

176. H. Leimeister and K. Dehnicke, *J. Organomet. Chem.*, **31**, C3 (1971).

177. R. G. Goel and D. R. Ridley, *Inorg. Chem.*, **13**, 1252 (1974).

178. J. R. Durig, J. F. Sullivan, and A. W. Cox, Jr. *J. Mol. Struct.*, **44**, 31 (1978).

179. J. E. Förster, M. Vargas, and H. Müller, *J. Organomet. Chem.*, **59**, 97 (1973).

180. J. Relf, R. P. Cooney, and H. F. Henneike, *Organomet. Chem.*, **39**, 75 (1972).

181. F. Weller, I. L. Wilson, and K. Dehnicke, *J. Organomet. Chem.*, **30**, C1 (1971).

182. K. Dehnicke, *Angew. Chem.*, **79**, 942 (1967).

183. M. Wada and R. Okawara, *J. Organomet. Chem.*, **8**, 261 (1967).

184. H. Bürger and U. Goetze, *J. Organomet. Chem.*, **10**, 380 (1967).

185. J. S. Thayer, *Inorg. Chem.*, **7**, 2599 (1968).

186. E. E. Aynsley, N. N. Greenwood, G. Hunter, and M. J. Sprague, *J. Chem. Soc. A*, 1344 (1966).

187. W. Beck and E. Schuierer, *J. Organomet. Chem.*, **3**, 55 (1965).

188. M. R. Booth and S. G. Frankiss, *Spectrochim. Acta*, **26A**, 859 (1970).

189. J. R. Durig, Y. S. Li, and J. B. Turner, *Inorg. Chem.*, **13**, 1495 (1974).

190. F. Watari, *J. Mol. Struct.*, **32**, 285 (1976).

191. R. Okawara, D. E. Webster, and E. G. Rochow, *J. Am. Chem. Soc.*, **82**, 3287 (1960).

192. M. P. Brown, R. Okawara, and E. G. Rochow, *Spectrochim. Acta*, **16**, 595 (1960).

193. M. Shindo and R. Okawara, *J. Organomet. Chem.*, **5**, 537 (1966).

194. R. Okawara and M. Ohara, *J. Organomet. Chem.*, **1**, 360 (1964).

195. K. Yasuda and R. Okawara, *J. Organomet. Chem.*, **3**, 76 (1965).

196. H. C. Clark and R. J. O'Brien, *Inorg. Chem.*, **2**, 740 (1963).

197. R. G. Goel, H. S. Prasad, G. M. Bancroft, and T. K. Sham, *Can. J. Chem.*, **54**, 711 (1976).

198. N. Q. Dao and D. Breitinger, *Spectochim. Acta*, **27A**, 905 (1971).

199. M. J. Almond, C. E. Jenkins, D. A. Rice, and K. Hagen, *J. Organomet. Chem.*, **439**, 251 (1992).

200. H. C. Clark, R. J. O'Brien, and A. L. Pickard, *J. Organomet. Chem.*, **4,** 43 (1965).

201. B. S. Ault, *J. Phys. Chem.*, **96,** 7908 (1992).

202. T. Tanaka. M. Komura, Y. Kawasaki, and R. Okawara, *J. Organomet. Chem.*, **1,** 484 (1964).

203. J. P. Clark and C. J. Wilkins, *J. Chem. Soc. A*, 871 (1966).

204. R. J. H. Clark, A. G. Davies, and R. J. Puddenphatt, *J. Chem. Soc. A*, 1828 (1968).

205. D. E. Clegg and J. R. Hall, *Spectrochim. Acta*, **23A,** 263 1967).

206. G. Ferguson, R. G. Goel, F. C. March, D. R. Ridley, and H. S. Prasad, *J. Chem. Soc., Dalton Trans.*, 1547 (1971).

207. R. Okawara and K. Yasuda, *J. Organomet. Chem.*, **1,** 356 (1964).

208. J. M. Brown, A. C. Chapman, R. Harper, D. J. Mowthorpe, A. G. Davies, and P. J. Smith, *J. Chem. Soc., Dalton Trans.*, 338 (1972).

209. P. I. Paetzold, *Z. Anorg. Allg. Chem.*, **326,** 53 (1963).

210. J. F. Helling and D. M. Braitsch, *J. Am. Chem. Soc.*, **92,** 7207 (1970).

211. R. S. Tobias and C. E. Freidline, *Inorg. Chem.*, **4,** 215 (1965).

212. R. S. Tobias and S. Hutcheson, *J. Organomet. Chem.*, **6,** 535 (1966).

213. R. S. Tobias, M. J. Sprague, and G. E. Glass. *Inorg. Chem.*, **7,** 1714 (1968).

214. P. A. Bulliner and T. G. Spiro, *Inorg. Chem.*, **8,** 1023 (1969).

215. H. Kriegsmann, *Z. Anorg. Allg. Chem.*, **294,** 113 (1958).

216. F. K. Butcher, W. Gerrard. E. F. Mooney, R. G. Rees, and H. A. Willis, *Spectrochim. Acta*, **20,** 51 (1964).

217. J. Lorberth and M. R. Kula, *Chem. Ber.*, **97,** 3444 (1964).

218. G. Mann, A. Haaland, and J. Weidlein, *Z. Anorg. Allg. Chem.*, **398,** 231 (1973).

219. T. Tanaka, *Bull. Chem. Soc. Jpn.*, **33,** 446 (1960).

220. E. W. Abel, *J. Chem. Soc.*, 4406 (1960).

221. K. A. Hooton and A. L. Allred, *Inorg. Chem.*, **4,** 671 (1965).

222. J. E. Drake, H. E. Henderson, and L. N. Khasrou, *Spectrochim. Acta*, **38A,** 31 (1982).

223. P. G. Harrison and S. R. Stobart, *J. Organomet. Chem.*, **47,** 89 (1973).

224. M. Wada and R. Okawara, *J. Organomet. Chem.*, **4,** 487 (1965).

225. P. L. Goggin and L. A. Woodward, *Trans. Faraday Soc.*, **58,** 1495 (1962).

226. D. E. Clegg and J. R. Hall, *J. Organomet. Chem.*, **17,** 175 (1969); *Spectrochim. Acta.*, **21,** 357 (1965).

227. V. B. Ramos and R. S. Tobias, *Spectrochim. Acta*, **29A,** 953 (1973).

228. Y. Kawasaki, T. Tanaka, and R. Okawara, *Spectrochim. Acta*, **22,** 1571 (1966).

229. Y. Kawasaki, T. Tanaka, and R. Okawara, *Bull. Chem. Soc. Jpn.*, **37,** 903 (1964).

230. H. A. Meinema, A. Mackor, and J. G. Noltes, *J. Organomet. Chem.*, **37,** 285 (1972).

231. M. G. Miles, G. E. Glass, and R. S. Tobias, *J. Am. Chem. Soc.*, **88,** 5738 (1966).

232. M. M. McGrady and R. S. Tobias, *Inorg. Chem.*, **3,** 1161 (1964); *J. Am. Chem. Soc.*, **87,** 1909 (1965).

233. C. Z. Moore and W. H. Nelson, *Inorg. Chem.*, **8,** 138 (1969).

234. G. A. Miller and E. O. Schlemper, *Inorg. Chem.*, **12,** 677 (1973).

235. D. F. Ball, T. Carter, D. C. McKean, and L. A. Woodward, *Spectrochim. Acta*, **20,** 1721 (1964).

236. D. F. Ball, P. L. Goggin, D. C. McKean, and L. A. Woodward, *Spectrochim. Acta*, **16,** 1358 (1960).

237. M. W. Mackenzie, *Spectrochim. Acta*, **38A,** 1083 (1982).

238. D. F. Van de Vondel and G. P. Van der Kelen, *Bull. Soc. Chim. Belg.*, **74,** 467 (1965).

239. H. Kimmel and C. R. Dillard, *Spectrochim. Acta*, **24A,** 909 (1968).

240. C. R. Dillard and L. May, *J. Mol. Spectrosc.*, **14,** 250 (1964).

241. E. Amberger, *Angew. Chem.*, **72,** 494 (1960).

242. A. J. F. Clark and J. E. Drake, *Spectrochim. Acta*, **34A,** 307 (1978).

243. A. J. F. Clark, J. E. Drake, and Q. Shen, *Spectrochim. Acta*, **34A,** 311 (1978).

244. W. J. Lehmann, C. O. Wilson, and I. Shapiro, *J. Chem. Phys.*, **32,** 1088, 1786 (1960); **33,** 590 (1960); **34,** 476, 783 (1961).

245. A. S. Grady, S. G. Puntambekar, and D. K. Russell, *Spectrochim. Acta*, **47A,** 47 (1991).

246. R. A. Burnham and S. R. Stobart, *J. Chem. Soc., Dalton Trans.*, 1269 (1973).

247. D. J. Cardin, S. A. Keppie, and M. F. Lappert, *Inorg. Nucl. Chem. Lett.*, **4,** 365 (1968).

248. A. Terzis, T. C. Strekas, and T. G. Spiro, *Inorg. Chem.*, **13,** 1346 (1974).

249. G. F. Bradley and S. R. Stobart, *J. Chem. Soc., Dalton Trans.*, 264 (1974).

250. N. A. D. Carey and H. C. Clark, *Chem. Commun.*, 292 (1967); *Inorg. Chem.*, **7,** 94 (1968).

251. B. Fontal and T. G. Spiro, *Inorg. Chem.*, **10,** 9 (1971).

252. R. J. H. Clark, A. G. Davies, R. J. Puddenphatt, and W. McFarlane, *J. Am. Chem. Soc.*, **91,** 1334 (1969).

253. H. Bürger and U. Goetze, *Spectrochim. Acta*, **26A,** 685 (1970).

254. G. Davidson, *Organomet. Rev. A*, **8,** 303 (1972).

255. J. Chatt, L. A. Duncanson, and R. G. Guy, *Nature (London)*, **184,** 526 (1959).

256. J. A. J. Jarvis, B. T. Kilbourn, and P. G. Owston, *Acta Crystallogr.*, **B27,** 366 (1971).

257. N. Rösch, R. P. Messmer, and K. H. Johnson, *J. Am. Chem. Soc.*, **96,** 3855 (1974).

258. M. J. Grogan and K. Nakamota, *J. Am. Chem. Soc.*, **88,** 5454 (1966); **90,** 918 (1968).

259. J. P. Sorzano and J. P. Fackler, *J. Mol. Spectrosc.*, **22,** 80 (1967).

260. J. Hiraishi, *Spectrochim. Acta*, **25A,** 749 (1969).

261. J. A. Crayston and G. Davidson, *Spectrochim. Acta*, **43A,** 559 (1987).

262. J. Mink, M. Gal, P. L. Goggin, and J. L. Spencer, *J. Mol. Struct.*, **142,** 467 (1986).

263. P. Csaszar, P. L. Goggin, J. Mink, and J. L. Spencer, *J. Organomet. Chem.*, **379,** 337 (1989).

264. G. Davidson and C. L. Davies, *Inorg. Chim. Acta*, **165,** 231 (1989).

265. D. C. Andrews and G. Davidson, *J. Organomet. Chem.*, **35,** 161 (1972).

266. M. A. Bennett, R. J. H. Clark, and D. L. Miller, *Inorg. Chem.*, **6,** 1647 (1967).

267. J. Howard and T. C. Waddington, *J. Chem. Soc., Faraday Trans. 2*, **74,** 1275 (1978).

268. D. J. Stufkens, T. L. Snoeck, W. Kaim, T. Roth, and B. Olbrich-Deussner, *J. Organomet. Chem.*, **409,** 189 (1991).

269. M. Bigorgne, *J. Organomet. Chem.*, **127,** 55 (1977).

270. J. A. Crayston and G. Davidson, *Spectrochim. Acta*, **42A,** 1385 (1986).

271. A. E. Smith, *Acta Crystallogr.*, **18,** 331 (1965).

272. K. Shobatake and K. Nakamota, *J. Am. Chem. Soc.*, **92,** 3339 (1970).

273. D. C. Andrews and G. Davidson, *J. Organomet. Chem.*, **55,** 383 (1973).

274. D. M. Adams and A. Squire, *J. Chem. Soc. A*, 1808 (1970).

275. D. C. Andrews and G. Davidson, *J. Organomet. Chem.*, **124,** 181 (1977).

276. G. Davidson and D. C. Andrews, *J. Chem. Soc., Dalton Trans.*, 126 (1972).

277. T. B. Chenskaya, L. A. Leites, and V. T. Aleksanyan, *J. Organomet. Chem.*, **148,** 85 (1978).

278. I. S. Butler and G. G. Barna, *J. Raman Spectrosc.*, **1,** 141 (1973).

279. J. Howard and T. C. Waddington, *Spectrochim. Acta*, **34A,** 807 (1978).

280. P. J. Hendra and D. B. Powell, *Spectrochim. Acta*, **17,** 909 (1961).

281. G. Davidson, *Inorg. Chim. Acta*, **3,** 596 (1969).

282. G. Davidson and D. A. Duce, *J. Organomet. Chem*, **44,** 365 (1972).

283. M. J. Grogan and K. Nakamoto, *Inorg. Chim. Acta*, **1,** 228 (1967).

284. M. A. Bennett and J. D. Saxby, *Inorg. Chem.*, **7,** 321 (1968).

285. B. Dickens and W. N. Lipscomb, *J. Am. Chem. Soc.*, **83,** 4062 (1961); *J. Chem. Phys.*, **37,** 2084 (1962).

286. R. T. Bailey, E. R. Lippincott, and D. Steele, *J. Am. Chem. Soc.*, **87,** 5346 (1965).

287. G. G. Barna and I. S. Butler, *J. Raman Spectrosc.*, **7,** 168 (1978).

288. C. Udovich, J. Takemoto, K. Shobatake, and K. Nakamoto, unpublished.

289. M. A. Coles and F. A. Hart, *J. Organomet. Chem.*, **32,** 279 (1971).

290. R. Nast and H. Schindel, *Z. Anorg. Allg. Chem.*, **326,** 201 (1963).

291. I. A. Garbusova, V. T. Alexanjan, L. A. Leites, I. R. Golding, and A. M. Sladkov, *J. Organomet. Chem.*, **54,** 341 (1973).

292. K. Yasufuku and H. Yamazaki, *Bull. Chem. Soc. Jpn.*, **45,** 2664 (1972).

293. J. Chatt, G. A. Rowe, and A. A. Williams, *Proc. Chem. Soc.*, 208 (1957); J. Chatt, R. Guy, and L. A. Duncanson, *J. Chem. Soc.*, 827 (1961).

294. Y. Iwashita, A. Ishikawa, and M. Kainosho, *Spectrochim. Acta*, **27A,** 271 (1971).

295. Y. Iwashita, F. Tamura, and A. Nakamura, *Inorg. Chem.*, **8,** 1179 (1969).

296. K. Stahl, U. Müller, and K. Dehnicke, *Z. Anorg. Allg. Chem.*, **527,** 7 (1985).

297. G. J. Kubas and T. G. Spiro, *Inorg. Chem.*, **12,** 1797 (1973).

298. W. M. Scovell and T. G. Spiro, *Inorg. Chem.*, **13,** 304 (1974).

299. W. J. Bland, R. D. Kemmitt, and R. D. Moore, *J. Chem. Soc., Dalton Trans.*, 1292 (1973).

300. D. H. Payne, Z. A. Payne, R. Rohmer, and H. Frye, *Inorg. Chem.*, **12,** 2540 (1973).

301. M. Moskovitz and G. A. Ozin, *Cryochemistry*, Wiley, New York, 1976, p. 263.

302. L. Manceron and L. Andrews, *J. Phys. Chem.*, **90,** 4514 (1986).

303. L. Manceron and L. Andrews, *J. Phys. Chem.*, **94,** 3513 (1990).

304. T. Merle-Méjean, C. Cosse-Mertens, S. Bouchareb, J. Mascetti, and M. Tranquille, *J. Phys. Chem.*, **96,** 9148 (1992).

305. T. Merle-Méjean, S. Bouchareb, and M. Tranquille, *J. Phys. Chem.*, **93,** 1197 (1989).

306. L. Manceron and L. Andrews, *J. Am. Chem. Soc.*, **107,** 563 (1985).

307. E. S. Kline, Z. H. Kafafi, R. H. Hauge, and J. L. Margrave, *J. Am. Chem. Soc.*, **109,** 2402 (1987).

308. G. A. Ozin and W. J. Power, *Inorg. Chem.*, **19,** 3860 (1980).

309. D. Tevault, D. P. Strommen, and K. Nakamoto, *J. Am. Chem. Soc.*, **99,** 2997 (1977).

310. A. Weselucha-Birczynska, I. R. Paeng, A. A. Shabana, and K. Nakamoto, *New J. Chem.*, **16,** 563 (1992).

311. H. P. Fritz, *Adv. Organomet. Chem.*, **1,** 239 (1964).

312. H. P. Fritz and L. Schäfer, *Chem. Ber.*, **97,** 1827 (1964).

313. E. O. Fischer and H. Fischer, *J. Organomet. Chem.*, **3,** 181 (1965).

314. E. O. Fischer and G. Stölzle, *Chem. Ber.*, **94,** 2187 (1961).

315. E. R. Lippincott, J. Xavier, and D. Steele, *J. Am. Chem. Soc.*, **83,** 2262 (1961).

316. J. Lusztyk and K. B. Starowieyski, *J. Organomet. Chem.*, **170,** 293 (1979).

317. K. A. Allan, B. G. Gowenlock, and W. E. Lindsell, *J. Organomet. Chem.*, **55,** 229 (1973).

318. O. G. Garkusha, I. A. Garbuzova, B. V. Lokshin, and G. K. Borisov, *J. Organomet. Chem.*, **336,** 13 (1987).

319. I. A. Garbuzova, O. G. Garkusha, B. V. Lokshin, G. K. Borisov, and T. S. Moro-zova, *J. Organomet. Chem.*, **279,** 327 (1985).

320. A. Almenningen, A. Haaland, and T. Motzfeldt, *J. Organomet. Chem.*, **7,** 97 (1967).

321. P. G. Harrison and M. A. Healy, *J. Organomet. Chem.*, **51,** 153 (1973).

322. S. J. Pratten, M. K. Cooper, and M. J. Aroney, *Polyhedron*, **3,** 1347 (1984).

323. K. W. Nugent, J. K. Beattie, T. W. Hambley, and M. R. Snow, *Aust. J. Chem.*, **37,** 1601 (1984).

324. K. W. Nugent and J. K. Beattie, *Inorg. Chem.*, **27,** 4269 (1988).

325. G. Wilkinson, P. L. Pauson, and F. A. Cotton, *J. Am. Chem. Soc.*, **76,** 1970 (1954).

326. R. T. Bailey, *Spectrochim. Acta*, **27A,** 199 (1971).

327. J. S. Bodenheimer and W. Low, *Spectrochim. Acta*, **29A,** 1733 (1973).

328. B. V. Lokshin, V. T. Aleksanian, and E. B. Rusach, *J. Organomet. Chem.*, **86,** 253 (1975).

329. E. R. Lippincott and R. D. Nelson, *Spectrochim. Acta*, **10,** 307 (1958).

330. I. Pavlik and J. Klilorka, *Collect. Czech. Chem. Commun.*, **30,** 664 (1965).

331. D. Hartley and M. J. Ware, *J. Chem. Soc. A*, 138 (1969).

332. V. T. Aleksanyan, B. V. Lokshin, G. K. Borisov, G. G. Devyatykh, A. S. Smirnov,

R. V. Nazarova, J. A. Koningstein, and B. F. Gachter, *J. Organomet. Chem.*, **124,** 293 (1977).

333. K. Nakamoto, C. Udovich, J. R. Ferraro, and A. Quattrochi, *Appl. Spectrosc.*, **24,** 606 (1970).

334. L. Phillips, A. R. Lacey, and M. K. Cooper, *J. Chem. Soc., Dalton Trans.*, 1383 (1988).

335. I. J. Hyams, *Chem. Phys. Lett.*, **15,** 88 (1972).

336. L. Schäfer, J. Brunvoll, and S. J. Cyvin, *J. Mol. Struct.*, **11,** 459 (1972).

337. J. Brunvoll, S. J. Cyvin, and L. Schäfer, *J. Organomet. Chem.*, **27,** 107 (1971).

338. K. Yokoyama, S. Kobinata, and S. Maeda, *Bull. Chem. Soc. Jpn.*, **49,** 2182 (1976).

339. I. J. Hyams, *Spectrochim. Acta*, **29A,** 839 (1973).

340. F. Rocquet, L. Berreby, and J. P. Marsault, *Spectrochim. Acta*, **29A,** 1101 (1973).

341. D. M. Adams and W. S. Fernado, *J. Chem. Soc., Dalton Trans.*, 2507 (1972).

342. E. Maslowsky, Jr. and K. Nakamoto, *Inorg. Chem.*, **8,** 1108 (1969).

343. F. A. Cotton and T. J. Marks, *J. Am. Chem. Soc.*, **91,** 7281 (1969).

344. G. Wilkinson and T. S. Piper, *J. Inorg. Nucl. Chem.*, **2,** 32 (1956).

345. E. Gallinella, B. Fortunato, and P. Mirone, *J. Mol. Spectrosc.*, **24,** 345 (1967).

346. A. Davison and P. E. Rakita, *Inorg. Chem.*, **9,** 289 (1970).

347. F. W. Siegert and H. J. de Liefde Meijer, *J. Organomet. Chem.*, **15,** 131 (1968).

348. F. W. Siegert and H. J. de Liefde Meijer, *J. Organomet. Chem.*, **20,** 141 (1969).

349. V. I. Kulishov, E. M. Brainina, N. G. Bokiy, and Yu. T. Struchkov, *Chem. Commun.*, 475 (1970).

350. J. L. Calderon, F. A. Cotton, B. G. DeBoer, and J. Takats, *J. Am. Chem. Soc.*, **93,** 3592 (1971).

351. R. D. Rogers, R. V. Bynum, and J. L. Atwood, *J. Am. Chem.Soc.*, **100,** 5238 (1978).

352. E. O. Fischer and Y. Hristidu, *Chem. Ber.*, **95,** 253 (1962).

353. J. L. Calderon, F. A. Cotton, and J. Takats, *J. Am. Chem. Soc.*, **93,** 3587 (1971).

354. A. E. Smith, *Inorg. Chem.*, **11,** 165 (1972).

355. R. A. Forder and K. Prout, *Acta. Crystallogr.*, **B30,** 491 (1974).

356. F. N. Tebbe and G. W. Parshall, *J. Am. Chem. Soc.*, **93,** 3793 (1971).

357. H. P. Fritz and E. F. Paulus, *Z. Naturforsch.*, **18B,** 435 (1963).

358. P. J. Fitzpatrick, Y. Le Page, J. Sedman, and I. S. Butler, *Inorg. Chem.*, **20,** 2852 (1981).

359. S. M. Barnett, F. Dicaire, and A. A. Ismail, *Can. J. Chem.*, **68,** 1196 (1990).

360. R. Feld, E. Hellner, A. Klopsch, and K. Dehnicke, *Z. Anorg. Allg. Chem.*, **442,** 173 (1978).

361. D. M. Adams and A. Squire, *J. Organomet. Chem.*, **63,** 381 (1973).

362. B. V. Lokshin, Z. S. Klemmenkova, and Yu. V. Makarov, **28A,** 2209 (1972).

363. J. R. Durig, R. B. King, L. W. Houk, and A. L. Marston, *J. Organomet. Chem.*, **16,** 425 (1969).

364. A. Davison, M. L. H. Green, and G. Wilkinson, *J. Chem. Soc.*, 3172 (1961).

365. B. F. Hallam and P. L. Pauson, *Chem. Ind. (London)*, **23,** 653 (1955).

366. T. S. Piper and G. Wilkinson, *Chem. Ind. (London)*, **23,** 1296 (1955); *J. Inorg. Nucl. Chem.*, **3,** 104 (1956).

367. M. J. Bennett, F. A. Cotton, A. Davison, J. W. Faller, S. J. Lippard, and S. M. Morehouse, *J. Am. Chem. Soc.*, **88,** 4371 (1966).

368. O. S. Mills, *Acta Crystallogr.*, **11,** 620 (1958); R. F. Bryan and P. T. Greene, *J. Chem. Soc. A*, 3064 (1970).

369. R. F. Bryan, P. T. Greene, M. J. Newlands, and D. S. Field, *J. Chem. Soc. A*, 3068 (1970).

370. F. A. Cotton and G. Yagupsky, *Inorg. Chem.*, **6,** 15 (1967).

371. R. D. Fischer, A. Vogler, and K. Noack, *J. Organomet. Chem.*, **7,** 135 (1967).

372. A. R. Manning, *J. Chem. Soc. A*, 1319 (1968).

373. J. G. Bullitt, F. A. Cotton, and T. J. Marks, *Inorg. Chem.*, **11,** 671 (1972).

374. A. Alich, N. J. Nelson, D. Strope, and D. F. Shriver, *Inorg. Chem.*, **11,** 2976 (1972); N. J. Nelson, N. E. Kime, and D. F. Shriver, *J. Am. Chem. Soc.*, **91,** 5173 (1969).

375. N. E. Kim, N. J. Nelson, and D. F. Shriver, *Inorg. Chim. Acta*, **7,** 393 (1973).

376. J. S. Kristoff and D. F. Shriver, *Inorg. Chem.*, **13,** 499 (1974).

377. P. McArdle and A. R. Manning, *J. Chem. Soc. A*, 717 (1971).

378. F. C. Wilson and D. P. Shoemaker, *J. Chem. Phys.*, **27,** 809 (1957).

379. G. Davidson and E. M. Riley, *J. Organomet. Chem.*, **51,** 297 (1973).

380. R. D. Adams and F. A. Cotton, *Inorg. Chem. Acta*, **7,** 153 (1973).

381. A. Davison, W. McFarlane, E. Pratt, and G. Wilkinson, *J. Chem. Soc.*, 3653 (1962).

382. M. A. Neuman, Trinh-Toan, and L. F. Dahl, *J. Am. Chem. Soc.*, **94,** 3382 (1972).

383. A. Terzis and T. G. Spiro, *Chem. Commun.*, 1160 (1970).

384. E. Maslowsky, Jr. and K. Nakamoto, *Appl. Spectrosc.*, **25,** 187 (1971).

385. E. Samuel, R. Ferner, and M. Bigorgne, *Inorg. Chem.*, **12,** 881 (1973).

386. P. M. Druce, B. M. Kingston, M. F. Lappert, and R. C. Srivastava, *J. Chem. Soc. A*, 2106 (1969).

387. G. Balducci, L. Bencivenni, G. DeRosa, R. Gigli, B. Martini, and S. Nunziante, *J. Mol. Struct.*, **64,** 163 (1980).

388. R. J. Haines, R. S. Nyholm, and M. H. B. Stiddard, *J. Chem. Soc. A*, 1606 (1966).

389. G. Paliani, R. Cataliotti, A. Poletti, and A. Foffani, *J. Chem. Soc., Dalton Trans*, 1741 (1972).

390. O. Crichton and A. J. Rest, *Chem. Commun.*, 407 (1973).

391. T. S. Piper and G. Wilkinson, *J. Inorg. Nucl. Chem.*, **2,** 38 (1956).

392. H. Brunner, *J. Organomet. Chem.*, **16,** 119 (1969).

393. R. C. Elder, F. A. Cotton, and R. A. Schunn, *J. Am. Chem. Soc.*, **89,** 3645 (1967).

394. M. J. D'Aniello, Jr. and E. K. Barefield, *J. Organomet. Chem.*, **76,** C50 (1974).

395. R. L. Cooper, M. L. H. Green, and J. T. Moelwyn-Hughes, *J. Organomet. Chem.*, **3,** 261 (1965).

396. M. P. Johnson and D. F. Shriver, *J. Am. Chem. Soc.*, **88,** 301 (1966).

397. M. Gerloch and R. Mason, *J. Chem. Soc.*, 296 (1965).

398. B. D. James, R. K. Nanda, and M. G. H. Wallbridge, *Inorg. Chem.*, **6,** 1979 (1967).

399. L. Banford and G. E. Coates, *J. Chem. Soc.*, 5591 (1964).

400. J. E. Bercaw and H. H. Brintzinger, *J. Am. Chem. Soc.*, **91,** 7301 (1969).

401. C. White, D. S. Gill, J. W. Kang, H. B. Lee, and P. M. Maitlis, *Chem. Commun.*, 734 (1971).

402. J. Müller and H. Dorner, *Angew. Chem., Int. Ed. Engl.*, **12,** 843 (1973).

403. J. L. Burmeister, E. A. Deardorff, A. Jensen, and V. H. Christiansen, *Inorg. Chem.*, **9,** 58 (1970).

404. A. Jensen, V. H. Christiansen, J. F. Hansen, T. Likowski, and J. L. Burmeister, *Acta Chem. Scand.*, **26,** 2898 (1972).

405. A. Schulz and T. M. Klapötke, *J. Organomet. Chem.*, **436,** 179 (1992).

406. A. C. Filippou, W. Grünleitner, and E. Herdtweck, *J. Organomet. Chem.*, **373,** 325 (1989).

407. A. C. Filippou, E. O. Fischer, and W. Grünleitner, *J. Organomet. Chem.*, **386,** 333 (1990).

408. M. L. H. Green and W. E. Silverthorn, *Chem. Commun.*, 557 (1971).

409. A. E. Fenster and I. S. Butler, *Can. J. Chem.*, **50,** 598 (1972).

410. P. M. Treichel and G. P. Werber, *J. Am. Chem. Soc.*, **90,** 1753 (1968).

411. D. C. Andrews and G. Davidson, *J. Organomet. Chem.*, **36,** 349 (1972).

412. J. Howard and T. C. Waddington, *Spectrochim. Acta*, **34A,** 445 (1978).

413. D. C. Andrews and G. Davidson, *J. Organomet. Chem.*, **76,** 373 (1974).

414. H. P. Fritz, *Z. Naturforsch.*, **16B,** 415 (1961).

415. M. L. H. Green, L. Pratt, and G. Wilkinson, *J. Chem. Soc.*, 3753 (1959).

416. P. M. Treichel and R. L. Shubkin, *Inorg. Chem.*, **6,** 1328 (1967).

417. G. Winkhaus, L. Pratt, and G. Wilkinson, *J. Chem. Soc.*, 3807 (1961).

418. H. P. Fritz and E. O. Fischer, *J. Organomet. Chem.*, **7,** 121 (1967).

419. H. P. Fritz, W. Lüttke, H. Stammreich, and R. Forneris, *Spectrochim. Acta*, **17,** 1068 (1961).

420. S. J. Cyvin, J. Brunvoll, and L. S. Schäfer, *J. Chem. Phys.*, **54,** 1517 (1971).

421. M. P. Andrews, S. M. Mattar, and G. A. Ozin, *J. Phys. Chem.*, **90,** 744 (1986).

422. D. W. Ball, Z. H. Kafafi, R. H. Hauge, and J. L. Margrave, *J. Am. Chem. Soc.*, **108,** 6621 (1986).

423. A. McCamley and R. N. Perutz, *J. Phys. Chem.*, **95,** 2738 (1991).

424. H. Saito, Y. Kakiuchi, and M. Tsutsui, *Spectrochim. Acta*, **23A,** 3013 (1967).

425. J. D. Laposa, N. Hao, B. G. Sayer, and M. J. McGlinchey, *J. Organomet. Chem.*, **195,** 193 (1980).

426. D. M. Adams, R. E. Christopher, and D. C. Stevens, *Inorg. Chem.*, **14,** 1562 (1975).

427. E. M. Bisby, G. Davidson, and D. A. Duce, *J. Mol. Struct.*, **48,** 93 (1978).

428. W. Wang, P. Jin, Y. Liu, Y. She, and K. Fu, *J. Phys. Chem.*, **96,** 1278 (1992).

429. R. D. Nelson, W. G. Fateley, and E. R. Lippincott, *J. Am. Chem. Soc.*, **78,** 4870 (1956).

430. P. D. Harvey, I. S. Butler, and D. F. R. Gilson, *Inorg. Chem.*, **26,** 32 (1987).

431. L. Hocks, J. Goffart, G. Duyckaerts, and P. Teyssié, *Spectrochim. Acta*, **30A,** 907 (1974).

432. J. Goffart and L. Hocks, *Spectrochim. Acta*, **37A,** 609 (1981).

433. K. O. Hodgson, F. Mares, D. F. Starks, and A. Streitwieser, Jr., *J. Am. Chem. Soc.*, **95,** 8650 (1973).

434. E. Samuel and M. Bigorgne, *J. Organomet. Chem.*, **19,** 9 (1969); **30,** 235 (1971).

435. F. Kaufman and D. O. Cowan, *J. Am. Chem. Soc.*, **92,** 6198 (1970).

436. T. J. Katz and N. Acton, *J. Am. Chem. Soc.*, **94,** 3281 (1972).

437. A. Salzer and H. Werner, *Angew. Chem., Int. Ed. Engl.*, **11,** 930 (1972).

438. L. A. Garbuzova, O. G. Garkusha, B. V. Lockshin, A. R. Kudinov, and K. I. Rybinskaya, *J. Organomet. Chem.*, **408,** 247 (1991).

V

APPLICATIONS IN BIOINORGANIC CHEMISTRY

Metal ions in biological systems are divided into two classes. The first class consists of ions such as K^+, Na^+, Mg^{2+}, and Ca^{2+} which are found in relatively high concentrations. These ions are important in maintaining the structure of proteins by neutralizing negative charges of peptide chains and in controlling the function of cell membranes which selectively pass certain molecules. In the second class, ionic forms of Mn, Fe, Co, Cu, Zn, Mo, and so on exist in small to trace quantities, and are often incorporated into proteins (metalloproteins). The latter class is divided into two categories: (A) transport and storage proteins and (B) enzymes. Type A includes oxygen-transport proteins such as hemoglobin (Fe), myoglobin (Fe), hemerythrin (Fe), and hemocyanin (Cu), electron-transfer proteins such as cytochromes (Fe), iron–sulfur proteins (Fe), blue-copper proteins (Cu), and metal-storage proteins such as ferritin (Fe) and ceruloplasmin (Cu). Type B includes hydrolases such as carboxypeptidase (Zn) and aminopeptidase (Zn, Mg), oxidoreductases such as oxidase (Fe, Cu, Mo) and nitrogenase (Mo, Fe), and isomerases such as vitamin B_{12} coenzyme (Co).

To understand the roles of these metal ions in metalloproteins, it is first necessary to know the coordination chemistry (structure and bonding) of metal ions in their active sites. Such information is difficult to obtain since these active sites are buried in a large and complex protein backbone. Although X-ray crystallography would be ideal for this purpose, its application is hampered by the difficulties in growing single crystals of large protein molecules and in analyzing diffraction data with high resolution. As will be discussed later, these difficulties have been overcome in some cases, and knowledge of precise geometries has made great contribution to our understandings of their biological functions in terms of molecular structure. In other cases where X-ray structural information is not available or definitive, a variety of physicochemical techniques have been employed to gain structural and bonding information about the metal and its environment. These include electronic, infrared, resonance Raman, ESR, NMR, ORD, CD, Mössbauer spectroscopy, EXAFS, and electrochemical, thermodynamic, and kinetic measurements.

Recently, resonance Raman (RR) spectroscopy (Sec. I-22 of Part A) has been used extensively for the study of active sites of metalloproteins. The reason for this is twofold:

1. Most metalloproteins have strong electronic absorptions in the uv-visible region which originate in a chromophore containing a metal center. By tuning the laser wavelength into these bands, it is possible to selectively enhance the vibrations localized in this chromophore without interference from the rest of the protein.

2. Owing to strong resonance enhancement of these vibrations, only a dilute solution is needed to observe their RR spectra. This enables one to obtain spectra from a small volume of dilute aqueous solution under biological conditions. This is particularly significant in assigning metal–ligand vibrations by using metal–isotope techniques, because isotopes such as ^{54}Fe and ^{68}Zn are expensive.

In some cases, however, the vibrations of interest may not be enhanced with sufficient intensity. A typical example is the $\nu(O_2)$ of oxyhemoglobin. Then, one must resort to IR spectroscopy which exhibits all vibrations allowed by IR selection rules. It should be noted, however, that IR measurements in aqueous media are generally limited to the regions where water does not absorb strongly (Sec. III-10). Furthermore, it is often necessary to use difference techniques to cancel out interfering bands due to the solvent and some solute bands.

In the following, we will review typical results to demonstrate the utility of vibrational spectroscopy in deducing structural and bonding information about large and complex biological molecules. Recently, marked progress has been made in biomimetic chemistry where the active site is modeled by relatively simple coordination compounds.

For example, a number of iron porphyrins have been prepared to mimic heme proteins, and the vibrational spectra of some of these compounds have been discussed in Secs. III-5 and 17–19. Thus, we compare vibrational spectra of biological molecules and their model systems whenever appropriate or necessary.

Since biospectroscopy is one of the most exciting areas of modern research, the volume of literature on biological compounds is increasing explosively. It is clearly not possible to cover all important topics in a limited space. Several excellent monographs[1–4a] and review articles cited in each section should be consulted for further information.

V-1. MYOGLOBIN AND HEMOGLOBIN

Myoglobin (Mb, MW ~ 16,000) is an oxygen-storage protein found in animal muscles. Figure V-1 shows the structure of sperm-whale myoglobin as determined by X-ray analysis. It is a monomer consisting of 153 amino acids, and its active site is an iron protoporphyrin (see Fig. III-26a) which is linked axially to the proximal histidine (F8). In the deoxy state, the iron is divalent and high spin, and the Fe atom is out of the porphyrin-core plane by ~0.6 Å as shown in Fig. V-2. Upon oxygenation, the dioxygen molecule coordinates to the vacant axial position, and the heme core becomes planar. The Fe atom in oxy-Mb is low spin, and its oxidation state is close to Fe(III) (vide infra).

Hemoglobin (Hb, MW ~ 64,000) is an oxygen-transport protein found in animal blood. It consists of four subunits (α_1, α_2, β_1, and β_2), each of which takes a structure similar to that of Mb. However, these four subunits are not

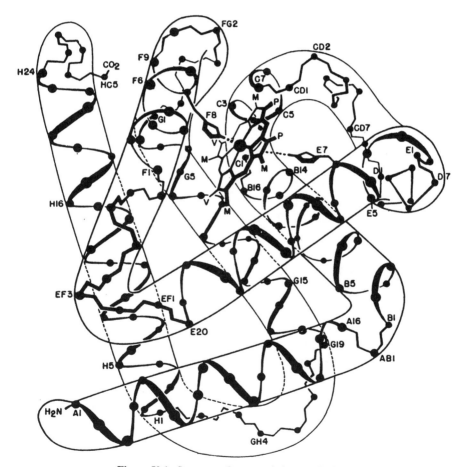

Figure V-1. Structure of sperm-whale myoglobin.

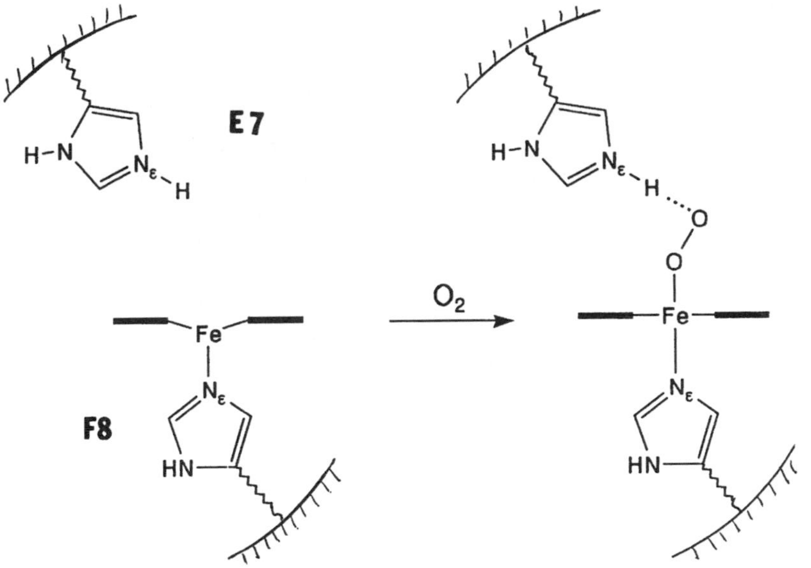

Figure V-2. Structures of deoxy- and oxymyoglobins.

completely independent of each other. Oxygen-uptake studies show that the oxygen affinity of each subunit depends upon the number of other subunits that are already oxygenated (cooperativity). This phenomenon has been explained in terms of two quaternary structures called the T and R states. Deoxy-Hb is in the T_0 (most tense) state. As it gradually absorbs dioxygen, the R state becomes more stable than the corresponding T state. Finally, oxy-Hb assumes the R_4 (most relaxed) state.[5]

Several review articles are available on RR[6–8] and IR[9] spectra of heme proteins.

(1) Selective Excitation of RR spectra

As stated previously, one of the great advantages of RR spectroscopy is its ability to selectively enhance chromophor vibrations by tuning the excitation wavelength to the electronic transition of a particular chromophor in a large and complex molecule such as myoglobin and hemoglobin. This is clearly demonstarated in Fig. V-3, obtained by Asher.[10] The absorption spectrum of the fluoride complex of sperm whale myoglobin is shown in the bottom (b). The absorption near 500 nm is due to the Fe–F (axial ligand) CT and/or the π–π^* transition of the porphyrin core. Thus, the RR spectrum obtained by excitation near 500 nm (inset f) shows strong enhancement of the porphyrin core as well as the ν(Fe–F) vibrations. Excitation at the Soret band near 400 nm (π–π^* transition) of the porphyrin core) produces the RR spectrum (inset e) in which totally symmetric porphyrin vibrations are strongly enhanced.

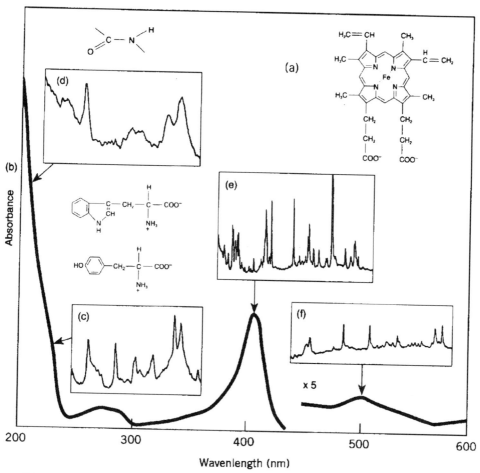

Figure V-3. (a) Structure of iron protoporphyrin. (b) Absorption spectrum of Mb fluoride. (c) UV excitation at 225 nm enhances tyrosine and tryptophan bands. (d) Excitation further in the UV region enhances amide vibrations. (e) Excitation near 400 nm enhances totally symmetric vibrations of the heme core. (f) The absorption bands in the 600–500 nm region are due to the porphyrin core as well as Fe–axial ligand CT transitions, and excitation in this region enhances nontotally symmetric porphyrin core vibrations as well as Fe–axial ligand vibrations (reproduced with permission from Ref. 10).

In contrast, excitation below 300 nm produces RR spectra which exhibit peptide chain vibrations with no major interference from porphyrin core vibrations. The absorption band in the 270–220 nm region originates in the π–π^* transitions of aromatic amino acids such as tyrosine and tryptophan (structures shown). Thus, their phenyl and indole ring vibrations are enhanced by excitation in this region (inset c). Strong enhancement of peptide chain vibrations can occur by excitation below 220 nm (inset d), since the electronic absorption in this region is due to the π–π^* transition of the peptide backbone.

In Mb(III)F, Hb(III)F, and their model compounds, the vinyl group vibrations, such as the $\nu(C=C)$ (~1620 cm^{-1}), are overlapped by strong porphyrin core vibrations when RR spectra are obtained by Soret excitation.[11] In the [Fe(PP)(CN)$_2$]$^-$ ion, however, the vinyl stretching as well as the vinyl–heme stretching (1125 cm^{-1}) vibrations are clearly observed without interference by porphyrin core vibrations if excitation is made in the UV region (225 nm).[12] Recently, UVRR techniques have been utilized to elucidate the structures of peptides and proteins in biological macromolecules.[13]

(2) Porphyrin Core Vibrations

"Structure-sensitive bands" of porphyrin core vibrations in heme proteins were first discovered by Spiro and Strekas[14] in 1974. Table V-1 lists four structure-sensitive bands reported by these workers. Bands I and IV are an oxidation-state marker and a spin-state marker, respectively, while Bands II and V are sensitive to both oxidation and spin states. Based on these results, they proposed that the Fe–O$_2$ bond in oxy–Hb should be formulated as Fe(III)—O$_2^-$. In Sec. III-5, we discussed structure-sensitive bands of model compounds such as Ni(OEP) and Ni(TPP) based on the results of normal coordinate analyses. The normal modes obtained for these model systems are not directly transferable to heme proteins since the effects of peripheral substutuents, axial ligands, and peptide chains on porphyrin core vibrations must be considered. Approximate correlations may be made, however between these two systems. Thus, bands I, II, IV, and V listed in Table V-1 are often referred to as the ν_4, ν_3, ν_{19}, and ν_{10} of the model compound, respectively (see Table III-10).

The oxidation-state sensitive bands (I, II, and V) contain $\nu(C_\alpha C_m)$ or $\nu(C_\alpha N)$ as the major contributors in their potential energy distribution. By lowering the oxidation state, back-donation of d-electrons to the porphyrin π^* orbitals increases. Thus, the porphyrin π-bonds are weakened, and their stretching frequencies are lowered. As seen in Table V-1, this is most clearly demonstrated by band I, which is a relatively pure oxidation-state marker. In general, axial coordination of π-acceptor ligands (CO, O$_2$, etc.) raises its frequency, while that of π-donor ligands (RS$^-$, etc.) lowers it. In fact, cytochrome P–450 (Fe(II), high

TABLE V-1. Structure-Sensitive Bands of Heme Proteins (cm^{-1})^a

Protein	Oxidation State	Spin State	Band I (p)	Band II (p)	Band IV (ap)	Band V (dp)
Ferricytochrome c	Fe(III)	Low spin	1374	1502	1582	1636
CN-Met-Hb	Fe(III)	Low spin	1374	1508	1588	1642
F-Met-Hb	Fe(III)	High spin	1374	1482	1555	1608
deoxy-Hb	Fe(II)	High spin	1358	1473	1552	1607
Ferrocytochrome c	Fe(II)	Low spin	1362	1493	1584	1620
oxy-Hb	Fe(III)	Low spin	1377	1506	1586	1640

^{a}The bands are numbered following the convention given by T. G. Spiro and J. M. Burke (*J. Am. Chem. Soc.*, **98**, 5482 (1976)). Bands I, II, IV, and V correspond approximately to ν_4, ν_3, ν_{19}, and ν_{10}, respectively, of metalloporphyrins (See. III-5(2)). For a more complete listing, see Ref. 6.

spin) exhibits band I at 1346 cm^{-1}, which is much lower than that of deoxy-Hb (1358 cm^{-1}) because its axial ligand is a mercaptide sulfur of a cysteinyl residue.[15]

As discussed in Sec. III-5, the sensitivity of RR bands to spin state is attributed to expansion or out-of-plane deformation of the porphyrin core. In high-spin iron, electrons populate the antibonding $d_{x^2-y^2}$ orbital, and the lengthened Fe–N bonds are accommodated by expansion of the porphyrin core or displacement of the Fe atom from the porphyrin-core plane. This results in weakening of the methine bridge bonds in high-spin complexes. Thus, the frequencies of spin-state-sensitive bands (ν_3, ν_{19}, and ν_{10}) are lower in high-spin than in low-spin complexes since all these vibrations contain $\nu(C_\alpha C_m)$ as the major contributor in their normal modes (see Table III-10). The spin-state-sensitive bands are also metal sensitive since electron occupation in the antibonding $d_{x^2-y^2}$ orbital is varied in a series of transition metals.[16]

(3) Fe–Histidine Vibrations[17]

As shown in Fig. V-2, the iron protoporphyrin is linked to the nitrogen (N$_\epsilon$) atom of the proximal histidine (F8) in Mb, Hb, and many other heme proteins. Thus, the ν(Fe–N$_\epsilon$(His)) vibration is highly important in understanding the nature of the T and R states mentioned previously. Originally, this vibration was assigned to the Raman line at 372 cm^{-1},[18] or 406 cm^{-1},[19] of deoxy Mb. However, Nagai et al.[20] have shown definitively that the ν(Fe–N$_\epsilon$(His)) is near 220 cm^{-1}, which is much lower than those proposed by others. Their assignments were confirmed by the observed ^{54}Fe/^{58}Fe isotope shifts (~2 cm^{-1}) of these bands. As seen in Fig. V-4, this band is at 215 cm^{-1} for the T state of deoxy Hb, and at 221 cm^{-1} for the R state of deoxy NES des–Arg$^{141\alpha}$ Hb.[20] It is the only band that can differentiate between the T and R states of deoxy Hb. The observed upshift in going from the T to R state indicates that the Fe–N$_\epsilon$ bond is stretched in the T state because of a strain exerted by the globin.[21] Stein et al.[22] proposed an alternative explanation that partial donation of the N$_\delta$H proton to an acceptor such as the COO$^-$ group of the peptide backbone would strengthen the Fe–N$_\epsilon$ bond, and that the degree of such partial donation might be less in the T than in the R state.

Finally, the ν(Fe–N$_\epsilon$(His)) of oxyMb (Fe(III), low spin) is much higher (271 cm^{-1})[7] than that of deoxyMb (Fe(II), high spin) (218 cm^{-1}). Extensive isotope labelling studies have shown that the latter should be regarded as a vibration of the whole imidazole moiety against the Fe center, and not as a simple Fe–N$_\epsilon$ diatomic vibrator.[23]

V-2. LIGAND BINDING TO MYOGLOBIN AND HEMOGLOBIN

When a diatomic (XY) ligand such as CO, NO, and O_2 binds to myoglobin and hemoglobin, the ν(XY) as well as the ν(Fe–XY) and δ(FeXY) vibrations

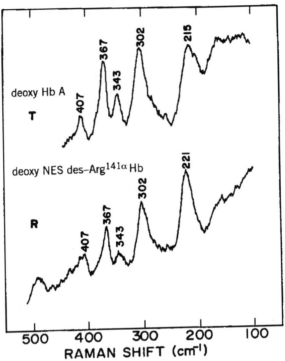

Figure V-4. The RR spectra (441.6 nm excitation) of deoxy-Hb at the T and R states; Hb A and NES des-Arg (141 α)Hb were used to represent these states, respectively. The 215 cm^{-1} band is asymmetric and broad because contributions from the α and β chains are not equivalent (reproduced with permission from Ref. 20).

are expected in IR and Raman spectra. In RR spectra, the origin of resonance enhancement of these axial vibrations is attributed to "direct coupling" between the ν(Fe–XY) vibration and the porphyrin π–π* electronic transition.[7] These axial vibrations provide valuable information about the steric and electronic effects of the heme cavity on the Fe–X–Y moiety.

(1) CO Adducts[24,25]

In general, the ν(CO) is strong in IR but weak in RR spectra. The opposite trend holds for the ν(Fe–CO) and δ(FeCO) in the low-frequency region. Tsubaki et al.[26] first assigned the ν(CO), δ(FeCO), and ν(Fe—CO) of HbCO at 1951, 578, and 507 cm^{-1}, respectively. Isotopic substitution experiments show:

	$^{12}C^{16}O$		$^{13}C^{16}O$		$^{12}C^{18}O$		$^{13}C^{18}O$
δ(FeCO) (cm^{-1})	578	>	563	<	576	>	560
ν(Fe—CO) (cm^{-1})	507	>	503	>	498	>	494

These assignments were supported by approximate normal coordinate analysis on a linear (tilted) Fe—C≡O model. The δ(FeCO) exhibits a "zigzag" pattern (Sec. III-17(7)), whereas the ν(Fe—CO) changes monotonously as the mass of CO increases. Although the trend δ(FeCO) > ν(Fe—CO) is somewhat unusual, it has been reported for some metal carbonyls (Table III-48). Recently, Hirota et al.[27] assigned the δ(FeCO) at 365 cm⁻¹ and attributed the band at 578 cm⁻¹ to a combination of the δ(FeCO) with a porphyrin or a Fe–C deformation mode (displacement of the C atom parallel to the porphyrin plane). Their assignments are based on the Raman difference spectra of HbCO obtained by using four isotopomers of CO. As seen in Fig. V-5, Hirota et al. observed three difference features at 577, 506, and 367 cm⁻¹ which were assigned to a combination band, ν(Fe—CO), and δ(FeCO), respectively. More recently, Hu et al.[27a] observed two strong bands at 574 and 495 cm⁻¹ in the IR spectrum of Fe(OEP)(py)(CO) which were downshifted by 17 and 5 cm⁻¹, respectively, by ¹²CO/¹³CO substitution. However, they could not observe any isotope-sensitive bands near 360 cm⁻¹. This observation led them to support the original assignment by Tsubaki et al.

As discussed in Sec. III-17, σ-donation from the CO to the metal tends to raise the ν(CO) while π-back donation from the metal to the CO tends to lower the ν(CO). It is expected, therefore, that the Fe—C bond order would increase, and the CO bond order would decrease, as π-back bonding increases. In fact, a negative linear relationship was found between the ν(CO) and ν(Fe—CO) in a series of heme proteins containing imidazole as the axial ligand.[28] Deviation from the straight-line relationship occurs when (1) electron-donating substituents are introduced in the porphyrin ring, (2) the donor strength of the *trans* axial ligand is increased, and (3) the coordinated CO interacts with the distal histidine.

In a protein-free environment, the Fe—C≡O bond is perpendicular to the porphyrin plane. In the heme cavity, however, it may be bent and/or tilted due to steric hindrance and/or electronic interaction with distal histidine (E7). Three probable geometries are illustrated in Fig. V-6. According to X-ray analysis,[29] the Fe—C≡O bond in MbCO is linear but tilted by 13° from the normal to the porphyrin plane. Such distortion is expected to raise the ν(CO) since it decreases the Fe(dπ) → CO(π*) back-bonding. However, this effect is over-

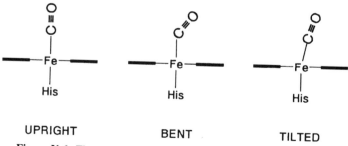

UPRIGHT   BENT   TILTED

Figure V-6. Three geometries of Fe—C≡O bond in heme proteins.

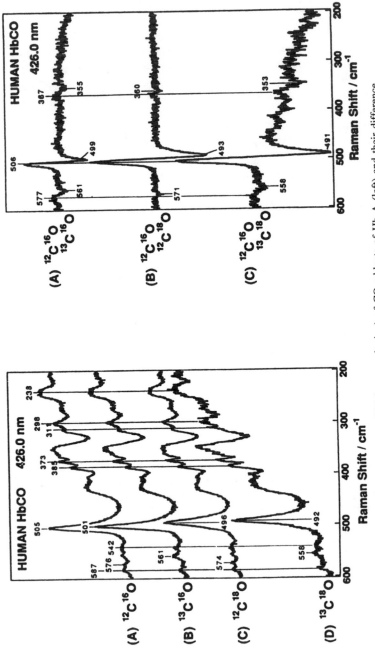

Figure V-5. The RR spectra (426.0 nm excitation) of CO adducts of Hb A (left) and their difference spectra as specified (right) (reproduced with permission from Ref. 27).

come by an increase in the pyrrole (π)–$CO(\pi^*)$ overlap resulting from such tilting. As a result, tilting lowers the $\nu(CO)$ and raises the $\nu(Fe—CO)$ as discussed for "strapped" porphyrins in Sec. III-17(7).

In heme proteins, the upright $\rightleftharpoons$ tilted conformational change is observed by changing the pH of the solution. Thus, the $\nu(Fe—CO)$ of the sperm whale MbCO at 507 cm^{-1} in neutral pH solution is shifted to 488 cm^{-1} in acidic solution.[30] This result corresponds to the previous IR observation that the $\nu(CO)$ of soybean legHbCO at 1947.5 cm in neutral pH is upshifted to 1967 cm^{-1} in acidic solution.[31] As illustrated in Fig. V-7, the N_ϵ atom of the distal histidine is protonated in acidic solution. This may induce the displacement of the histidine to accommodate the upright geometry of the Fe—C$\equiv$O bond. Similar upshifts of the $\nu(CO)$ are observed when the β-chain distal histidine of Hb is replaced by nonpolar residues such as glycine and valine,[32] and when the degree of hydration of hydrated films of Hb and Mb is changed.[33]

The IR spectra of sperm whale MbCO exhibit three $\nu(CO)$ near 1967, 1944, and 1933 cm^{-1} in solution as well as in the crystalline state. Three different environments of the Fe—C$\equiv$O moiety were proposed to account for this observation.[34] In human HbCO, two $\nu(CO)$ bands of α and β subunits overlap to give a single band at 1951 cm^{-1}. In contrast, rabbit HbCO exhibits two $\nu(CO)$ at 1951 (β) and 1928 (α) cm^{-1}; the latter frequently is unusually low, and its intensity is about half that of the former. It has been suggested that the distal histidine acts as a nucleophilic donor to the CO in the α subunit.[35] All the observations of $\nu(CO)$ mentioned above were made by using aqueous IR techniques (Sec. III-10).

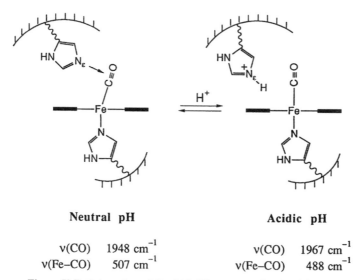

Neutral pH

$\nu(CO)$ 1948 cm^{-1}
$\nu(Fe–CO)$ 507 cm^{-1}

Acidic pH

$\nu(CO)$ 1967 cm^{-1}
$\nu(Fe–CO)$ 488 cm^{-1}

Figure V-7. Orientation of distal histidine at neutral and acidic pH.

(2) O$_2$ Adducts[24,25]

As discussed in Sec. III-19, dioxygen coordinated to metalloporphyrins can take end-on, side-on, and bridging structures. However, the bridging structure is too bulky to occur in a heme cavity. Although the side-on coordination is stereo-chemically possible, it may be too unstable under biological conditions. Thus, the end-on coordination, such as shown in Fig. V-2, is most probable. In fact, this structure was found by X-ray[36] and neutron diffraction[37] on MbO$_2$ and X-ray diffraction on HbO$_2$.[38] These studies also revealed the presence of hydrogen bonding between the bound O$_2$ and the N$_\epsilon$ atom of the distal imidazole (Fig. V-2). The N—H$\cdots$O$_2$ distance in MbO$_2$ is 2.97 Å, whereas in HbO$_2$ it is 3.7 and 3.2–3.4 Å, respectively, for the α- and β-subunits.

The coordinated dioxygen of the end-on type exhibits the $\nu(O_2)$, $\nu(Fe—O_2)$, and $\delta(FeOO)$. Thus far, the $\nu(O_2)$ of heme proteins have been observed only in IR spectra. Attempts to resonance-enhance this mode have been unsuccessful because the "direct-coupling" mechanism invoked for CO adducts does not work or because the oscillator strength of the Fe $\rightarrow$ O$_2$ CT transition is too small.[7] Exceptions are found in five-coordinate Fe(TPP)O$_2$ (Sec. III-19) and O$_2$ adducts of cytochrome P–450 (Sec. V-3). In contrast, the $\nu(Fe—O_2)$ and $\delta(FeOO)$ in the low-frequency region have been observed exclusively by RR spectroscopy. Thus, the $\nu(Fe—O_2)$ vibration of HbO$_2$ was first observed at 567 cm^{-1} by Brunner,[39] and the end-on geometry was confirmed by $^{16}O^{18}O$ experiments which showed two $\nu(Fe—O_2)$ vibrations due to mixing of the Fe–^{16}O–^{18}O and Fe–^{18}O–^{16}O bonds.[40] Recently, Hirota et al.[41] were able to locate the $\delta(FeOO)$ of HbO$_2$ at 425 cm^{-1} using Raman difference techniques. As shown in Fig. V-8, difference features are observed at 568 and 425 cm^{-1}. The same results were obtained independently by Jeyarajah et al.[41a]

The $\nu(O_2)$ of HbO$_2$ was first located at 1107 cm^{-1} by IR spectroscopy.[42] This was followed by similar work on MbO$_2$ which gave almost the same frequency.[43] Later IR studies revealed, however, that HbO$_2$ exhibits two $\nu(O_2)$ bands at 1156 and 1107 cm^{-1}, although a single $\nu(^{18}O_2)$ was observed at 1065 cm^{-1}. Therefore, the observed splitting was attributed to Fermi resonance between $\nu(^{16}O_2)$ near 1130 cm^{-1} and the first overtone of the $\nu(Fe—O_2)$ at 567 cm^{-1}.[44]

The interpretation presented above was challenged by Tsubaki and Yu,[45] who observed the $\nu(O_2)$ of cobalt(II)-reconstituted HbO$_2$ (CoHbO$_2$) using Soret excitation, which is in resonance with the Co–O$_2$ CT transition. These workers observed three oxygen–isotope sensitive bands at 1152 (weak), 1137 (strong), and 1107 cm^{-1} (very weak). The origin of this multiple-band structure was attributed to the presence of two conformers; conformer I is responsible for the bands at 1137 and 1107 cm^{-1}, which result from Fermi resonance between the unperturbed $\nu(O_2)$ (~1122 cm) and the porphyrin mode at 1121 cm^{-1}, whereas conformer II is responsible for the 1152 cm^{-1} band. This interpretation is based on X-ray analysis of MbO$_2$ in which the Fe–O–O plane can take two orientations relative to the porphyrin plane.[36] Thus, in conformer I, the Co–O–O plane

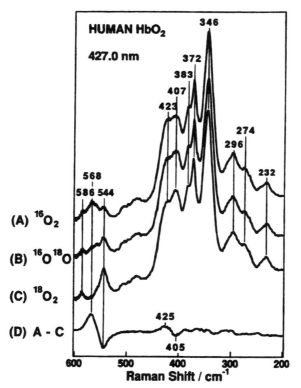

Figure V-8. The RR spectra (427.0 nm excitation) of HbO$_2$ and their difference spectrum as specified (reproduced with permission from Ref. 41).

is in the direction which permits the formation of the N—H··O$_2$ bond mentioned earlier. In conformer II, this plane is rotated by about 40° from that of conformer I so that the O$_2$ is free from hydrogen bonding. As a result, the ν(O$_2$) of the latter (1152 cm^{-1}) is much higher than that of the former (~1122 cm^{-1}). The observed upshift of the 1137 cm^{-1} band (2 cm^{-1}) by D$_2$O/H$_2$O (solvent) exchange was regarded as evidence to support their interpretation.[46] More recent IR studies by Potter et al.[47] confirmed the presence of the three bands mentioned above. These workers noted, however, that the observed difference in ν(O$_2$) (30 cm^{-1}) between the two conformers is too large to attribute it to the effect of hydrogen bonding alone, and proposed a structure of conformer I in which the Fe–O–O and imidazole planes are eclipsed on the N–Fe–N axis of the porphyrin ring since π-electron donation from the imidazole to the O$_2$ mediated through the metal would cause a marked reduction in the ν(O$_2$).

Quite contrary to these investigations, Bruha and Kincaid[48] interpret the RR spectra of CoMbO$_2$ and CoHbO$_2$ in terms of a single conformer. Figure V-9 shows the RR spectra of CoHbO$_2$ obtained by these authors. The complicated features arise because of two reasons. First, several porphyrin vibrations appear

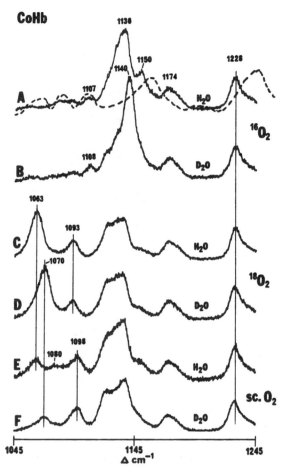

Figure V-9. The RR spectra (406.7 nm excitation) of CoHbO$_2$. The dashed line in (A) is the Raman spectrum of L-histidine (reproduced with permission from Ref. 48).

in this region. They are easily identified at 1228, 1174, ~1136, and ~1123 cm^{-1} because they show no oxygen–isotope shifts and appear in all the compounds studied. Second, the remaining oxygen–isotope sensitive bands are analyzed by considering the possibilities of vibrational couplings between the $\nu(O_2)$ fundamental and internal modes of imidazole (proximal or distal). In this case, vibrational coupling occurs between the $\nu(^{16}O_2)$ near 1136 cm^{-1} and the imidazole mode near 1160 cm^{-1}. Similar coupling occurs between the $\nu(^{18}O_2)$ near 1063 cm^{-1} and the second imidazole mode near 1100 cm^{-1} (these imidazole bands are seen in the spectrum of histidine shown by the dotted line in trace A). Thus, the $\nu(^{16}O_2)$ and $\nu(^{18}O_2)$ of HbO$_2$ are assigned near 1136 and 1063 cm^{-1}, respectively. As stated in Sec. III-19(5), these vibrational couplings have been analyzed quantitatively by using the Fermi resonance scheme.

Vibrational spectra of O_2 adducts of heme protein model compounds such as "picket-fence" and "strapped" porphyrins have also been discussed in Sec. III-19(5).

(3) NO Adducts[24,25]

Similar to the case of O_2 adducts, the NO groups bonded to ferrous Mb and Hb take a bent end-on geometry. The bent Fe–N–O group is expected to show the $\nu(NO)$ at 1700–1600 cm^{-1}, and the $\nu(Fe\!-\!NO)$ and $\delta(FeNO)$ in the 600–400 cm^{-1} region. The NO has been used to probe conformational changes of the heme moiety when the quaternary structure of Hb is switched from the R- to the T-state. Human HbNO in the R-state has four six-coordinate hemes, whereas the T-state is a hybrid of five- and six-coordinate NO moieties. Using IR spectroscopy, Maxwell and Caughey[49] observed the $\nu(NO)$ of six-coordinate heme (R-state) at 1618 cm^{-1} and that of five-coordinate heme (T-state induced by adding IHP (inositol hexaphosphate) at 1668 cm^{-1}. The lack of discernible pH effects on these frequencies suggested that a polar (donor–acceptor) interaction is more likely than hydrogen bonding between the NO and the distal imidazole. Spiro and co-workers[50,51] observed the $\nu(Fe\!-\!NO)$ of these six- and five-coordinate hemes near 550 and 590 cm^{-1}, respectively (413.1 and 454.5 nm excitation). However, Yu and co-workers[52,53] could not detect the latter by 406.7 nm excitation. According to Benko and Yu,[54] the band near 554 cm^{-1} in ferrous MbNO is the $\delta(FeNO)$ and not $\nu(Fe\!-\!NO)$. Their assignment is based on the zigzag isotopic shift pattern in the order of NO (554 cm^{-1}) > ^{15}NO (545 cm^{-1}) < N^{18}O (554 cm^{-1}).

Figure V-10 shows the RR spectra of NO adducts of ferrous Mb obtained by Hu and Kincaid.[55] These workers assigned the bands at 554 and 449 cm^{-1} to the $\nu(Fe\!-\!NO)$ and $\delta(FeNO)$, respectively, although substantial mode mixing was noted. The former shows a zigzag isotope shift pattern, whereas the frequency of the latter decreases monotonously as the total mass of the NO ligand increases. Thus, the observation of a zigzag isotope shift pattern does not necessarily indicate a bending mode.

The NO can also bind to ferric heme proteins, although ferric nitrosyl complexes have a tendency toward spontaneous autoreduction. Since the Fe(III)–NO is isoelectronic with the Fe(II)–CO, it may take a linear geometry which would be distorted in a heme cavity. The $\nu(Fe(III)\!-\!NO)$ and $\delta(Fe(III)NO)$ of MbNO are observed at 595 and 573 cm^{-1}, respectively.[54] In this case, the latter shows a zigzag isotope shift pattern.

The NO group vibrations are also reported for the NO adducts of Co(II)-[56,57] and Mn(II)-reconstituted[58] Mb and Hb. The following trend is found in the Mb series:

	$\nu(Mn(II)\!-\!NO)$	>	$\nu(Fe(III)\!-\!NO)$	>	$\nu(Co(II)\!-\!NO)$	>	$\nu(Fe(II)\!-\!NO)$
$\tilde{\nu}(cm^{-1})$	627		595		576		554

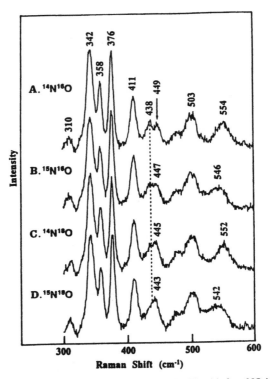

Figure V-10. The RR spectra (406.7 nm excitation) of MbNO with four NO isotopomers (reproduced with permission from Ref. 55).

The RR spectra of NO adducts of Mn(II) complexes of "unprotected" and "strapped" porphyrins (Fig. III-62*b*) have been compared to study the steric effects of the "strap."[59]

(4) Adducts of Other Axial Ligands[24,44]

The CN^- ion binds strongly to ferric heme proteins. The $\nu(CN)$ of ferric MbCN and HbCN are observed at 2125 cm^{-1} in IR spectra.[60] This frequency is higher than that of free CN^- ion (2083 cm^{-1}) for the reason discussed in Sec. III-15. Since the $Fe(III)—C{\equiv}N$ bond is linear, it may be distorted in a heme cavity. Although an X-ray diffraction study on ferric (or met) HbCN confirmed such a distortion, the exact geometry has not yet been known because of poor resolution (2.8 Å).[61] Yu et al.[62] suggest that the linear $Fe—C{\equiv}N$ bond is only tilted because both the $\nu(Fe—^{13}C{\equiv}N)$ and $\nu(Fe—C{\equiv}^{15}N)$ bands appear at 450 cm^{-1}. The CN^- ion can also bind to ferrous heme proteins, but its affinity is much lower and the corresponding complexes are readily photodissociated.

The N_3^- ion binds to ferric heme proteins to form a mixture of high-spin (hs) and low-spin (ls) complexes at room temperature. Thus, the IR spectrum of metMbN$_3$ exhibits two $\nu_a(N_3)$ bands at 2045 and 2023 cm^{-1} which were assigned to the hs and ls complexes, respectively. Similar bands were observed at 2047 (hs) and 2025 (ls) cm^{-1} for metHbN$_3$.[60] In RR spectra, Tsubaki et al.[63] observed two sets of porphyrin vibrations corresponding to the low- and high-spin states of metMbN$_3$. They also assigned the RR band at 411 cm^{-1} to the ν(Fe—N$_3$) of the low-spin complex, although the band at 413 cm^{-1} was previously assigned to the high-spin complex.[63a]

The ν(Fe—OH) of metHbOH is observed at 495 cm^{-1},[64]. In the abnormal subunit of Hb M Boston, the heme iron is bonded to the phenolate oxygen of tyrosine (E7)[65] instead of the proximal histidine (F8). Nagai et al.[66] assigned the band at 603 cm^{-1} to the ν(Fe—O(tyrosine)).

In metMbF, the ν(Fe—F) vibrations were observed at 461 and 421 cm^{-1} which were attributed to the nonhydrogen-bonded and hydrogen-bonded (to water) Fe–F moieties, respectively.[67]

(5) Photochemistry of HbCO and HbO$_2$

Recently, TR^3 spectroscopy (Sec. III-4(2)) has been utilized extensively to study the structures and dynamics of extremely short-lived transient species (in the order of nano (10^{-9}) and pico (10^{-12}) sec) which are created by photolysis of HbCO, HbO$_2$, and other proteins. The results obtained have been reviewed by Rousseau and Friedman[68] and Kincaid.[69] Here we discuss typical results obtained by Terner et al.,[70] who employed TR^3 spectroscopy to monitor structural changes in the porphyrin core which follows the photolysis of HbCO (low spin). The TR^3 spectra obtained with ~30 picosecond and 20 nanosecond pulses (576 nm) exhibit the ν_{10}, ν_{19}, and ν_{11} at 1603, 1552, and 1542 cm^{-1}, respectively. The frequencies of these structure-sensitive bands are very close to but slightly lower (2–4 cm^{-1}) than those of deoxy-Hb (high spin). These observations indicate that a high-spin Fe(II) species has been produced by the photolysis of HbCO within ~30 picoseconds. This spin conversion is ~10^3 times faster than typical spin-conversion rates in the ground state of Fe(II) complexes. These workers suggest that the photolysis pathway involves intersystem crossing for the initially excited singlet π–π* state to a low-lying excited state of HbCO. The observed small downshifts of the structure-sensitive bands indicate a slightly larger core size of the photoproduct relative to that of deoxy-Hb. Namely, the Fe atom in the photoproduct is closer to the heme plane than that of deoxy-Hb (~0.06 Å). These downshifts are observed even when the laser pulses are lengthened to ~20 nanoseconds. The observed slow relaxation to the structure of deoxy-Hb is associated with changes in the globin tertiary structure.

Similar work on HbO$_2$[71] shows that the photoproduct obtained by ~30 picosecond pulses (532 nm) exhibits the ν_{10} and ν_{11} at frequencies lower by 10 and 5 cm^{-1}, respectively, than the HbCO photoproduct. These large shifts were attributed tentatively to the formation of an electronically excited deoxy-Hb.

More recently, Kaminaka et al.[72] studied the dynamics of quaternary structural changes of HbCO after the photolysis by using UV TR^3 spectroscopy (218 nm). Finally, ultrafast (femto (10^{-15}) sec) IR spectroscopy was used to characterize ^{13}CO bonded to the α- and β-subunits of Hb M Boston.[73]

V-3. CYTOCHROMES AND OTHER HEME PROTEINS

(1) Cytochrome c[74]

Cytochromes (a, b, and c) are electron carriers in the mitochondrial respiratory chain. Among them, cytochromes c are relatively small (MW ~ 13,000) and relatively easily crystallized. The structures of cytochromes c from various sources have been determined by X-ray crystallography.[75] These studies show that the prosthetic group of cytochrome c is a heme in which the vinyl side chains of iron protoporphyrin are replaced by cysteinyl thioether bonds and to which the imidazole (His 18) nitrogen and the methionine (Met 80) sulfur (thioether) atoms are coordinated axially. One of the structural features of cytochrome c is the presence of an "opening" at the edge of the heme cavity through which the electron transfer may occur. In most cytochrome c, the iron atoms are in the low-spin state, and the basic structure of the heme is unchanged by changing the oxidation state of the iron.[76] As shown in Table V-1, bands I (ν_4), II (ν_3), and V (ν_{10}) are shifted markedly to higher frequencies in going from the ferrous to ferric state.[14]

Cytochrome c takes five different structures depending on the pH with

I		II		III		IV		V
0.42		2.50		9.35		12.76		

pK values shown above.[77] As stated above, the heme iron is axially bonded to the imidazole nitrogen (His 18) and the methionine sulfur (Met 80) at neutral pH (III). However, these axial ligands are replaced by water at acidic pH (I and II). At alkaline pH, the Fe–S (Met 80) bond is cleaved and may be replaced by another ligand (Lys 79) although the Fe—N (His 18) bond is intact (IV). At extremely alkaline pH, both of these axial ligands may be replaced by other ligands. Thus, vibrational studies of cytochrome c as a function of pH are of particular interest.

The RR spectra of ferri-cytochrome c as a function of pH were first studied by Kitagawa et al.[78] These workers noted that the bands at 1375 (ν_4), 1504 (ν_3), 1563 (ν_{11}), and 1637 cm (ν_{10}) are shifted by 2–3 cm^{-1} to higher frequencies when the pH is increased from 7 to 10.8. This result is expected since a weak π-back donation from the Met 80 to the porphyrin (π^*) via the Fe($d\pi$) orbital is disrupted at alkaline pH. As mentioned above, both axial ligands are replaced by water at pH = 2.5. Lanir et al.[79] observed that the bands at 1563 (ν_{11}), 1585 (ν_{19}), and 1637 cm (ν_{10}) are downshifted to 1556, 1569, and 1623 cm^{-1}, respectively, by decreasing the pH from 7.0 to 2.0. Thus, these workers concluded that structure II mentioned above is high spin.

The RR spectra of ferro-cytochrome c at pH = 7 to 11.2 exhibit many bands below 500 cm^{-1}. At pH = 13.6, however, this feature is replaced by a much simpler spectrum in this region. Valance and Strekas[80] interpret this result as follows: In neutral to alkaline solution, the heme is rigidly held by the peptide chain, and the resulting asymmetric heme activates many Raman bands. At pH over 13, however, the protein structure is relaxed (unfolded) and the symmetry of the heme becomes effectively higher, resulting in fewer Raman bands. Thus far, not much information is available on axial vibrations. The ν(Fe—N (His 18)) is observed at 182 cm^{-1}, and the ν(Fe—S (Met 80)) is estimated to be near 344 cm^{-1}[,81].

X-Ray analysis has been reported on several model compounds of cytochrome c such as [Fe(TPP)(THT)$_2$]ClO$_4$ and [Fe(TPP)(PMS)$_2$]ClO$_4$ where THT and PMS denote tetrahydrothiophene and pentamethylene sulfide, respectively.[82,83] Oshio et al.[84] assigned the ν_a(S—Fe—S) of these compounds at 328 and 323.5 cm^{-1}, respectively, based on ^{54}Fe/^{56}Fe isotope shifts observed in IR spectra.

(2) Cytochrome P-450[85]

Cytochromes P-450 (MW ~ 50,000) are monooxygenase enzymes that catalyze hydroxylation reactions of substrates such as drugs, steroids, pesticides, and carcinogens:

$$\underset{\diagup}{\overset{\diagdown}{}}\!C\!-\!H + O_2 + 2H^+ + 2e^- \longrightarrow \underset{\diagup}{\overset{\diagdown}{}}\!C\!-\!OH + H_2O$$

One of the microbial species in which cytochrome P-450 is found is *Pseudomonas Putida*. When this bacteria is grown in air with camphor as the substrate, cytochrome P-450$_{cam}$ can be isolated in a crystalline form. Thus far, most spectroscopic studies have been made on this compound.

The active site of cytochrome P-450 is an iron protoporphyrin with the iron center axially bound to the mercaptide sulfur of a cysteinyl residue. The axial Fe–S linkage is retained throughout its reaction cycle shown in Fig. V-11.[86] Recent X-ray analysis[87] on cytochrome P-450$_{cam}$ (B state) confirmed this structure. The term P-450 was derived from the position of the Soret band at 450 nm of its CO adduct.

As stated in Sec. V-1, Ozaki et al.[15] observed the oxidation-state marker band of cytochrome P-450$_{cam}$ in the C state at 1346 cm^{-1} which is much lower than those of other Fe(II) porphyrins. Similar observations have been made for cytochromes P-450 from other sources.[88] This anomaly was attributed to the extra negative charge transmitted to the porphyrin $\pi^*(e_g)$ orbital from the mercaptide sulfur (RS$^-$) which has two lone-pair electrons. Champion et al.[89] first observed the ν(Fe–S) of cytochrome P-450$_{cam}$ (B state) at 351 cm^{-1} in the RR spectrum (364 nm excitation). This band is shifted by ^{54}Fe/^{56}Fe and ^{32}S/^{34}S

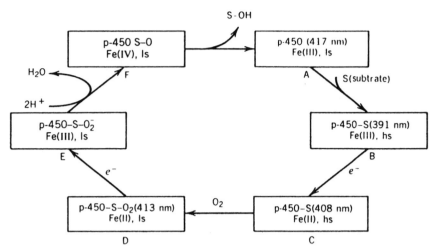

Figure V-11. Reaction cycle of cytochrome P-450.

substitutions by 2.5 ± 0.2 and 4.9 ± 0.3 cm^{-1}, respectively. Later, excitation profile studies on the ν(Fe–S) and other modes were carried out by Bangcharoenpaurpong et al.[90] The ν(Fe–S) of model compounds such as Fe(TPP)(SC$_6$H$_5$) are observed at 345–335 cm^{-1} in IR spectra.[84]

According to IR studies by O'Keefe et al.,[91] cytochrome P-450$_{cam}$—CO exhibits the ν(CO) at 1940 cm^{-1} (nonlinear FeCO bond), whereas the camphor-free compound exhibits two ν(CO) at 1963 (linear) and 1942 cm^{-1} (nonlinear). The ν(CO) of model compounds such as Fe(T$_{piv}$PP)(CH$_3$S)(CO) is observed at 1945 cm^{-1},[92]. Uno et al.[93] assigned the bands at 1940, 558, and 481 cm^{-1} to the ν(CO), δ(FeCO), and ν(Fe—CO), respectively. This ν(Fe—CO) is markedly lower than that of HbCO (507 cm^{-1}) due to the *trans* effect of the mercaptide sulfur discussed earlier.

The ν(O$_2$) of cytochrome P-450$_{cam}$—O$_2$ (D state) was first observed at 1140 cm^{-1} in RR spectra (420 nm excitation) by Bangcharoenpaurpong et al.[94] This frequency is very close to the ν(O$_2$) of a model compound, [Fe(T$_{piv}$PP)(SC$_6$HF$_4$)(O$_2$)]$^-$ (1139 cm^{-1}), observed in IR[95] as well as in RR spectra.[96] In a Co(II)-substituted model compound, [Co(TPP)(SC$_6$H$_5$)(O$_2$)]$^-$, the ν(O$_2$) is at 1122 cm^{-1}, which is 22 cm^{-1} lower than that of Co(TPP)(py-d$_5$)(O$_2$).[97]

The ν(Fe—O$_2$) of cytochrome P-450$_{cam}$—O$_2$ was first located at 541 cm^{-1} in RR spectra by Hu et al.[98] As seen in Fig. V-12, this band is rather weak, but its presence is confirmed by the difference spectrum (trace C). These workers also noted that two ν(O$_2$) are observed at 1139 and 1147 cm^{-1} when camphor is replaced by adamantanone. This may indicate the existence of two conformers which have different types of interactions between the bound O$_2$ and the substrate.

Hu and Kincaid[99] studied the effect of changing the substrate structure on

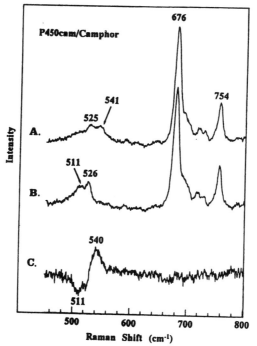

Figure V-12. The RR spectra (441.6 nm excitation) of O_2 adducts of cytochrome P-450$_{cam}$: (A) $^{16}O_2$, (B) $^{18}O_2$ and their difference spectrum (C) (reproduced with permission from Ref. 98).

the Fe(III)–NO bond of cytochrome P-450. Table V-2 summarizes their results together with those obtained for the Fe(II)–NO series. The Fe(III)–NO bond is expected to be linear since it is isoelectronic with the Fe(II)–CO bond. It is seen that the ν(Fe–NO) of ferri-cytochrome P-450–NO is by 70 cm^{-1} lower than that of ferri–Mb–NO because of the thiolate ligand in the former. Furthermore, this band is shifted sensitively by changing the substrate. These results have been

TABLE V-2. Effect of Substrates on FeNO Group Vibrations of Cytochrome P-450–NO (cm^{-1})[99]

Substrate	Fe(III)–NO (Linear)		Fe(II)–NO (Bent)	
	ν(Fe–NO)	δ(FeNO)	ν(Fe–NO)	δ(FeNO)
Substrate-free	528	—	547	444
Norcamphor	524	—	545	441
Camphor	522	546	553	445
Adamantanone	520	542	554	446
Mb–NO	594[a]	573[a]	554[b,c]	450[b]

[a]Ref. 54.
[b]Ref. 55.
[c]This band was previously (Ref. 54) assigned to the δ(FeNO).

explained by considering electronic and kinematic effects for a slightly bent Fe(III)–NO bond in the substrate-bound form. Recently, their work has been extended to the cyanide adducts of cytochrome P-450$_{cam}$.[100]

According to Fig. V-11, hydroxylation of the substrate molecule is accomplished by the activated oxygen released from the oxoferrylporphyrin (F state). Although such a state has not yet been characterized spectroscopically, oxoferryl stretching (Fe(IV)=O) vibrations have been observed for model compounds such as FeO(TPP) (852 cm^{-1}) (Sec. III-20(3)) and for Horseradish Peroxidase Compound II (HRP–II) at ~780 cm^{-1}.

(3) Horseradish Peroxidase[101,102]

Peroxidases are enzymes that catalyze the oxidation of a substrate, AH$_2$, by H$_2$O$_2$:

$$AH_2 + H_2O_2 \rightarrow A + 2H_2O$$

Among them, reaction mechanisms of horseradish peroxidase (HRP) (MW ~ 40,000) have been studied most extensively. The active site of HRP is the same as that of Mb, namely, iron protoporphyrin, which is axially bonded to the proximal histidine. However, there are marked differences between the two; HRP binds O$_2$ irreversibly, whereas Mb does so reversibly. Also, HRP is active biologically in the ferric state, whereas Mb is active in the ferrous state. This may be due to the difference in the heme environment; the proximal histidine in HRP is strongly hydrogen-bonded to nearby amino acid residues and this hydrogen bonding increases σ-basicity of the proximal histidine. As a result, the ν(Fe–N (His)) of HRP is at 244 cm^{-1}, which is much higher than that of Mb (220 cm^{-1}).[103]

The reaction cycle of HRP involves two intermediates, HRP–I and HRP–II.

$$HRP \text{ (ferric)} + H_2O \rightarrow HRP\text{–I} + H_2O$$
$$HRP\text{–I} + AH_2 \rightarrow HRP\text{–II} + AH$$
$$HRP\text{–II} + AH \rightarrow HRP(\text{ferric}) + A + H_2O$$

Thus, HRP–I (green) and HRP–II (red) have oxidation states higher than the native Fe(III) state by two and one oxidizing equivalents, respectively. It has been found that both intermediates are oxoferryl (Fe(IV)=O) porphyrins and that HRP–II is low-spin Fe(IV), whereas HRP–I is its π-cation radical, which is one electron deficient in the porphyrin π-orbital of HRP–II.

As expected from its high oxidation state, HRP–II exhibits the ν_4 at 1381 cm^{-1}, which is the highest among heme proteins.[104] The ν(Fe=O) of HRP–II was reported by Hashimoto et al.[105] and Terner et al.[106] almost simultaneously. Figure V-13 shows the RR spectra of HRP–II obtained by the former work-

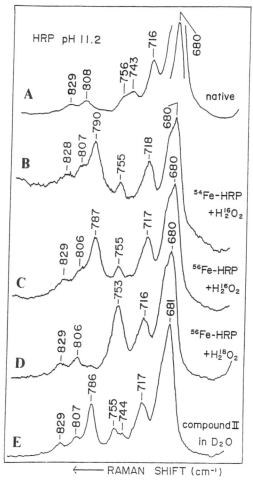

Figure V-13. The RR spectra (406.7 nm excitation) of HRP–II (pH = 11.2) containing isotopomers as indicated (reproduced with permission from Ref. 101).

ers. Upon reacting HRP with H_2O_2 at alkaline pH, a new band appears at 787 cm^{-1} (trace C) which is shifted to 790 cm^{-1} by $^{56}Fe/^{54}Fe$ substitution (trace B), and to 753 cm^{-1} by $H_2^{16}O_2/H_2^{18}O_2$ substitution (trace D). Thus, this band was assigned to the $\nu(Fe{=}O)$ of HRP–II. In neutral solution, the $\nu(Fe{=}O)$ band was observed at 774 cm^{-1} which was shifted to 740 cm^{-1} by $H_2^{16}O_2/H_2^{18}O_2$ substitution. The observed downshift (787 to 774 cm^{-1}) in going from alkaline to neutral solution has been attributed to the formation of a hydrogen bond between the oxoferryl oxygen and the NH group of the distal histidine which disappears in alkaline pH.

As discussed in Sec. III-20(3), the $\nu(Fe{=}O)$ of model compounds such as FeO(TPP) were first observed near 852 cm^{-1} in O_2 matrices. These frequencies are much higher than that of HRP–II because the former is a five-coordinate complex. In fact, the $\nu(Fe{=}O)$ of six-coordinate model compounds such as FeO(TPP)(1-MeIm) (820 cm^{-1}) are lower than that of five-coordinate complexes.

HRP–I is much more unstable and photolabile than HRP–II. The RR spectrum of a model compound of HRP–I, FeO(TMP$^{\cdot+}$), was first measured by Hashimoto et al.[107] who located the $\nu(Fe{=}O)$ at 828 cm^{-1} in CH_2Cl_2/CH_3OH mixed solvents at $-80°C$. However, Kincaid et al.[108] observed the $\nu(Fe{=}O)$ of the same compound at 802 cm^{-1} in pure CH_2Cl_2 at $-78°C$, and suggested that the 828 cm^{-1} band observed by the former workers is due to a photoproduct. According to Kincaid et al., the $\nu(Fe{=}O)$ of FeO(TMP$^{\cdot+}$) (802 cm^{-1}) is 41 cm^{-1} lower than that of the nonradical, FeO(TMP) (843 cm^{-1}). Correspondingly, the $\nu(Fe{=}O)$ of HRP–I (737 cm^{-1}) observed by Paeng and Kincaid[109] is 37 cm^{-1} lower than that of HRP–II (774 cm^{-1}). Later, Hashimoto et al.[110] presented evidence to show that the bands observed at 828 and 802 cm^{-1} by these two groups are due to CH_3OH-bound and CH_3OH-free species, respectively. However, this interpretation contradicts the general rule that the Fe–L (axial) stretching frequencies of five-coordinate species are higher than those of six-coordinate species. Clearly, further experiments are needed to solve this controversy.

When one electron is lost from the porphyrin π-orbital, two types of π-cation radicals, $^2A_{1u}$ and $^2A_{2u}$, are formed, and the distinction between them can be made based on the directions and magnitudes of the shifts of porphyrin core vibrations such as ν_2 and ν_4 (Sec. III-20(3)). However, these criteria are applicable to porphyrins having no or weak axial ligands. Furthermore, the frequencies (ν_2, ν_{11}, and ν_4) observed by four research groups[108,111–113] are markedly different among them. Thus, the radical character of HRP–I is still undetermined.

(4)   Other Heme Proteins

There are many other heme proteins on which IR and RR studies have been made. These include cytochrome c oxidase (CCO), cytochrome c peroxidase (CCP), myeloperoxidase (MPO), lactoperoxidase (LPO), and catalase (CAT). Several review articles mentioned previously should be consulted for these and other heme proteins.

V-4.   HEMERYTHRINS[114–118]

Hemerythrins (Hr) are molecular oxygen carriers found in invertebrate plyla. Different from Hb and Mb, Hr have no heme groups. Thus far, spectroscopic investigations have been concentrated on hemerythrin isolated from *Golfingia gouldii*, a sipunculan worm (MW 108,000) which consists of eight identical subunits. Each unit contains 113 amino acids and two Fe atoms, and each pair of Fe atoms binds one molecule of dioxygen. However, its oxygen affinity is slightly lower than hemoglobin, and no cooperativity is found in its oxygenation reaction. Deoxy-Hr (colorless) turns to pink upon oxygenation ("pink blood").

Figure V-14 shows the primary structure of Hr obtained from *G. gouldii*, while Fig. V-15 shows the tertiary structure of monomeric myohemerythrin obtained by low-resolution X-ray analysis[15]; it consists of four nearly parallel helical segments, 30–40 Å long, connected by sharp nonhelical turns.

Figure V-16 shows the electronic spectra of deoxy-, oxy-, and Met-Hr obtained by Dunn et al.[120] The oxy form exhibits a band at 500 nm which does not exist in the deoxy form. When the laser wavelength falls under this electronic absorption, two bands are resonance enhanced at 844 and 500 cm^{-1}

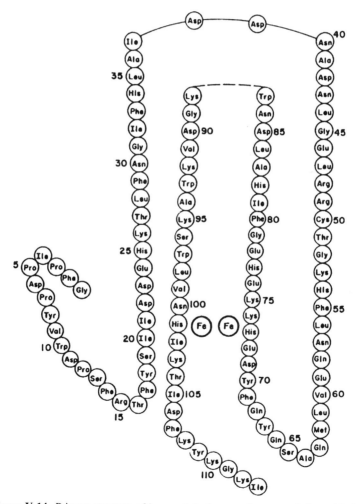

Figure V-14. Primary structure of hemerythrin from erythrocytes of *G. gouldii*.[119]

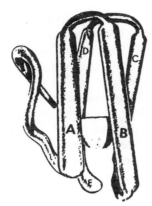

Figure V-15. Tertiary structure of monomeric myohemerythrin.[119]

which are shifted to 798 and 478 cm^{-1}, respectively, by $^{16}O_2$-$^{18}O_2$ substitution (Fig. V-17). These two bands are assigned to the $\nu(O_2)$ and $\nu(Fe-O_2)$ of the oxy form, respectively. Apparently, the electronic transition at 500 nm is due to $Fe \rightarrow O_2$ charge transfer. Also, the observed frequency of $\nu(O_2)$ ($844\ cm^{-1}$) suggests that the dioxygen is not of "superoxo" but of "peroxo" type (Sec. III-19).

In order to gain more information about the geometry of O_2 binding, Kurtz et al.[121] measured the RR spectra of the oxy-Hr with isotopically scrambled oxygen ($^{16}O_2$/$^{16}O^{18}O$/$^{18}O \approx 1/2/1$). Figure V-18A shows that the central band

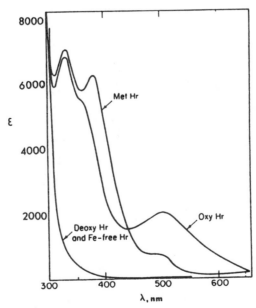

Figure V-16. Electronic spectra of hemerythrin in the deoxy, oxy, and Met forms.[120]

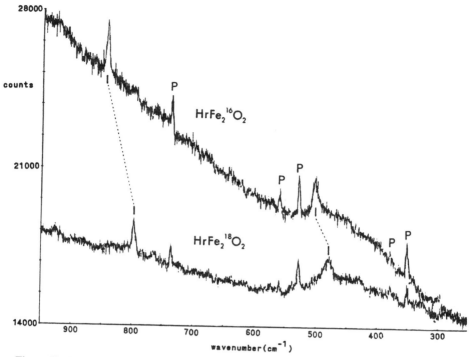

Figure V-17. The RR spectra (488.0 nm excitation) of oxy-hemerythrin ($^{16}O_2$ and $^{18}O_2$)[120]; p denotes laser plasma lines.

due to the $^{16}O^{18}O$ adduct clearly splits into two peaks, indicating the nonequivalence of the two oxygen atoms. This conclusion is also supported by the RR spectrum in the $\nu(Fe-O_2)$ region. As is seen in Fig. V-18B, the spectrum consists of two composite bands, one near 502 cm^{-1} and the other near 483 cm^{-1}. Simple normal coordinate calculations on models I and II indicate

that the $\nu(Fe-O_2)$ of the Fe–$^{16}O^{16}O$ (a) and Fe–$^{16}O^{18}O$ (b) adducts nearly overlap, as do those of the Fe–$^{18}O^{16}O$ (c) and Fe–$^{18}O^{18}O$ (d) adducts (a–d refer to the vertical lines in Fig. V-18B). If the two oxygen atoms were equivalent as shown below,

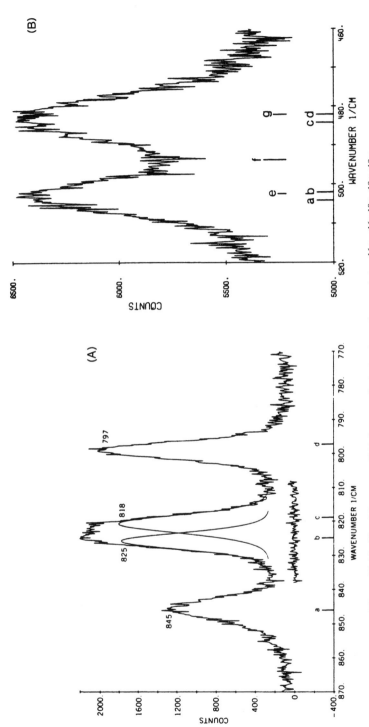

Figure V-18. The RR spectra (514.5 nm excitation) of oxy-hemerythrin ($^{16}O_2/^{16}O^{18}O/^{18}O^{18}O_2$ = 1/2/1).[121] (A) The $\nu(O_2)$ region. The smooth curves represent deconvolution of the 822 cm^{-1} feature into two components. The difference between the observed and fitted curves is shown below the spectrum near 822 cm^{-1}. Lines a–d show the calculated peak positions for models I and II of Fe-$^{16}O_2$ (845 cm^{-1}), Fe-$^{16}O^{18}O$ (825 cm^{-1}), Fe-$^{18}O^{16}O$ (818 cm), and Fe-$^{18}O_2$ (797 cm^{-1}), respectively. (B) The ν(Fe–O$_2$) region. Lines a–d represent calculated positions for the isotopic species defined in (A). Lines e–g show, for models III and IV, the calculated peak positions and estimated relative intensities for $^{16}O_2$ (502 cm^{-1}), $^{16}O^{18}O$ (495 cm^{-1}), and $^{18}O_2$ (489 cm^{-1}), respectively.

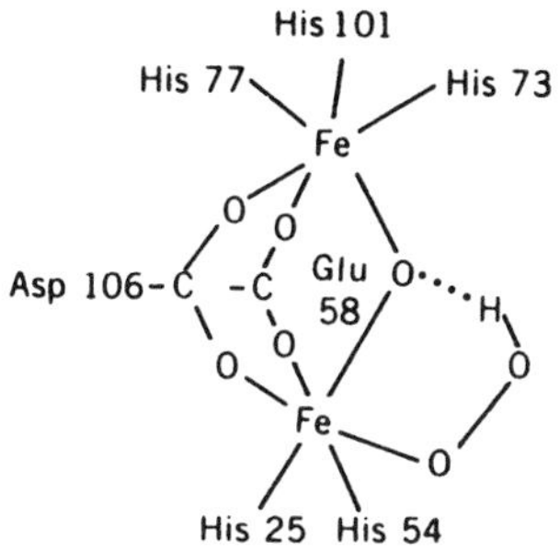

Figure V-19. Structure of the active site of hemerythrin in the oxy form.[123]

a three-peak spectrum with the $1:2:1$ intensity ratio would have appeared in the positions indicated by e–g in Fiv. V-18B.

Later, X-ray analyses were carried out on *met*-azidohemerythrin[122] and oxyhemerythrin.[123] Figure V-19 shows the structure of the active site of the latter; the two Fe atoms are separated by 3.25 Å, and bridged by an oxo atom and two carboxylate groups of the peptide chain. The structure of the former is similar except that the protonated peroxide ion is replaced by the azide ion. Shiemke et al.[124] observed the ν_a(FeOFe) and ν_s(FeOFe) of the oxo bridge at 753 and 486 cm^{-1}, respectively, in the RR spectrum (363.8 nm excitation). They also noted that both $\nu(O_2)$ (844 cm^{-1}) and ν(Fe–O_2) (503 cm^{-1}) of oxy-Hr in H_2O are shifted by +4 and −3 cm^{-1}, respectively, in D_2O solution. These shifts are consistent with the protonated peroxide structure shown in Fig. V-19.

Recently, Kaminaka et al.[125] found via RR studies that, in a cooperative hemerythrin (*Lingula unguis*), hydrogen bonding between bound O_2 and a nearby amino acid residue is responsible for cooperativity in oxygen affinity.

As shown in Fig. V-19, the two iron centers in hemerythrin are bonded via one oxo and two carboxylato bridges. Similar structures are found in enzymes such as ribonucleotide reductase, purple acid phosphatase and methane monooxygenase.[126] In Sec. III-20(4), the ν_a(FeOFe) and ν_s(FeOFe) are listed for a number of model compounds which mimic these oxo-bridged dinuclear iron centers.

V-5. HEMOCYANINS[114,127]

Hemocyanins (Hc) are oxygen-transport nonheme proteins (MW$10^5 \sim 10^7$) which are found in the blood of some insects, crustaceans, and other invertebrates. One of the smallest Hc (MW 450,000) extracted from spiny lobster *Panulirus interruptus* consists of six subunits each containing two Cu atoms. Upon oxygenation, the deoxy form (Cu(I), colorless) turns to blue (Cu(II), "blue blood") by binding one O_2 molecule per two Cu atoms.

Oxy-Hc extracted from *Cancer magister* (Pacific crab) and *Busycon canaliculatum* (channeled whelk) exhibit absorption bands near 570 and 490 nm.

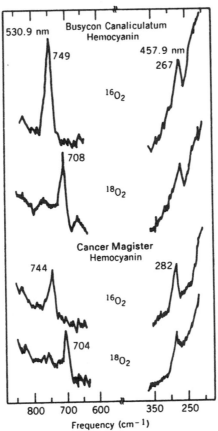

Figure V-20. The RR spectra of oxy-hemocyanins ($^{16}O_2$ and $^{18}O_2$) with 530.9 and 457.9 nm excitation.[128]

Freedman et al.[128] measured the RR spectra of these compounds with 530.9 and 457.9 nm excitations. The results shown in Fig. V-20 clearly indicate that the bands near 747 cm^{-1} are sensitive to $^{16}O_2$-$^{18}O_2$ substitution and must be assigned to the $\nu(O_2)$ characteristic of the peroxo(O_2^{2-}) type. Excitation profiles of the $\nu(O_2)$ consist of two components and indicate that the absorption bands near 570 and 490 nm are due to $O_2^{2-} \rightarrow Cu(II)$ charge transfer. These workers proposed a nonplanar (C_2) structure to account for the appearance of the two CT bands:

$$Cu \blacktriangleright O$$
$$ O \blacktriangleleft Cu$$

The equivalence of the two oxygen atoms in this structure was confirmed by the RR spectrum of oxy-Hc which exhibits a single $\nu(O_2)$ band at 728 cm^{-1}

for the $^{16}O^{18}O$ adduct.[129] This is a marked contrast to oxy-Hr discussed in the preceding section. Larrabee and Spiro[130] observed $\nu(Cu-N(Im))$ below 350 cm^{-1} in the RR spectra of oxy-Hc with 363.8 nm excitation. Their assignments were confirmed by $^{63}Cu-^{65}Cu$ and H_2O-D_2O frequency shifts.

In 1980, Brown et al.[131] carried out an EXAFS study on oxy- and deoxy-Hc of *B. canaliculatum*, and proposed the structure shown in Fig. V-21A for the oxy-form; the two Cu atoms are bound to the protein via three histidine ligands each, and bridged by the O_2^{2-} and an X atom from a protein, possibly tyrosine. Later, Gaykema et al.[132] carried out X-ray analysis (3.2 Å resolution) on colorless single crystals of Hc extracted from *Panulirus interruptus*. This molecule consists of six subunits (MW 75,000), and each subunit is folded into three domains. The structure of the second domain in which two Cu atoms are located is shown in Fig. V-22. The Cu—Cu distance is 3.8 Å, and each Cu atom is coordinated by three histidyl residues as suggested by Brown et al.[131] for the deoxy form. No evidence for a bridging protein ligand was found although it was not possible to rule out such a possibility from low-resolution X-ray analysis.

Recent X-ray analysis by Magnus et al.[132a] revealed that the two Cu(II) atoms in oxy-Hc (from *Limulus polyphemus*) are bridged by a side-on peroxide as shown in Fig. V-21B. Here, each Cu(II) atom takes a square-pyramidal structure with four equatorial bonds (two Cu–N and two Cu–O bonds) and one axial Cu–N bond so as to obtain the overall C_{2h} symmetry. Ling et al.[133] have measured the RR spectra of oxy-Hc from several sources, and made complete band assignments via normal coordinate analysis using isotopic shift data ($^{16}O/^{18}O$, $^{63}Cu/^{65}Cu$, and H/D). The $\nu_a(Cu_2O_2)$ and its first overtone are located at 542 and 1085 cm^{-1}, respectively, for oxy-Hc from *Octopus dofleini*. The $\nu(Cu-N(His))$ of oxy-Hc (*L. polyphemus*) appear in the 370–190 cm^{-1} region although some of these are coupled with the $\nu(Cu-O)$ modes.

Fager and Alben[134] studied the FTIR spectra of HcCO using $^{13}C^{16}O$ and $^{12}C^{18}O$, and proposed a structure in which the CO is coordinated to one Cu via the O atom in a trigonal-planar fashion while the second Cu is free from such interaction. Pate et al.[135] proposed the μ-1,3 bridging structure for *met*-HcN$_3$ based on RR spectra obtained by using the isotopic $^{14}N_2^{15}N^-$ ligand.

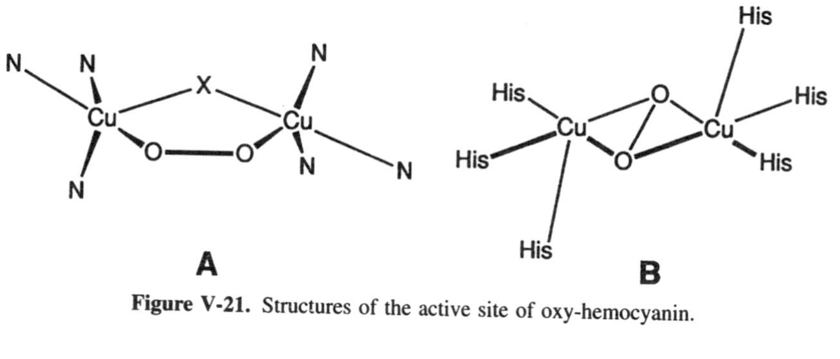

Figure V-21. Structures of the active site of oxy-hemocyanin.

$$^{14}N-^{14}N-^{15}N \qquad\qquad ^{15}N-^{14}N-^{14}N$$
$$\text{Cu} \qquad\qquad \text{Cu} \qquad\qquad \text{Cu} \qquad\qquad \text{Cu}$$

For the $^{14}N-^{14}N-^{15}N$ ion, two $\nu(N_3)$ bands were observed at 2035 and 2024 cm^{-1}. This observation suggests nonequivalence in the two Cu—N interactions which originates in differences between the two Cu environments in the protein.

Karlin et al.[136] first prepared a model compound of Hc which is shown in Fig. V-23A. This complex performs reversible oxygenation at $-70°$C. As seen in Fig. V-24, the $^{16}O^{18}O$ adduct exhibits a broad $\nu(O_2)$ centered at 780 cm^{-1} (peroxide-type) and two $\nu(Cu—O)$ bands at 486 and 465 cm^{-1} in RR spectra. Through normal coordinate analyses and computer simulations of the observed band shapes, Pate et al.[137] have shown that the peroxide is asymmetrically bonded to the Cu atoms, although the nature of asymmetry is not clear.

Model compounds that mimic HcO$_2$ more closely were prepared by Kitajima et al.[138] As shown in Fig. V-23B, their compounds contain two Cu centers

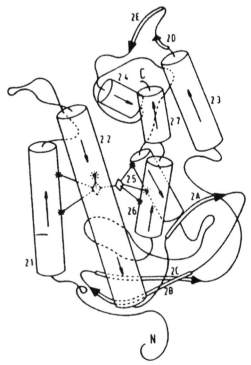

Figure V-22. Structure of the second domain of Hc extracted from *P. interruptus*. The Cu atoms are indicated by diamonds. The cylinders (2.1–2.7) indicate the α-helical structure while the strips (2A–2E) represent the β-structure of the peptide chain.[132]

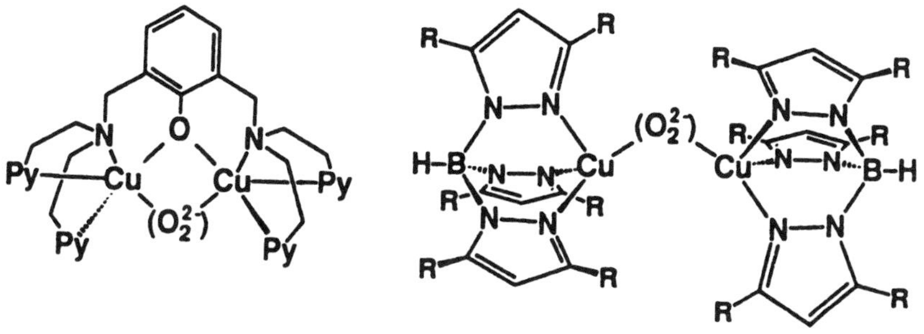

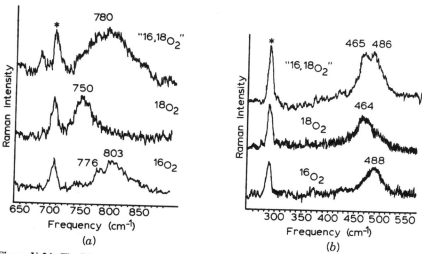

Figure V-23. Structures of model compounds of oxy-hemocyanin. (A) Py denotes the 2-pyridyl group. (B) R denotes the Me, *i*-Pr, or Ph group.

which are bonded via the peroxo bridge without a phenoxo bridge. Figure V-25 shows the RR spectrum of their complex (R = Pr, in Fig. V-23B), which was reacted with isotopically scrambled dioxygen ($^{16}O_2/^{16}O^{18}O/^{18}O_2 = 1/2/1$) at $-40°C$. It is seen that the intensity ratios of the three $\nu(O_2)$ peaks are close to 1/2/1 and their band widths are nearly identical. These results confirm that the peroxide is symmetrically coordinated as shown in Fig. V-21B. Electronic and vibrational spectra of the model compound shown in Fig. V-23A were also studied by Baldwin et al.[138a] Karlin[139] reviewed the reactions of O_2 with copper complexes.

Figure V-24. The RR spectra of model compound A (Fig. V-23) with $^{16}O_2$, $^{18}O_2$, and $^{16}O^{18}O$ (a mixture of $^{16}O_2$, $^{16}O^{18}O$, and $^{18}O_2$ in 1/2/1 ratio). (a) The $\nu(O_2)$ region (488.0 nm excitation). The asterisk indicates the peak due to CH_2Cl_2. (b) The $\nu(Cu–O)$ region (647.1 nm excitation) (reproduced with permission from Ref. 136).

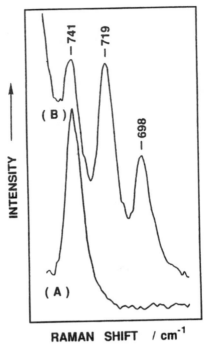

Figure V-25. The RR spectra (514.5 nm excitation) of model compound B (Fig. V-23) in acetone at $-40°C$. (A) $^{16}O_2$. (B) A mixture of $^{16}O_2$, $^{16}O^{18}O$, and $^{18}O_2$ in $1/2/1$ ratio. (reproduced with permission from Ref. 138).

V-6. BLUE COPPER PROTEINS[140,127]

Blue (Type I) copper proteins are found widely in nature. Two groups of blue copper proteins are known: single-copper proteins such as plastocyanin, azurin, and stellacyanin and multicopper proteins such as ceruloplasmin and ascorbate oxidase. For example, plastocyanin (MW ~ 10,500) contains one Cu atom per protein and serves as a component of the electron transfer chain in plant photosynthesis, while ascorbate oxidase (MW ~ 150,000) contains eight Cu atoms per protein and is involved in the oxidation of ascorbic acid. The oxidized forms of these proteins are characterized by intense blue color due to electronic absorption near 600 nm. In addition, blue copper proteins exhibit unusual properties such as extremely small hyperfine splitting constants (0.003 ~ 0.009 cm^{-1}) in ESR spectra and rather high redox potential (+0.2 ~ 0.8 V) compared to the Cu(II)/Cu(I) couple in aqueous solution.

The structure of the active site of blue copper proteins has been a subject of many physicochemical studies.[127] Siiman et al.[141] were the first to measure the RR spectra of blue copper proteins using excitation at 647.1 and 488.0 nm. This was followed by several groups of investigators[142–145] who proposed vibrational assignments and possible structures based on their RR spectra. Figure V-26

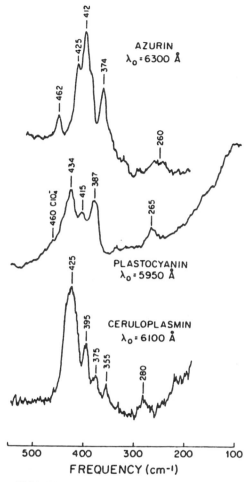

Figure V-26. The RR spectra of three blue copper proteins.[142]

shows the RR spectra of azurin, plastocyanin, and ceruloplasmin obtained by Miskowski et al.[142] Several bands observed between 470 and 350 cm^{-1} were tentatively assigned to the ν(Cu—N) or ν(Cu—O) while a weak band near 270 cm^{-1} was suggested to be the ν(Cu—S). Based on these results and other information, they proposed an approximately trigonal–bipyramidal structure with a sulfur and two nitrogen ligands in the equatorial plane and less strongly bound nitrogen or oxygen ligands at the axial positions.

Siiman et al.[144] also made similar assignments for RR spectra of five blue copper proteins, and proposed a distorted four-coordinate structure involving one cysteinyl sulfur and three nitrogens, at least one of which is an amide nitrogen. These workers assigned the electronic bands near 600 as well as 450 nm to the S → Cu CT transition. Ferris et al.[145] compared the ν(Cu–S) of natural proteins with those of Cu(II) complexes of macrocyclic thiaether (280 ~ 247 cm^{-1}) and mercaptide ligands (~300 cm^{-1}), and concluded that the RR bands near 260 cm^{-1} in blue copper proteins must be assigned to the ν(Cu–S(Met)). Then, the strong bands in the 470–350 cm^{-1} region must be attributed to the ν(Cu–S(Cys)) and ν(Cu–N(His)).[146]

In 1978, the crystal structure of poplar plastocyanin was first determined by X-ray diffraction with 2.7 Å resolution,[147] and later refined to 1.6 Å resolution.[148] Figures V-27A and B show the location of the Cu atom in the peptide chain and the environment around the Cu atom, respectively. It was found that the Cu atom is coordinated by two histidyl nitrogens (His 37 and 87), one cysteinyl sulfur (Cys 84), and one methionyl sulfur (Met 92) in a distorted tetrahedral environment. The two Cu—N(His) distances are 2.10 and 2.04 Å, and the Cu—S(Cys) distance is 2.13 Å, while the Cu—S(Met) is 2.90 Å. In 1986, the crystal structure of azurin was determined by X-ray diffraction with 1.8 Å resolution.[149] The Cu site takes a distorted trigonal–planar structure with three short equatorial bonds (two Cu—N(His) and one Cu—S(Cys)) and two long axial bonds (Cu—S(Met) and Cu—O (Gly)).

Since then, several groups of investigators[150–153] have proposed new assignments based on the results of X-ray diffraction studies mentioned above. These workers assigned the group of bands in the 470–350 cm^{-1} region to the ν(Cu—S(Cys)) mixed with other modes such as ν(Cu—N(His)) and inter-

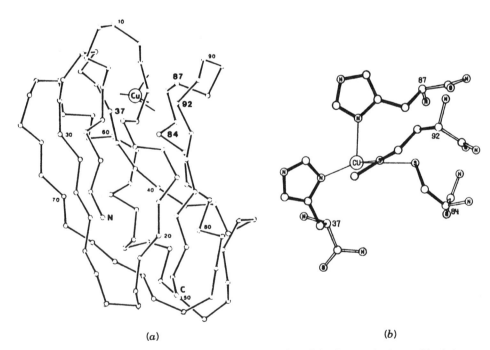

(a)   (b)

Figure V-27. Crystal structure of plastocyanin. (a) Location of the Cu atom in the peptide chain. (b) Environment around the Cu atom.[147]

nal modes of Cys, and the band near 260 cm^{-1} to the ν(Cu—N(His)). Normal coordinate analyses on model compounds[150,152] confirmed their assignments. More recently, vibrational couplings in the former region were examined via isotopic substitutions (^{63}Cu/^{65}Cu, ^{32}S/^{34}S, and ^{14}N/^{15}N).[154,155] Urushiyama and Tobari[156] carried out normal coordinate calculations using 485 internal coordintes around the Cu site of azurin (*Alcaligenes denitrificans*). Their results show that the ν(Cu—S(Cys)) couples not only with the ν(Cu—N(His)) and ν(Cu—S(Met)) but also with many other bending modes of the protein, and that the 417 cm^{-1} vibration possesses the highest ν(Cu—S(Cys)) character (32.7%).

Metal complexes of the type LCuSR (L = hydrotris(3,5-diisopropyl-1-pyrazolyl)borate, R = *t*-Bu, etc.), are known to mimic the Cu sites of blue copper proteins. The nature of vibrational couplings mentioned above has been studied via normal coordinate analysis on the RR spectra of these model complexes.[157]

V-7. IRON–SULFUR PROTEINS[158,159]

Iron–sulfur proteins are found in a variety of organisms, bacteria, plants, and animals, and serve as electron transfer agents via one-electron oxidation–reduction step [redox potential (E_m), -0.43 V in chloroplasts to $+0.35$ V in photosynthetic bacteria]. For example, ferredoxin in green plants (chloroplasts) is involved in the electron transfer system of photosynthesis. The molecular weights of iron–sulfur proteins range from 5600 (rubredoxin from *Clostridium Pasteurianum, Cp*) to 83,000 (beef heart aconitase). All these compounds show strong absorptions in the visible and near-uv regions which are due to Fe $\leftarrow$ S CT transitions. Thus, laser excitation in these regions is expected to resonance-enhance ν(Fe–S) vibrations of iron–sulfur proteins.

The most simple iron–sulfur protein is rubredoxin (Rd), which contains one Fe atom per protein. The Fe atom is coordinated by four sulfur atoms of cysteinyl residues in a tetrahedral environment. Figure V-28 shows the crystal structure of a model compound, [Fe(S$_2$-*o*-xyl)$_2$]$^-$ (S$_2$-*o*-xyl = *o*-xylene-α,α'-dithiolate).[160] Long et al.[161] first obtained the RR spectrum of oxidized rubredoxin, and assigned two bands at 368(ν_3) and 314(ν_1) cm^{-1} to the ν(Fe–S) and those at 150(ν_4) and 126(ν_2) cm^{-1} to the δ(FeS$_4$) of the FeS$_4$ tetrahedron. Later, Yachandra et al.[162] attributed three bands observed near 371, 359, and 325 cm^{-1} of oxidized rubredoxins to the splitting components of the ν_3(F_2) vibration. Figure V-29 shows the RR spectra of oxidized rubredoxin from *Desulfovibrio gigas (Dg)* at 77 K obtained by Czernuszewicz et al.[163] The split components are clearly seen at 376, 363, and 348 cm^{-1} with the ν_1(A_1) at 314 cm^{-1}. These workers were able to assign all the fundamentals as well as overtones and combination bands as indicated in Fig. V-29. Recently, Czernuszewicz et al.[164] studied the origins of the F_2 mode splitting by using model compounds, [FeL$_4$]$^-$ (L = SMe$^-$ and SEt$^-$; (L)$_2$ = S$_2$-*o*-xyl). Saito et al.[165] carried out normal coordinate analysis on rubredoxins, considering over 1000 internal coordinates around the FeS$_4$ site. Their results show the presence of extensive vibrational couplings between the ν(Fe–S) and bending modes of the peptide skeleton.

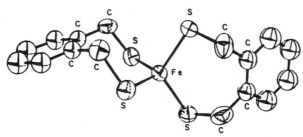

Figure V-28. ORTEP drawing of $[Fe(S_2\text{-}o\text{-xyl})_2]^-$ viewed down the C_2 axis.[160]

The RR spectra of rubrerythrin (from *Desulfovibrio vulgaris*) demonstrates the presence of a rubredoxin-type FeS_4 site as well as a (μ-oxo) diiron (III) cluster.[166]

Two-iron proteins are found in ferredoxin from chloroplast (MW ~ 10,000) and in adrenodoxin from adrena cortex of mammals (MW ~ 13,000), and so on. These proteins contain the $Fe_2S_2(cysteinyl)_4$ cluster in which two Fe atoms are bridged by two "labile" (inorganic) sulfur atoms and each Fe atom is tetrahedrally coordinated by two bridging and two cysteinyl sulfur atoms (2Fe–2S

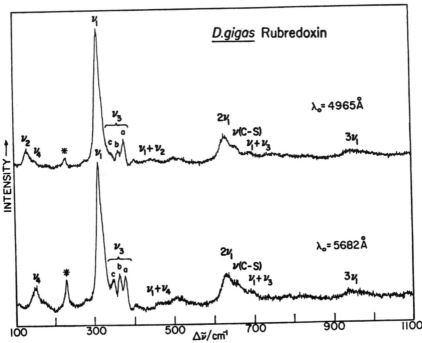

Figure V-29. The RR spectra of oxidized *D. gigas* rubredoxin obtained in a liquid N_2 Dewar using the excitation lines indicated. The asterisk indicates a band due to ice (reproduced with permission from Ref. 163).

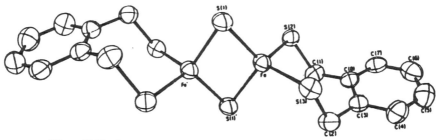

Figure V-30. ORTEP drawing of $[Fe_2S_2(S_2\text{-}o\text{-xyl})_2]^{2-}$ in its Et_4N^+ salt.[168]

cluster). This structure was confirmed by X-ray analysis of the ferredoxin from *Spirulina platensis* (*Sp* Fd).[167] The $Fe_2S_2S_4'$ core (**D**$_{2h}$ symmetry) is modeled by the $[Fe_2S_2(S_2\text{-}o\text{-xyl})_2]^{2-}$ ion whose structure is shown in Fig. V-30.[168] Yachandra et al.[169] measured the RR spectra of oxidized spinach ferredoxin and its ^{34}S-enriched analog containing such a 2Fe–2S cluster. Later, Han et al.[170] remeasured the RR spectra of bovine adrenodoxin (Ado) and ferredoxin (Fd) from *Porphyra umbilicalis* with ^{34}S substituted at the bridge positions. Figure V-31

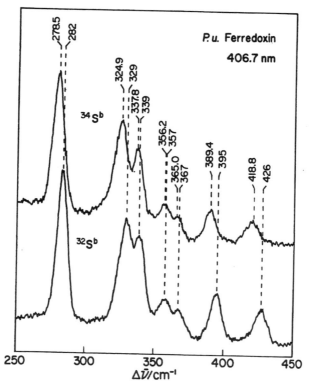

Figure V-31. The RR spectra (77 K) of native and $^{34}S_b$-reconstituted *P. umbilicalis* Fd (reproduced with permission from Ref. 170).

TABLE V-3. RR frequencies and Vibrational Assignments of the $Fe_2S_2(SR)_4$ Core (cm^{-1})[165]

Vibrational Mode[a]	Symmetry[b]	Fd[c]	$[Fe_2S_2(S_2\text{-}o\text{-}xyl)_2]^{2-,c}$
$\nu(Fe-S_b)$	B_{2u}	426(7.2)	415(6.0)
$\nu(Fe-S_b)$	A_g	395(5.6)	391(5.9)
$\nu(Fe-S_b)$	B_{3u}	367(2.0)	342(3.2)
$\nu(Fe-S_t)$	B_{1u} ⎫	357(0.8)	—
$\nu(Fe-S_t)$	B_{2g} ⎭		
$\nu(Fe-S_t)$	A_g	339(1.2)	323(2.0)
$\nu(Fe-S_b)$	B_{1g}	329(4.1)	313(3.2)
$\nu(Fe-S_t)$	B_{3u}	282(3.5)	276(3.2)

[a] S_b and S_t denote the bridging and terminal sulfur atoms, respectively.
[b] D_{2h} symmetry.
[c] The number in parentheses indicates the $^{32}S-^{34}S$ isotopic shift.

shows the RR spectra of the native and $^{34}S_b$-reconstituted Fd measured at 77 K, and Table V-3 lists the band assignments for Fd and its model compound which have been confirmed by normal coordinate calculations. These workers also carried out normal coordinate analysis on model compounds for the $[Fe_2S_2]S_4'$-type proteins to study vibrational couplings between $\nu(Fe-S)$ and bending modes.[171] Kuila et al.[172] measured the RR spectra of *Thermus(thermophilus)* Rieske protein (TRP) and phthalate dioxygenase (PDO) from *Pseudomonas cepacia* and discussed possible structures for the $[Fe_2S_2]S_2'N_2$-type core.

One of the most common Fe–S clusters in iron–sulfur proteins is the 4Fe–4S cube containing interpenetrating Fe_4 and S_4 tetrahedra, the Fe corners of which are bound to cysteinyl sulfur atoms. Figure V-32 shows the X-ray crystal structure of a bacterial ferrodoxin from *Peptococcus aerogenes* (MW ~ 6,000) containing two such clusters.[173] The geometry of this 4Fe–4S cluster is in good agreement with that of the synthetic analog, $[Fe_4S_4(SR)_4]^{2-}$ (R = $CH_2C_6H_5$, C_6H_5, etc.) prepared by Berg and Holm.[174] In both cases, the 4Fe–4S cube is slightly squashed with four short and eight long Fe–S bonds (approximately D_{2d} symmetry).

The RR spectra of 4Fe–4S proteins have been reported by several investigators.[175–178] Figure V-33 shows the RR spectra of the high-potential iron protein (HiPIP) from *Chromatium vinosum* (Cv), ferredoxin from *Clostridium pasteurianum* (Cp), and their model compounds, $(Et_4N)_2$ $[Fe_4S_4(SCH_2Ph)_4]$, in the solid state and in solution obtained by Czernuszewicz et al.[177] Table V-4 lists the observed frequencies and band assignments for these compounds. Based on normal coordinate calculations using $^{32}S/^{34}S$ isotopic shift data, these workers confirmed that the symmetry of the model compound above is T_d in solution but D_{2d} in the solid state. If the symmetry of the $(Fe_4S_4)S_4'$ core shown in Fig. V-34 is T_d, it should give five $\nu(Fe-S_b)$ ($A_1 + E + F_1 + 2F_2$) and two $\nu(Fe-S_t)$ ($A_1 + F_2$) modes. Here, S_b and S_t denote the bridging and terminal S atoms, respectively. If it is D_{2d}, all the degenerate vibrations under T_d symmetry should split into two bands. As a result, a total of 12 vibrations are expected.

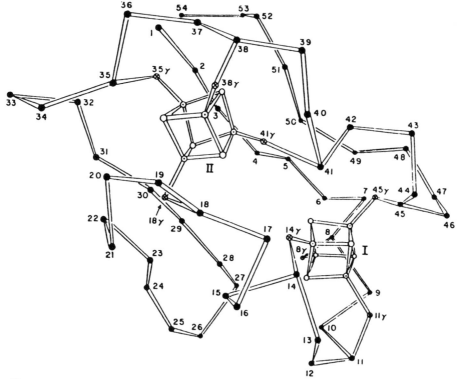

Figure V-32. Structure of bacterial ferredoxin. $\odot$, $\bigcirc$, $\otimes$, and $\bullet$ indicate Fe, S (inorg.), S (cysteinyl), and C atoms, respectively.[173]

Thus, the results shown in Table V-4 and Fig. V-33 support their conclusions. Furthermore, the RR spectra of HiPIP and *Cp* Fd are similar to that of the model compound in solution and in the solid state, respectively. Thus, the symmetries of the Fe_4S_4 cores of these proteins must be $\mathbf{T}_d$ and $\mathbf{D}_{2d}$, respectively.

In the active form of aconitase which catalyzes the isomerization of citrate to isocitrate, one of the Fe atoms in the Fe_4S_4 cluster is coordinated by an OH_x ($x = 1$ or 2) group instead of a cysteinyl residue. Kilpatrick et al.[179] assigned the ν(Fe–S) of aconitase and its model compounds via normal coordinate analysis.

A number of Fe–S proteins contain 3Fe centers. In some cases, the 3Fe center can be converted to the 4Fe center, and vice versa. The structures of 3Fe clusters have been controversial. In 1980, Stout et al.[180] determined the crystal structure of ferredoxin I extracted from *Azotobacter vinelandii* (*Av* Fd I; MW, 14,000) which contains a 3Fe–3S cluster in addition to a 4Fe–4S cluster. For the former, they proposed a novel Fe_3S_3 planar ring structure. However, Beinert et al. showed that four labile sulfides, not three, are associated with the 3Fe center.[181] Thus, these workers as well as Johnson et al.[175] proposed cubane-like structures in which one of the corner Fe atoms is lost (Fig. V-35).

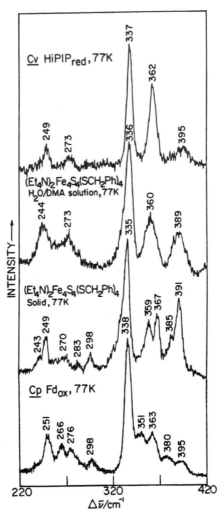

Figure V-33. The RR spectra (77 K) of *Cv* HiPIP$_{red}$ (457.9 nm excitation), *Cp* Fd$_{ox}$ (488.0 nm excitation), and (Et$_4$N)$_2$[Fe$_4$S$_4$(SCH$_2$Ph)$_4$] in solution (room temp., 457.9 nm excitation) and solid state (KCl pellet, 488.0 nm excitation) (reproduced with permission from Ref. 177).

Johnson et al.[175] first measured the RR spectra of *Av* Fd I and *Tt* Fd (from *Thermus thermophilus*), both of which contain 3Fe as well as 4Fe clusters. As seen in Fig. V-36, their RR spectra are dominated by the 3Fe spectra which exhibit bands at 390, 368, 347, 285, and 266 cm^{-1}. The weak band at 334 cm^{-1} is attributed to the 4Fe–4S cluster. Oxidized *Dg* Fd II[182] and ferricyanide-treated *Cp* Fd,[183] which are known to contain only 3Fe clusters, show no such bands. The ^{34}S sulfide substitution in *Tt* Fd and ferricyanide-treated *Cp* Fd produced downshifts of the bands near 266, 285, and 347 cm^{-1}. Therefore, these bands must be assigned to the bridging ν(Fe–S). The strong band at 347 cm^{-1} is due

TABLE V-4. RR Frequencies (cm^{-1}) and Band Assignments for the [Fe$_4$S$_4$(SCH$_2$Ph)$_4$]$^{2-}$ Ion, Oxidized *Cp Fd* and Reduced *Cv* HiPIP^a

D$_{2d}$	[Fe$_4$S$_4$(SCH$_2$Ph)$_4$]$^{2-}$ (Solid)	*Cp Fd*	*Cv* HiPIP	[Fe$_4$S$_4$(SCH$_2$Ph)$_4$]$^{2-}$ (Solution)	**T$_d$**
		Mainly Terminal ν(Fe–S)			
A$_1$	391(1)	395(3.9)	395	384(1)	A$_1$
B$_2$(F$_2$)	367(1)	351(0.7) ⎤			
E(F$_2$)	359(2)	363(2.0) ⎦ 362		358(1)	F$_2$
		Mainly Bridging ν(Fe–S)			
B$_2$(F$_2$)	385(6)	380(5.6) ⎤			
E(F$_2$)	—	— ⎦ 395		384(1)	F$_2$
A$_1$	335(8)	338(7.0)	337	333(7)	A$_1$
A$_1$(E)	298(5)	298(4.9) ⎤			
B$_1$(E)	283(4)	276(4.5) ⎦ 273		268(3)	E
E(F$_1$)	283(4)	276(4.5) ⎤			
A$_2$(F$_1$)	270(3)	266(4.0) ⎦ 273		268(3)	F$_1$
B$_2$(F$_2$)	249(6)	251(6.2) ⎤			
E(F$_2$)	243(5)	— ⎦ 249		241(6)	F$_2$

^{a}Numbers in parentheses indicate downshifts due to S substitution for the bridging S atoms.

to the totally symmetric breathing-cluster mode, while the remaining bands near 390 and 368 cm^{-1} are assigned to the terminal ν(Fe–S).

Normal coordinate calculations by Johnson et al.[175] have shown that the RR spectra of *Av* Fd I crystals and 3Fe-bacterial ferredoxins (*Cp* Fd and *Tt* Fd) are compatible with cubane-like 3Fe–4S structures shown in Fig. V-35, but not with the 3Fe–3S structure reported by Stout et al. The RR spectra of aconitase (inactive form) and *Desulfovibrio desulfuricans*[184] are also very similar to those mentioned above, indicating the possibility of the cubane-like 3Fe–4S structures in these proteins. Recently, the planar Fe$_3$S$_3$ structure originally proposed by Stout et al.[180] was found to be in error.[185]

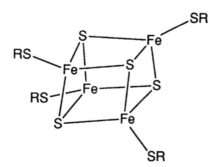

Figure V-34. Structure of the (Fe$_4$S$_4$)S$_4'$ cluster.

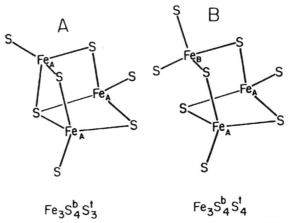

$$Fe_3S_4^b S_3^t \qquad\qquad Fe_3S_4^b S_4^t$$

Figure V-35. Structures proposed for 3Fe–4S clusters.[181]

V-8. INTERACTIONS OF METAL COMPLEXES WITH NUCLEIC ACIDS[186,187]

Nucleic acids interact with a variety of compounds including simple metal ions, metal complexes, antibiotics, carcinogens, and biological stains. In general, their interactions involve three types; intercalation, groove binding, and covalent bonding, and these are often reinforced by hydrogen bonding and/or coulombic interaction.

The mode of interaction can be studied by a variety of physicochemical techniques. Although X-ray crystallography provides the most definitive and precise structural information, its application is limited by the difficulties in growing single crystals of large molecules such as drug–nucleic acid complexes. Recently, vibrational spectroscopy has become a powerful tool in elucidating the mode of interaction in solution as well as in the solid state. In particular, resonance Raman spectroscopy is ideal for such studies because of the reasons mentioned earlier. In the following, we discuss two examples which involve interactions of metal complexes with nucleic acids.

(1) *cis*-Platin

cis-Platin, cis-$Pt(NH_3)_2Cl_2$, is the most well known metal-containing antitumor drug. Alix et al.[188] first compared the interaction of *cis*-platin and its *trans*-isomer with DNA by using Raman spectroscopy (488 nm excitation). Under a low concentration of DNA, with a Pt/DNA(phosphate) ratio of 0.06, they observed the spectral changes listed in Table V-5. Based on these observations, they proposed two possible structures for the *cis*-platin–DNA interaction shown in Fig. V-37. In (A), the Pt atom is chelated via the N-7 and O-6 atoms of the same guanine (G) ligand, whereas in (B), the Pt atom is bonded to two N-7 atoms

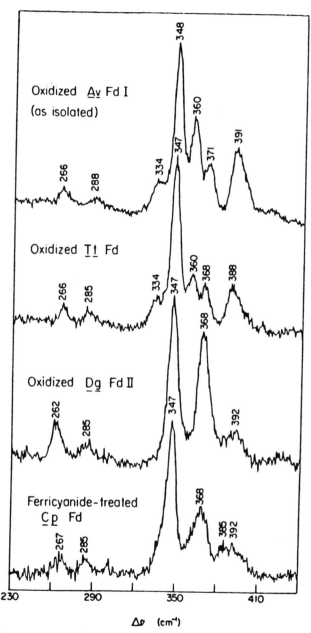

Figure V-36. Low-temperature RR spectra (488.0 nm excitation) of oxidized *Av* Fd I, oxidized *Tt* Fd, oxidized *Dg* Fd II, and ferricyanide-treated *Cp* Fd.[182]

TABLE V-5. Interactions of DNA with *cis*- and *trans*-Pt(NH₃)₂Cl₂ as Observed by Raman Spectroscopy (cm⁻¹)

TABLE V-5. Interactions of DNA with *cis*- and *trans*-Pt(NH$_3$)$_2$Cl$_2$ as Observed by Raman Spectroscopy (cm^{-1})

DNA[a]	DNA + *cis*-Platin	DNA + *trans*-Platin
1628 (G, C)	Shift to 1596	No change
1578 (A, G)	Shift to 1584	No change
—	1540[b]	1540[b]
1490 (G, A)	Large decrease	Small decrease and shift to 1510 cm
—	1412[b]	—
1376 (T, A, G)	Small decrease	No change
684 (G)	Small decrease	No change

[a]G, C, A, and T denote guanine, cytosine, adenine, and thymine, respectively.
[b]Indicates a new band due to platinum-bound G.

of different G ligands in the same strand of DNA. In both cases, the Pt–N(G) bonds are thought to be covalent. FTIR studies by Theophanides[189] also supported these proposals. Later, X-ray analysis on *cis*-Pt(NH$_3$)$_2$ (*d*(GpG))[190] confirmed the intrastrand cross-linking structure shown by Fig. V-37B. As seen in Table V-5, *trans*-Pt(NH$_3$)$_2$Cl$_2$ reacts differently with DNA. Thus far, the Pt–N$_7$ (G) stretching vibration has not been observed.

(2) Water-Soluble Metalloporphyrins

Currently, hematoporphyrin derivatives (HPDs) are used for photoradiation therapy of malignant tumors. Water-soluble metalloporphyrins such as tetrakis (4-N-methylpyridyl) porphyrin, H$_2$(TMpy-P4), and its metal derivatives (Fig.

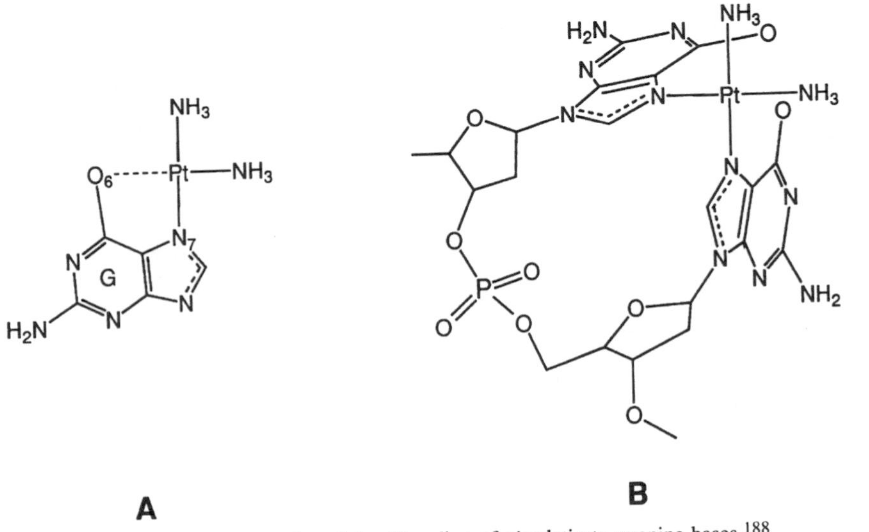

A   **B**

Figure V-37. Proposed models of bonding of *cis*-platin to guanine bases.[188]

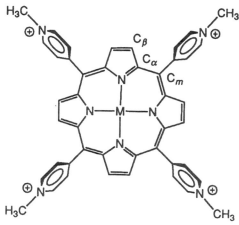

Figure V-38. Structure of M(TMpy-P4).

V-38), which can be obtained in a pure form, are regarded as attractive alternatives to HPDs. In fact, some of these porphyrins are known to cause DNA strand scissions in the presence of oxygen and visible light.

Interactions of water-soluble porphyrins with nucleic acids have been studied by using a variety of physicochemical techniques. The results of these studies show that H_2(TMpy-P4) and its Cu(II) and Ni(II) derivatives with no axial ligands intercalate at the G–C sites, whereas the Zn(II), Co(III), Fe(III), and Mn(III) derivatives with axial water coordination form "outside-bound" and/or "groove-bound" complexes at the A–T regions of DNA.[191]

The porphyrin–DNA system is ideal for RR studies because only porphyrin vibrations can be resonance-enhanced by using excitation lines in the 400–500 nm region. Schneider et al.[192] were the first to measure band shifts of metalloporphyrins resulting from interaction with nucleic acids by using Raman difference techniques. Figure V-39 shows the RR spectra of Cu(TMpy-P4) (trace A) and its mixture with poly(dG–dC)$_2$ (porphyrin/phosphate ratio = 0.04) in dilute solution (trace B). Although nine bands are observed in this region, only six (I, II, V, VI, VIII, and IX) are shifted by interaction with poly(dG–dC)$_2$. Among them, band II near 1100 cm^{-1} shows the largest shift (+6.8 cm). This band corresponds to the ν_9 of Ni(TPP) (Table III-10), and is largely due to the $\delta(C_\beta\text{–H})$ mode. In free Cu(TMpy-P4), the N-methylpyridyl (pyr) rings are nearly perpendicular to the porphyrin plane. In order to form an "intercalated" complex with poly(dG–dC)$_2$, however, it is necessary to rotate the pyr ring toward the porphyrin plane. This would increase repulsion between the C$_\beta$–H and the hydrogen of the pyr ring at the ortho position, resulting in an upshift of the $\delta(C_\beta\text{–H})$ vibration. Conversely, the observation of such a trend signals "intercalation" of the metalloporphyrin between base pairs of nucleic acids. In fact, a mixture of Cu(TMpy-P4) with poly(dA–dT)$_2$, which is known to be "groove-bound," shows only a small upshift (0.2 cm^{-1}) of this band. Similar observa-

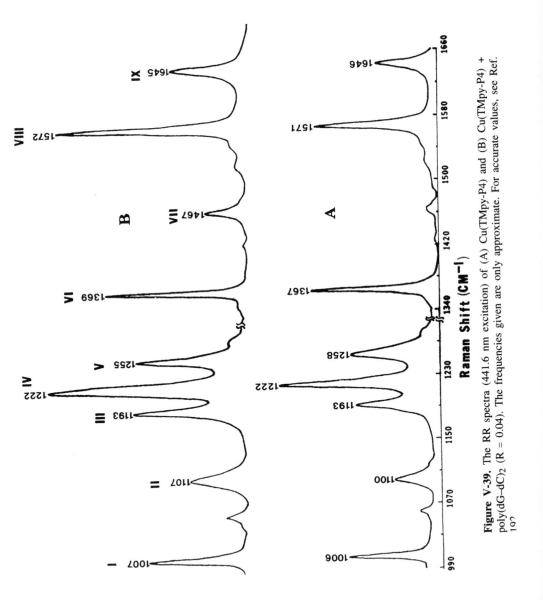

Figure V-39. The RR spectra (441.6 nm excitation) of (A) Cu(TMpy-P4) and (B) Cu(TMpy-P4) + poly(dG–dC)$_2$ (R = 0.04). The frequencies given are only approximate. For accurate values, see Ref. 197

tion is made for Co(III)(TMpy-P4) mixed with poly(dG–dC)$_2$ or poly(dA–dT)$_2$. Bands V (~1258 cm^{-1}) and IX (~1646 cm^{-1}) originate in the N-methylpyridyl group, and show small downshifts regardless of the mode of interaction. Later, Nonaka et al.[193] observed that the ν_a(PO$_2$) at 1221 cm^{-1} and ν_s(PO$_2$) at 1086 cm^{-1} of DNA are upshifted by 17–12 cm^{-1} and downshifted by 26–18 cm^{-1},

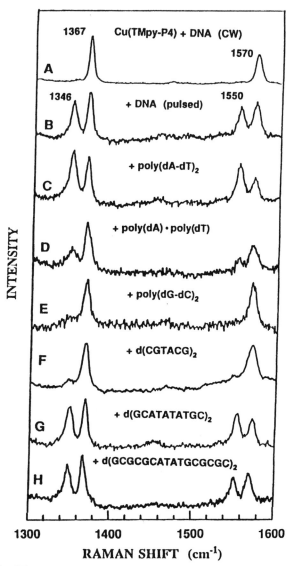

Figure V-40. The RR spectra of Cu(TMpy-P4)–nucleic acid complexes. All the spectra were obtained by using a pulsed-laser excitation (416 nm) except for the top spectrum (CW 406.7 nm excitation).

respectively, when DNA is mixed with M(TMpy-P4). These results indicate the presence of strong coulombic interaction between the N^+–CH_3 groups of M(TMpy-P4) and the PO_2 groups of DNA which strengthens their interaction.

More recently, Strahan et al.[194] found that the Cu(TMpy-P4) intercalated between the GC/CG sequence of DNA is translocated to the ATAT site upon electronic excitation by a pulsed laser. As seen in Fig. V-40, the RR spectrum of Cu(TMpy-P4) obtained by high-power pulsed laser exhibits new bands at 1550 and 1346 cm^{-1} (trace B) which are not observed by CW laser excitation (trace A). These new bands do not appear with low-power pulsed-laser excitation. They are observed with poly(dA–dT)$_2$ (trace C) but not with poly(dG-dC)$_2$ (trace E). These new bands have been attributed to an electronically excited Cu(TMpy-P4) that is stabilized by forming a π-cation radical exciplex, $(Cu(TMpy\text{-}P4))^{+*} (AT)^-$, at an AT site.[195] If oligonucleotides contain GC/CG as well as ATAT or a longer A/T sequence, the exciplex bands are observed

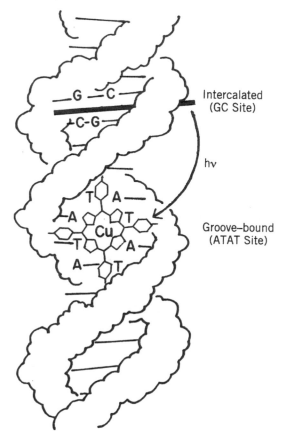

Figure V-41. Schematic diagram showing translocation of Cu(TMpy-P4) from GC to ATAT site upon electronic excitation.

as shown in traces G and H. More elaborate experiments show that, in these cases, some of the intercalated porphyrin at the GC/CG site is translocated to the ATAT site (major groove binding) as illustrated in Fig. V-41. Strahan et al.[194] have estimated the time scale of this translocation and discussed the biological significance of such exciplex formation by laser irradiation.

REFERENCES

1. T. G. Spiro, ed., *Biological Applications of Raman Spectroscopy*, Vols. 1, 2, and 3, Wiley, New York, 1987–1988.

2. A. T. Tu, *Raman Spectroscopy in Biology*, Wiley, New York, 1982.

3. P. B. Carey, *Biochemical Applications of Raman and Resonance Raman Spectroscopies*, Academic Press, New York, 1982.

4. F. S. Parker, *Application of Infrared, Raman and Resonance Raman Spectroscopy in Biochemistry*, Plenum Press, New York, 1983.

4a. T. G. Spiro and R. S. Czernuszewicz, "Resonance Raman Spectroscopy of Metalloproteins," in *Methods in Enzymology*, (K. Sauer, ed.) Vol. 246, Academic Press, San Diego, CA, 1995, p. 416.

5. R. G. Shulman, J. J. Hopfield, and S. Ogawa, *Q. Rev. Biophys.*, **8**, 325 (1975).

6. R. H. Felton and N.-T. Yu, "Resonance Raman Scattering from Metalloporphyrins and Hemopreteins," in D. Dolphin, ed., *The Porphyrins*, Vol. 3, Part A, Academic Press, New York, 1978, p. 347.

7. T. G. Spiro, "The Resonance Raman Spectroscopy of Metalloporphyrins and Heme Proteins," in A. B. P. Lever and H. B. Gray, eds., *Iron Porphyrins*, Addison-Wesley, Reading, MA, 1983.

8. S. A. Asher "Resonance Raman Spectroscopy of Hemoglobin," in S. P. Colowick and N. O. Kaplan, eds., *Methods in Enzymology*, Vol. 76, Academic Press, New York, 1981, p. 371.

9. J. O. Alben, "Infrared Spectroscopy of Porphyrins," in D. Dolphin, ed., *The Porphyrins*, Vol. 3, Part A, Academic Press, New York, 1978, p. 323.

10. S. A. Asher, *Anal. Chem.*, **65**, 59A, 201A (1993).

11. S. Choi, T. G. Spiro, K. C. Langry, K. M. Smith, D. L. Budd, and G. N. La Mar, *J. Am. Chem. Soc.*, **104**, 4345 (1982).

12. V. L. DeVito and S. A. Asher, *J. Am. Chem. Soc.*, **111**, 9143 (1989).

13. P. G. Hildebrandt, R. A. Copeland, T. G. Spiro, J. Otlewski, M. Laskowski, Jr., and F. G. Prendergast, *Biochemistry*, **27**, 5426 (1988).

14. T. G. Spiro and T. C. Strekas, *J. Am. Chem. Soc.*, **96**, 338 (1974).

15. Y. Ozaki, T. Kitagawa, Y. Kyogoku, H. Shimada, T. Iizuka, and Y. Ishimura, *J. Biochem.*, **80**, 1447 (1976).

16. H. Oshio, T. Ama, T. Watanabe, J. Kincaid, and K. Nakamoto, *Spectrochim. Acta*, **40A**, 863 (1984).

17. T. Kitagawa, "Heme Protein Structure and the Iron-Histidine Stretching Mode," in T. G. Spiro, ed., *Biological Applications of Raman Spectroscopy*, Vol. 3, Wiley, New York, 1988, Chapter 3, p. 97.

18. J. Kincaid, P. Stein, and T. G. Spiro, *Proc. Natl. Acad. Sci. U.S.A.*, **76**, 549, 4156 (1979).

19. A. Desbois and M. Lutz, *Biochim. Biophys. Acta*, **671**, 168 (1981).

20. K. Nagai, T. Kitagawa, and H. Morimoto, *J. Mol. Biol.*, **136**, 271 (1980).

21. S. Matsukawa, K. Mawatari, Y. Yoneyama, and T. Kitagawa, *J. Am. Chem. Soc.*, **107**, 1108 (1985).

22. P. Stein, M. Mitchell, and T. G. Spiro, *J. Am. Chem. Soc.*, **102**, 7795 (1980).

23. A. V. Wells, J. T. Sage, D. Morikis, P. M. Champion, M. L. Chiu, and S. G. Sligar, *J. Am. Chem. Soc.*, **113**, 9655 (1991).

24. N.-T. Yu, "Vibrational Modes of Coordinated CO, CN$^-$, O$_2$ and NO, in, T. G. Spiro, ed., *Biological Applications of Raman Spectroscopy*, Vol. 3, Wiley, New York, 1988, Chapter 2, p. 39.

25. K. Nakamota and R. S. Czernuszewicz, "Infrared Spectroscopy," in J. F. Riordan and B. L. Vallee, eds., *Methods in Enzymology*, Part C, Vol. 226, Academic Press, San Diego, CA, 1993.

26. M. Tsubaki, R. B. Srivastava, and N.-T. Yu, *Biochemistry*, **21**, 1132 (1982).

27. S. Hirota, T. Ogura, K. Shinzawa-Itoh, S. Yoshikawa, M. Nagai, and T. Kitagawa, *J. Phys. Chem.*, **98**, 6652 (1994).

27a. S. Hu, K. M. Vogel, and T. G. Spiro, *J. Am. Chem. Soc.*, **116**, 11187 (1994).

28. X.-Y. Li and T. G. Spiro, *J. Am. Chem. Soc.*, **110**, 6024 (1988).

29. J. M. Baldwin, *J. Mol. Biol.*, **136**, 103 (1980).

30. J. Ramsden and T. G. Spiro, *Biochemistry*, **28**, 3125 (1989).

31. W. H. Fuchsman and G. A. Appleby, *Biochemistry*, **18**, 1309 (1979).

32. R. Hoffman, M. M. L. Chen, and D. L. Thorn, *Inorg. Chem.*, **16**, 503 (1977).

33. W. E. Brown, III, J. W. Sutcliffe, and P. D. Pulsinelli, *Biochemistry*, **22**, 2914 (1983).

34. M. W. Makinen, R. A. Houtchens, and W. S. Caughey, *Proc. Natl. Acad. Sci. U.S.A.*, **76**, 6042 (1979).

35. J. D. Sattwellee, M. Teintze, and J. H. Richards, *Biochemistry*, **27**, 1456 (1978).

36. S. E. V. Philips, *J. Mol. Biol.*, **142**, 521 (1980).

37. S. E. V. Philips and B. P. Schoenborn, *Nature*, **292**, 81 (1981).

38. B. Shaanan, *Nature*, **296**, 683 (1982).

39. H. Brunner, *Naturwissenschaflen*, **61**, 129 (1974).

40. L. L. Duff, E. H. Appilelman, D. F. Shriver, and I. M. Klotz, *Biochem. Biophys. Res. Commun.*, **90**, 1098 (1979).

41. S. Hirota, T. Ogura, E. H. Appelman, K. Shinzawa-Itoh, S. Yoshikawa, and T. Kitagawa, *J. Am. Chem. Soc.*, **116**, 10564 (1994).

41a. S. Jeyarajah, L. M. Proniewicz, H. Bronder, and J. R. Kincaid, *J. Biol. Chem.*, **269**, 31047 (1994).

42. C. H. Barlow, J. C. Maxwell, W. J. Wallace, and W. S. Caughey, *Biochem. Biophys. Res. Commun.*, **55**, 91 (1973).

43. J. C. Maxwell and W. S. Caughey, *Biochem. Biophys. Res. Commun.*, **60**, 1309 (1974).

44. J. O. Alben, in D. Dolphin, ed., *The Porphyrins*, Vol. 3, Part A, Academic Press, New York, 1978, p. 323.

45. M. Tsubaki and N.-T. Yu, *Proc. Natl. Acad. Sci. U.S.A.*, **78,** 3581 (1981).

46. T. Kitagawa, M. R. Ondrias, D. L. Rousseau, M. Ikeda-Saito, and T. Yonetani, *Nature*, **298,** 869 (1982).

47. W. T. Potter, M. P. Tucker, R. A. Houtchens, and W. S. Caughey, *Biochemistry*, **26,** 4699 (1987).

48. A. Bruha and J. R. Kincaid, *J. Am. Chem. Soc.*, **110,** 6006 (1988).

49. J. C. Maxwell and W. S. Caughey, *Biochemistry*, **15,** 388 (1976).

50. J. D. Stong, J. M. Burke, P. Daly, P. Wright, and T. G. Spiro, *J. Am. Chem. Soc.*, **102,** 5815 (1980).

51. M. A. Walters and T. G. Spiro, *Biochemistry*, **21,** 6989 (1982).

52. M. Tsubaki and N.-T. Yu, *Biochemistry*, **21,** 1140 (1982).

53. H. C. Mackin, B. Benko, N.-T. Yu, and K. Gersonde, *FEBS Lett.*, **158,** 199 (1983).

54. B. Benko and N.-T. Yu, *Proc. Natl. Acad. Sci. U.S.A.*, **80,** 7042 (1983).

55. S. Hu and J. R. Kincaid, *J. Am. Chem. Soc.*, **113,** 9760 (1991).

56. N.-T. Yu, H. M. Thompson, H. Mizukami, and K. Gersonde, *Eur. J. Biochem.*, **159,** 129 (1986).

57. S. Hu, *Inorg. Chem.*, **32,** 1081 (1993).

58. N. Parthasarathi and T. G. Spiro, *Inorg. Chem.*, **26,** 2280, 3792 (1987).

59. N.-T. Yu, S. H. Lin, C. K. Chang, and K. Gersonde, *Biophys. J.*, **55,** 1137 (1989).

60. S. McCoy and W. S. Caughey, *Biochemistry*, **9,** 2387 (1970).

61. J. F. Deatherage, R. S. Loe, C. M. Anderson, and K. Moffat, *J. Mol. Biol.*, **104,** 687 (1976)

62. N.-T. Yu, B. Benko, E. A. Kerr, and K. Gersonde, *Proc. Natl. Sci. U.S.A.*, **81,** 5106 (1984).

63. M. Tsubaki, R. B. Srivastava, and N.-T. Yu, *Biochemistry*, **20,** 946 (1981).

63a. S. A. Asher and T. M. Schuster, *Biochemistry*, **18,** 5377 (1979).

64. S. A. Asher, L. E. Vickery, T. M. Schuster, and K. Sauer, *Biochemistry*, **16,** 5849 (1977).

65. D. D. Pulsinelli, M. F. Perutz, and R. L. Nagel, *Proc. Natl. Acad. Sci. U.S.A.*, **70,** 3870 (1973).

66. K. Nagai, T. Kagimoto, A. Hayashi, F. Taketa, and T. Kitagawa, *Biochemistry*, **22,** 1305 (1983).

67. S. A. Asher, M. L. Adams, and T. M. Schuster, *Biochemistry*, **20,** 3339 (1981).

68. D. L. Rousseau and J. M. Friedman, "Transient and Cryogenic Studies of Photodissociated Hemoglobin and Myoglobin," in T. G. Spiro, ed., *Biological Applications of Raman Spectroscopy*, Vol. 3, Wiley, New York, 1988, Chapter 4, p. 133.

69. J. R. Kincaid, "Structure and Dynamics of Transient Species Using Time-Resolved Resonance Raman Spectroscopy," in *Methods in Enzymology*, (K. Sauer, ed.) Vol. 246, Academic Press, San Diego, CA, 1995, p. 460.

70. J. Terner, J. D. Stong, T. G. Spiro, M. Nagumo, M. Nicol, and M. A. El-Sayed, *Proc. Natl. Sci. U.S.A.*, **78,** 1313 (1981).

71. J. Terner, D. F. Voss, C. Paddock, R. B. Miles, and T. G. Spiro, *J. Phys. Chem.*, **86,** 859 (1982).

72. S. Kaminaka, T. Ogura, and T. Kitagawa, *J. Am. Soc.*, **112,** 23 (1990).

73. T. Lian, B. Locke, T. Kitagawa, M. Nagai, and R. M. Hochstrasser, *Biochemistry*, **32**, 5809 (1993).

74. B. Cartling, "Cytochrome c," in T. G. Spiro, ed., *Biological Applications of Raman Spectroscopy*, Vol. 3, Wiley, New York, 1988, Chapter 5, p. 217.

75. R. Timkovich, "Cytochrome c," in D. Dolphin, ed., *The Porphyrins*, Vol. 7, Part B, Academic Press, New York, 1979.

76. T. Takano and R. E. Dickerson, *Proc. Natl. Acad. Sci. U.S.A.*, **77**, 6371 (1980).

77. R. E. Dickerson and R. Timkovich, "Cytochromes c," in P. D. Boyer ed., *The Enzymes*, Vol. 11, Academic Press, New York, 1975, p. 397.

78. T. Kitagawa, Y. Ozaki, J. Teraoka, Y. Kyogoku, and T. Yamanaka, *Biochim. Biophys. Acta*, **494**, 100 (1977).

79. A. Lanir, N.-T. Yu, and R. H. Felton, *Biochemistry*, **18**, 1656 (1979).

80. W. G. Valance and T. C. Strekas, *J. Phys. Chem.*, **86**, 1804 (1982).

81. B. Carling, *Biophys. J.*, **43**, 191 (1983).

82. T. Mashiko, J.-C. Marchon, D. T. Musser, C. A. Reed, M. E. Kastner, and W. R. Scheidt, *J. Am. Chem. Soc.*, **101**, 3653 (1979).

83. T. Mashiko, C. A. Reed, K. J. Haller, M. E. Kastner, and W. R. Scheidt, *J. Am. Chem. Soc.*, **103**, 5758 (1981).

84. H. Oshio, T. Ama, T. Watanabe, and K. Nakamoto, *Inorg. Chim. Acta*, **96**, 61 (1985).

85. P. M. Champion, "Cytochrome P450 and the Transform Analysis of Heme Protein Raman Spectra," in T. R. Spiro, ed., *Biological Applications of Raman Spectroscopy*, Vol. 3, Wiley, New York, 1988, Chapter 6, p. 249.

86. For example, see L. S. Alexander and H. M. Goff, *J. Chem. Educ.*, **59**, 179 (1982).

87. T. L. Poulos, B. C. Finzel, I. C. Gunsalus, G. C. Wagner, and J. Kraut, *J. Biol. Chem.*, **260**, 16122 (1985).

88. Y. Ozaki, T. Kitagawa, Y. Kyogoku, Y. Imai, C. Hashimoto-Yutsudo, and R. Sato, *Biochemistry*, **17**, 5826 (1978).

89. P. M. Champion, B. R. Stallard, G. C. Wagner, and I. C. Gunsalus, *J. Am. Chem. Soc.*, **104**, 5469 (1982).

90. O. Bangcharoenpaupong, P. M. Champion, S. A. Martinis, and S. G. Sligar, *J. Chem. Phys*, **87**, 4273 (1987).

91. D. H. O'Keefe, R. E. Ebel, J. A. Peterson, J. C. Maxwell, and W. S. Caughey, *Biochemistry*, **17**, 5845 (1978).

92. J. P. Collman and T. N. Sorrell, *J. Am. Chem. Soc.*, **97**, 4133 (1975).

93. T. Uno, Y. Nishimura, R. Makino, T. Iizuka, Y. Ishimura, and M. Tsuboi, *J. Biol. Chem.*, **260**, 2023 (1985).

94. O. Bangcharoenpaurpong, A. K. Rizos, P. M. Champion, D. Jollie, and S. G. Sligar, *J. Biol. Chem.*, **261**, 8089 (1986).

95. M. Schappacher, L. Richard, R. Weiss, R. Montiel-Montoya, E. Bill, U. Gonser, and A. Trautwein, *J. Am. Chem. Soc.*, **103**, 7646 (1981).

96. G. Chottard, M. Schappacher, L. Richard, and R. Weiss, *Inorg. Chem.*, **23**, 4557 (1984).

97. K. Nakamoto and H. Oshio, *J. Am. Chem. Soc.*, **107**, 6518 (1985).

98. S. Hu, A. J. Schreider, and J. R. Kincaid, *J. Am. Chem. Soc.*, **113**, 4815 (1991).

99. S. Hu and J. R. Kincaid, *J. Am. Chem. Soc.*, **113**, 2843 (1991).

100. M. C. Simianu and J. R. Kincaid, *J. Am. Chem. Soc.*, **117**, 4628 (1995).

101. T. Kitagawa, "Resonance Raman Spectra of Reaction Intermediates of Heme Proteins," in *Raman Spectroscopy of Biological Systems*, (R. J. H. Clark and R. E. Hester, eds.) Vol. 13, Wiley, New York, 1986, p. 443.

102. T. Kitagawa and Y. Mizutani, *Coord. Chem. Rev.*, **135/136**, 685 (1994).

103. J. Teraoka and T. Kitagawa, *J. Biol. Chem.*, **256**, 3969 (1981).

104. G. Rakhit, T. G. Spiro, and M. Uyeda, *Biochem. Biophys. Res. Commun.*, **71**, 803 (1976).

105. S. Hashimoto, Y. Tatsuno, and T. Kitagawa, *Proc. Jpn. Acad.*, **60B**, 345 (1984); *Proc. Natl. Acad. Sci. U.S.A.*, **83**, 2417 (1986).

106. J. Terner, A. J. Sitter, and M. Reczek, *Biochem. Biophys. Acta*, **828**, 73 (1985); *J. Biol. Chem.*, **260**, 7515 (1985).

107. S. Hashimoto, Y. Tatsuno, and T. Kitagawa, *J. Am. Chem. Soc.*, **109**, 8096 (1987).

108. J. R. Kincaid, A. J. Schneider, and K.-J. Paeng, *J. Am. Chem. Soc.*, **111**, 735 (1989).

109. K.-J. Paeng and J. R. Kincaid, *J. Am. Chem. Soc.*, **110**, 7913 (1988).

110. S. Hashimoto, Y. Mizutani, Y. Tatsuno, and T. Kitagawa, *J. Am. Chem. Soc.*, **113**, 6542 (1991).

111. T. Ogura and T. Kitagawa, *Rev. Sci. Instrum.*, **59**, 1316 (1988).

112. V. Palaniappan and J. Terner, *J. Biol. Chem.*, **264**, 16046 (1989).

113. W.-J. Chuang and H. E. Van. Wart, *J. Biol. Chem.*, **267**, 13293 (1992).

114. T. M. Loehr and A. K. Shiemke, "Nonheme Respiratory Proteins," in T. G. Spiro, ed., *Biological Applications of Raman Spectroscopy*, Vol. 3, Wiley, New York, 1988, Chapter 10, p. 439.

115. D. M. Kurtz, Jr., *Chem. Rev.*, **90**, 585 (1990).

116. J. B. Vincent, G. L. Oliver-Lilley, and B. A. Averill, *Chem. Rev.*, **90**, 1447 (1990).

117. I. M. Klotz and D. M. Kurtz, Jr., *Acc. Chem. Res.*, **17**, 16 (1984).

118. D. M. Kurtz, Jr., D. F. Shriver, and I. M. Klotz, *Coord. Chem. Rev.*, **24**, 145 (1977).

119. W. A. Henderickson, G. L. Klippenstein, and K. B. Ward, *Proc. Natl. Acad. Sci. U.S.A.*, **72**, 2160 (1975).

120. J. B. R. Dunn, D. F. Shriver, and I. M. Klotz, *Proc. Natl. Acad. Sci. U.S.A.*, **70**, 2582 (1973).

121. D. M. Kurtz, Jr., D. F. Shriver, and I. M. Klotz, *J. Am. Chem. Soc.*, **98**, 5033 (1976).

122. R. E. Stenkamp, L. C. Sieker, and L. H. Jensen, *J. Am. Chem. Soc.*, **106**, 618 (1984).

123. R. E. Stenkamp, L. C. Sieker, L. H. Jensen, J. D. McCallum, and J. Sanders-Loehr, *Proc. Natl. Acad. Sci. U.S.A.*, **82**, 713 (1985).

124. A. K. Shiemke, T. M. Loehr, and J. Sanders-Loehr, *J. Am. Chem. Soc.*, **106**, 4951 (1984).

125. S. Kaminaka, H. Takizawa, T. Handa, H. Kihara, and T. Kitagawa, *Biochemistry*, **31**, 6997 (1992).

126. J. Sanders-Loehr, W. D. Wheeler, A. K. Shiemke, B. A. Averill, and T. M. Loehr, *J. Am. Chem. Soc.*, **111**, 8084 (1989).

127. "Copper Proteins," in T. G. Spiro, ed., *Metal Ions in Biology*, Vol. 3, Wiley, New York, 1981.

128. T. B. Freedman, J. S. Loehr, and T. M. Loehr, *J. Am. Chem. Soc.*, **98**, 2809 (1976); J. S. Loehr, T. B. Freedman, and T. M. Loehr, *Biochem. Biophys. Res. Commun.*, **56**, 510 (1974).

129. T. J. Thamann, J. S. Loehr, and T. M. Loehr, *J. Am. Chem. Soc.*, **99**, 4187 (1977).

130. J. A. Larrabee and T. G. Spiro, *J. Am. Chem. Soc.*, **102**, 4217 (1980).

131. J. M. Brown, L. Powers, B. Kincaid, J. A. Larrabee, and T. G. Spiro, *J. Am. Chem. Soc.*, **102**, 4210 (1980).

132. W. P. J. Gaykema, W. G. J. Hol, J. M. Vereijken, N. M. Soeter, H. J. Bak, and J. J. Beintema, *Nature*, **309**, 23 (1984).

132a. K. A. Magnus, B. Hazes, H. Ton-That, C. Bonaventura, J. Bonaventura, and W. G. J. Hol, *Proteins*, **19**, 302 (1994).

133. J. Ling, L. P. Nestor, R. S. Czernuszewicz, T. G. Spiro, R. Fraczkiewicz, K. D. Sharma, T. M. Loehr, and J. Sanders-Loehr, *J. Am. Chem. Soc.*, **116**, 7682 (1994).

134. L. Y. Fager and J. O. Alben, *Biochemistry*, **11**, 4786 (1972).

135. J. E. Pate, T. J. Thamann, and E. I. Solomon, *Spectrochim. Acta*, **42A**, 313 (1986).

136. K. D. Karlin, R. W. Cruse, Y. Gultneh, J. C. Hayes, and J. Zubieta, *J. Am. Chem. Soc.*, **106**, 3372 (1984).

137. J. E. Pate, R. W. Cruse, K. D. Karlin, and E. I. Solomon, *J. Am. Chem. Soc.*, **109**, 2624 (1987).

138. N. Kitajima, K. Fujisawa, C. Fujimoto, Y. Moro-oka, S. Hashimoto, T. Kitagawa, K. Toriumi, K. Tatsumi, and A. Nakamura, *J. Am. Chem. Soc.*, **114**, 1277 (1992).

138a. M. J. Baldwin, D. E. Root, J. E. Pate, K. Fujisawa, N. Kitajima, and E. I. Solomon, *J. Am. Chem. Soc.*, **114**, 10421 (1992).

139. K. D. Karlin, *Prog. Inorg. Chem.*, **35**, 219 (1987).

140. W. H. Woodruff, R. B. Dyer, and J. R. Schoonover, "Resonance Raman Spectroscopy of Blue Copper Proteins," in T. G. Spiro, ed., *Biological Applications of Raman Spectroscopy*, Vol. 3, Wiley, New York, 1988, Chapter 9, p. 413.

141. O. Siiman, N. M. Young, and P. R. Carey, *J. Am. Chem. Soc.*, **96**, 5583 (1974).

142. V. Miskowski, S.-P. W. Tang, T. G. Spiro, E. Shapiro, and T. H. Moss, *Biochemistry*, **14**, 1244 (1975).

143. L. Tosi, A. Garnier, M. Hervé, and M. Steinbuch, *Biochem. Biophys. Res. Commun.*, **65**, 100 (1975).

144. O. Siiman, N. M. Young, and P. R. Carey, *J. Am. Chem. Soc.*, **98**, 744 (1976).

145. N. S. Ferris, W. H. Woodruff, D. L. Tennent, and D. R. McMillin, *Biochem. Biophys. Res. Commun.*, **88**, 288 (1979).

146. N. S. Ferris, W. H. Woodruff, D. B. Rorabacher, T. E. Jones, and L. A. Ochrymowycz, *J. Am. Chem. Soc.*, **100**, 5939 (1978).

147. P. M. Colman, H. C. Freeman, J. M. Guss, M. Morata, V. A. Norris, J. A. M. Ramshaw, and M. P. Venkatappa, *Nature*, **272**, 319 (1978).

148. J. M. Guss and H. C. Freeman, *J. Mol. Biol.*, **169**, 521 (1983).

149. G. E. Morris, B. G. Anderson, and E. N. Baker, *J. Mol. Biol.*, **165**, 501 (1983); *J. Am. Chem. Soc.*, **108**, 2784 (1986).

150. T. J. Thaman, P. Frank, L. J. Willis, and T. M. Loehr, *Proc. Natl. Acad. Sci. U.S.A.*, **79**, 6396 (1982).

151. W. H. Woodruff, K. A. Norton, B. I. Swanson, and H. A. Fry, *J. Am. Chem. Soc.*, **105**, 657 (1983).

152. L. Nestor, J. A. Larrabee, G. Woolery, B. Reinhammer, and T. G. Spiro, *Biochemistry*, **23**, 1084 (1984).

153. D. F. Blair, G. W. Campbell, J. R. Schoonover, S. I. Chan, H. B. Gray, B. G. Malmstrom, I. Pecht, B. I. Swanson, W. H. Woodruff, W. K. Cho, A. M. English, H. A. Fry, V. Lum, and K. A. Norton, *J. Am. Chem. Soc.*, **107**, 5755 (1985).

154. D. Qiu, S. Dong, J. A. Ybe, M. H. Hecht, and T. G. Spiro, *J. Am. Chem. Soc.*, **117**, 6446 (1995).

155. B. C. Dave, J. P. Germanas, and R. S. Czernuszewicz, *J. Am. Chem. Soc.*, **115**, 12175 (1993).

156. A. Urushiyama and J. Tobari, *Bull Chem. Soc. Jpn.*, **63**, 1563 (1990).

157. D. Qiu, L. Kilpatrick, N. Kitajima, and T. G. Spiro, *J. Am. Chem. Soc.*, **116**, 2585 (1994).

158. T. G. Spiro, J. Hare, V. Yachandra, A. Gowirth, M. K. Johnson, and E. Remsen, "Resonance Raman Spectra of Iron-Sulfur Proteins and Analogs," in T. G. Spiro, ed., *Iron-Sulfur Proteins*, Wiley-Interscience, New York, 1982, Chapter 11, p. 409.

159. T. G. Spiro, R. S. Czernuswicz, and S. Han, "Iron-Sulfur Proteins and Analog Complexes," in T. G. Spiro, ed., *Biological Applications of Raman Spectroscopy*, Vol. 3, Wiley, New York, 1988, Chapter 12, p. 523.

160. R. W. Lane, J. A. Ibers, R. B. Frankel, R. H. Holm, and G. C. Papaefthymiou, *J. Am. Chem. Soc.*, **99**, 84 (1977).

161. T. V. Long and T. M. Loehr, *J. Am. Chem. Soc.*, **92**, 6384 (1970); T. V. Long, T. M. Loehr, J. R. Alkins, and W. Lovenberg, *ibid.*, **93**, 1809 (1971).

162. V. K. Yachandra, J. Hare, I. Moura, and T. G. Spiro, *J. Am. Chem. Soc.*, **105**, 6455 (1983).

163. R. S. Czernuszewicz, J. LeGall, I. Moura, and T. G. Spiro, *Inorg. Chem.*, **25**, 696 (1986).

164. R. S. Czernuszewicz, L. K. Kilpatrick, S. A. Koch, and T. G. Spiro, *J. Am. Chem. Soc.*, **116**, 7134 (1994).

165. H. Saito, T. Imai, K. Wakita, A. Urushiyama, and T. Yagi, *Bull. Chem. Soc. Jpn.*, **64**, 829 (1991).

166. B. C. Dave, R. S. Czernuszewicz, B. C. Prickril, and D. M. Kurtz, Jr., *Biochemistry*, **33**, 3572 (1994).

167. K. Fukuyama, T. Hase, S. Matsumoto, T. Tsukihara, Y. Katsube, N. Tanaka, M. Kakudo, K. Wada, and H. Matsubara, *Nature*, **286**, 522 (1980).

168. J. J. Mayerle, S. E. Denmark, B. V. DePamphilis, J. A. Ibers, and R. H. Holm, *J. Am. Chem. Soc.*, **97,** 1032 (1975).

169. V. K. Yachandra, J. Hare, A. Gewirth, R. S. Czernuszewicz, T. Kimura, R. H. Holm, and T. G. Spiro, *J. Am. Chem. Soc.*, **105,** 6462 (1983).

170. S. Han, R. S. Czernuszewicz, T. Kimura, M. W. W. Adams, and T. G. Spiro, *J. Am. Chem. Soc.*, **111,** 3505 (1989).

171. S. Han, R. S. Czernuszewicz, and T. G. Spiro, *J. Am. Chem. Soc.*, **111,** 3496 (1989).

172. D. Kuila, J. A. Fee, J. R. Schoonover, W. H. Woodruff, C. J. Batie, and D. P. Ballou, *J. Am. Chem. Soc.*, **109,** 1559 (1987).

173. E. T. Adman, L. C. Sieker, and L. H. Jensen, *J. Biol. Chem.*, **248,** 3987 (1973).

174. T. M. Berg and R. H. Holm, "Iron-Sulfur Proteins," in T. G. Spiro, ed., *Metal Ions in Biology*, Vol. 4, Wiley, New York, 1982, Chapter 1.

175. M. K. Johnson, R. S. Czernuszewicz, T. G. Spiro, J. A. Fee, and W. V. Sweeney, *J. Am. Chem. Soc.*, **105,** 6671 (1983).

176. J.-M. Moulis, J. Meyer, and M. Lutz, *Biochem. J.*, **219,** 829 (1984); *Biochemistry*, **23,** 6605 (1984).

177. R. S. Czernuszewicz, K. A. Macor, M. K. Johnson, A. Gewirth, and T. G. Spiro, *J. Am. Chem. Soc.*, **109,** 7178 (1987).

178. G. Backes, Y. Mino, T. M. Loehr, T. E. Meyer, M. A. Cusanovich, W. V. Sweeney, A. T. Adman, and J. Sanders-Loehr, *J. Am. Chem. Soc.*, **113,** 2055 (1991).

179. L. K. Kilpatrick, M. C. Kennedy, H. Beinert, R. S. Czernuszewicz, D. Qiu, and T. G. Spiro, *J. Am .Chem. Soc.*, **116,** 4053 (1994).

180. C. D. Stout, D. Ghosh, V. Pattabhi, and A. H. Robbins, *J. Biol. Chem.*, **255,** 1797 (1980); D. Ghosh, W. Furey, Jr., S. O'Donnell, and C. D. Stout, *ibid.*, **256,** 4185 (1981); D. Ghosh, S. O'Donnell, W. Furey, Jr., A. H. Robinson, and C. D. Stout, *J. Mol. Biol.*, **158,** 73 (1982).

181. H. Beinert, M. H. Emptage, J. L. Dryer, R. A. Scott, J. E. Kahn, K. O. Hodgson, and A. Y. Thompson, *Proc. Natl. Acad. Sci. U.S.A.*, **80,** 393 (1983).

182. M. K. Johnson, J. W. Hare, T. G. Spiro, J. J. G. Moura, A. V. Xavier, and J. Legall, *J. Biol. Chem.*, **256,** 9006 (1981).

183. M. K. Johnson, T. G. Spiro, and L. E. Mortenson, *J. Biol. Chem.*, **257,** 2447 (1982).

184. M. K. Johnson, R. S. Czernuszewicz, T. G. Spiro, R. R. Ramsey, and T. P. Singer, *J. Biol. Chem.*, **258,** 12771 (1983).

185. C. D. Stout, *J. Biol. Chem.*, **263,** 9256 (1988); *J. Mol. Biol.*, **205,** 545 (1989).

186. M. J. Waring, *Annu. Rev. Biochem.*, **50,** 159 (1981).

187. M. Manfait and T. Theophanides, "Drug-Nucleic Acid Interactions," in R. J. H. Clark and R. E. Hester, eds., *Spectroscopy of Biological Systems*, Vol. 13, Wiley, New York, 1986, Chapter 6, p. 311.

188. A. J. P. Alix, L. Bernard, M. Manfait, P. K. Ganguli, and T. Theophanides, *Inorg. Chim. Acta*, **55,** 147 (1981).

189. T. Theophanides, *Appl. Spectrosc.*, **35,** 461 (1981).

190. S. E. Sherman, D. Gibson, A. H.-J. Wang, and S. J. Lippard, *Science*, **230,** 412 (1985).

191. R. F. Pasternack, E. J. Gibbs, and J. J. Villafranca, *Biochemistry*, **22,** 2406 (1983).

192. J. H. Schneider, J. Odo, and K. Nakamoto, *Nucleic Acids Res.*, **16,** 10323 (1988).

193. Y. Nonaka, D. S. Lu, A. Dwivedi, D. P. Strommen, and K. Nakamoto, *Biopolymers*, **29,** 999 (1990).

194. G. D. Strahan, D. S. Lu, M. Tsuboi, and K. Nakamoto, *J. Phys. Chem.*, **96,** 6450 (1992).

195. P. Y. Turpin, L. Chinsky, A. Laigle, M. Tsuboi, J. R. Kincaid, and K. Nakamoto, *Photochem. Photobiol.*, **51,** 519 (1990).

Index

Since the number of compounds included in this book is numerous, entries for most of individual compounds are collected under general entries such as ammine complexes and metalloporphyrins. Infrared and Raman spectra are listed separately under respective entries. Boldface numbers refer to figures and tables.

INFRARED SPECTRA:

RAMAN SPECTRA